Communications
in Computer and Information Science   348

Ana Fred   Jan L.G. Dietz   Kecheng Liu
Joaquim Filipe (Eds.)

Knowledge Discovery, Knowledge Engineering and Knowledge Management

Third International Joint Conference, IC3K 2011
Paris, France, October 26-29, 2011
Revised Selected Papers

 Springer

Volume Editors

Ana Fred
IST - Technical University of Lisbon, Portugal
E-mail: afred@lx.it.pt

Jan L.G. Dietz
Delft University of Technology, The Netherlands
E-mail: j.l.g.dietz@tudelft.nl

Kecheng Liu
Henley Business School, University of Reading, UK
E-mail: k.liu@henley.reading.ac.uk

Joaquim Filipe
INSTICC and IPS, Estefanilha, Setúbal, Portugal
E-mail: joaquim.filipe@estsetubal.ips.pt

ISSN 1865-0929   e-ISSN 1865-0937
ISBN 978-3-642-37185-1   e-ISBN 978-3-642-37186-8
DOI 10.1007/978-3-642-37186-8
Springer Heidelberg Dordrecht London New York

Library of Congress Control Number: 2013935259

CR Subject Classification (1998): H.3.1, H.3.3-7, I.2.4, H.2.4, H.2.8, I.2.8, J.1, K.4.3

Typesetting: Camera-ready by author, data conversion by Scientific Publishing Services, Chennai, India

Printed on acid-free paper

Springer is part of Springer Science+Business Media (www.springer.com)

Preface

The present book includes extended and revised versions of a set of selected papers from the Third International Joint Conference on Knowledge Discovery, Knowledge Engineering and Knowledge Management (IC3K 2011), held in Paris, France, during October 26–29, 2011. IC3K was sponsored by the Institute for Systems and Technologies of Information Control and Communication (INSTICC) and it was hosted by the University PARIS-EST Créteil (UPEC) at the campus of Sénart.

This conference was held in cooperation with the Association for the Advancement of Artificial Intelligence (AAAI), the ACM Special Interest Group on Management Information Systems (ACM SIGMIS), the LISSI Laboratory and the Information & Knowledge Management Society (IKMS).

The purpose of IC3K is to bring together researchers, engineers, and practitioners in the areas of Knowledge Discovery, Knowledge Engineering and Knowledge Management, fostering scientific and technical advances in these areas.

IC3K is composed of three concurrent and co-located conferences, each specialized in at least one of the afore-mentioned main knowledge areas, namely:

- KDIR (International Conference on Knowledge Discovery and Information Retrieval). Knowledge Discovery is an interdisciplinary area focusing on methodologies for identifying valid, novel, potentially useful and meaningful patterns from data, often based on underlying large data sets. A major aspect of Knowledge Discovery is data mining, i.e., applying data analysis and discovery algorithms that produce a particular enumeration of patterns (or models) over the data. Knowledge Discovery also includes the evaluation of patterns and identification of those that add to knowledge. This has proven to be a promising approach for enhancing the intelligence of software systems and services. The ongoing rapid growth of online data due to the Internet and the widespread use of large databases created an important need for knowledge discovery methodologies. The challenge of extracting knowledge from data draws upon research in a large number of disciplines including statistics, databases, pattern recognition, machine learning, data visualization, optimization, and high-performance computing to deliver advanced business intelligence and Web discovery solutions. Information retrieval (IR) is concerned with gathering relevant information from unstructured and semantically fuzzy data in texts and other media, searching for information within documents and for metadata about documents, as well as searching relational databases and the Web. Automation of information retrieval enables the reduction of what has been called "information overload."

Information retrieval can be combined with knowledge discovery to create software tools that empower users of decision support systems to better understand and use the knowledge underlying large data sets.

- KEOD (International Conference on Knowledge Engineering and Ontology Development). Knowledge Engineering (KE) refers to all technical, scientific, and social aspects involved in building, maintaining, and using knowledge-based systems. KE is a multidisciplinary field, bringing in concepts and methods from several computer science domains such as artificial intelligence, databases, expert systems, decision support systems, and geographic information systems. From the software development point of view, KE uses principles that are strongly related to software engineering. KE is also related to mathematical logic, as well as strongly involved in cognitive science and socio-cognitive engineering where the knowledge is produced by humans and is structured according to our understanding of how human reasoning and logic works. Currently, KE is strongly related to the construction of shared knowledge bases or conceptual frameworks, often designated as ontologies.

Ontology Development aims at building reusable semantic structures that can be informal vocabularies, catalogs, glossaries as well as more complex finite formal structures representing the entities within a domain and the relationships between those entities. Ontologies have been gaining interest and acceptance in computational audiences: formal ontologies are a form of software, thus software development methodologies can be adapted to serve ontology development. A wide range of applications are emerging, especially given the current Web emphasis, including library science, ontology-enhanced search, e-commerce, and configuration.

- KMIS (International Conference on Knowledge Management and Information Sharing). Knowledge Management (KM) is a discipline concerned with the analysis and technical support of practices used in an organization to identify, create, represent, distribute, and enable the adoption and leveraging of good practices embedded in collaborative settings and, in particular, in organizational processes. Effective KM is an increasingly important source of competitive advantage, and a key to the success of contemporary organizations, bolstering the collective expertise of its employees and partners.

There are several perspectives on KM, but all share the same core components, namely: people, processes, and technology. Some take a techno-centric focus, in order to enhance knowledge integration and creation; some take an organizational focus, in order to optimize organization design and workflows; some take an ecological focus, where the important aspects are related to people interaction, knowledge, and environmental factors as a complex adaptive system similar to a natural ecosystem.

Information Sharing (IS) is a term that has been used for a long time in the information technology (IT) lexicon, related to data exchange, communication protocols, and technological infrastructures. Although standardization is indeed an essential element for sharing information, IS effectiveness requires going beyond the syntactic nature of IT and delving into the human functions involved in the semantic, pragmatic, and social levels of organizational semiotics. The two areas are intertwined, as information sharing is the foundation for knowledge management.

IC3K received 429 paper submissions from 59 countries in all continents. To evaluate each submission, a double-blind paper review was performed by the Program Committee. After a stringent selection process, 47 papers were accepted to be published and presented as full papers.

From these, we further selected 30 papers, based not only on the reviewers classifications and comments but also on the Session Chairs assessment of oral presentations, whose extended and revised versions are included in this volume.

On behalf of the conference Organizing Committee, we would like thank all participants. First of all the authors, whose quality work is the essence of the conference, and the members of the Program Committee, who helped us with their expertise and diligence in reviewing the papers. As we all know, realizing a conference requires the effort of many individuals. We wish to thank also all the members of our Organizing Committee, whose work and commitment were invaluable.

September 2012   Ana Fred
                                                  Jan L. G. Dietz
                                                   Kecheng Liu
                                                 Joaquim Filipe

Organization

Conference Chair

Joaquim Filipe — Polytechnic Institute of Setúbal / INSTICC, Portugal

Program Co-chairs

KDIR

Ana Fred — Technical University of Lisbon/IT, Portugal

KEOD

Jan L. G. Dietz — Delft University of Technology, The Netherlands

KMIS

Kecheng Liu — University of Reading, UK

Organizing Committee

Sérgio Brissos	INSTICC, Portugal
Helder Coelhas	INSTICC, Portugal
Vera Coelho	INSTICC, Portugal
Andreia Costa	INSTICC, Portugal
Patrícia Duarte	INSTICC, Portugal
Bruno Encarnação	INSTICC, Portugal
Liliana Medina	INSTICC, Portugal
Carla Mota	INSTICC, Portugal
Raquel Pedrosa	INSTICC, Portugal
Vitor Pedrosa	INSTICC, Portugal
Daniel Pereira	INSTICC, Portugal
Cláudia Pinto	INSTICC, Portugal
José Varela	INSTICC, Portugal
Pedro Varela	INSTICC, Portugal

KDIR Program Committee

Anjana Kakoti Mahanta, India
Edson T. Matsubara, Brazil
Eduarda Mendes Rodrigues, Portugal

Marcos Gonçalves Quiles, Brazil
Hisham Al-Mubaid, USA
Francisco Martínez Álvarez, Spain

KDIR Auxiliary Reviewers

Apoorv Agarwal, USA
Laurence Amaral, Brazil
Patricia Fernandes, Brazil
Alexandre Levada, Brazil
Eneldo Loza Mencia, Germany
Tiemi Sakata, Brazil
Edimilson Santos, Brazil
Pattaraporn Warintarawej, France
Gene P.K. Wu, Hong Kong

KEOD Program Committee

Alia Abdelmoty, UK
Carlo Allocca, UK
Yuan An, USA
Francisco Antunes, Portugal
Pasquale Ardimento, Italy
Sören Auer, Germany
Janaka Balasooriya, USA
Claudio De Souza Baptista, Brazil
Jean-Paul Barthes, France
Teresa M. A. Basile, Italy
Nicola Boffoli, Italy
Patrick Brezillon, France
Giacomo Bucci, Italy
Vladimír Bureš, Czech Reublic
Ladislav Burita, Czech Republic
Doina Caragea, USA
Núria Casellas, USA
Sharma Chakravarthy, USA
Ricardo Choren, Brazil
Magdalini Eirinaki, USA
Anna Fensel, Austria
Johannes Fuernkranz, Germany
David Fonseca, Spain
Raul Garcia-Castro, Spain
Serge Garlatti, France
Rosario Girardi, Brazil
Matteo Golfarelli, Italy
Stephan Grimm, Germany
Sven Groppe, Germany
Christopher Hogger, UK
Ronghuai Huang, China

Asanee Kawtrakul, Thailand
Katia Lida Kermanidis, Greece
Pavel Kordik, Czech Republic
Patrick Lambrix, Sweden
Ming Li, China
Antoni Ligeza, Poland
Matthias Loskyll, Germany
Rocio Abascal Mena, Mexico
Riichiro Mizoguchi, Japan
Malgorzata Mochol, Germany
Juan Bautista Mocholi, Spain
Benjamin Molina, Spain
Ralf Möller, Germany
Andrés Montoyo, Spain
Claude Moulin, France
Kazumi Nakamatsu, Japan
Keiichi Nakata, UK
Roberto Navigli, Italy
Erich Neuhold, Austria
Jørgen Fischer Nilsson, Denmark
Nan Niu, USA
Judit Olah, USA
Jivka Ovtcharova, Germany
Salvatore Flavio Pileggi, Spain
Enric Plaza, Spain
Mihail Popescu, USA
Violaine Prince, France
Juha Puustjärvi, Finland
Amar Ramdane-Cherif, France
Domenico Redavid, Italy
M. Teresa Romá-Ferri, Spain

François Rousselot, France
Martin Serrano, Ireland
Jorge Sá Silva, Portugal
Nuno Silva, Portugal
Derek Sleeman, UK
Deborah Stacey, Canada
Anna Stavrianou, France
Heiner Stuckenschmidt, Germany
Mari Carmen Suárez-Figueroa, Spain
Christos Tatsiopoulos, Greece
Orazio Tomarchio, Italy
Shengru Tu, USA

Rafael Valencia-Garcia, Spain
Iraklis Varlamis, Greece
Cristina Vicente-Chicote, Spain
Bruno Volckaert, Belgium
Sebastian Wandelt, Germany
Yi-dong Wei, China
Franz Wotawa, Austria
Yue Xu, Australia
Gian Piero Zarri, France
Jinglan Zhang, Australia
Zehua Zhang, China
Catherine Faron Zucker, France

KEOD Auxiliary Reviewers

Thomas Hubauer, Germany
Nuno Luz, Portugal
Paulo Maio, Portugal
Karsten Martiny, Germany
Giuseppe Di Modica, Italy
Heiko Paulheim, Germany
Thomas Riechert, Germany
Sebastian Tramp, Germany

KMIS Program Committee

Marie-Helene Abel, France
Miriam C. Bergue Alves, Brazil
Bernd Amann, France
Ioannis Anagnostopoulos, Greece
Alessio Bechini, Italy
Steve Beitzel, USA
Milton Borsato, Brazil
Malgorzata Bugajska, Switzerland
Marcello Castellano, Italy
Xiaoyu Chen, China
Ying Chen, USA
Reynold Cheng, Hong Kong
Giulio Concas, Italy
Bruce Cronin, UK
Dominique Decouchant, France
William Dixon, USA
Mariagrazia Dotoli, Italy
Zamira Dzhusupova, Macau

Elsa Estevez, Macau
Susana Falcão, Portugal
Joan-Francesc Fondevila-Gascón,
    Spain
Anna Goy, Italy
Song Han, Australia
Jan Hidders, The Netherlands
Jiankun Hu, Australia
Anca Daniela Ionita, Romania
Dan Kirsch, USA
Elise Lavoué, France
Alessio Malizia, Spain
Anthony Masys, Canada
Nada Matta, France
Michael May, Denmark
Mauro Mazzieri, Italy
Sonia Mendoza, Mexico
Christine Michel, France

Owen Molloy, Ireland
Fei Nan, USA
Augusta Maria Paci, Italy
Hye-young Paik, Australia
Marina Ribaudo, Italy
John Rohrbaugh, USA
Wendy HuiWang, USA

Robert Warren, Canada
Leandro Krug Wives, Brazil
Andreas Wombacher, The Netherlands
Clement T. Yu, USA
Wei Zhang, USA
Yonggang Zhang, USA

KMIS Auxiliary Reviewers

Federica Cena, Italy
Kimberly García, Mexico
Wenjing Ma, USA
Avila Mora Ivonne Maricela, Mexico

Invited Speakers

Jay Liebowitz

Carole Goble
Mayer Aladjem
Guus Schreiber
Jan L. G. Dietz

University of Maryland University
    College, USA
University of Manchester, UK
Ben-Gurion University of the Negev, Israel
VU University Amsterdam, The Netherlands
Delft University of Technology,
    The Netherlands

Table of Contents

Part II: Knowledge Engineering and Ontology Development

Part III: Knowledge Management and Information Sharing

Invited Paper

Accelerating Scientists' Knowledge Turns

Carole Goble[1], David De Roure[2], and Sean Bechhofer[1]

[1] School of Computer Science, The University of Manchester, Manchester, U.K.
{carole.goble,sean.bechhofer}@manchester.ac.uk
[2] Oxford e-Research Centre, University of Oxford, U.K.
david.deroure@oerc.ox.ac.uk

Abstract. A "knowledge turn" is a cycle of a process by a professional, including the learning generated by the experience, deriving more good and leading to advance. The majority of scientific advances in the public domain result from collective efforts that depend on rapid exchange and effective reuse of results. We have powerful computational instruments, such as scientific workflows, coupled with widespread online information dissemination to accelerate knowledge cycles. However, turns between researchers continue to lag. In particular method obfuscation obstructs reproducibility. The exchange of "Research Objects" rather than articles proposes a technical solution; however the obstacles are mainly social ones that require the scientific community to rethink its current value systems for scholarship, data, methods and software.

Keywords: Reproducible Research, Scientific Workflow, Research Object, Digital Scholarship, Open Science.

1 Introduction

A "knowledge turn" in manufacturing enterprises is the cycle of a process that derives more good leading to new or better products and competitive advantage [43]. One turn corresponds to a single trial-and-error cycle by a professional in a focused area of knowledge, and includes the learning generated by the experience. Long-term sustainable competitive advantage is advanced if a business learns faster than its competitors: the experiences of its people recorded, efficiently harnessed, communicated and built upon. Scientific research is the business of knowledge turning. The classical scientific method turns observations and hypothesis – through experimentation, comparison, and analysis – into new knowledge and improved "scientific products" such as more observations and hypotheses, experimental methods, models, techniques, data, protocols, and publications. A "Hypothesis-Prediction-Observation-Analysis" cycle operates over a knowledge pool of "know-what, know-how, know-why and know-who" populated by the publication of peer-reviewed articles, the gathering of scientists at symposia, sharing of skills and experience in collaborations, and the exchange of datasets, methods and, increasingly, computational tools and software.

As an example of knowledge turns and clear motivation for accelerating them, Josh Sommer is an impressive young man suffering from Chordoma, a form of bone cancer. His condition is uncommon and the research scattered and piecemeal. A cure was

A. Fred et al. (Eds.): IC3K 2011, CCIS 348, pp. 3–25, 2013.

being held back by lack of funding, but it was also being hindered by the scattered researchers getting hold of resources, restrictions on information flow between the researchers and poor coordination and collaboration [54].

Three labs are researching the condition (Fig. 1). Lab1 produces experimental results that Lab3 uses to test a hypothesis. Data produced by Lab3 generates a hypothesis that when combined with a locally innovative technique in Lab2 produces new insights that, unexpectedly, is just the missing piece that Lab1 needs to confirm an earlier experiment. Each step is a knowledge turn, to turn prior results into new results. Restricted flows between the labs slow down the whole process. Years can pass between the emergence of a result and its availability to another researcher, if it ever becomes available, wasting time and wasting opportunities.

Josh co-founded the Chordoma Foundation (www.chordomafoundation.org) as a resource and knowledge broker, to remove the barriers on information flow, particularly addressing the social influences operating on scientific communication practices that make them slow and ineffective: e.g. the jealous guarding of pre-published results and ignorance of other research. The foundation has maximized productivity and reduced lags by providing resources, by creating, collecting, storing, and distributing information and biological materials, and by facilitating information exchange and collaboration among researchers.

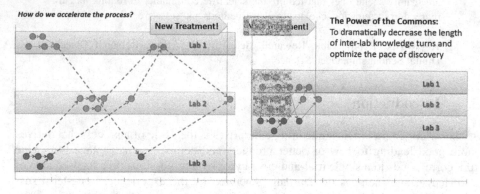

Fig. 1. Rapidly contributing results in a commons increases knowledge flow, decreases the length of inter-lab knowledge turns and optimizes the pace of discovery [54]

In this paper we will discuss what affects knowledge turns, the practices and challenges of science in the web era, and how we might accelerate the turns of scientific knowledge. In Section 2 we introduce today's scientific research practice through digital instruments, information technology and community. In Section 3 we consider reproducibility and the primacy of method, and offer the important social perspective in Section 4. In Section 5 we introduce *Research Objects* and our current work in the "Workflow Forever" project, and we conclude in Section 6.

2   The Mechanics of Fact Making

Shapin [49] highlights three key aspects in the "mechanics of fact-making" that have revolutionized scientific research: (i) The *instrument technologies/experimental equipment* used to undertake science, and produce, process and analyze knowledge; (ii) the *information technologies* used for recording and spreading knowledge; and (iii) the composition and *social organization of scientific communities*, and their conventions in cooperation, collaboration and dealing with scientific claims. How are advances in these components revolutionizing science in the web era?

2.1   Digital Scientific Instruments

Scientific instruments are devices for observing, measuring and handling natural or experimentally produced phenomena. Instrument technology relates to their design, production and use. We are in an age of digital instruments; observations and measurements are born digitally or the instruments themselves are *in silico*. To paraphrase McLuhan: "scientists shape tools and thereafter the tools shape science" [35]. Gray defines four paradigms of science: *empirical*, observing and describing natural phenomena; *theoretical*, using models and forming generalizations; *computational*, simulating complex phenomena; and *data exploration*, unifying theory, experimentation, and simulation [24]. How are these shaped by some of today's instrument technologies?

Revolutions in technology have increased the rate of empirical observations and decreased the cost of ownership. Genomes can now be sequenced ~50,000 times faster than in 2000 yet Next Generation Genome Sequencing machines are cheap enough to be local laboratory commodities potentially generating terabytes of data [40]. Acceleration in data scales is accompanied by a proliferation in the varieties, with derived, distilled and predictive datasets alongside raw data, and online accessibility. The Molecular Biology Database Collection lists 1380 databases [17] ranging across genomes, proteomes, structures, pathways, etc. PubMed Central (a citation index for the biomedical literature, www.ncbi.nlm.nih.gov/pubmed) indexes over 20 million published articles, a new one every 30 seconds. Generating data is no longer the bottleneck – rather, it is what we do with it, and scientists can now turn to data first where they used to turn to an instrument.

Today software is an instrument – it makes it possible to predict, simulate, and generate/confirm a hypothesis on the basis of datasets and models alone [29]. Increased automation copes with the scale, repetition, accuracy and complexity of processing. Our bottlenecks are curation, comparison, validation, filtering, mining, integration, visualization and analytics; the issue is not the $1000 genome but the $100,000 analysis. Publishing software and data as Web Services opens up an ecosystem of interacting services; i.e. "service-oriented science". Commodity based elastic compute and data commons cloud platforms opens up the prospect of a pay-as-you-go "science as a service" delivery model: a lab generates the data and the questions; a public or commercial service manages the data, processing and even the processing pipeline, spreading the cost and outsourcing routine but scarce analysis expertise [3].

Example: Science as a Service Using Taverna Workflows. Some cattle breeds in sub-Saharan Africa are highly tolerant of sleeping sickness infection (African bovine trypanosomiasis), but the potentially more productive breeds are more susceptible. Increasing their tolerance could have a major impact on food production. A multi-institution, multi-disciplinary team sampled many cattle specimens and repeatedly used computational analysis techniques that accessed online tools, drew data from public datasets, and compared to findings in the literature. They succeeded in finding the candidate genes that could hold the key to the differences between the cattle [42]. To accelerate the analysis the team automated it using the Taverna Workflow Management system [45] (Fig. 2). A scientific workflow combines data and processes into a configurable, structured sequence of steps. Taverna provides software to combine datasets systematically and automatically, and integrate analytical software tools and services even when they are independently made and incompatible [26].

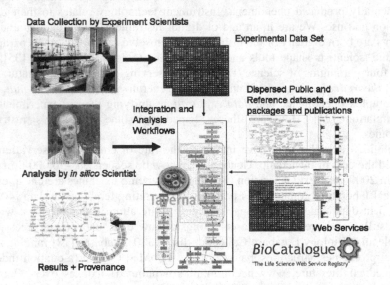

Fig. 2. Improving research productivity using automated Taverna workflows

A visual interface enables skilled computational scientists to assemble pipelines and access services shielded from low-level programming concerns. The execution platform accesses the scattered datasets and tools remotely hosted by providers so consumers do not need to download resources or learn their codes [15]. The Taverna workflows ensure that the process is comprehensive, systematic and unbiased, important for protocols often repeated. They represent an accurate record of the computational method and automate recording the provenance of the computed results which is crucial for experimental and data reproducibility (discussed in Section 3). They liberate the scientists from routine informatics, reducing the manual analysis task from weeks to hours, so the scientists can concentrate on scientific discovery.

Computational workflows are effective because the data are in machine-readable formats and the databases and tools are accessible through machine-processable interfaces (APIs) such as Web Services. For example, the BioCatalogue [5], a public catalogue of Web Services in the life sciences, is integrated into Taverna. It currently registers 2000+ services from 160+ providers. Next Generation Sequencing makes it possible for this consortium to sequence a genome every few days. By putting the analysis workflows on Amazon Cloud, alongside reference datasets and tools, we create an "Analysis as a Service", enabling non-informatics skilled experimentalists to rerun "pre-cooked" computational protocols repeatedly against new samples and changes in the reference services, without having to manage complex infrastructure. This opens up the processes and data resources to a wider group of scientists and scientific application developers. Elastic compute, pay-as-you-go Clouds enable scientists to pay on-demand. The costs are spread by making the service available to the whole community when the analytical process is routine.

2.2   Digital Scientific Information Technology

Information technology is the means used to accumulate knowledge; to record, transmit, analyze, verify and discuss claims; and to "virtually witness" experiments and observations. The traditional method is the peer-reviewed scientific publication, the traditional technology is print, and the traditional archive is the library. The Web, originated as a research communication tool, is *the* information technology that has most changed the dispersal of scholarship: publications and most results and analyses are "born digital", and anyone can be a publisher or a librarian.

However, mainstream online publishing is chiefly putting facsimiles of print on the Web. Alternatives to written articles, like videos, are seen as exotic and time-consuming with little benefit for the author (though this may change as tools improve and new digitally native researchers gain influence). Data and methods are frequently unavailable for peers to test claims [34], or inaccessibly embedded in documents in unprocessable forms (tables, graphs, pictures, text) [44]. Journals with data policies that demand that data be available in public archives or supplementary material have accelerated discovery in the Life Sciences, but frequently turn out to neither enforce their own policies nor ensure long-term compliance [1]. Open Access enables the flow of research – if we count citations as a metric [18] – and are actively promoted within disciplines (e.g. ArXive.org in Physics) and by funders (e.g. the US NIH publicaccess.nih.gov). However, Open Access does not guarantee reusability [52], and though it is free for consumers it is not for providers, restricting publication to those who can afford it. As the majority of publishers remain subscription-based and legislative efforts in the USA aim to block open access mandates, open access for all content is some way off.

The Web supports the technological and social innovations to circumvent publishers and libraries. Scientists can now assemble their own publishing environment, organize agile peer review and create their own scholarly research productivity platforms [27]. Such agility is enabled by:

• *Mass Market Resources*: Free publishing platforms (e.g. wikis and blogs) and cloud-based services (LinkedIn, YouTube, SlideShare, Twitter, Wikipedia, Google-Docs, DropBox, GitHub, though not so much Facebook);

• *Specialised Resources*: data commons (FigShare, Sage Bionetworks); methods (nanoHUB, myExperiment, OpenWetWare); models (BioModels), reference management (Mendeley, CiteULike, ReadCube); publishing analysis (Google Citations); review management (EasyChair); LIMS (YourLabData); Lab management (LabGuru); social networking (ResearchGate, NatureNetworks, SciLink etc);

• *Cross-platform Search and Indexing*: free and powerful search tools (Google, Yahoo) and services (Google Scholar) enable cross-library, cross-publisher and cross-repository discovery, impact metrics, and science mappings;

• *Open Standards*: open licenses, non-proprietary standards, open data formats and APIs that enable platforms to interoperate, and added-value services to proliferate. Monolithic publishing systems are giving way to ecosystems of services.

• *Open Metadata*: common reporting standards and ontologies allow different systems to represent, exchange and link information on communities (e.g. SIOC), citations (e.g. CiTO [51]), and artefacts (e.g. Dublin Core). Ontologies representing claims and discourse enable recorded discussions across publications, datasets and communities [8].

These resources improve the flow of results and provide the specialized management services needed for different data types, but they also act to fragment information.

Embedded Science and Scattered Science. The scholarly article as a monolithic, single document is no longer fit for purpose, especially for online scholars using digital instruments. Readers need (i) easy access to underlying data, software, and methods and (ii) the ability to consume content on a variety of electronic devices and by digital instruments [44]. From one perspective articles are compound "collages", incorporating data (in tables and charts) and algorithms/methods (in incomplete written sketches). For articles to be actionable and verifiable scholarship, these need to be disinterred from computationally inaccessible forms into links to the actual data and executable codes − to re-link the scholarship with the instruments. Besides, the data and methods are crucial scholarly results that should be published, cited and credited in their own right.

From another viewpoint, articles are structured reports of claims related to evidence and prior work. But these claims are deeply embedded and computationally inaccessible. To surface such content requires authors to structure their claims more systematically, expert readers to explicitly recover and structure them [9], and text mining to assist discovery and recovery. For example, components of the workflow in Fig. 2 mine the literature available in PubMed to surface links to genes that are then tracked in the datasets. Semantic Publishing [50] and Semantic Publishing systems such as Utopia Documents [44] are attempts to tackle computational accessibility through explicit metadata.

Accelerating knowledge turns between laboratories is not just about flow and availability of results. Researchers must cope with the piecemeal and fragmentary nature of publishing, gathering results before they can begin to use them. Scattering occurs in several forms:

• *Stewardship*: On one hand we have massive centralization in the various general data centers, specialist data collections and publishers' supplementary repositories. On the other we have massive decentralization in private lab books, group wikis, project web sites and institutional repositories. The original experimental data is on a disk under a post-doc's desk whereas the analysis results are embedded in a table in PDF supplementary file held by the publisher behind a pay-wall.

- *Asset Type*: For one experiment the analysis software is on GitHub, the workflow is on myExperiment, and the tutorial is on Slideshare. Scattering results onto specialist service providers delegates their stewardship to experts, but it also risks the demise of a "rented" resource from which content is hard to recover; for example, Google shut down its medical and scientific data vault. On the other hand research labs are even less likely to maintain accessible archives.

- *Multi-part Datasets*: Although biological science is integrative its data ecosystem is "siloed" and dispersed. Separate resources hold information on gene sequences, structures, interactions, proteins, etc. These scattered complex datasets can only be reassembled by consumers equipped to navigate the various reporting standards and technologies used. Unified frameworks for metadata that ensure biomedical research datasets become interoperable are urgently needed [47].

Value-added data integration platforms like Galaxy [21] and topic-specific *commons* like Sage Commons (sagebase.org/commons) and Pathway Commons (www.pathwaycommons.org) attempt to improve productivity for the hard-pressed researchers who otherwise have to track down the diverse components of an experiment for themselves.

Fig. 3. myExperiment showing an entry from the cattle workflow suite by its author

Example: The myExperiment e-Laboratory for Workflow Exchange. An e-Laboratory is a set of (usually) scattered online components – workflows, resources, data, algorithms, texts, queries – that circulate in a collaborative space used by (usually) scattered scientists. They enable the planning, execution, analysis and publication of *in silico* experiments and their results. As we have already seen, scientific workflows capture a computational process so that results can be reproduced, the method accurately reviewed and validated and know-how shared, reused and adapted.

myExperiment [11] is an e-Laboratory for the circulation, conservation, preservation and publishing of workflows. Communication flow is encouraged by social tagging, comments, ratings and recommendations, social network analysis and reuse mining (what is used with what, for what, and by whom). myExperiment encourages the flow and reuse of methods, and supports workflow citation and "altmetrics" for its contributors. The open public myExperiment.org currently has 2000+ workflows from 21+ workflow systems and 5000+ registered members. Workflow design is a skill often beyond a researcher. By establishing shared collections of workflows that contain reusable processing pipelines we help avoid the reinvention of techniques and propagate best practices: some workflows have access statistics of 1000+. The workflows developed by Fisher for the cattle investigation (Fig. 3), have been successfully reused for chronic colitis [31] and bacterial responses to oxygen [33].

myExperiment directly addresses scattering through the notion of "packs". Essentially a pack is a single sharable collection of items which could include anything from workflows and data to slides and papers, and may be on the myExperiment site or elsewhere. Packs are one of the "social objects" of myExperiment, but equally they are a technical object in that they can be accessed programmatically though myExperiment's Linked Data interface, a semantic interface using the Object Reuse and Exchange representation. Packs form the basis of the Research Objects discussed in Section 5.

2.3   The Online Scientific Community – Collaboration and the Crowd

Scientific investigation is a social activity. Each scientific community and sub-community operates within conventions shaped by the roles of its members, their activities and the perceived value of their contributions. We classify the "crowd" into:

• *Suppliers/Consumers of Instruments*: Experimental, theoretical scientists and modelers are usually thought of as *consumers*. *Suppliers* include scientific informaticians, computational scientists, specialist tool developers, service and resource providers, content curators, infrastructure developers and system administrators. A bioinformatician is a consumer of software from a programmer who in turn is a consumer of a dataset supplied by an experimental scientist.

• *Suppliers/Consumers of Information*: All players consume. Librarians, publishers, data centres, the Web, supply technology. Scientists supply content, i.e. *results*.

• *Governance Regulators*: funders, employers, and peer-reviewers define expectations, confer value on scientific investigations and evaluate the performance of members of all the scientific community through metrics such as paper citations, software download figures or scales of datasets.

How does this social "ego-system" (i) collaborate and (ii) peer-produce? Networks of people construct scientific instruments (e.g. Uniprot Protein Sequence Database, the Large Hadron Collider), conduct and discuss experiments, and record and publish experimental results. Almost all original natural science papers have multiple authors; in January-October 2008, Nature published only six single-author papers, out of a total of 700 reports [58]. Traditionally networks may be organized scientific societies or projects. Thanks to the Web virtual communities are easy to set up and can span geographical, institutional, discipline and time zone boundaries. Across a spectrum of formal, coordinated and funded collaborations to informal self-organizing, volunteer-based networks, we are in an era of "Networked Science".

The use of online tools has the potential to dramatically speed up the rate of scientific discovery across all of science but also to amplify the *collective* intelligence of science, and expand the range of scientific problems which can be tackled [36]. Virtual communities of scholars in all fields and all team sizes produce higher impact and more highly cited work than comparable co-located teams or solo scientists [61]. Moreover, increasing specialization of all fields and the need to tackle problems beyond the scope of a single discipline or research practice necessitates interdisciplinary team research. Thus collaborative research needs *shared understanding* in addition to shared expertise and shared resources, and hence clear and regular communication.

Although face-to-face networking is essential, all scientists work in online cooperative spaces. Just using email makes us part of an "invisible college" on a scale that was not feasible 30 years ago. This completely changes the speed and range of conversation. First, cheap and instant communication – instant twitter broadcasting, instant message conversations and virtual Skype meetings – make complex cooperation feasible across a scattered community. Word of mouth recommendation has been raised to web-scale by micro-blogging with twitter, feeds, blogs, social bookmarking and auto citation tracking: to cope with an information deluge, science gossip is now electronic. Papers that are tweeted appear to be more highly cited [14]. Second, "limited focus" social networking improves the flow of expertise between people who already collaborate and between people in the "long tail" of small but expert research labs and singleton graduate students. They create "hang outs" for isolated and/or distributed researchers to crowd around a focus: people (e.g. ResearchGate); resources, such as citation recommendations (e.g. CiteULike, Mendeley); digital instruments such as workflows (myExperiment) or simulations (nanoHUB); and scientific problems such as SysMO-SEEK for Systems Biology in MicroOrganisms and Scratchpads for BioDiversity data (scratchpads.eu). Finally, collaboration tools enable two-way communication and collaboration with practitioners and beneficiaries, such as patients and policy makers [9].

Science has always relied on crowd-sourcing, building upon a cumulative pool of other's results. Traditionally, publishing houses crowd-source experimental results, peer-review harnesses the wisdom of the crowd and citations are crowd-voting metrics. Public datasets like Uniprot, and data commons like Sage Bionetwork, are repositories for community-contributed results. Now we have the means to widen the crowd to scientists in different disciplines, to scientists in the "long tail" including amateurs, and to citizens too (e.g. the Galaxy Zoo citizen contributions to Astronomy [32]). Curated public databases like Chemspider (www.chemspider.org) for chemistry, and wikis such as Specieswiki (species.wikimedia.org), are a game changer. Inspired by Wikipedia they are also mechanisms for concentrating information, for quality-improving mass-curated distillations of scientific knowledge otherwise dispersed [57]. Wikipedia itself is used as a way of improving "official" datasets by mass curation [10]. The same phenomenon holds with open source software: it is popular because it is free and collective ownership gives confidence in its durability, quality and continued availability.

myExperiment supports a mixed crowd including a "long tail" of isolated computational scientists who are strangers working independently. The site allows them to find, share and brag about expertise and get/offer help. Interestingly, tagging and reviewing other's content is rare, and collaborations sparked by the site are continued outside it privately. Groups, typically from specific projects or disciplines, establish their own collaboration and sharing norms. SysMO-SEEK (www.sysmo-db.org) is a more elaborate version of a group-based e-Laboratory: a specialist, private resource supporting 15 multi-partner consortiums (around 300 scientists) for a European-wide research program in Systems Biology. The collaborations are within established groups and between the groups, focused on one discipline, and focused on exchanging data, models, and procedures. The social dynamics of the two systems have commonalities and differences, and we will refer to them both in this paper.

3   Reproducibility and the Primacy of Method

A scientific communication has two prime goals: to announce a result so it can pass into use by the reader and to convince the reader that the result is correct [34]. Reproducibility underpins the scientific method and is fundamental to the flow of knowledge between researchers [55]. Productivity is strongly linked to the ease with which a result can be reproduced and sufficiently validated to be reused. Consequently, a communication must present its experimental materials and methods. We are now familiar with the notion that the data (materials) described should be available for inspection. However, to truly reproduce results we also need methodological transparency. The methods need to be precisely described to stand up to critical examination and reproduction, and the provenance of the data needs to be precisely reported so that it can be accurately cited, safely compared and correctly interpreted. This is not just for the consumers' benefit but also the producers'. Misunderstanding the subtle contexts of data and methods leads to misuse and misinterpretation of results when used by other disciplines, policy makers, journalists and the public.

In experimental science, methods include laboratory protocols and Standard Operating Procedures (SOP). These are detailed, written instructions to achieve uniformity of the methodology used and without them it is impossible to compare or validate results. SOP repositories in biology include Nature Protocols, OpenWetWare.org and MolMeth.org. The earliest request for SysMO-SEEK by its users was for a Standard Operating Procedures registry. Media publishing sites like Jove and SciVee publish videos of experimental methods that can be linked to written protocols or used in teaching. In computational science methods include: (i) the *in silico* instrument – algorithms, models, workflows, scripts, web services, tools, software, and so on; (ii) the configuration of variables, parameters, error bounds and thresholds; and (iii) the protocols used to choose and apply them. Digital method public repositories include: BioModels for System Biology models; myExperiment for computational workflows; NanoHUB for simulation codes; BioCatalogue for cataloguing Web Services; MethodBox.org for statistical scripts over social science surveys.

Access to (and preservation of) software is a particular issue with the reproducibility of computational experiments. Open software, foundations and open software repositories such as GitHub are technical mechanisms to try to make software source codes available and sustainable. Virtualization platforms such as SHARE [22] try to preserve the software binaries so they can be rerun as they were when a result was announced. SHARE was a prizewinner at the Elsevier's Executable Paper Grand Challenge 2011 (www.executablepapers.com) which set out to stimulate innovations that make it easy to rerun and reuse methods from publications. Other prizewinners adapted existing publication models to include data and analyses, by embedding executable code in papers [41], and facilitated the validation, citation and tracking information through identification schemes for verifiable computational results [19]. The challenge reflects a rather incremental view of scholarly communication in that the hub is still the written paper. Similarly, the *Open Research Computation* journal (www.openresearchcomputation.com) uses traditional reviewed articles to describe software in much the same way as data journals such as GigaScience use traditional articles to publish data. This traditionalism is largely due to credit being tied up with articles, which we discuss in the next section.

The absence of explicit, comprehensive and accurate descriptions of methods is recognized as a serious problem, leading to "black box" science [37], difficulties in peer review and exposure to allegations of fraud. Scientific papers conform to conventions for presenting results rather than accurately and comprehensively describing what really happened. Tacit knowledge and shorthand references are prevalent. Furthermore software might not be available to be executed or examined to accurately assess the algorithm it encodes. We will address the various social reasons for this and why this obfuscation of method is sometimes unintended but often intentional.

Workflows give an accurate and transparent record of the method used to comprehend, justify and compare resultant data products. For example a Taverna workflow uses a mixture of data and software instruments which may be code embedded in the workflow logic, calls to locally provisioned datasets, or calls to services supplied by public providers. The workflow is deposited in myExperiment so it has a unique identifier for citations. As a freestanding method it is an asset that can be reused – retrieved, referenced, and repeatedly used on its original data or new data. For a specific execution bundled in a pack with the configuration of input data and variable settings, the provenance of its run can be examined to "replay" its execution.

However, even if a computational method is recorded and shared it might cease to be useful. "Reproducibility by re-run" is sensitive to the stewardship of components, especially web services hosted by third parties. Components alter, become incompatible or unavailable. Step decay leads to workflow decay, equivalent to an instrument becoming obsolete. Techniques are available to reduce this decay; e.g. Taverna workflows deposited in myExperiment that use web services deposited in BioCatalogue are monitored for such circumstances.

Where actual reproducibility cannot be achieved, partial reproducibility is a means to play back workflow execution based on the provenance of previous executions

[13]. Even in cases where the workflow cannot be executed and no provenance trace is included, the details of the workflow description may be enough to justify and explain a result. To review research requires retrieval of workflows which are transparent but not necessarily executable. However, to re-run these workflows as "black boxes" they could be executable but not transparent. It is useful to distinguish the former case as *preservation* of a workflow and the latter as *conservation*, whereby a method is restored or repaired so as to be re-executable. Workflows also lend themselves to *repurposing*; i.e. reuse of a method against different settings, for example different datasets or parameter configurations, but also *structural reuse*, for example substituting or reordering steps; and *fragment reuse*, using parts or combinations of parts of the workflow [60]. The descriptions of the relationships between these parts and how they are assembled inform their re-use. As the majority of experimental designs are variants, reusing scientific workflows leads to better procedures and spreading best practice. Reproducibility in computational science turns out to have many subtle aspects that warrant further investigation: the Wf4Ever project (www.wf4ever-project.org) is investigating workflow reproducibility, preservation and conservation, proposing a framework based on "Research Objects" which we discuss in Section 5.

4   Open Knowledge Flow: The Common Good vs. Self-interest

In general, we are moving towards an era of greater transparency in all of these topics (methodology, data, communication and collaboration). We have the instrument and information technology to deliver, but do we have the necessary social conventions? Mismatched motivations, value placed on knowledge and social capital, reward schemes, poor reciprocity and distrust together conspire to block the circulation of knowledge [38] [6].

Open Science [46][48] encompasses the ideals of transparency in experimental methodology, observation, and collection of data coupled with the public availability and reusability of scientific data and public accessibility and transparency of scientific communication. Openness is a means of achieving accelerated knowledge transfer and networked science [36]. Underlying this open ideal is a notion of voluntary sharing of methods, results, and scholarship, and the objects of scholarship belonging not to the individual scientist but to the larger community [55]. However, scientists are people working within their social norms and as self-interested as any other group of people. Their prime motivations include funding, building reputation, and getting sufficient time, space, and resources do their research. Sharing results is not a motivation in itself, so has to be placed within a context of maximizing reward, minimizing risk and optimizing costs:

• *Reward for Sharing*: to gain competitive advantage over rivals by establishing a claim on priority of a result; to establish public reputation and recognition through credit; to accelerate the widespread adoption/acceptance of a result; to gain access to otherwise unavailable instruments, data, techniques or expertise.

• *Risk of Sharing*: the threat of rivals gaining a competitive advantage; damage to public reputation through scrutiny or misinterpretation; not getting credit; a sense that others will get a "free-ride".

• *Cost of Sharing*: the time and resources needed to prepare; the inconvenience and/or difficulty in preparing to share or sharing; potential long-term sustainability obligations.

Stodden [55] gives an excellent sociological account of scientific sharing practices, broadly summed up as: scientists do not like to share and when they have to they prefer to share with those they trust and be rewarded for it. Although they do not always trust data other than their own, they like to be shared *with*. This mismatch leads to "Data-Mine-ing" – your data is mine and my data is mine. Liebowitz also argues that the intrinsic worth of knowledge is a sharing factor related to the scarcity of prized commodity in local asset economies. Researchers will trade when there is a local unavailability of an asset such as specialist data. If assets are expensive and have to be collectively obtained, like the Large Hadron Collider or a telescope, the consortium typically obliges the results to be collectively shared. However, if assets are local investments, like a Next Generation Genome Sequencer, and data is locally available, then there is no need to share to acquire, and if the data is scarce and prized it will be protected.

The behavior of the members of the SysMO-SEEK consortium highlights the value placed on knowledge capital and the distrust that lies between rivals, manifested as incremental sharing that widens the availability of content as its local value proposition changes. At first (or perhaps only) an individual or laboratory uses the e-Laboratory as a private, preserved repository. This is useful when scientists are mobile, moving from grant to grant and institution to institution. Next, *trusted* collaborators within each project may exchange pre-published content. Results shared outside a trusted group prior to publication are rare. When a scientific article is finally published publicly we could expect its associated data/method/model to be deposited publicly. However, if the investigator thinks they can wring another paper out of some data they will not share it even if it is the basis of an announced result. Data are only made widely available when their local capital is exhausted. We also observe that (i) models, procedures and workflows are more likely to be openly shared than data, suggesting that the scientific community places greater value on data than experimental method; (ii) formal consortia are less likely to publicly share than individuals; and (iii) young researchers and very senior well established researchers are more willing to share than mid-career researchers in the midst of establishing their reputations.

The e-Laboratories we have built deal exclusively with non-personalized data. In the social and clinical sciences a common objection to data commons is the risk to personal privacy. Although of course important, in practice the Chordoma Foundation and other patient groups have found that patients are far more positive towards data sharing than their self-appointed guardians in the clinical profession. Yakowitz argues that risks from anonymized data rarely materialize, and current privacy policies overtax valuable research without reducing any realistic risks [62]. A discussion about the throttling of clinical knowledge exchange by well meaning but ill-informed ethics committees is a topic for another paper.

4.1   Reciprocity and Flirting

The tendency to protect data is sometimes called "data hugging". In myExperiment and SysMO-SEEK we observe "data flirting" where scientists strategically (or maybe tacitly) hold back information, communicating *just enough* to interest their community and publish *just enough* to preserve their claim for priority on the findings but *not enough* in practice for competitors to be able to take advantage. Specialized knowledge on experimental details is withheld. Scientific jargon is used to frustrate competitors. This "counterfeit sharing" is prevalent when funders make data sharing directives. The data is deposited, and thus "shared", but it is hard to find and impossible to reuse or reproduce [38].

Borgman [6] highlights the underlying mismatches in the motivations to share by data producers and data consumers. myExperiment highlights some examples. Providers of workflows want credit and, sometimes, control of who benefits from their work or how it is used. Consumers tend to follow a "search-download-logoff" pattern, wanting to easily reuse the workflow without constraint. However, they often fail to credit the provider, or contribute comment or review. They do not feedback results that arose from using the workflow and it is hard to track the workflow from published research unless it was explicitly cited. This lack of feedback fosters a sense of "free riding".

Reciprocity is fundamental to effective knowledge exchange. Where myExperiment is used by an organized group, good citizenship is more governed. However, when it is used by individuals who do not even have to be registered members the social pressure for reciprocity is absent or at best tacit. To close the reciprocity feedback loop, and secure/preserve their reputation, workflows are described by their producers such that the consumer must enter into a dialogue to reuse them. The author can then secure credit or negotiate a beneficial collaboration. They can also protect their reputation by guarding against misuse, or maybe even withhold access altogether. Interestingly, this dialogue is usually conducted outside myExperiment making it invisible to other consumers (and the service provider).

4.2   Reusability: The Burden of Curation

To be reusable an asset must be sufficiently well described that it can be understood, so that the consumers and producers have a shared understanding [30]. However, the cost of preparing metadata (known as *curation* and *annotation*) combined with data flirting conspires against data being sufficiently well described to be reusable [38]. The workflow in Fig. 2 integrates several datasets. Its complexity is due to mapping between various identifiers, nomenclatures, schemas, structures, formats and terminologies. Reuse of data is hard. Consumers need assets described as well as possible by understandable and explicit reporting standards: insufficient contextual information about methods means they cannot really be trusted or validated, and if results cannot be understood they will be reinvented rather than reused. Furthermore if they cannot be redone then there is a risk of misunderstanding and misuse. Reuse is correlated with familiarity (we find this in myExperiment) but this reduces the opportunity for innovation and cross-discipline sharing. To overcome the many esoteric formats and nonstandard terminologies that face consumers, biology has developed a range of *reporting standards:* 150+ minimum reporting guidelines, 260+ terminologies and 50+ exchange formats listed by the BioSharing.org initiative, mostly targeted for specific data types.

For providers, curating is a burden. The range of standards is bewildering and many are difficult to adopt. A survey of 160 major biology data providers revealed that although 74% used controlled vocabularies only 26% used community standards, and although 31% used minimum reporting checklists; only 8% used those recognized by the community [53]. It takes knowledge, skill, effort and time to curate, especially when using new or combinations of technologies [47], and consistent description – particularly when contributors self-curate – is hard, especially in the absence of tool support. In reality, quality curation is only evident when professional, dedicated curators, such as members of the International Society for Biocuration, are paid to do it.

The general view is that the best time to describe something is at the time it comes into being. For data that would be when it is collected. For a workflow it would be when it is designed. However, experimental data may be acquired with the expectation it be thrown away or never published. A workflow may be created for private or temporary use, and it is uneconomic to curate results that are anticipated to be disposable or private. So curation is left until later when the reporting information is difficult to recover retrospectively. Online, public e-Laboratories are particularly vulnerable to poor curation. Self-curation is commonly sketchy as it is usually only intended for the author or close colleagues. Contributions by non-authors, for example reviews and ratings, require effort. Our experiences indicate that the technical incorporation of reporting standards is relatively straightforward but getting contributors to curate against them is very difficult.

4.3   Easing the Curation Burden: Ramps for Knowledge Transfer

Convenient knowledge mechanisms are a crucial component of successful knowledge transfer between people [30], and hence knowledge turning. Providers, and consumers, need convenient *automated and manual curation ramps*. A "ramp" is a mechanism embedded in a routine practice or familiar tool that eases a user to use a technology. For example, a common instrument for data collection is the Microsoft Excel spreadsheet. The RightField tool [59] wires acceptable terms from community controlled vocabularies into Excel templates. Data is thus collected using selections against the correct terminologies by the experimentalist without changing their work practices or tools. *Stealth ramps* attempt to gather metadata in the appropriate tool at the right time in a familiar work practice, a kind of knowledge acquisition by stealth. *Automated ramps* use instrumentation for knowledge collection, for example the Taverna Workbench is instrumented to record usage and responses of workflows (from myExperiment) and services (from BioCatalogue) to feedback on operational and usage profiles. *Collective ramps* (re)assemble the component parts of an investigation scattered across databases. Commons, wikis, our e-Laboratories and special programs like ENCODE in genomics, are examples.

Collective ramps are vulnerable to a kind of "tragedy of the commons". People use them but do not contribute to or maintain them, instead relying on other curators to integrate and validate descriptions, and other data providers to submit and check data. Recent work in identification of co-author groups and formally declared consortia are first steps in establishing responsibilities for stewardship over complex datasets spanning multiple institutions, journals, databases and funders. The suggestion is that more complete and granular information about the people who generate knowledge will contribute to sustainable access to the datasets in perpetuity [39]; however, fundamentally we need to ensure contributors are rewarded.

4.4   Credit Where Credit Is Due

Liebowitz [30] highlights reciprocity, the intrinsic worth of knowledge, and interpersonal trust and respect as factors for successful knowledge sharing. Scientific reputation is the key measure of worth and respect, and the giving and gaining of credit is the way we express reciprocity and measure reputation. Incentivizing through credit is needed to accelerate the sharing and adoption of results and reward the burden of curation. Scientists fight to get their names onto papers because currently credit is based on peer-citation of articles. However, now web-scale information technology can build credit and attribution networks at the *article* (not journal) level and for *all* digital instruments. Altmetrics (altmetrics.org) and Scientometrics widen credit metrics to all the commodities of science essential for communication (blogs, wikis), and reproducibility (software, data, models, methods), the better for measuring impact. Downloads and views, service calls, expert opinion in public peer review, links, bookmarks, citations in twitter or blogs all contribute to a richer picture. myExperiment download and visit statistics, cross-attributions and trackable derivations, ratings and favorite bookmarking, and references in papers are all bragging material for their authors. Technologies for data citation, like Datacite.org, and for tracking the attribution and acknowledgement of researchers, like Orcid.org, needs to be wired into our data commons.

Technically, we have to rethink what we mean when we cite data and methods. Published articles do not change but database entries improve and software evolves. Citing resources that are both archival and active is an open question. Citing web pages that are in flux is similarly challenging [56]. Provenance tracking transparently records where results came from, but we also need accurate propagation of attributions on method variation and data derivation using shared knowledge models for citation [51] and provenance, such as that proposed by the W3C's Provenance Working Group (www.w3.org/TR/prov-dm).

Sociologically, we need community governance regulators to recognize the value of alternative metrics and to build reputation and asset economies for data, method and code. For example, software, because it can be copied and distributed at essentially no cost, opens the door to unprecedented levels of sharing and collaborative innovation. However reputation economies of software production are not well rewarded through the traditional reputation economy of science leading to over-production of independent scientific software packages, and the under-production of collaborative projects based on pre-existing codes [25]. Paid service professionals, scientists who produce intrinsic software as a by-product of their work, and scientists who are already well established produce software because they do not need academic reward. Where the software *is* the academic credit then recognition, respect and sustainability is essential or else it will not happen. Similar findings are made for data [6]. Data and software journals are temporary measures between old and new credit systems.

Curators are skilled people, required to be experts in both instrument and information technologies, and are motivated by many different drivers that are not just financial reward. The most significant incentives are reputation building, altruistic improvement of the quality of information for their field and the not-so-altruistic promotion of their ideas. However, curators are undervalued and low worth is applied to their contribution, even the professional ones. Curators, and scientific informaticians, need to be first class citizens rather than the "blue collar" scientists as they are often viewed, without a recognized career path. Crowd-sourced curation is needed to match the scales of data but we know that voluntary peer review does not just happen [2]. Reviewers, commentators and curators of third party, openly published data need tangible recognition and social kudos. An excellent review should be citable.

5   Research Objects and Workflow Preservation

We have argued that scientific workflows are computational *instrument technologies* for executing methods and *information technologies* for recording and disseminating computational method. However, they are not only "technical objects" which can be interpreted and executed by machines. They are simultaneously "social objects" that are shared by people as part of the flow in scientific knowledge turns in collaborative environments like myExperiment. As *information objects* they have reusable and repurposable knowledge capital in their own right. They encode scientific methods and know-how and are thus worth exchanging and preserving, subject to the same stresses as data for sharing, credit, attribution and curation. As *digital instruments* they are components in the reproducibility of computational experiments, components of a greater experimental design, linked with publications, more data and other computational and experimental methods, and aggregations combining data, configurations and executable steps (services, codes, tools, databases).

We need workflows to be both embedded and non-embedded within scholarly communications, to behave as independently and collectively, and to be both a record and an executable. We need a form of information currency that allows workflows to: (i) be, and be part of, an aggregated and richly annotated complex scholarly communication; (ii) be an accurate, citable preserved record of method; and (iii) be an actively conserved executable method.

The printing press gave us the academic paper, but a picture of a workflow in a PDF is not going to do the job. Force11 [16] calls for "a new, enriched form of scholarly publication that enables the creation and management of relationships between knowledge, claims and data" and the need for "a full record of the research undertaken requires preservation of these processing steps and software tools employed, in addition to the datasets upon which they acted".

The Research Object (RO) [4], depicted in Fig. 4, aggregates potentially scattered resources that are collected in order to support research activity, investigation and experimentation. A Workflow Research Objects bundles a workflow, together with provenance traces obtained by its enactment, and annotations that semantically describe the domains of the workflow parameters, its operational semantics, its version history, author attributions, citation credit, license etc. An Experiment Research Object bundles a Workflow Research Object with others and with data, ideas, people, a description of the experiment, publications and so on, and the semantic interrelationships between them.

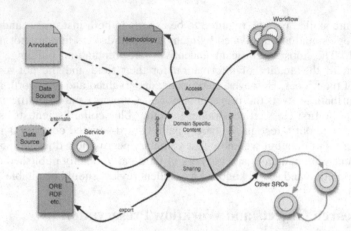

Fig. 4. A notional schematic of a Research Object

The notion of collecting or aggregating resources is not new. Hunter proposes the idea of Scientific Publication Packages (SPP) to describe "the selective encapsulation of raw data, derived products, algorithms, software and textual publications" [28]. SPPs are motivated primarily by the need to create archives for the variety of artifacts produced during the course of a scientific investigation; they ideally contain data, methods, software and documents, but also their provenance as well. The LiquidPub project introduces Scientific Knowledge Objects [20], which describe aggregation structures intended to describe scientific papers, books and journals. A key aspect of this approach is consideration of the *lifecycle* of publications in terms of "states": Gas, Liquid and Solid, which represent early, tentative and finalized work respectively.

Wf4Ever (www.wf4ever-project.org) is an EU-funded STREP project that aims to develop technological infrastructure for the preservation and efficient retrieval and reuse of scientific workflows, in particular through definition of suitable Research Objects and the services that support their creation and management. The focus of the project is on workflows as concrete descriptions of *method*, and of workflow-centric Research Objects, particularly examining the duality of both preserving computational workflows and conserving them in order to support reproducibility. The specific user domains targeted in the project are Astrophysics and Genomics.

From a social perspective ROs must support different roles. For the creator and contributor they should gather credit metrics and provide citation mechanisms; for the reader they should gather and provide quality measures; for the re-user they should provide execution and repair mechanisms; for the reviewer they should provide a means to compare reruns and for the curator they should provide mechanisms to gather, maintain and validate annotations. From a technical point of view ROs that use external, specific or fragile services, data or execution platform are susceptible to *decay*. Changes to, or unavailability of, external resources may compromise the possibility of re-executing a workflow and "reproducibility by rerun" is likely to be problematic. ROs need to carry sufficient provenance information in order to support the replay of an execution [13] or an examination of the processes enacted in order to allow validation.

The aggregated structure of the RO also supports the repurposing of constituent parts. The RO provides the container within which information relating to the workflow and its use or execution can be maintained. If this information is of sufficient detail to allow a refactoring of the process (e.g. via the substitution of an alternate appropriate service or dataset), method *conservation* can be achieved. In terms of the social role of Research Objects, *understanding* is the key to useful preservation. Again, the inclusion of information regarding the provenance of results is facilitating the understanding of those results that then supports reusability.

The Workflow Research Object approach in Wf4Ever defines (i) an abstract RO data model and concrete encodings; (ii) core RO services (such as credit management and execution capability management) and added-value RO services (such as recommendation systems and quality control); and (iii) the protocols for interoperating services and managing the model. Using standardized web infrastructure makes the approach backwards compatible, adoptable by publishers and libraries and future enabled. We specifically propose encoding ROs into Linked Data [23]. Linked Data has uptake in scientific domains [7] and neatly fits scholarly and semantic publishing [50]. We intend that the RO model itself be small and will extensively reuse community ontologies for citation, discourse, provenance and so on. Similarly we aim to leverage standard models for aggregation and harvesting, such as OAI-ORE and OAI-PMH. Linked Data publication is complemented and enriched by ROs [4]. The ROs not only aggregate resources, but add additional annotations and metadata supporting both the technical and social roles that those objects play.

The view of workflows (and associated aggregations) as social objects has also been observed in the myExperiment platform [12] with packs demonstrating a role in workflow reuse and curation. The Wf4Ever RO reference implementation of services is being built using myExperiment, Taverna, BioCatalogue and the dLibra Digital Repository. However the aim is not to create a monolithic system but lightweight components that can be incorporated into data services such as DataVerse (www.thedata.org), digital repositories and platforms such as Galaxy.

6   Conclusions

A "knowledge turn" is the cycle of a process that derives more good leading to an advance. The majority of scientific advances in the public domain result from collective efforts that depend on rapid exchange and effective reuse of results. We have powerful computational instruments, such as scientific workflows, coupled with widespread online information dissemination to accelerate knowledge cycles.

However knowledge turns between researchers continue to lag. Open science and open data are still movements in their infancy and method obfuscation continues to obstruct reproducibility. An ecosystem of "Research Objects" is a possible technical approach towards reusable results. In this paper we have suggested that the shared objects of scientific practice, which underlie knowledge turns, are both technical and social – and indeed that the real obstacles are social.

Strategically, we want to do our best to circulate results and methods so we can attack the big goals of science, like curing Chordoma and protecting the cattle of Africa. But operationally the metrics, processes and norms developed over the past 50 years need serious revision to meet these goals. Until transparent, open science is rewarded it will remain elusive and safer to hug than share. Until curation is recognized as a necessity rather than a luxury results will remain un-reusable. The whole scientific community – from the lab to the publisher and policy makers – needs to rethink and re-implement its value systems for scholarship, data, methods and software. Otherwise we are seriously letting down Josh Sommer.

Acknowledgements. We acknowledge the many members of the myGrid team. We thank Tim Clark for introducing us to Shapin, and acknowledge Tim's insights along with those of Marco Roos, Jose Enrique Ruiz, Josh Sommer, Chris Taylor, Sweitze Roffel, Robert Stevens, Andy Brass, Paul Fisher, Katy Wolstencroft, Jay Liebowitz, Dawn Field, James Howison, Heather Piwowar, Victoria Stodden, Susanna-Assunta Sansone, Phil Bourne, Scott Edmunds, Anita De Waard and Chris Borgman. We thank Ian Cottam for his unswerving support. The work was supported by EU FP7 270192 Wf4Ever, EPSRC EP/G026238/1 myGrid Platform Grant and the BBSRC BB/I004637/1 SysMO-DB2.

References

1. Adie, E.: Commenting on scientific articles (PLoS edition), `http://blogs.nature.com/nascent/2009/02/commenting_on_scient ific_artic.html` (last accessed February 11, 2009)
2. Alsheikh-Ali, A.A., Qureshi, W., Al-Mallah, M.H., Ioannidis, J.P.A.: Public Availability of Published Research Data in High-Impact Journals. PLoS One 6(9) (2011)
3. Baker, M.: Next-generation sequencing: adjusting to data overload. Nature Methods 7, 495–499 (2010)
4. Bechhofer, S., Buchan, I., et al.: Why linked data is not enough for scientists. Future Generation Computer Systems 29(2), 599–611 (2012), doi:10.1016/j.future.2011.08.004
5. Bhagat, J., Tanoh, F., et al.: BioCatalogue: a universal catalogue of web services for the life sciences. Nucleic Acids Research 38(suppl. 2), W689–W694 (2010)
6. Borgman, C.L.: The Conundrum of Sharing Research. Journal of the American Society for Information Science and Technology, 1–40 (2011)
7. Chen, B., et al.: Chem2Bio2RDF: a semantic framework for linking data and mining chemogenomic and systems chemical biology data. BMC Bioinformatics 11, 255 (2010)
8. Ciccarese, P., Wu, E., Wong, G., Ocana, M., Kinoshita, J., Ruttenberg, A., Clark, T.: The SWAN biomedical discourse ontology. J. Biomed. Inform. 41(5), 739–751 (2008)
9. Clark, T., Kinoshita, J.: Alzforum and SWAN: The Present and Future of Scientific Web Communities. Briefings in Bioinformatics 8(3), 163–171 (2007)
10. Daub, J., Gardner, P.P., Tate, J., et al.: The RNA WikiProject: Community annotation of RNA families. RNA 14(12), 2462 (2008)
11. De Roure, D., Goble, C., Stevens, R.: The Design and Realisation of the myExperiment Virtual Research Environment for Social Sharing of Workflows. Future Generation Computer Systems 25(5), 561–567 (2009)
12. De Roure, D., Bechhofer, S., Goble, C., Newman, D.: Scientific Social Objects: The Social Objects and Multidimensional Network of the myExperiment Website. In: IEEE Third International Conference on Social Computing (socialcom), October 9-11, pp. 1398–1402 (2011), doi:10.1109/PASSAT/SocialCom.2011.245

13. De Roure, D., Belhajjame, K., et al.: Towards the Preservation of Scientific Workflows. In: Proc. 8th Intl. Conf. on Preservation of Digital Objects (2011)

14. Eysenbach, G.: Can Tweets Predict Citations? Metrics of Social Impact Based on Twitter and Correlation with Traditional Metrics of Scientific Impact. J. Med. Internet Res. 2011 13(4), e123 (2012)

15. Fisher, P., et al.: A systematic strategy for large-scale analysis of genotype–phenotype correlations: identification of candidate genes involved in African trypanosomiasis. Nucleic Acids Research 35(16), 5625–5633 (2007)

16. Future of Research Communications and e-Scholarship (FORCE 11) Force11 Manifesto (2011), http://force11.org/white_paper (last accessed February 2, 2012)

17. Galperin, M.Y., Fernandez-Suarez, X.M.: The 2012 Database Issue and the online Molecular Biology Database Collection. Nucleic Acids Research 40(D1), D1–D8 (2012)

18. Gargouri, Y., Hajjem, C., Larivière, V., et al.: Self-Selected or Mandated, Open Access Increases Citation Impact for Higher Quality Research. PLoS One 5(10) (2010)

19. Gavish, M., Donoho, D.: A universal identifier for computational results. Procedia Computer Science 4, 637–647 (2011)

20. Giunchiglia, F., ChenuAbente, R.: Scientific Knowledge Objects V.1, Technical Report DISI-09-006, University of Trento (2009)

21. Goecks, J., Nekrutenko, A., Taylor, J.: The Galaxy Team. Galaxy: a comprehensive approach for supporting accessible, reproducible, and transparent computational research in the life sciences. Genome Biol. 11(8), R86 (2010)

22. Gorp, P.V., Mazanek, S.: Share: a web portal for creating and sharing executable research papers. Procedia Computer Science 4, 589–597 (2011); Proc. Intl. Conf. on Comp. Sci.

23. Heath, T., Bizer, C.: Linked Data: Evolving the Web into a Global Data Space. Synthesis Lectures on the Semantic Web: Theory and Technology 1(1), 1–136 (2011)

24. Hey, T., Tansley, S., Tolle, K. (eds.): The Fourth Paradigm: Data-Intensive Scientific Discovery, Microsoft (2009)

25. Howison, J., Herbsleb, J.D.: Scientific software production: incentives and collaboration. In: Proc. ACM 2011 Conf. Computer Supported Cooperative Work, pp. 513–522 (2011)

26. Hull, D., Wolstencroft, K., Stevens, R., Goble, C., Pocock, M., Li, P., Oinn, T.: Taverna: A tool for building and running workflows of services. Nucleic Acids Research 34(Web Server Issue), 729–732 (2006)

27. Hull, D., et al.: Defrosting the digital library: bibliographic tools for the next generation web. PLoS Comput. Biol. 4(10), e1000204 (2008)

28. Hunter, J.: Scientific Publication Packages – A Selective Approach to the Communication and Archival of Scientific Output. Intl. J. of Digital Curation 1(1) (2006)

29. Kell, D.B., Oliver, S.G.: Here is the evidence, now what is the hypothesis? BioEssays 26(1), 99–105 (2004)

30. Liebowitz, J., Ayyavoo, N., Nguyen, H., Carran, D., Simien, J.: Cross-generational knowledge flows in edge organizations. Industrial Management & Data Systems 107(8), 1123–1153 (2007)

31. Levison, S.F., et al.: Colonic transcriptional profiling in resistance and susceptibility to Trichuriasis: phenotyping a chronic colitis and lessons for iatrogenic helminthosis. Inflammatory Bowel Diseases 16(12), 2065–2079 (2010)

32. Lintott, C.J., Schawinski, et al.: Galaxy Zoo: morphologies derived from visual inspection of galaxies from the Sloan Digital Sky Survey. Monthly Notices of the Royal Astronomical Society 389, 1179–1189 (2008)

33. Maleki-Dizaji, S., Rolfe, M., Fisher, P., Holcombe, M.: A Systematic Approach to Under-standing Bacterial Responses to Oxygen Using Taverna and Webservices. In: 13th International Conference on Biomedical Engineering, pp. 77–80 (2009)
34. Mesirov, J.: Accessible Reproducible Research. Science 327(5964), 415–416 (2010)
35. McLuhan, M.: Understanding Media: The Extensions of Man. McGraw Hill (1964)
36. Nielson, M.: Reinventing Discovery: The New Era of Networked Science. Princeton University Press (2011)
37. Nature Editorial "Illuminating the black box". Nature 442, 1 (2006)
38. Nature Editorial "Data's shameful neglect". Nature 461, 145 (2009)
39. Nature Genetics Editorial "It's not about the data". Nature Genetics 44, 111 (2012)
40. Nature Special Issue on Big Data. Nature 455 (2008)
41. Nowakowski, P., Ciepiela, E., et al.: The collage authoring environment. Procedia Computer Science 4, 608–617 (2011); Proc. of the Intl. Conf. on Comp. Science
42. Noyes, H., et al.: Genetic and expression analysis of cattle identifies candidate genes in pathways responding to Trypanosoma congolense infection. PNAS 108(22), 9304–9309 (2011)
43. Orr, J.: One Good Turn,
http://cofes.com/About/OneGoodTurn/tabid/57/Default.aspx
(last accessed January 31, 2012)
44. Pettifer, S., McDermott, P., Marsh, J., Thorne, D., Villeger, A., Attwood, T.K.: Ceci n'est pas un hamburger: modelling and representing the scholarly article. Learned Publishing 24(3), 207–220 (2011)
45. Rennie, C., Hulme, H., Fisher, P., Halp, L., Agaba, M., Noyes, H., Kemp, S., Brass, A.: A systematic, data-driven approach to the combined analysis of microarray and QTL data. Developments in Biologicals 132, 293–299 (2008)
46. The Research Information Network. Open science case studies (2010),
http://www.rin.ac.uk/
47. Sansone, S.-A., et al.: Toward interoperable bioscience data. Nat. Gen. 44, 121–126 (2012)
48. Schroeder, R.: e-Research Infrastructures and Open Science: Towards a New System of Knowledge Production? Prometheus: Critical Studies in Innovation 25(1), 1–17 (2007)
49. Shapin, S.: Pump and Circumstance: Robert Boyle's Literary Technology. Social Studies of Science 14(4), 481–520 (1984)
50. Shotton, D., Portwin, K., Klyne, G., Miles, A.: Adventures in semantic publishing: exemplar semantic enhancement of a research article. PLoS Comp. Bio. 5(4) (2009)
51. Shotton, D.: CiTO, the Citation Typing Ontology. J. of Biomed. Sem. 1(suppl. 1), S6 (2010)
52. Shotton, D.: The Five Stars of Online Journal Articles — a Framework for Article Evaluation. D-Lib Magazine 18(1/2) (January/February 2012)
53. Southan, C.: Elixir Database Provider Survey, report from ELIXIR Preparatory Phase (2009), http://www.elixir-europe.org/prep (last accessed January 31, 2011)
54. Sommer, J.: Sage Commons: Josh Sommer, Chordoma Foundation. Video available on, http://fora.tv/2010/04/23/Sage_Commons_Josh_Sommer_Chordoma_Foundation
55. Stodden, V.: The Scientific Method in Practice: Reproducibility in the Computational Sciences. MIT Sloan Research Paper No. 4773-10 (2010), Available at SSRN, http://ssrn.com/abstract=1550193, doi:10.2139/ssrn.1550193

56. Van de Sompel, H., et al.: An HTTP-Based Versioning Mechanism for Linked Data. In: Proc. of Linked Data on the Web, LDOW 2010 (2010), http://arxiv.org/abs/1003.3661v1
57. Waldrop, M.: Big data: Wikiomics. Nature 455, 22–25 (2008)
58. Whitfield, J.: Collaboration: Group Theory. Nature 455, 720–723 (2008)
59. Wolstencroft, K., Owen, S., et al.: RightField: Embedding ontology annotation in spreadsheets. Bioinformatics 27(14), 2012–2022 (2011)
60. Wroe, C., Goble, C., et al.: Recycling workflows and services through discovery and reuse. Concurrency and Computation: Practice and Experience 19(2), 181–194 (2007)
61. Wuchty, S., Jones, B.F., Uzzi, B.: The increasing dominance of teams in production of knowledge. Science 316(5827), 1036–1039 (2007)
62. Yakowitz, J.: Tragedy of the Data Commons. Harvard J. of Law and Tech. 25 (2011)

Part I

Knowledge Discovery
and Information Retrieval

Mining Graphs of Prescribed Connectivity

Natalia Vanetik

Shamoon College of Engineering, Beer Sheva, Israel
natalyav@sce.ac.il

Abstract. Many real-life data sets, such as social, biological and communication networks are naturally and easily modeled as large labeled graphs. Finding patterns of interest in these graphs is an important task, but due to the nature of the data not all of the patterns need to be taken into account. Intuitively, if a pattern has high connectivity, it implies that there is a strong connection between data items. In this paper, we present a novel algorithm for finding frequent graph patterns with prescribed connectivity in large single-graph data sets. We also show how this algorithm can be adapted to a dynamic environment where the data changes over time. We prove that the suggested algorithm generates no more candidate graphs than any other algorithm whose graph extension procedure we employ.

Keywords: Mining graphs, Graph connectivity.

1 Introduction

Representing large complex naturally occurring data structures as labeled graphs has gained popularity in the last decade due to the simplicity of the translation process and because such a representation is intuitive to users. The graph is now a standard format for representing social and biological networks, biochemical and genetic data, and Web and document structure. Frequent subgraphs that represent substructures of the dataset, which are characteristic to that dataset, are considered important and useful indicator of the nature of the dataset. Frequent subgraphs are used to build indices for graph datasets [16] that improve search efficiency, to facilitate classification or clustering for machine learning tasks [14], and to determine normal and abnormal structures within the data [7].

Not all of the frequent subgraphs are usually of interest to the user performing a specific search task, both because of the subgraph meaning in the particular database and because of the high complexity of the graph mining problem. This obstacle becomes especially disturbing when the dataset in question is represented by a single very large labeled graph, such as a Web or a DNA sequence. In this paper, we concentrate on the problem of finding frequent subgraphs that satisfy a user-defined constraint of minimum edge connectivity, that determines how many edges should be removed from a graph in order to separate in into two parts. A minimal edge connectivity requirement allows us to discard frequent graphs that do not characterize strong relations between data items in the native dataset. Moreover, the edge connectivity of a graph can be verified fairly easily (in polynomial time and space), unlike some of the other constraints, such as symmetry, maximum clique size etc.

A. Fred et al. (Eds.): IC3K 2011, CCIS 348, pp. 29–44, 2013.

While there is a number of algorithms for the task of general frequent subgraph mining exist (see, for instance [11]), the issue of finding frequent graphs that are subject to connectivity constraints has rarely been addressed in the literature. In [15], the authors address the issue of mining all closed frequent graphs with predefined edge connectivity and propose two algorithms that handle this problem. The algorithms do not address the issue of frequent patterns that have high connectivity but are not closed.

The authors of [13] have proposed the 'CODENSE' algorithm that finds coherent dense subgraphs – all edges in a coherent subgraph exhibit correlated occurrences in the whole graph set; these graphs naturally have high connectivity.

In this paper, we propose a novel graph mining algorithm that finds frequent subgraphs with a user-specified constraint on edge connectivity. Our algorithm uses the minimum cut structure of a graph in order to perform the task efficiently; this structure can be computed in low polynomial time (even linear, if one uses the randomized algorithm of [9]), which makes our algorithm especially suitable for databases consisting of a single large graph. The mincut structure of a graph also allows us to increase frequent patterns by more than just a node or an edge at a time than the standard approach. We also prove the optimality of this algorithm by showing that every frequent subgraph produced by our algorithm (even if it is only used as a building block for a supergraph satisfying edge connectivity constraints) has to be produced by a competing algorithm.

We present two extensions of our algorithm: the first one focuses on weaker connectivity constraint where frequent subgraphs with at least the required connectivity need to be found. This process can be performed effectively using a simple optimization of the CactusMiningAlgorithm and it is proven to be optimal as well. The second extension allows incremental maintenance of frequent subgraphs with prescribed connectivity in dynamic datasets where edge deletions may happen. Our approach eliminates the need repetitive mining and focuses on candidate subgraph structure instead, saving precious computational time.

This paper is organized as follows. Section 2 contains the basic definitions and graph theoretic facts required for our approach. Section 3 describes the algorithm and contains proofs of the algorithm's correctness. Section 4 contains the proof of algorithm's optimality. Section 5 describes the algorithm for weaker connectivity constraint and Section 6 describes the algorithm for incremental frequent subgraph maintenance.

2   Statement of the Problem

2.1   Basic Definitions

In this paper, we deal with undirected labeled graphs. In a graph $G = (V, E)$, V denotes the node set, $E \subseteq V \times V$ denotes the edge set, and each node $v \in V$ has a label $l(v)$. A graph $G' = (V', E')$ is called a *subgraph* of G, denoted by $G' \subseteq G$, if $V' \subseteq V$, $E' \subseteq E$ and every edge in E' has both ends in V'. G' is an *induced subgraph* of G if it is a subgraph of G and for every pair of nodes $v, u \in V'$ such that (u, v) is an edge of G, (u, v) is also an edge of G'.

A graph $G = (V, E)$ is *disconnected* if there exists a partition V_1, V_2 of V so that no edge in E has one end in V_1 and another in V_2. If no such partition exists, G is called *connected*. G is called *k-edge-connected*, $k \in \mathbb{N}$, if G is connected and one has to remove at least k edges from E to make G disconnected.

A partition of edge set E into $X \subset E$ and $\overline{X} := E \setminus X$ is called a *cut*. Removing all edges having one end in X and another in $\overline{X}$ (called $(X, \overline{X})$-*edges*) from G disconnects the graph. The *size* of a cut $(X, \overline{X})$ is the number of $(X, \overline{X})$-edges, denoted $|(X, \overline{X})|$. The $(X, \overline{X})$-edges whose removal disconnects the graph are often also called a cut. A cut of minimum size is called a *minimum cut* or a *mincut*. The least size of a cut in a graph is the *edge connectivity* of a graph. In general, for two foreign subsets $X, Y \subset V$ we denote by $|(X, Y)|$ the number of edges in G with one end in X and another in Y.

We study the problem of graph mining in the following setting: our database is a single large undirected labeled graph G. We are given a user-supplied support threshold $S \in \mathbb{N}$ and a connectivity constraint k and we are looking for all k-edge-connected subgraphs of G with a count of at least S (these subgraphs are called *frequent*). The *count* of a graph in a database is determined by a function *count*() that satisfies the *downward closure property*: for all subgraphs g_1, g_2 of any database graph D such that $g_1 \subseteq g_2$ we always have $count(g_1, D) \geq count(g_2, D)$. The main idea of our approach is to employ the special structure of mincuts in the database graph in order to make the search for frequent k-edge-connected subgraphs faster.

2.2   The Cactus Structure of Mincuts

An unweighted undirected multigraph is called a *cactus* if each edge is contained in exactly one cycle (i.e., any pair of cycles has at most one node in common). Dinitz, Karzanov and Lomonosov showed in [3] that all minimum cuts in a given graph with n vertices can be represented as a cactus of size $0(n)$. This cactus representation plays an important role in solving many connectivity problems, and we use it here for the efficient mining of graphs with connectivity constraints.

Formally, let $G = (V, E)$ be an undirected multigraph and let $\{V_1, ..., V_n\}$ be a partition of V. We denote the set of all minimum cuts of G by $Cuts(G)$. Let $R = (V_R, E_R)$ be a multigraph with node set $V_R := \{V_1, ..., V_n\}$ and edge set $E_R := \{(V_i, V_j) \mid (v_i, v_j) \in E, v_i \in V_i, v_j \in V_j\}$.

Definition 1. *R is a **cactus representation** of $Cuts(G)$ if there exists a one-to-one correspondence $\rho : Cuts(G) \rightarrow Cuts(R)$ such that for every mincut $(X, \overline{X}) \in Cuts(G)$ holds $\rho((X, \overline{X})) \in Cuts(R)$ and for every mincut $(X, \overline{X}) \in Cuts(R)$ $\rho^{-1}((X, \overline{X})) \in Cuts(G)$.*

Dinitz, Karzanov and Lomonosov [3] have proved that for any undirected multigraph, there exists a cactus representation (in fact, they showed that this is always true for any weighted multigraph). A *dual graph* to any cactus representation, if the cactus cycles are taken as nodes, is a tree. The size of a cactus tree is linear in the number of vertices in the original graph, and any cut can be retrieved from the cactus representation in time linearly proportional to the size of the cut. In addition, the cactus displays explicitly all nesting and intersection relations among minimum cuts. Note that a graph can have at most $\binom{n}{2}$ mincuts, where n is the size of graph's node set. The following definition and a fundamental lemma entirely describe the structure of a cactus representation.

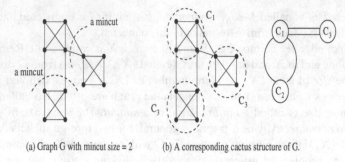

(a) Graph G with mincut size = 2   (b) A corresponding cactus structure of G.

Fig. 1. A cactus structure of a graph

Definition 2. *Let λ be the size of a mincut in graph $G = (V, E)$. A **circular partition** is a partition of V into $k \geq 3$ disjoint subsets $\{V_1, ..., V_k\}$ such that*

1. *$|(V_i, V_j)| = \lambda/2$ when $j - i = 1 \mod k$.*
2. *$|(V_i, V_j)| = 0$ when $j - i \neq 1 \mod k$.*
3. *For $1 \leq a < b \leq k$, $\cup_{i=a}^{b-1} V_i$ is a mincut. Moreover, if any mincut $(X, \overline{X})$ is not of this form, then either X or $\overline{X}$ is contained in some V_i.*

Lemma 1. *[3,1] If X_1 and X_2 are crossing cuts in G (have a non-trivial intersection as sets), then G has a circular partition $\{V_1, ..., V_k\}$ such that each of $X_1 \cap X_2$, $\overline{X_1 \cup X_2}$, $X_1 \setminus X_2$ and $X_2 \setminus X_1$ equal $\cup_{i=a}^{b-1} V_i$ for appropriate choices of a and b.*

Corollary 1. *[3] Every graph has a cactus representation.*

Corollary 2. *[5] Every graph on n vertices has a cactus representation with no more than $2n - 2$ vertices.*

Figure 1 shows a 2-edge-connected multigraph and its cactus representing all three mincuts that exist in the graph. In this example, there is a one-to-one correspondence between the cycles of the cactus and the circular partitions of G.

2.3 Cactus Construction Algorithms

The earliest well-defined algorithm for finding all minimum cuts in a graph uses maximum flows to compute mincuts for all pairs of vertices (see [6]). Karzanov and Timofeev [10] outlined the first algorithm to build a cactus for an unweighted graph. A randomized algorithm by Karger [8] finds all minimum cuts in $O(n^2 \log n)$ time. Fleischer in [5] describes an algorithm that arranges the minimum cuts into an order suitable for a cactus algorithm which runs in $O(nm + n^2 \log n)$ time. Finally, Karger and Panigrahi proposed a near-linear time randomized algorithm in [9].

3 Finding Frequent k-connected Graphs

In this section, we present the CactusMining algorithm for finding all frequent k-connected graphs in a graph database. For simplicity, we assume here that the database is a single large graph; when a database consists of two or more disconnected graphs, graph decomposition and support counting should be performed once for each transaction. The CactusMining algorithm searches for all frequent k-connected subgraphs in a bottom-up fashion and relies on the search space reduction that is implied by the cactus structure of the database.

3.1 Computing the Cactus Structure of a Graph

To compute the cactus structure for a given graph G and a connectivity bound k, we employ a cactus-constructing algorithm, denoted as $BuildCactus()$ (for instance, the one described in [5]).

3.2 Basic Properties

In this section, we describe several useful properties of a cactus mincut structure.

Property 1. Let g be a $(k+1)$-connected subgraph of G. Then g is entirely contained in some V_i, $1 \leq i \leq k$. The converse is not true, i.e. non-k-edge-connected subgraphs of V_i may exist. □

Property 2. Let g be a k-edge-connected subgraph of G. Let C be a minimal subcactus of G with circular partition $\{V_1, ..., V_k\}$ containing g as a subgraph. Then either g contains all the (V_i, V_j)-edges or it contains no such edges.

Proof: This property is trivial since removing a (V_i, V_j)-edge decreases the edge connectivity of a subcactus C containing g.

Corollary 3. *In Property 2, subgraph $g \cap V_i$ contains all the nodes incident to the (V_i, V_j)-edges of a circular partition.*

Proof: Follows from the fact that g contains all (V_i, V_j)-edges.

3.3 Growing Subgraphs

In this section, we describe how an instance of a candidate subgraph can be grown from an existing frequent subgraph instance without violating connectivity constraints.

The intuition behind our subgraph extension approach relies on properties of its location within the cactus structure. Let $T = (V_T, E_T)$ be a dual cactus structure of $(k+1)$-cuts in database D with nodes V_T being the cactus cycles and the adjacency relation E_T determining whether two cactus cycles share a node. Each cactus cycle $C \in V_T$ is a graph, denoted by $C = (V_C, E_C)$, with the structure of a simple cycle. In C, the nodes of V_C are the basic $(k+1)$-connected components of D, i.e. the components that contain no edge of a k-cut in D. Two such components $c_1, c_2 \in V_C$ are adjacent if there exist edges of D that belong to a k-cut and are incident to nodes in c_1 and c_2 (there are precisely $\lceil \frac{k}{2} \rceil$ such edges). To simplify the notation, we say that (c_1, c_2) denotes the set of these edges.

Our goal is to extend instances of frequent graphs gradually, while complying with the following rule:

– do not produce an extension whose cactus structure in C does not ensure k-connectivity.

In order to achieve the objective, our extension procedure depends strongly of the location of an instance within the database cactus structure. Moreover, our approach allows to extend an instance by more than one node.

Let $f \subset D$ be a frequent subgraph instance that we are currently extending. We introduce several additional parameters of f that are updated by our mining algorithm:

1. $f.type$ can assume the values $node$, $cycle$ and $tree$,
2. $f.cycle$ denotes the node $t \in V_T$ containing f as a subgraph (if one exists),
3. $f.tree$ denotes the subtree of T containing f as a subgraph; $|f.tree|$ denotes the number of nodes in the said subtree.

For each value of $f.type$, we propose a separate extension procedure. The first two procedures extend the subgraph instance within its own type; they can fail to extend either because no extension is possible at all or because the type of extension needs to be changed. For $f.type = cycle$ and $f.type = tree$, additional precaution needs to be taken in order to ensure a better search space reduction. In this case, a subgraph of such an instance contained within a $(k+1)$-connected component of the database graph may cause a cut of size less than k to appear in the extended instance. We apply the Contraction() procedure (see Section 3.3) that determines exactly if these subgraphs produce a smaller than required edge cut or not.

The Contraction Procedure. The Contraction() procedure, described in Algorithm 1, receives as an input a subgraph g contained in a $(k+1)$-connectivity component of a cactus structure T, and contracts all the parts of T that are not adjacent or incident to g into single nodes. For each subtree $t \in T$ adjacent to g, $t \setminus g$ is turned into a single node. In fact, every such subtree is turned into a two-node cycle with t and $t \setminus g$ as nodes, and the cycle edges incident to g as edges.

Algorithm 1. Contraction()

Input: subgraph g,
        subcactus T containing g.
Output: contraction of g
 1: $K :=$ a node of T containing g;
 2: **for all** subtrees $t \in T$ incident to K **do**
 3:    replace $t \setminus \{g, \text{cycle edges incident to } g\}$
      with a single node;
 4: **end for**
 5: **return** T;

Figure 2 gives two examples of applying the Contraction() procedure. The following claim ensures correctness of the procedure.

Claim. Let $T = (V_T, E_T)$ be a cactus structure of k-cuts in a graph G and let a subgraph $f \subset G$ span the $(k+1)$-connectivity components of T and contain all the cycle

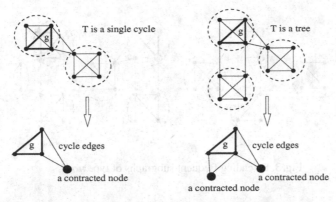

Fig. 2. Contracting subgraphs in $(k + 1)$-connected components

edges of the cactus structure. Then there exists a $(k + 1)$-connectivity component K of G such that Contraction$(f \cap K, T)$ is not k-connected if and only if f is not k-connected.

Proof: The "only if" direction is trivial since the contraction described in Algorithm 1 does not reduce the connectivity of f.

For the "if" direction, let us assume an f that is not k-connected. Then there exists a partition V_1, V_2 of its node set so that $|(V_1, V_2)| < k$. Since the number m of cactus components $C_1, ..., C_m$ in T is at least 2, there exists $i \in [1, m]$ s.t. $|(V_1, V_2) \cap C_i < \frac{k}{2}$. Thus, we have a partition U_1, U_2 of a node set of $f \cap C_i$ so that $|(U_1, U_2)| < \frac{k}{2}$.

Let us denote by $f' :=$ Contraction$(f \cap C_i, T)$. Since there are at most k cactus edges incident to $f \cap C_i$, w.l.o.g. there are at most $\frac{k}{2}$ cactus edges incident to U_1 in f', which we denote by E'. Then $(V_1, V_2) \cup E'$ is an edge cut of f' of size less than k, and f' is not k-connected.

Algorithm 2. ExtendNodeType()

Input: Cactus $T = (V_T, E_T)$ of $(k + 1)$-cuts in D,
           frequent subgraph f.
Output: extensions of f.
 1: $Ext(f) :=$ basic-extend(g);
 2: **for all** $h \in Ext(f)$ **do**
 3:     **if** $h \cap f.cycle \neq f.cycle$ **then**
 4:        $Ext(f) := Ext(f) \setminus \{h\}$;
 5:     **else**
 6:        $h.type = node$;
 7:        $h.cycle = f.cycle$;
 8:     **end if**
 9: **end for**
10: **return** $Ext(f)$;

Extension Procedures. The procedure for $f.type = node$ is described in Algorithm 2. This procedure simply adds a node or an edge to an existing subgraph instance within a $(k + 1)$-connectivity component of D. It uses a basic pre-existing extension procedure

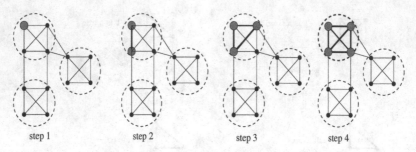

step 1   step 2   step 3   step 4

Fig. 3. Extending frequent subgraphs of type *node*

basic-extend(), for instance such as the one in FSG [11]. An example of ExtendNode-
Type() procedure is given in Figure 3 (extended subgraphs in bold).

The procedure for $f.type = cycle$ is described in Algorithm 3. It extends an instance
f of a frequent graph within a single cycle $C = (V_C, E_C)$ that forms a node of the
cactus dual tree structure. The main concern is to extend a subgraph so as not to create
a not-k-connected instance, and for this purpose the edges E_C must be present in an
extension. Therefore, a frequent graph instance that is a subgraph of V_C must be added
to f. An example of joining subgraphs of type *node* into subgraphs of type *cycle* is
given in Figure 4 (extended subgraphs in bold).

Algorithm 3. ExtendCycleType

Input: frequent subgraph instances F
        of type *node*, frequent subgraph f.
Output: extensions of f.
  1: **let** $C = (\{c_1, ..., c_n\}, E_C) := f.cycle$;
  2: **let** $f \subseteq c_j$;
  3: $Ext(f) := \emptyset$;
  4: $E =: \cup_{i,j}\{(c_i, c_j) \in E_C\}$ (as edge sets);
  5: $F_{\text{good}} := F \cap C$; {frequent subgraphs in C}
  6: **for all** $f \in F_{\text{good}}$ **do**
  7:     $f' :=$ Contraction(f, C);
  8:     **if** f' **is not** k-connected **then**
  9:         $F_{\text{good}} := F_{\text{good}} \setminus \{f'\}$;
 10:     **end if**
 11: **end for**
 12: **for all** $f_i \in F_{\text{good}} \cap c_i$,
        $1 \le i \le n, i \ne j$ **do**
 13:     $h := \cup_{1 \le i \le n, \, i \ne j} f_i \cup E$;
 14:     **if** h is a graph **then**
 15:         $Ext(f) := Ext(f) \cup \{h\}$;
 16:         $h.type = cycle$; $h.cycle = f.cycle$;
 17:         $h.tree = h.cycle$;
 18:     **end if**
 19: **end for**
 20: **return** $Ext(f)$;

The final procedure for $f.type = tree$ is described in Algorithm 4. In this case, f is contained in a subtree of the dual cactus tree structure and it is extended by an instance of a frequent subgraph of type *cycle*. In order not to generate the same instance twice, we assume that the tree T is a directed out-tree and that extending a subtree of T by a node is possible only in the direction of T's edges. An example of joining two subgraphs

---

Algorithm 4. ExtendTreeType

Input: database cactus structure
$\qquad T = (V_T, E_T)$,
$\qquad$ frequent subgraph instances F
$\qquad$ of type *cycle*, frequent subgraph f.
Output: extensions of f.
1: $T' = (V_{T'}, E_{T'}) := f.tree$;
2: $Ext(f) := \emptyset$;
3: **for all** $C \in V_T$ and $C' \in V_{T'}$ **do**
4: $\quad$ **if** $(C', C) \in E_T$ **then**
5: $\qquad$ **for all** $g \in F$ such that $g \cap C = f \cap C$ **do**
6: $\qquad\quad$ $T'' := T' \cup C$;
7: $\qquad\quad$ $g' := \text{Contraction}(g \cap C, T'')$;
8: $\qquad\quad$ **if** g' is k-connected **then**
9: $\qquad\qquad$ $h := f \cup g$;
10: $\qquad\qquad$ $Ext(f) := Ext(f) \cup \{h\}$;
11: $\qquad\qquad$ $h.type = tree$;
12: $\qquad\qquad$ $h.tree = (V_{T'} \cup \{C\}, E_{T'} \cup \{(C', C)\})$;
13: $\qquad\quad$ **end if**
14: $\qquad$ **end for**
15: $\quad$ **end if**
16: **end for**
17: **return** $Ext(f)$;

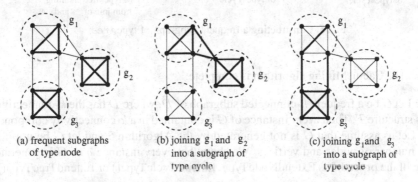

(a) frequent subgraphs   (b) joining g_1 and g_2 (c) joining g_1 and g_3
$\quad$ of type node $\quad$ into a subgraph of $\quad$ into a subgraph of
$\qquad$ $\quad$ type cycle $\quad$ type cycle

Fig. 4. Constructing frequent subgraphs of type *cycle*

of type *cycle* into a subgraph of type *tree* is given in Figure 5 (extended subgraphs in bold).

3.4   The CactusMining Algorithm

The CactusMining algorithm extends each frequent graph until it spans beyond connectivity component of the database, at which point the extension must include a nontrivial subcactus. If such an extension is not possible, the frequent subgraph must be abandoned. The existence of a counting procedure for subgraphs, denoted count(), is assumed.

3.5   Proof of Correctness

The aim of this section is to show that every maximal frequent k-edge-connected subgraph g of G is generated by the above algorithm at some point (completeness), and no subgraph that is not k-edge-connected is added to a candidate set (soundness).

Claim. The CactusMining algorithm is sound.

Proof: This claim is trivial since step 39 of Algorithm 5 filters out all not-k-connected graphs.   □

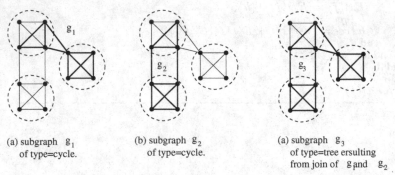

(a) subgraph g_1
of type=cycle.

(b) subgraph g_2
of type=cycle.

(a) subgraph g_3
of type=tree ersulting
from join of g and g_2.

Fig. 5. Constructing a frequent subgraph of type *tree*

Claim. The CactusMining algorithm is complete.

Proof: Let G be a frequent k-connected subgraph of D, where D has the k-connectivity cactus structure T. Then every instance of G is contained in a k-connectivity component of D. Let us assume that G is not generated by the Algorithm 5 and let G be minimal in the number of nodes and vertices. We show that every instance g of G is generated by one of the procedures ExtendNodeType, ExtendCycleType() or ExtendTreeType().

- If $g.type = node$, it is generated by the ExtendNodeType() procedure that filters out nothing. Otherwise, $g.type \in \{cycle, tree\}$ and by Claim 3.3 g is not generated only if g is not-k-connected - a contradiction.   □

Algorithm 5. CactusMining()

Input: graph database D, support S,
   connectivity bound k.
Output: frequent k-edge-connected graphs.
 1: $F_1 :=$ frequent nodes of D;
 2: $D := D \setminus \{\text{non-frequent nodes}\}$;
 3: $BuildCactus(D, k - 1)$;
 4: $D \leftarrow k$-connectivity components of D;
 5: $T := BuildCactus(D, k)$;
 6: $F_1 :=$ frequent nodes; {type=node}
 7: $i := 1$;
 8: **while** $F_i \neq \emptyset$ **do**
 9:    $C_{i+1} := \emptyset$;
10:    **for all** $f \in F_i$ **do**
11:       $C_{i+1} := C_{i+1} \cup \text{ExtendNodeType}(T, f)$;
12:    **end for**
13:    $i := i + 1$;
14:    $F_i :=$ frequent graphs from C_i;
15: **end while**
16: $F_I := \cup_{j=1}^{i-1} F_j$;
17: $F_1 := F_I$; {type=cycle}
18: $i := 1$;
19: **while** $F_i \neq \emptyset$ **do**
20:    $C_{i+1} := \emptyset$;
21:    **for all** $f \in F_i$ **do**
22:       $C_{i+1} := C_{i+1} \cup \text{ExtendCycleType}(F_I, f)$;
23:    **end for**
24:    $i := i + 1$;
25:    $F_i :=$ frequent graphs from C_i;
26: **end while**
27: $F_{II} := \cup_{j=1}^{i-1} F_j$;
28: $F_1 := F_{II}$; {type-tree}
29: $i := 1$;
30: **while** $F_i \neq \emptyset$ **do**
31:    $C_{i+1} := \emptyset$;
32:    **for all** $f \in F_i$ **do**
33:       $C_{i+1} := C_{i+1} \cup \text{ExtendTreeType}(T, F_{II}, f)$;
34:    **end for**
35:    $i := i + 1$;
36:    $F_i :=$ frequent graphs from C_i;
37: **end while**
38: $F_{III} := \cup_{j=1}^{i-1} F_j$;
39: remove $< k$-connected graphs from $F_I \cup F_{II} \cup F_{III}$;
40: **return** $F_I \cup F_{II} \cup F_{III}$

---

4   Candidate Graphs Generated by the Algorithm

In this section, we prove that the CactusMining algorithm is optimal w.r.t. the set of candidate subgraphs generated by it. We show that any algorithm based on pattern extension must produce a superset of candidate subgraphs generated by the Cactus-Mining algorithm.

Theorem 1. *Let AnyAlgorithm be a graph mining algorithm based on pattern extension. Then CactusMining algorithm produces no more candidates than AnyAlgorithm.*

Proof: We show that any frequent candidate subgraph g produced by the CactusMining algorithm has to be produced by AnyAlgorithm. We denote the database graph by D and its cactus tree structure by $T = (V_T, E_T)$. We denote the $(k+1)$-connectivity of the cactus structure by $C_1, ..., C_n$. Let us assume that g is produced by the CactusMining algorithm but not by AnyAlgorithm as a candidate subgraph.

If $g \subseteq g'$, where g' is a k-connected frequent graph in D, then g is produced by AnyAlgorithm (we can assign labels to g''s nodes so that g is the lexicographically minimal extension of g) – a contradiction. Therefore, g is neither k-connected nor is it a subgraph of a frequent k-connected graph in D. Let therefore V_1, V_2 be a partition of g's nodes so that $|(V_1, V_2)| < k$ (i.e. an edge cut of size $< k$ separates V_1 and V_2). We assume first that g has a non-empty intersection with cactus nodes $C_1, ..., C_m$, where $m > 1$. Then g contains all the cactus edges connecting $C_1, ..., C_m$, for otherwise it is not produced by the CactusMining algorithm (as $g.type = cycle$ or $g.type = node$). We denote the edge set (V_1, V_2) separating g by $E_{1,2}$. Since $m > 1$ and $|E_{1,2}| < k$, there exists i s.t. $|C_i \cap E_{1,2}| < \frac{k}{2}$. Thus, we have a partition U_1, U_2 of a node set of $g \cap C_i$ so that $|(U_1, U_2)| < \frac{k}{2}$. Let $g' :=$Contraction(g, C_i). Since there are at most k cactus edges incident to $g \cap C_i$, w.l.o.g. there are at most $\frac{k}{2}$ cactus edges incident to U_1 in g', which we denote E'. Then $E_{1,2} \cup E'$ is an edge cut of g' of size less than k – in contradiction to step 8 of Algorithm 3 or to step 8 of Algorithm 4. Then g is not produced by the CactusMining algorithm – a contradiction.

Let us assume now that $g \subseteq C_i$ for some $i \in [1, m]$. Since g is minimal in the node and edge set, there exists a node or an edge x such that $g - x$ is either a k-connected frequent graph in D or is a subset of a k-connected frequent graph g' in D. We can assume that AnyAlgorithm is locally optimal, and, since Algorithm 2 can use any function as basic-extend(), that the same function is used. Therefore, g is produced by both AnyAlgorithm and the CactusMining algorithm – a contradiction.   □

5   Weaker Connectivity Constraints

Sometimes, users of graph mining systems (network analysts or biologists) may not know the precise connectivity of frequent subgraphs in a large dataset that they look for. Often, only the minimal requirement of connectivity equal to or bigger than the specified constraint is known. In his case, the CactusMining algorithm can be easily modified into an algorithm that finds all such patterns. In fact, if we are given a connectivity bound k and we are looking for all frequent subgraphs that are *at least k-connected* in dataset D, all these subgraphs fall into two categories:

  – subgraphs can be $(k + 1)$-connected and are located inside cactus nodes of k-connectivity cactus structure of D;

– subgraphs that are k-connected but not $(k+1)$-connected that span the cactus edges of D's cactus structure. This property follows from the fact that any frequent subgraph of type *tree* or *node* has a k-cut.

In order to find patterns that are at least $(k + 1)$-connected we modify the Cactus-Mining algorithm in order to focus the search process on subgraphs within cactus nodes of dataset cactus structure. The algorithm ExtendedCactusMining is presented in Algorithm 6.

The ExtendedCactusMining algorithm, much like CactusMining algorithm, produces the subset of candidate subgraphs of any other mining algorithm. Indeed, if a subgraph has connectivity at least $(k + 1)$ that is can only be of type=*node* for otherwise it contains a k-cut. Inside cactus nodes, both algorithms employ the most efficient extension procedure than it available. It is necessary to produce all the candidate patterns inside cactus nodes as such a node can be, for instance, a complete graph of large order. We obtain the following simple corollary from Theorem 1.

Corollary 4. *Let AnyAlgorithm be a graph mining algorithm based on pattern extension. Then ExtendedCactusMining algorithm produces no more candidates than AnyAlgorithm.*   □

Algorithm 6. ExtendedCactusMining()

Input: graph database D, support S, connectivity bound $k + 1$.
Output: frequent graphs that are at least $(k + 1)$-connected.
 1: $F_1 :=$ frequent nodes of D;
 2: $D := D \setminus \{\text{non-frequent nodes}\}$;
 3: $T := BuildCactus(D, k)$;
 4: $D \leftarrow (k + 1)$-connectivity components of D;
 5: $F_1 :=$ frequent nodes; {type=node}
 6: $i := 1$;
 7: **while** $F_i \neq \emptyset$ **do**
 8:     $C_{i+1} := \emptyset$;
 9:     **for all** $f \in F_i$ **do**
10:         $C_{i+1} := C_{i+1} \cup \text{ExtendNodeType}(T, f)$;
11:     **end for**
12:     $i := i + 1$;
13:     $F_i :=$ frequent graphs from C_i;
14: **end while**
15: $F = \bigcup_i F_i$;
16: remove $< (k + 1)$-connected graphs from F;
17: **return** F

---

6   Connectivity Constraints and Dynamic Datasets

The next issue we address is the fact that many single-graph datasets have dynamic nature, i.e. edges may appear and disappear over time. Social networks, web and communication/routing networks are natural examples; protein structure also changes over time while protein folds or moves from one conformation to another. A trivial approach to handling the task of finding all frequent k-connected subgraphs in a dynamic dataset would be to run the mining algorithm anew every time a change occurs. This approach, however, is computationally infeasible. We suggest an easier solution, keeping in mind the fact that computing the cut structure of a pattern is an easy polynomial task. Our approach consists of following steps.

- We run the CactusMining algorithm once and keep the set of candidate subgraphs and their instances in the database for future use. One should notice that the instances are usually kept in the system anyway since they are used by data miners and analysts for insight and verification.
- For each k-connected candidate subgraph we compute its k-connectivity structure (the cactus).
- When a change occurs in the dataset, we look at the exact place this change happened and act accordingly. Our algorithm handles edge deletion only, since addition of an edge may cause subgraphs previously discarded by the mining process to become valid again. In this case, a re-computation may be required.

The DynamicCactusMining algorithm depicted in Algorithm 7 describes a simple procedure for incremental maintenance of frequent k-connected patterns in case of edge deletion. The DynamicCactusMining algorithm performs no mining; it uses previously computed cactus structure of frequent subgraph instances, denoted by $f.cactus$, in order to determine whether the change is dangerous or not. In the former case, the instance

needs to be discarded while in the latter case the instance is still k-connected. However, the subgraph represented by this instance may change. If this happens, the count of subgraph in question may need to be updated. Thus, even when k-connectivity is preserved, frequency may change and so can the cactus structure of an instance.

Algorithm 7. DynamicCactusMining()

Input: instances of frequent k-connected subgraphs I,
          deleted edge $e = (v, u)$, connectivity bound k, support S.
Output: frequent k-edge-connected graphs.
 1: $NewI := \emptyset$;
 2: **for all** $f \in I$ **do**
 3:    $C := f.cactus$;
 4:    **if** $u, v \in$ same node of C **then**
 5:       $f := f \setminus e$;
 6:       $count(f)$++;
 7:       $count(f \cup e)$- -;
 8:       $NewI := NewI \cup \{f\}$;
 9:       $f.cactus := BuildCactus(f, k)$;
10:    **end if**
11: **end for**
12: $F' := \emptyset$.
13: **for all** $f \in NewI$ **do**
14:    **if** $count(f) \geq S$ **then**
15:       $F' := F' \cup \{f\}$;
16:    **end if**
17: **end for**
18: **return** F'.

Claim. DynamicCactusMining algorithm finds all k-connected patterns in the dataset with edge e removed.

Proof: Let our dataset be denoted D. Each k-connected candidate subgraph in $D \setminus e$ is a candidate subgraph in D since edge deletion can only decrease connectivity. The case where a candidate subgraph f is k-connected but $f \setminus e$ is not can only happen when the edge e is a cactus edge. Indeed, if $f \setminus e$ is not k-connected, then e lies in a k-cut X, which by definition contains only edges from $f.cactus$. Moreover, since every edge $g \in f.cactus$ lies in some k-cut X of f, removing e from f creates a cut $X \setminus e$ in $f \setminus e$ of size $(k - 1)$. Therefore, lines 5–9 of the DynamicCactusMining algorithm keep a subgraph instance if and only if it remains k-connected. Since the graph changes after edge deletion, instance counts need to be updated as it happens in lines 6–7 of the algorithm. Finally, due to line 14 only frequent subgraphs are kept in the final set.   □

7   Conclusions

In this paper, we have presented the CactusMining algorithm for mining frequent k-edge connected subgraphs in a graph database, where k is a user-defined integer constant. The method presented here is defined and described for a single graph database case, but is adapted trivially to multiple graph databases. Our method relies on the Dinitz-Karzanov-Lomonosov cactus minimum cut structure theory and on the existence of efficient polynomial algorithms that compute this structure. Our algorithm employs the pattern-growth approach, and the cactus structure of mincuts allows us to grow frequent subgraphs by more than a node or an edge at a time. We have proved that the CactusMining algorithm is sound and correct, and have also shown that the set of frequent patterns it produces is the least possible, i.e. a competing graph mining algorithm will produce all the candidate patterns that our algorithm produces. We demonstrated how our approach can be adapted to the case of weaker connectivity constraints (ExtendedCactusMining algorithm) and the case of dynamic dataset where edge deletion happen (DynamicCactusMining algorithm).

References

1. Bixby, R.E.: The minimum number of edges and vertices in a graph with edge connectivity n and m n-bonds. Networks 5, 253–298 (1975)
2. De Vitis, A.: The cactus representation of all minimum cuts in a weighted graph. Technical Report 454, IASI-CNR (1997)
3. Dinitz, E.A., Karzanov, A.V., Lomonosov, M.V.: On the structure of a family of minimal weighted cuts in a graph. In: Fridman, A.A. (ed.) Studies in Discrete Optimization, pp. 290–306. Nauka, Moscow (1976)
4. Fiedler, M., Borgelt, C.: Support computation for mining frequent subgraphs in a single graph. In: International Workshop on Mining and Learning with Graphs (2007)
5. Fleischer, L.: Building Chain and Cactus Representations of All Minimum Cuts from Hao-Orlin in the Same Asymptotic Run Time. In: Bixby, R.E., Boyd, E.A., Ríos-Mercado, R.Z. (eds.) IPCO 1998. LNCS, vol. 1412, pp. 294–309. Springer, Heidelberg (1998)

6. Gomory, R.E., Hu, T.C.: Multi-terminal network flows. J. Soc. Indust. Appl. Math. 9(4), 551–570 (1991)
7. Horváth, T., Ramon, J.: Efficient frequent connected subgraph mining in graphs of bounded tree-width. Theor. Comput. Sci. 411(31-33), 2784–2797 (2010)
8. Karger, D.R., Stein, C.: A new approach to the minimum cut problem. Journal of the ACM 43(4), 601–640 (1996)
9. Karger, D.R., Panigrahi, D.: A near-linear time algorithm for constructing a cactus representation of minimum cuts. In: SODA 2009, pp. 246–255 (2009)
10. Karzanov, A.V., Timofeev, E.A.: Efficient algorithms for finding all minimal edge cuts of a nonoriented graph. Cybernetics 22, 156–162 (1986); Translated from Kibernetika 2, 8–12 (1986)
11. Kuramochi, M., Karypis, G.: Frequent Subgraph Discovery. In: ICDM 2001, pp. 313–320 (2001)
12. Nagamochi, H., Kameda, T.: Canonical cactus representation for minimum cuts. Japan Journal of Industrial Appliel Mathematics 11, 343–361 (1994)
13. Papadopoulos, A., Lyritsis, A., Manolopoulos, Y.: Skygraph: an algorithm for important subgraph discovery in relational graphs. Journal of Data Mining and Knowledge Discovery 17(1) (2008)
14. Seeland, M., Girschick, T., Buchwald, F., Kramer, S.: Online Structural Graph Clustering Using Frequent Subgraph Mining. In: Balcázar, J.L., Bonchi, F., Gionis, A., Sebag, M. (eds.) ECML PKDD 2010, Part III. LNCS, vol. 6323, pp. 213–228. Springer, Heidelberg (2010)
15. Yan, X., Zhou, X.J., Han, J.: Mining Closed Relational Graphs with Connectivity Constraints. In: ICDE 2005, pp. 357–358 (2005)
16. Zhang, S., Li, S., Yang, J.: GADDI: distance index based subgraph matching in biological networks. In: EDBT 2009, pp. 192–203 (2009)

Concept Discovery and Automatic Semantic Annotation for Language Understanding in an Information-Query Dialogue System Using Latent Dirichlet Allocation and Segmental Methods

Nathalie Camelin[1], Boris Detienne[2], Stéphane Huet[2],
Dominique Quadri[2], and Fabrice Lefèvre[2]

[1] Université du Maine, LIUM, Le Mans, France
[2] Université d'Avignon, LIA-CERI, Avignon, France
forename.surname@lium.univ-lemans.fr,
forename.surname@univ-avignon.fr

Abstract. Efficient statistical approaches have been recently proposed for natural language understanding in the context of dialogue systems. However, these approaches are trained on data semantically annotated at the segmental level, which increases the production cost of these resources. This kind of semantic annotation implies both to determine the concepts in a sentence and to link them to their corresponding word segments. In this paper, we propose a two-step automatic method for semantic annotation. The first step is an implementation of the latent Dirichlet allocation aiming at discovering concepts in a dialogue corpus. Then this knowledge is used as a bootstrap to infer automatically a segmentation of a word sequence into concepts using either integer linear optimisation or stochastic word alignment models (IBM models). The relation between automatically-derived and manually defined task-dependent concepts is evaluated on a spoken dialogue task with a reference annotation.

Keywords: Concept discovery, Segmental semantic annotation, Language understanding, Latent Dirichlet analysis, Dialogue systems.

1 Introduction

Spoken dialogue systems in the field of information query are basically used to interface a database with users using speech. When probabilistic models are used in such systems, good performance is at the price of collecting a large set of field data, which must be transcribed and annotated at the semantic level. These data allow then efficient models to be trained in a supervised way. However, the annotation process is costly and as a consequence represents a real difficulty hindering the widespread development of these systems. Therefore devising automatic methods for the annotation would be highly beneficial for portability to new tasks, domains or languages.

To give a full description of the architecture of a dialogue system is out of the scope of this paper. Instead we limit ourselves to briefly recall the main information pipeline: once a speech recognizer has transcribed the signal it is common (though avoidable

A. Fred et al. (Eds.): IC3K 2011, CCIS 348, pp. 45–59, 2013.

words	concept	normalized value
donnez-moi	null	
le	refLink-coRef	singular
tarif	object	payment-amount-room
puisque	connectProp	imply
je voudrais	null	
une chambre	number-room	1
qui coûte	object	payment-amount-room
pas plus de	comparative-payment	less than
cinquante	payment-amount-integer-room	50
euros	payment-unit	euro

Fig. 1. Semantic concept representation for the query *"give me the rate since I'd like a room charged not more than fifty euros"*

for very simple tasks) to use a module dedicated to extract the meaning of the user's queries. This meaning representation is then conveyed to an interaction manager that decides upon the next best action to perform considering the current user's input and the dialogue history (defining the dialogue state all together). One of the very first steps to build the spoken language understanding (SLU) module is the identification of literal concepts in the word sequence hypothesised by the speech recogniser. An example of a semantic representation in terms of literal concepts is given in Figure 1. Once the concepts have been identified they can be further composed to form the overall meaning of the sentence, for instance by means of a tree representation based on hierarchical semantic frames [12].

To address the issue of concept tagging several techniques are available. Some of these techniques now classical rely on probabilistic models, that can be either discriminative or generative. Among these, the following models have been studied in the last decade: hidden Markov models, finite state transducers, maximum entropy Markov models, support vector machines, dynamic Bayesian networks and conditional random fields (CRF). In [6] it is shown that CRFs obtain the best performance on a tourist information retrieval task in French (MEDIA [2]), but also in two other comparable corpora in Italian and Polish.

To be able to apply any such technique, basic concept units have to be defined by an expert. In the best case, most of these concepts can be derived straightforwardly from the pieces of information lurking in the database tables (mainly table fields but not exclusively). Some others are general (dialogue units but also generic entities such as number, dates etc.). However, to provide efficient and usable information to the reasoning modules (the dialogue manager in our case) concepts have to be fine-grained enough and application-dependent (even general concepts might have to be tailored to peculiar uses). To that extent it seems out of reach to derive the concept definitions using a fully automatic procedure. Anyhow the process can be bootstrapped, for instance by induction of semantic classes such as in [17] or [8]. Our assumption here is that the most time-consuming parts of concept inventory and data tagging could be obtained in an unsupervised way, even though a final (but hopefully minimal) manual procedure is still required to tag the classes so as to manually correct automatic annotation.

Unlike the previous attempts cited above which developed *ad hoc* approaches, we investigate here the use of broad-spectrum knowledge extraction methods. Topic as used in information retrieval systems is a notion related to that of concept in SLU. For a long time, the topic detection task was limited to associate a single topic with a document and thus was not fitted to the SLU task requirements. The recently proposed LDA technique allows a document to have a probabilistic representation as a mixture of topics. Then multiple topics can co-occur inside a document and the same topic can be repeated. From these characteristics it is possible to consider the application of LDA to unsupervised concept inventory and concept tagging for SLU. A shortcoming is that LDA does not take into account at all the sequentiality of the data. To address this issue we propose to conclude the procedure with a final step introducing specific constraints for a correct segmentation of the data: the assignments of topics proposed by LDA are modified to be more segmentally coherent. Heretwo variants are evaluated to derive an optimal concept segmentation using the probabilities output by the LDA inference step. First the recourse to a technique used in the Machine Translation field is considered: stochastic word alignment with IBM models. Then a more general framework for global optimisation under constraints is also investigated: integer linear programming.

The paper is organised as follows. Principles of automatic induction of semantic classes are presented in Section 2, followed by the presentation of an induction system based on LDA. The additional segmentation step is presented in Section 3 with two variants: stochastic word alignment with IBM models (GIZA) and integer linear programming (ILP). Then evaluations and results are reported in Section 4 on the French MEDIA dialogue task.

2 Automatic Induction of Semantic Classes

2.1 Context Modelling

The idea of automatic induction of semantic classes is based on the assumption that concepts often share the same context (syntactic or lexical). Implemented systems are based on the observation of co-occurring words according to two different ways. The observation of consecutive words (bigrams or trigrams) allows for the generation of lexical compounds supposed to follow syntactic rules. The comparison of right and left contexts considering pairs of words allows for the clustering of words and compound words into semantic classes.

In [17] and [14], iterative systems are presented. Their implementations differ in the metrics chosen to evaluate the similarity during the generation of syntactic rules and semantic classes, but also in the number of words taken into account in a word context and the order of successive steps (which ones to generate first: syntactic rules or semantic classes?). An iterative procedure is executed to obtain a sufficient set of rules in order to automatically extract knowledge from the data.

While there may be still room for improvement in these techniques we decided to skip them and to focus instead on general knowledge extraction approaches in order to evaluate their potential. For that purpose a global strategy based on an unsupervised machine learning technique is adopted in our work to produce semantic classes.

2.2 Implementation of an Automatic Induction System Based on Latent Dirichlet Allocation

Several approaches are available for topic detection in the context of knowledge extraction and information retrieval. They all more or less rely on the projection of the documents of interest in a semantic space to extract meaningful information. However, as the considered spaces (initial document words and latent semantics) are discrete performance of the proposed approaches for the topic extraction tasks are pretty unstable, and also greatly depend on the quantity of data available. In this work we were motivated by the recent development of a very attractive technique with major distinct features such as the detection of multiple topics in a single document. LDA [1] is the first principled description of a Dirichlet-based model of mixtures of latent variables. LDA will be used in our work to annotate the dialogue data in terms of topics in an unsupervised manner.

Basically LDA is a generative probabilistic model for text documents. LDA follows the assumption that a set of observations can be explained by latent variables. More specifically documents are represented by a mixture of topics (latent variables) and topics are characterized by distributions over words. The LDA parameters are $\{\alpha, \beta\}$. α represents the Dirichlet parameters of K latent topic mixtures as $\alpha = [\alpha_1, \alpha_2, \ldots, \alpha_K]$. β is a matrix representing a multinomial distribution in the form of a conditional probability table $\beta_{k,w} = P(w|k)$, where k and w represent resp. a topic and a word. Based on this representation, LDA can estimate the probability of a new document d of N words $d = [w_1, w_2, \ldots, w_N]$ using the following procedure.

A topic mixture vector θ is drawn from the Dirichlet distribution (with parameter α). The corresponding topic sequence $\kappa = [k_1, k_2, \ldots, k_N]$ is generated for the whole document accordingly to a multinomial distribution (with parameter θ). Finally each word is generated by the word-topic multinomial distribution (with parameter β, that is $p(w_i|k_i, \beta)$). After this procedure, the joint probability of θ, κ and d is then:

$$p(\theta, \kappa, d | \alpha, \beta) = p(\theta|\alpha) \prod_{i=1}^{N} p(k_i|\theta) p(w_i|k_i, \beta) \qquad (1)$$

To obtain the marginal probability of d, a final integration over θ and a summation over all possible topics considering a word is necessary:

$$p(d | \alpha, \beta) = \int p(\theta|\alpha) \left(\prod_{i=1}^{N} \sum_{k_i} p(k_i|\theta) p(w_i|k_i, \beta) \right) d\theta \qquad (2)$$

The framework is comparable to that of probabilistic latent semantic analysis, but the topic multinomial distribution in LDA is assumed to be sampled from a Dirichlet prior and is not linked to training documents. This approach is illustrated in Figure 2.

Training of the α and β parameters is possible using a corpus of M documents, with a fixed number K of topics to predict. A variational inference procedure is described in [1] which alleviates the intractability due to the coupling between θ and β in the summation over the latent topics. Once the parameters for the Dirichlet and multinomial distributions are available, topic scores can be derived for any given document or word sequence.

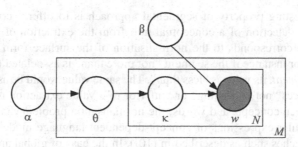

Fig. 2. Graphical representation for LDA variables (from [1]). The grey circle is the only observable variable.

In recent years, several studies have been carried out in language processing based on LDA. For instance, [18] worked on unsupervised language model adaptation; [4] ranked candidate passages in a question-answering system; [15] implemented LDA to classify short and sparse web texts.

In our work, LDA is employed to annotate each user's utterance of a dialogue corpus with topics. Utterances longer than one word are included in the training set as its sequence of *words*. Once the model has been trained, inference on data corpus assigns the topic with the highest probability to each word in a document. This probability is computed from the probability of the topic to appear in the document and the probability of the word to be generated by the topic. As a consequence we obtain a full topic annotation of the utterance.

Notice that LDA considers a user utterance as a bag of words. This implies that each topic is assigned to a word without any consideration for its immediate context. An additional segmental process is required if we want to introduce some context information in the topic assignment.

3 Segmental Annotation

3.1 Benefits of a Segmental Annotation

The segmental annotation of the data is not a strict requirement for language understanding. Up to quite recently, most approaches for literal interpretation were limited to lexical-concept relations; for instance this is the case of the PHOENIX system [20] based on the detection of keywords. However, in an NLP perspective, the segmental approach allows the various levels of sentence analysis (lexical, syntactic and semantic) to be connected. Even though, in order to simplify its application, segments are generally designed specifically for the semantic annotation and do not have any constraint on their relation with the actual syntactic units (chunks, phrasal groups etc.). Ignoring such constraints not only simplifies the annotation process but also increases the robustness of the interpretation module, which will ultimately be used inside a spoken dialogue system and will have to deal with noisy data. In particular, highly spontaneous and ungrammatical utterances from the users, combined with errors from the speech recognizer are likely to severely decrease the performance of syntactic analysers.

Another interesting property of segmental approach is to offer a convenient way to dissociate the detection of a conceptual unit from the extraction of its associated value. The value corresponds to the normalisation of the surface form (see last column in Fig. 1); for instance if the segment "not more than" is associated to the concept *comparative-payment*, its value is "less than". The same value would be associated with the word sequences "not exceeding" or "inferior to". Value extraction requires therefore a link between concepts and words; the normalisation problem can be addressed by means of regular expressions or concept-dependent language models (even using integrated approaches such as described in [10]). In the case of global approaches (i.e. not segmental), value extraction must be done directly at the level of the conceptual unit tagging, as in [11]. This additional level is very complex (as some values may not be enumerable, such as numbers and dates) and is only affordable when the number of authorised values (for the enumerable cases) is low.

To refine the LDA output, the topic-to-word alignment proposed by LDA is discarded and an automatic procedure is used to derive the best alignment between topics and words. While the underlying probabilistic models are pretty comparable, the major interest of this approach is to separate the tasks of detecting topics and aligning topics with words. It is then possible to introduce additional constraints (such as locality, total number of segments, limits on repetitions etc.) in the latter task which would otherwise hinder topic detection. Conversely the alignment is self-coherent and able to question the associations proposed during topic detection with respect to its own constraints only. Two approaches were designed to this purpose: one based on IBM alignment models and another one based on integer linear optimisation.

3.2   Alignment with IBM Models (GIZA)

LDA is able to generate one topic k_i by word w_i from the probability $p(k_i|w_i, d)$ that k_i is associated with w_i of the document d. However, this method has the drawback of considering a document as a bag-of-words and it ignores the effect of k_i on the choice of the topics of the surrounding words w_{i-1} and w_{i+1}. This independence assumption between consecutive topics tends to generate too many distinct topics for each utterance. To tackle this problem, we resorted to a statistical word-alignment method based on the so-called IBM models, widely used in Machine Translation [3].

The use of the IBM models in our context offers several interests. First, the output of LDA generates a topic sequence for each word sequence and can be thus seen as a translation pair of a word-topic parallel corpus. Contrary to LDA, the HMM model [19] as well as IBM models 2-4 take into account the word order in the translation pairs to align. Finally, the IBM models can be trained in an unsupervised way without providing any alignment between words and topics.

To refine the assignment of topics in the documents with the IBM models, we adopted the following approach. Since LDA tends to produce a too high number of different topics for each utterance, a filtering process is applied on the LDA outputs to keep only the most relevant topics for each document. We keep the χ_{max} topics with the highest probabilities $p(k_i|w_i, d)$ computed by LDA, where χ_{max} is a value empirically fixed from the expected set of topics according to the number of words in an utterance. The IBM models are expected to correct some errors made by LDA, and in particular to assign words associated with the discarded topics to more likely topics. Two additional processing steps are applied on the topic sequences. Only the first occurrence is kept for consecutive repetitions of the same topic, as the IBM models can align several words with a single topic. Besides, the *null* concept (for out-of-domain or irrelevant word segments) is removed since this concept disturbs the alignments between words and concepts, as shown by a previous experiment done in a similar setup [7]. The IBM models already incorporate an empty word to account for source words that are not aligned with any target word; this empty word can be considered as the *null* concept for our task. This of course supposes that a manual tagging of LDA topics has allowed to define task-dependent concepts from topics (but can also be limited to identify the topics corresponding to the *null* concept).

In our experiments the statistical word alignment toolkit GIZA++ [13] is used to train the IBM models 1-4 as well as the HMM model. The IBM models are built using the expectation-maximization algorithm from a word-topic parallel corpus, where the topic sequences generated from LDA are disconnected from the words. To be able to train the most informative IBM model 4, the following training pipeline was considered: 5 iterations of IBM1, 5 iterations of HMM, 3 iterations of IBM3 and 3 iterations of IBM4. The IBM4 model obtained at the last iteration is finally used to align words and topics. In order to improve alignment, IBM models are usually trained in both directions (words towards concepts and *vice versa*) then symmetrised by combining them. For this purpose, we resorted to the default symmetrization heuristics used in MOSES, a widely used machine translation system toolkit [9].

3.3   Alignment with Integer Linear Programming (ILP)

Another approach to the re-alignment of LDA outputs is based on a general optimisation technique. Integer Linear Programming (ILP) is a widely used tool for modelling and solving combinatorial optimisation problems. It broadly aims at formulating a decision process as a set of equations or inequations (called *constraints*) which are linear with regards to some *decision variables*. An ILP is also composed of a linear *objective function* which represents the criterion (for instance a cost, a profit or like in this study a probability). Solving an ILP consists in assigning values to decision variables, such that all constraints are satisfied and the objective function is optimized. Readers may refer to [5] for an overview of applications and methods of ILP.

We provide two ILP formulations for solving the topic assignment problem related to a given document. They both take as input data an ordered set d of words w_i, $i = 1...N$, a set of K available topics and, for each word $w_i \in d$ and topic $k = 1...K$, the natural logarithm of the probability $p(k|w_i, d)$ that k is assigned to w_i in the considered document d. Model $[ILP]$ simply finds the highest-probability assignment of one topic to each word in the document, such that at most χ_{max} different topics are assigned.

$$[ILP] : \max \sum_{i=1}^{N} \sum_{k=1}^{K} \ln(p(k|w_i, d)) \, x_{ik} \qquad (3)$$

$$\sum_{k=1}^{K} x_{ik} = 1 \qquad i \qquad (4)$$

$$y_k - x_{ik} \geq 0 \qquad i, k \qquad (5)$$

$$\sum_{k=1}^{K} y_k \leq \chi_{max} \tag{6}$$

$$x_{ik} \in \{0,1\} \qquad i,k$$

$$y_k \in \{0,1\} \qquad k$$

In this model, decision variable x_{ik} is equal to 1 if topic k is assigned to word w_i, and equal to 0 otherwise. Constraints 4 ensure that exactly one topic is assigned to each word. Decision variable y_k is equal to 1 if topic k is used. Constraints 5 force variable y_k to take a value of 1 if at least one variable x_{ik} is not null. Moreover, constraints 6 limit the total number of topics used. The objective function 3 merely states that we want to maximize the total probability of the assignment. Through this model, our assignment problem is identified as a *p-centre* problem (see [16] for a survey on such location problems).

Numerical experiments show that $[ILP]$ tends to give sparse assignments: most of the time, adjacent words are assigned to different topics even if the total number of topics is correct. To prevent this unnatural behaviour, we modified $[ILP]$ to consider groups of consecutive words instead of isolated words. Model $[ILP\_seg]$ partitions the document into segments of consecutive words, and assigns one topic to each segment, such that at most χ_{max} segments are created. For the sake of convenience, we denote by $\bar{p}(k|w_{ij}, d) = \sum_{l=i}^{j} \ln(p(k|w_l, d))$ the logarithm of the probability that topic k is assigned to all words from i to j in the current document.

$$[ILP\_seg] : \max \sum_{i=1}^{N} \sum_{j=i}^{N} \sum_{k=1}^{K} \bar{p}(k|w_{ij}, d)\, x_{ijk} \tag{7}$$

$$\sum_{j=1}^{i} \sum_{l=i}^{N} \sum_{k=1}^{K} x_{jlk} = 1 \qquad i \tag{8}$$

$$\sum_{i=1}^{N} \sum_{j=i}^{N} \sum_{k=1}^{K} x_{ijk} \leq \chi_{max} \tag{9}$$

$$x_{ijk} \in \{0,1\} \qquad i,j,k$$

In this model, decision variable x_{ijk} is equal to 1 if topic k is assigned to all words from i to j, and 0 otherwise. Constraints 8 ensure that each word belongs to a segment that is assigned a topic. Constraints 9 limit the number of segments. Due to the small size of the instances considered in this paper, both $[ILP]$ and $[ILP\_seg]$ are well solved by a direct application of an ILP solver.

4 Evaluation and Results

4.1 MEDIA Corpus

The MEDIA corpus is used to evaluate the proposed approach and to compare the various configurations. MEDIA is a French corpus related to the domain of tourism information and hotel booking [2]. 1,257 dialogues were recorded from 250 speakers with

Table 1. Proportion of user utterances as a function of the number of concepts in the utterance

# concepts	1	2	3	[4,72]
% utterances	49.4	14.1	7.9	28.6

Table 2. Number of concepts according to their occurrence range

# concept supports	[1,100[	[100,500[	[500,1k[	[1k,9k[	[9k,15k]
# concepts	33	21	6	14	1 (*null*)

a wizard of Oz technique (a human agent mimics an automatic system). This dataset contains 17k user utterances and 123,538 words, for a total of 2,470 distinct words.

The MEDIA data have been manually transcribed and semantically annotated. The semantic annotation uses 75 concepts (e.g. *location, hotel-state, time-month...*). Each concept is supported by a sequence of words, the *concept support*. The *null* concept is used to annotate every words segment that does not support any of the 74 other concepts (and does not bear any information wrt the task). On average, a concept support contains 2.1 words, 3.4 concepts are included in an utterance and 32% of utterances are restrained to a single word (generally "yes" or "no"). Table 1 gives the proportions of utterances according to the number of concepts in the utterance.

Notice that each utterance contains at least one concept (the *null* label being considered as a concept). As shown in Table 2, some concepts are supported by few segments. For example, 33 concepts are represented by less than 100 concept supports. Considering that, we can foresee that finding these poorly represented concepts will be extremely hard if not impossible for LDA.

4.2 Evaluation Protocol

Unlike previous studies, we chose a fully automatic way to evaluate the systems. In [17], a manual process is introduced to reject induced classes or rules that are not relevant to the task and also to name the semantic classes with the appropriate label. Thus, they were able to evaluate their semi-supervised annotation on the ATIS corpus. In [14], the relevance of the generated semantic classes was manually evaluated giving a mark to each induced semantic rule.

To evaluate the unsupervised procedure it is necessary to associate each induced topic with a MEDIA concept. To that purpose, the reference annotation is used to align topics with MEDIA concepts at the word level. A co-occurrence matrix is computed and each topic is associated with its most co-occurring concept.

As MEDIA reference concepts are very fine-grained, we also define a *high-level* concept hierarchy containing 18 clusters of concepts. For example, a high-level concept *payment* is created from the 4 concepts *payment-meansOfPayment, payment-currency, payment-total-amount, payment-approx-amount*; a high-level concept *location* corresponds to 12 concepts (*location-country, location-district, location-street, ...*). Thus, two levels of concepts are considered for the evaluation: *high-level* and *fine-level*.

The evaluation is presented in terms of the classical F-measure, defined as a combination of precision and recall measures. Two levels are also considered to measure topic assignment quality:

- *alignment* corresponds to a full evaluation where each word is considered and associated with one topic/concept;
- *generation* corresponds to the set of topics/concepts generated for a turn (no order, no word-alignment).

4.3 System Descriptions

Four systems are evaluated in our experiments.

$[LDA]$ is the result of the unsupervised learning of LDA models using GIBBSLDA++ tool[1]. It assigns the most probable topic to each word occurrence in a document as described in Section 2.2. This approach requires prior estimation of the number of clusters that are expected to be found in the data. To find an optimal number of clusters, we varied the number K of topics around 75, which is the number of reference concepts. 2k training iterations were made using default values for α and β.

$[GIZA]$ is the system based on the GIZA++ toolkit[2] which re-aligns for each utterance the topic sequence assigned by $[LDA]$ to word sequence as described in Section 3.2.

$[ILP]$ and $[ILP\_seg]$ systems are the results of the ILP solver IBM ILOG CPLEX[3] applied to the models described in Section 3.3.

For the three last systems, the χ_{max} value has to be fixed according to the desired concept annotation. As on average a concept support contains 2.1 words, χ_{max} is defined empirically according to the number of words in the sequence: if $i \in [2, 4]$: $\chi_{max} = i$; if $i \in [5, 10]$ words: $\chi_{max} = i - 2$; and for utterances containing more than 10 words: $\chi_{max} = i/2$.

For the sake of simplicity, single-word utterances are processed separately with prior knowledge. City names, months, days or answers (e.g. "yes", "no", "yeah") and numbers are identified in these one-word utterances.

4.4 Generated Topic Observations

In Table 3, six topics generated by $[LDA]$ are represented by their 8 highest probability words. For topic 13, it is interesting noting that words have quite similar weights. The most represented words are "du" ("from") and "au" ("to") and other words are numbers or months, which *a priori* corresponds to a "time-date" concept. For topic 43, the word "oui" ("yes") is given a 0.62 probability, other words are "absolutely" or "okay" leading to an *a priori* "answer-yes" concept.

---

[1] http://gibbslda.sourceforge.net/

[2] http://code.google.com/p/giza-pp/

[3] http://www-01.ibm.com/software/integration/optimization/
   cplex-optimizer/

Table 3. Examples of topics discovered by LDA ($K = 100$)

Topic 0 information		Topic 13 time-date		Topic 18 sightseeing		Topic 35 politeness		Topic 33 location		Topic 43 answer-yes	
words	prob	words	prob	words	prob	words	prob	words	prob	words	prob.
d'	0.28	du	0.16	de	0.30	au	0.31	de	0.30	oui	0.62
plus	0.17	au	0.11	la	0.24	revoir	0.27	Paris	0.12	et	0.02
infor-										abso-	
mations	0.16	quinze	0.08	tour	0.02	madame	0.09	la	0.06	lument	0.008
autres	0.10	dix-huit	0.07	vue	0.02	merci	0.08	près	0.06	autre	0.008
détails	0.03	décembre	0.06	Eiffel	0.02	bonne	0.01	proche	0.05	donc	0.007
obtenir	0.03	mars	0.06	sur	0.02	journée	0.01	Lyon	0.03	jour	0.005
										Notre-	
alors	0.01	dix-sept	0.04	mer	0.01	villes	0.004	aux	0.02	Dame	0.004
souhaite	0.003	nuits	0.04	sauna	0.01	bientôt	0.003	gare	0.02	d'accord	0.004

Table 4. Topic repartitions among the high-level or fine-level concepts for $[LDA]$ and $[ILP]$ ($K = 100$)

	Topic 18 sightseeing			Topic 33 location			Topic 43 answer-yes		
	#occ	concept	$Ent(k)$	#occ	concept	$Ent(k)$	#occ	concept	$Ent(k)$
					[LDA]				
high	292	Location		571	Location		705	Answer	
	258	null	2.25	156	null	1.72	107	null	1.10
	94	Name		87	Comparative		27	Location	
fine	258	null		226	loc.-distRel.		705	answer	
	136	loc.-placeRel.	2.78	190	loc.-city	2.57	107	null	1.19
	100	loc.-distRel.		156	null		17	object	
					[ILP]				
high	300	Location		661	Location		846	Answer	
	200	null	2.19	123	null	1.52	109	null	0.76
	102	Name		115	Comparative		24	Location	
fine	200	null		234	loc.-distRel.		846	answer	
	163	loc.-placeRel.	2.64	223	loc.-city	2.44	109	null	0.80
	98	name-hotel		129	loc.-placeRel.		16	name-hotel	

To observe which MEDIA concept is associated with these topics, the list of the 3 most co-occurring concepts and the number of co-occurrences are shown in Table 4. The 2 most co-occurring concepts in a topic are the same in $[LDA]$ and $[LDA+ILP]$. However, the number of co-occurrences is higher in $[ILP]$ than in $[LDA]$. An entropy measure $Ent(k)$ is computed for each topic k in order to evaluate the reliability of the topic-concept association over all the possible concepts. It is computed as follows:

$$Ent(k) = - \sum_{concepts\ c} p(c|k) \log p(c|k) \qquad (10)$$

$$\text{with} \quad p(c|k) = \frac{\#(c \cap k)}{\#k}$$

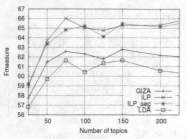
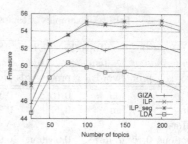

Fig. 3. F-measure of the high-level concept generation as a function of the number of topics

Fig. 4. F-measure of the high-level concept alignment as a function of the number of topics

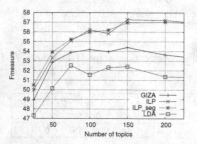
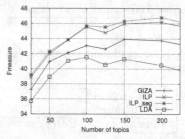

Fig. 5. F-measure of the fine-level concept generation as a function of the number of topics

Fig. 6. F-measure of the fine-level concept alignment as a function of the number of topics

The topic entropy is always smaller considering $[ILP]$ than $[LDA]$. This indicates that the re-assignment due to ILP alignment improves the reliability of the topic-concept association. Entropies measured with *high-level* concepts is always lower than with *fine-level* concepts, mainly because less classes are considered (18 instead of 75). Topic 18 is associated with the *Location* concept using the *high-level* configuration instead of the *null* one using the *fine-level* configuration, but the entropy is quite high. On the over hand, topic 43 has a low entropy, specifically with $[ILP]$. This shows that word "yes" is strongly associated with the concept "Answer". Other topics representing the *null* concept can show entropies as low as 0.47 for topic 6 ("il", "y", "a", "qu'", "est-ce"...).

4.5   Results

Plots comparing the different systems implemented wrt the different evaluation levels in terms of F-measure are reported in Figures 3, 4, 5 and 6 (*high-level* vs *fine-level*, *alignment* vs *generation*).

The $[LDA]$ system generates topics which are correctly correlated with the *high-level* concepts. It can be observed that the bag of 75 topics reaches an F-measure of 61.5% (Fig. 3). When not enough topics are required from $[LDA]$, induced topics are too wide to fit the fine-grained concept annotation of MEDIA. On the other hand if too many high-level topics are required, the generation performance stays the same while a substantial decrease of the F-measure is observed in the *alignment* evaluation (Fig. 4). This effect can be explained by the automatic process to associate generated topics with reference concepts. Indeed, the increase of the number of topics makes them co-occur with many concepts, which often leads to assign them to the most frequent concept (*null*) in the studied corpus.

From the *high-level* to *fine-level* concept evaluations, results globally decrease by 10%. An additional global loss of 10% is also observed between the *generation* and *alignment* scoring methods. In the *fine-level* evaluation, a maximum F-measure of 52.2% is observed for the *generation* of 75 topics (Fig. 5) whereas the F-measure decreases to 41.5% in the *alignment* evaluation (Fig. 6).

To conclude on the [*LDA*] system, we can see that it generates topics having a good correlation with the *high-level* concepts, seemingly the best representation level between topics and concepts. From these results it seems obvious that an additional step is needed to obtain a more accurate segmental annotation, which is expected with the following systems.

The [*GIZA*] system improves the [*LDA*] results. It is very likely that the filtering process helps to discard the irrelevant topics. Therefore, the automatic alignment between words and the filtered topics induced by [*LDA*] with IBM models seems more robust when more topics (a higher value for K) is required from [*LDA*], specifically in *high-level* concept *alignment* (Fig. 4).

Systems based on the ILP technique perform better than other systems whatever the evaluation. Considering [*LDA*] as the baseline, we can expect significant gains of performance. For example, an F-measure of 66% is observed for the ILP systems considering the *high-level* concept *generation* for 75 topics (Fig. 3), in which case the maximum for [*LDA*] was 61.5%, and an F-measure of 55% is observed (instead of 50.5% for [*LDA*]) considering the *high-level* concept *alignment* (Fig. 4).

No significant difference was finally measured between both ILP models for the concept generation evaluations. Even though [*ILP_seg*] seems to obtain slightly better results in the *alignment* evaluation. This could be expected since [*ILP_seg*] intrinsically yields alignments with grouped topics, closer to the reference alignment used for the evaluation.

It is worth noticing that unlike [*LDA*] system behaviour, the results of [*ILP*] are not affected when more topics are generated by [*LDA*]. A large number of topics enables [*ILP*] to pick up the best topic for a given segment in a longer selection list. As for [*LDA*], the same losses are observed between *high-level* and *fine-level* concepts and the *generation* and *alignment* paradigms. Nevertheless, an F-measure of 54.8% is observed with *high-level* concepts in the *alignement* evaluation (Fig. 4) that corresponds to a precision of 56.2% and a recall of 53.5%, which is not so low considering a fully-automatic high-level annotation system.

5   Conclusions and Perspectives

In this paper an approach has been presented for concept discovery and segmental semantic annotation of user's turns in an information-query dialogue system. An evaluation based on an automatic association between generated topics and expected concepts has shown that topics induced by LDA are close to *high-level* task-dependent concepts. The segmental annotation process increases performance both for the generation and alignment evaluations. On the whole these results confirm the applicability of the technique to practical tasks with expected gains in data production.

Future work will investigate the use of n-grams to extend LDA and to increase its accuracy for providing better hypotheses to the following segmentation techniques. Besides, other levels of data representation will be examined (use of lemmas, *a priori* semantic classes like city names...) in order to better generalise on the data.

Acknowledgements. This work is supported by the ANR funded project PORT-MEDIA (www.port-media.org) and the LIA OptimNLP project (www.lia.univ-avignon.fr).

References

1. Blei, D.M., Ng, A.Y., Jordan, M.I.: Latent dirichlet allocation. The Journal of Machine Learning Research 3, 993–1022 (2003)
2. Bonneau-Maynard, H., Rosset, S., Ayache, C., Kuhn, A., Mostefa, D.: Semantic annotation of the french MEDIA dialog corpus. In: Proceedings of the ISCA 9th European Conference on Speech Communication and Technology (2005)
3. Brown, P.F., Della Pietra, S.A., Della Pietra, V.J., Mercer, R.L.: The mathematics of statistical machine translation: Parameter estimation. Computational Linguistics 19(2), 263–311 (1993)
4. Celikyilmaz, A., Hakkani-Tur, D., Tur, G.: LDA based similarity modeling for question answering. In: Proceedings of the NAACL HLT 2010 Workshop on Semantic Search (2010)
5. Chen, D.-S., Batson, R., Dang, Y.: Applied Integer Programming: Modeling and Solution. Wiley (2010)
6. Hahn, S., Dinarelli, M., Raymond, C., Lef'evre, F., Lehnen, P., De Mori, R., Moschitti, A., Hermann Ney, H., Riccardi, G.: Comparing stochastic approaches to spoken language understanding in multiple languages. IEEE Transactions on Audio, Speech and Language Processing 19(6), 1569–1583 (2010)
7. Huet, S., Lefèvre, F.: Unsupervised alignment for segmental-based language understanding. In: Proceedings of the EMNLP 1st Workshop on Unsupervised Learning in NLP, UNSUP (2011)
8. Iosif, E., Tegos, A., Pangos, A., Fosler-Lussier, E., Potamianos, A.: Unsupervised combination of metrics for semantic class induction. In: Proceedings of the IEEE/ACL Spoken Language Technology Workshop (2006)
9. Koehn, P., Hoang, H., Birch, A., Callison-Burch, C., Federico, M., Bertoldi, N., Cowan, B., Shen, W., Moran, C., Zens, R., Dyer, C., Bojar, O., Constantin, A., Herbst, E.: Moses: Open source toolkit for statistical machine translation. In: Proceedings of ACL, Companion Volume (2007)
10. Lefèvre, F.: Dynamic bayesian networks and discriminative classifiers for multi-stage semantic interpretation. In: Proceedings of the IEEE International Conference on Audio, Signal and Speech Processing (2007)
11. Mairesse, F., Gašić, M., Jurčíček, F., Keizer, S., Thomson, B., Yu, K., Young, S.: Spoken language understanding from unaligned data using discriminative classification models. In: Proceedings of the IEEE International Conference on Audio, Signal and Speech Processing (2009)

12. Meurs, M.-J., Lefèvre, F., de Mori, R.: Spoken language interpretation: On the use of dynamic bayesian networks for semantic composition. In: Proceedings of the IEEE International Conference on Audio, Signal and Speech Processing (2009)
13. Och, F., Ney, H.: A systematic comparison of various statistical alignment models. Computational Linguistics 29(1), 19–51 (2003)
14. Pargellis, A., Fosler-Lussier, E., Potamianos, A., Lee, C.: Metrics for measuring domain independence of semantic classes. In: Proceedings of the 7th ISCA European Conference on Speech Communication and Technology (2001)
15. Phan, X., Nguyen, L., Horiguchi, S.: Learning to classify short and sparse text & web with hidden topics from large-scale data collections. In: Proceeding of the 17th ACM International Conference on World Wide Web (2008)
16. ReVelle, C.S., Eiselt, H.A.: Location analysis: A synthesis and survey. European Journal of Operational Research 165(1), 1–19 (2005)
17. Siu, K., Meng, H.: Semi-automatic acquisition of domain-specific semantic structures. In: Proceedings of the ISCA 6th European Conference on Speech Communication and Technology (1999)
18. Tam, Y., Schultz, T.: Unsupervised language model adaptation using latent semantic marginals. In: Proceedings of ISCA INTERSPEECH (2006)
19. Vogel, S., Ney, H., Tillmann, C.: HMM-based word alignment in statistical translation. In: Proceedings of COLING (1996)
20. Ward, W.: Understanding spontaneous speech. In: Proceedings of the IEEE International Conference on Audio, Signal and Speech Processing (1991)

Spectral Clustering: An Explorative Study of Proximity Measures

Nadia Farhanaz Azam and Herna L. Viktor

School of Electrical Engineering and Computer Science, University of Ottawa
800 King Edward Avenue, Ottawa, Ontario K1N6N5, Canada
{nazam085,hlviktor}@site.uottawa.ca
http://www.eecs.uottawa.ca

Abstract. Spectral clustering algorithms recently gained much interest in research community. This surge in interest is mainly due to their ease of use, their applicability to a variety of data types and domains as well as the fact that they very often outperform traditional clustering algorithms. These algorithms consider the pair-wise similarity between data objects and construct a similarity matrix to group data into natural subsets, so that the objects located in the same cluster share many common characteristics. Objects are then allocated into clusters by employing a proximity measure, which is used to compute the similarity or distance between the data objects in the matrix. As such, an early and fundamental step in spectral cluster analysis is the selection of a proximity measure. This choice also has the highest impact on the quality and usability of the end result. However, this crucial aspect is frequently overlooked. For instance, most prior studies use the Euclidean distance measure without explicitly stating the consequences of selecting such measure. To address this issue, we perform a comparative and explorative study on the performance of various existing proximity measures when applied to spectral clustering algorithm. Our results indicate that the commonly used Euclidean distance measure is not always suitable, specifically in domains where the data is highly imbalanced and the correct clustering of boundary objects are critical. Moreover, we also noticed that for numeric data type, the relative distance measures outperformed the absolute distance measures and therefore, may boost the performance of a clustering algorithm if used. As for the datasets with mixed variables, the selection of distance measure for numeric variable again has the highest impact on the end result.

Keywords: Spectral clustering, Proximity measures, Similarity measures, Boundary detection.

1 Introduction

In cluster analysis, objects that share similar characteristics are placed into the same cluster, whereas, the objects that are very different from one another are located in two different clusters. Therefore, the main objective of a clustering algorithm is to maximize the within cluster similarity and minimize the between cluster similarity. The difference between the objects is often measured by a proximity measure, such as, similarity, dissimilarity or distance measure. There have been many clustering algorithms

A. Fred et al. (Eds.): IC3K 2011, CCIS 348, pp. 60–78, 2013.

proposed in the literature. One such family of algorithms, collectively known as the spectral clustering algorithms, recently gained much interest in the research community. This popularity is due to, amongst other, their ease of use, the fact that it can be solved efficiently by standard linear algebra software and that these algorithms very often outperforms traditional clustering algorithms such as the k-means algorithm. The algorithms from this family have been successfully deployed across numerous domains, ranging from image segmentation to clustering protein sequences. Spectral clustering algorithms have been successfully applied in a wide range of domains, ranging from oil spill detection [17] to speech separation [3].

One of the main strengths of the spectral clustering algorithm is that the algorithm may be applied to a wide range of data types (i.e. numeric, categorical, binary, and mixed) as they are not sensitive to any particular data type. These algorithms consider the pair-wise similarity between the data objects to construct a *similarity matrix*. Some of the other alternative names are *proximity*, *affinity*, or *weight matrix*. The eigenvectors and eigenvalues of this similarity matrix or the *Laplacian matrix* (a matix manipulated and constructed from the similarity matrix) are then used to find the clusters ([1], [2], [5]). The various algorithms from this family mainly differ with respect to how the similarity matrix is manipulated and/or which eigenvalue(s) and eigenvector(s) are used to partition the objects into disjoint clusters. Significant theoretical progress has been made regarding the improvement of the spectral clustering algorithms as well as the proposal of new methods, or the application in various domains. However, little research has been performed on the selection of proximity measures, which is a crucial step in constructing the similarity matrix. In this paper, we evaluate the performance of a number of such proximity measures and perform an explorative study on their behavior when applied to the spectral clustering algorithms.

Proximity measures, i.e. similarity, dissimilarity and distance measures, often play a fundamental role in cluster analysis [7]. Early steps of the majority of cluster analysis algorithms often require the selection of a proximity measure and the construction of a similarity matrix (if necessary). Most of the time, the similarity matrix is constructed from an existing similarity or distance measure, or by introducing a new measure specifically suitable for a particular domain or task. It follows that the selection of such measures, particularly when existing measures are applied, requires careful consideration as the success of these algorithms relies heavily on the choice of the proximity function ([1], [8], [9]).

Most of the previous studies on the spectral clustering algorithm use the *Euclidean distance* measure, a distance measure based on linear differences, to construct the similarity matrix for numeric feature type ([2], [5], [6]) without explicitly stating the consequences of selecting the distance measure. However, there are several different proximity measures available for numeric variable types. Each of them has their own strengths and weaknesses. To our knowledge, no in-depth evaluation of the performance of these proximity measures on spectral clustering algorithms, specifically showing that the *Euclidean distance* measure outperforms, has been carried out. As such, an evaluation and an exploratory study that compares and analyzes the performance of various proximity measures may potentially provide important guideline for researchers when selecting a proximity measure for future studies in this area. This paper endeavors to

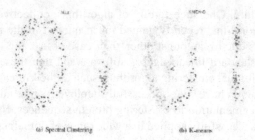

(a) Spectral Clustering   (b) K-means

Fig. 1. Comparison of results from spectral clustering and the K-means algorithm. (a) Results from the spectral clustering algorithm, (b) Results from the K-means algorithm.

evaluate and compare the performance of these measures and to imply the conditions under which these measures may be expected to perform well.

This paper is organized as follows. In Section 2, we discuss the two spectral clustering algorithms that we used in our experiment. Section 3 presents an overview of several proximity measures for numeric, and mixed variable types. This is followed by Section 4, where we present our experimental approach and evaluate and analyze the results obtained from our experiments. We conclude the paper in Section 5.

2   Spectral Clustering

Spectral clustering algorithms originated from the area of graph partitioning and manipulate the eigenvalue(s) and eigenvector(s) of the similarity matrix to find the clusters. There are several advantages, when compared to other cluster analysis methods, to applying the spectral clustering algorithms ([1], [5], [25], [26]). Firstly, the algorithms do not make assumption on the shape of the clusters. As such, while spectral clustering algorithms may be able to find meaningful clusters with strongly coherent objects, algorithms such as K-means or K-medians may fail to do so. One such example is depicted in Figure 1 [25]. Figure 1(a) shows the ring clusters obtained from the spectral clustering algorithm and Figure 1(b) depicts the results from the K-means algorithm when applied on the same sample dataset [25]. Secondly, the algorithms do not suffer from local minima. Therefore, it may not be necessary to restart the algorithm with various initialization options. Thirdly, the algorithms are also more stable than some algorithms in terms of initializing the user-specific parameters (i.e. the number of clusters). As such, the user-specific parameters may often be estimated accurately with the help of theories related to the algorithms. Prior studies also show that the algorithms from this group thus often outperform traditional clustering algorithms, such as, K-means and Single Linkage [1]. Importantly, the algorithms from the spectral family are able to handle different types of data (i.e. numeric, nominal, binary, or mixed) and one only needs to convert the dataset into a similarity matrix to be able to apply this algorithm on a given dataset [1].

The spectral clustering algorithms are divided into two types, namely *recursive algorithms* and *multi-way algorithms* [6]. The algorithms in the first group, as the name suggest, recursively bi-partition the data at each step until a stopping criterion is satisfied. The most representative algorithm from this group is, the *Normalized Cut Spectral*

Clustering by [2]. In contrast, multi-way spectral clustering algorithms directly partition the data into k number of groups. The best-known algorithms from this group are the *Ng, Jordan and Weiss algorithm* [5] and the *Meila - Shi algorithm* [11].

In this paper, we consider two algorithms, one from each group. From the first group, we select the normalized cut spectral clustering algorithm as this algorithm proved to have had several practical successes in a variety of fields [2]. We refer to this algorithm as *SM (NCut)* in the remainder of the paper. The Ng, Jordan and Weiss algorithm is an improvement to the algorithm proposed by Meila and Shi[11] and therefore, we select this algorithm (refer to as *NJW(K-means)*) from the second group. In the following section we present several algorithm-specific notations before we discuss the algorithms themselves.

2.1 Notations

Similarity Matrix or Weight Matrix, W: Let W be an $N \times N$ symmetric, non-negative matrix where N is the number of objects in a given dataset. Let i and j be any two objects in a given dataset, located at row i and row j, respectively. If the similarity (i.e. calculated from a proximity measure) between these two objects is $w_{i,j}$, then it will be located at the cell at row i and column j in the weight matrix.

Degree Matrix, D: Let d be an $N \times 1$ matrix with $d_i = \sum_{j=1}^{n} w_{i,j}$ as the entries which denote the total similarity value from object i to the rest of the objects. Therefore, the degree matrix D is an $N \times N$ diagonal matrix which contains the elements of d on its main diagonal.

Laplacian Matrix, L: The Laplacian matrix is constructed from the weight matrix W and the degree matrix D. The main diagonal of this matrix is always non-negative. In graph theory, the eigenvector(s) and eigenvalue(s) of this matrix contain important information about the underlying partitions present in the graph. The spectral clustering algorithms also use the same properties to find the clusters from a given dataset.

2.2 The SM (NCut) Algorithm

The SM (NCut) spectral clustering algorithm [2], as depicted in Figure 2, is one of the most widely used recursive spectral clustering algorithm ([1], [6]). The main intuition behind this algorithm is the optimization of an objective function called the *Normalized Cut*, or *NCut*. Minimizing the NCut function is the same as finding a cut such that the total connection in between two groups is weak, whereas the total connection within each group is strong. The algorithm uses the eigenvector associated with the second smallest eigenvalue of the generalized eigenvalue system which is considered as the real valued solution to the Normalized Cut problem. The partitions are obtained by thresholding this eigenvector. There are a number of ways this grouping may be performed. One may use a particular point (i.e. zero, mean, median) as the splitting criteria or use an existing algorithm such as the K-means or K-medians algorithms for this purpose. Components with similar values usually reside in the same cluster. Since, this algorithm bi-partition

---
Algorithm 1: The SM (NCut) Algorithm

Input: A dataset with N number of objects.
Output: A hierarchical tree of clusters.

1. Form a $N \times N$ symmetric weight matrix W and a degree matrix D.
2. Construct the $N \times N$ Laplacian matrix $(D - W)$.
3. Solve the eigensystem $(D - W) = \lambda Dx$ for eigenvectors with smallest eigenvalues.
4. Find the second smallest eigenvalue and use the eigenvector associated with this eigenvalue to bipartition the dataset.
5. Recursively re-partition the partitions, if necessary.

Fig. 2. The algorithm for SM (NCut) spectral clustering algorithm

---
Algorithm 2: The NJW (K-means) Algorithm

Input: A dataset with N number of objects.
Output: A set of k number of clusters.

1. Form a $N \times N$ symmetric weight matrix W and a degree matrix D.
2. Construct the $N \times N$ Laplacian matrix $D^{-\frac{1}{2}} W D^{-\frac{1}{2}}$.
3. Find the k eigenvectors associated with k largest eigenvalues, let these eigenvectors be $x_1, x_2 \ldots x_k$.
4. Form the $N \times k$ matrix $X = [x_1 x_2 \ldots x_k]$ by stacking the eigenvectors in columns.
5. Form another $N \times k$ matrix Y from X by normalizing each row of X so that it has a unit length. This is done by calculating $Y_{ij} = \dfrac{x_{ij}}{\left(\sum_j x_{ij}^2\right)^{\frac{1}{2}}}$ for each row of X.
6. Use K-means on Y to find k clusters, by treating each row of Y as a point in k-dimensions.
7. Assign the original points to the clusters formed from Y such that if a point i from Y belongs to cluster j then the original point i from the input dataset will also belong to cluster j.

Fig. 3. The algorithm for NJW(K-means) spectral clustering algorithm

the data, we get two disjoint clusters. To find more clusters we need to re-partition the segments by recursively applying the algorithm on each of the partitions.

2.3 The NJW (K-means) Algorithm

In contrast to the SM (NCut) algorithm that minimizes the *NCut* objective function and recursively bi-partitions the data, this algorithm directly partitions the data into k groups. The algorithm manipulates the normalized Laplacian matrix (as given in Figure 3) to find the clusters. The algorithm as shown in Figure 3, relates various theories from the *Random Walk Problem* and *Matrix Perturbation Theory* to theoretically motivate the partitioning solution ([1], [5], [11]). Once the eigensystem is solved and the k largest eigenvectors are normalized, the algorithm uses the *K-means* algorithm to find the k partitions.

3 Proximity Measures

Proximity measures quantify the distance or closeness between two data objects. They may be sub-categorized into three types of measures, namely *similarity*, *dissimilarity*, and *distance*.

Table 1. Proximity measures for numeric variables

Name	Function	Discussion
Euclidean Distance (EUC)	$d_{x_i,x_j} = \sqrt{\sum_{k=1}^{n}(x_{ik} - x_{jk})^2}$	Works well for compact or isolated clusters; Discovers clusters of spherical shape; Any two objects may not be influenced by the addition of a new object (i.e. outliers); Very sensitive to the scales of the variables; Not suitable for clusters of different shapes; The variables with the largest values may always dominate the distance.
Manhattan Distance (MAN)	$d_{x_i,x_j} = \sum_{k=1}^{n} |x_{ik} - x_{jk}|$	Computationally cheaper than the Euclidean distance; Scale dependent.
Minkowski Distance (MIN)	$d_{x_i,x_j} = \left(\sum_{k=1}^{n} |x_{ik} - x_{jk}|^{\lambda}\right)^{\frac{1}{\lambda}}$	One may control the amount of emphasis given on the larger differences; The Minkowski distance may cost more than the Euclidean and Manhattan distance when $\lambda > 2$.
Chebyshev Distance (CHEB)	$d_{x_i,x_j} = max_k |x_{ik} - x_{jk}|$	Suitable for situations where the computation time is very crucial; Very sensitive to the scale of the variables.
Canberra Distance (CAN)	$d_{x_i,x_j} = \sum_{k=1}^{n} \frac{|x_{ik}-x_{jk}|}{|x_{ik}+x_{jk}|}$	Not scale sensitive; Suitable for non-negative values; Very sensitive to the changes near the origin; Undefined when both the coordinates are 0.
Mahalanobis Distance (MAH)	$d(x,y) = \sqrt{(x - y)C^{-1}(x - y)^T}$	Considers the correlation between the variables; Not scale dependent; Favors the clusters of hyper ellipsoidal shape; Computational cost is high; May not be suitable for high-dimensional datasets.
Angular Distance (COS)	$d_{x_i,x_j} = 1 - \frac{\sum_{k=1}^{n} x_{ik} \cdot x_{jk}}{\left(\sum_{k=1}^{n} x_{ik}^2 \cdot \sum_{k=1}^{n} x_{jk}^2\right)^{\frac{1}{2}}}$	Calculates the relative distance between the objects from the origin; Suitable for semi-structured datasets (i.e. Widely applied in Text Document cluster analysis where data is highly dimensional); Does not depend on the vector length; Scale invariant; Absolute distance between the data objects is not captured.
Pearson Correlation Distance (COR)	$d_{ij} = 1 - \frac{\sum_{k=1}^{n}(x_{ik} - \bar{x}_i) \cdot (x_{jk} - \bar{x}_j)}{\left(\sum_{k=1}^{n}(x_{ik}-\bar{x}_i)^2 \cdot \sum_{k=1}^{n}(x_{jk}-\bar{x}_j)^2\right)^{\frac{1}{2}}}$	Scale invariant; Considers the correlation between the variables; Calculates the relative distance between the objects from the mean of the data; Suitable for semi-structured data analysis (i.e. applied in microarray analysis, document cluster analysis); Outliers may affect the results.

Similarity is a numerical measure that represents the similarity (i.e. how alike the objects are) between two objects. This measure usually returns a non-negative value that falls in between 0 and 1. However, in some cases similarity may also range from -1 to $+1$. When the similarity takes a value 0, it means that there is no similarity between the objects and the objects are very different from one another. In contrast, 1 denotes complete similarity, emphasizing that the objects are identical and possess the same attribute values.

The **dissimilarity** measure is also a numerical measure, which represents the discrepancy or the difference between a pair of objects [12]. If two objects are very similar then the dissimilarity measure will have a lower value, and visa versa. Therefore, this measure is reversely related to the similarity measure. The dissimilarity value also usually fall into the interval $[0, 1]$, but it may also take values ranging from -1 to $+1$.

The term **distance**, which is also commonly used as a synonym for the dissimilarity measure, computes the distance between two data points in a multi-dimensional space. Let $d(x, y)$ be the distance between objects x and y. Then, the following four properties hold for a distance measure ([13], [14]):

1. $d(x, y) = d(y, x)$, for all points x and y.
2. $d(x, y) = 0$, if $x = y$.
3. $d(x, y) \geq 0$, for all points x and y.
4. $d(x, y) \leq d(x, z) + d(z, y)$, for all points x, y and z. This implies that introducing a third point may never shorten the distance between two other points.

There are many different proximity measures available in the literature. One of the reasons for this variety is that these measures differ on the data type of the objects in a given dataset. Next we present the proximity measures that are used in this paper.

3.1 Proximity Measures for Numeric Variables

Table 1 presents the measures for numeric, real-valued or continuous variables used in our paper ([12], [27]). These measures may be categorized into three groups. The first group contains the functions that measure the absolute distance between the objects and are scale dependent. This list includes the Euclidean (EUC), Manhattan (MAN), Minkowski (MIN), and Chebyshev (CHEB) distances. The second group contains only the Canberra distance (CAN) which also calculates the absolute distance, however, the measure is not scale dependent. In the third group we have three distance measures, namely the Angular or Cosine (COS), Pearson Correlation (COR), and Mahalanobis (MAH) distances. These measures consider the correlation between the variables into account and are scale invariant.

3.2 Proximity Measures for Mixed Variables

In the previous section, we concentrated our discussion on datasets with numeric values. Nevertheless, in practical applications, it is often possible to have more than one type of attribute in the same dataset. It follows that, in such cases, the conventional proximity measures for those data types may not work well. A more practical approach is to process all the variables of different types together and then perform a single cluster analysis [10]. The Gower's General Coefficient and Laflin's General Coefficient are two such functions that incorporate information from various data types into a single similarity coefficient. Table 2 provides the equations and additional information about these two coefficients.

4 Experiments

This section discusses our experimental methodology and the results obtained for each of the data types. In order to compare the performance of the proximity measures for a particular data type, we performed ten-fold cross validation [15] and classes to clusters evaluation on each of the datasets. In this paper, we consider the external cluster evaluation measures to measure the goodness or the quality of the clusters obtained from

Table 2. Proximity measures for mixed variables

Name	Function	Discussion
Gower's General Coefficient (GOWER)	$d(i,j) = \frac{\sum_{f=1}^{p} \delta_{ij}^{(f)} d_{ij}^{(f)}}{\sum_{f=1}^{p} \delta_{ij}^{(f)}}$	For this coefficient $\delta_{ij}^{(f)} = 0$, if one of the values is missing or the variable is of type asymmetric binary variable. Otherwise, $\delta_{ij}^{(f)} = 1$. For numeric variables distance is calculated using the formula: $d_{ij}^{(f)} = \frac{|x_{if} - x_{jf}|}{max_h x_{hf} - min_h x_{hf}}$. For binary and nominal variables the dissimilarity is the number of un matched pairs. This measure may be extended to incorporate other attribute types (e.g. ordinal, ratio-scaled) [14].
Laflin's General Similarity Coefficient (LAFLIN)	$s(i,j) = \frac{N_1 \cdot s_1 + N_2 \cdot s_2 + \ldots + N_n \cdot s_n}{N_1 + N_2 + \ldots + N_n}$	In this function, $N_1 \ldots N_n$ represent the total number of attributes of each of the variable type, whereas, $s_1 \ldots s_n$ represent the total similarity measure calculated for each of the attribute type. The function uses existing similarity measures to calculate the similarity scores $s_1 \ldots s_n$.

spectral clustering algorithms, as the external class labels for each of the datasets used were available to us. Moreover, this allows us to perform a fair comparison against the known true clusters for all the proximity measures. We used two such measures namely, F-measure ([4], [16]) and G-means [17]. These measures have been previously used in numerous studies regarding cluster analysis and have proved to be successful in representing the quality of clusters numerically. In addition, we used the Friedman Test [19] to test the statistical significance of our results. The test is best suited for situations, like ours, where multiple comparisons are performed against multiple datasets. It returns the *p-value* that helps us to determine whether to accept or reject the null hypothesis. For us the null hypothesis was *"the performance of all the proximity measure is equivalent"*. For example, a p-value less than 0.05 signify that the result is only 5% likely to be extraordinary ([19], [28]).

4.1 Implementation and Settings

For all the experiments in this paper, the data preprocessing (e.g. replacing missing values, standardization) was performed using WEKA [18], an open-source Java-based machine learning software, developed at the University of Waikato in New Zealand. The spectral cluster analysis algorithms are implemented in MATLAB®. Since these algorithms manipulate the similarity matrix of a dataset, computation of eigenvalues and eigenvectors of the similarity matrix may be inefficient for a large matrix. However, MATLAB efficiently solves the eigensystem of large matrices. The cluster evaluation measures have been implemented in Java.

4.2 Datasets

Six datasets are used for each of the data types. All the datasets varied in size and are based on real-world problems representing various domains and areas. As for numeric data type, several proximity measures are scale-dependent. As such, we *standardized* [14] the attribute values to ensure the accuracy of our results. In addition, we used Equation 1 to convert the distance measure into a similarity measure in order to create the similarity or weight matrix W [2]. It is also important to note that the selection of

Table 3. Dataset Information

Numeric Datasets			
Dataset	No. of Tuples	No. of True Clusters	No. of Attributes
Body	507	2	21
Iris	150	3	4
Wine	178	3	13
Glass	214	6	8
Ecoli	336	5	7
SPECT	267	2	44
Mixed Datasets			
Dataset	No. of Tuples	No. of True Clusters	No. of Attributes
Automobile	205	6	25
CRX	690	2	15
Dermatology	366	6	33
Hepatitis	155	2	19
Post-Operative	90	3	8
Soybean	290	15	35

sigma is very crucial to the success of spectral clustering algorithm and the value of sigma varies depending on the proximity measure and the dataset used. We adopted the method proposed by Shi and Malik in [2] to select the sigma value. The sigma returned through this method was used as a starting point. Each of our experiments were performed on a range of values surrounding that sigma value and the result for which we achieved the best scores are included in this paper.

$$s(x, y) = \exp(\frac{-d(x, y)^2}{2 \times \sigma^2})$$ (1)

Five of our datasets are from UCI repository [21] and one of them (Body dataset) is from an external source [20]. In Table 3, we provide a summary of the numeric datasets. The six datasets used for the mixed variable type are also obtained from the UCI repository. A summary of the mixed datasets is presented in Table 3.

4.3   Experimental Results for Numeric Datasets

In Table 4 and Table 5 we present the F-measure and G-means scores obtained for the datasets with numeric variables when SM (NCut) algorithm is used. Our results show that the COR distance and the COS distance measure often scored higher than the rest of the distance measures (Figure 4). These two distance measures performed well in four out of six datasets. The datasets are Body, Iris, Wine, and Glass. We also notice that most of the time, these two coefficients achieved similar values for both the evaluation measures. The overall average difference for these two distance measures is 0.02, irrespective of the dataset or the splitting method used. In contrast, the MAH distance measure performed poorly in four out of six datasets. The datasets for which this distance measure scored the lowest are Body, Iris, Wine, and Glass. The MAN distance performed well for the Ecoli dataset and the CAN distance performed best for

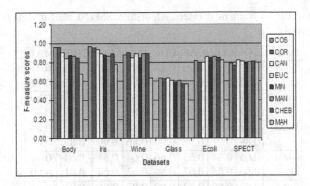

Fig. 4. F-measure scores for the numeric dataset when tested on the SM (NCut) algorithm

Table 4. F-measure scores for Numeric datasets. Algorithm: SM (NCut), Splitting points: Zero and Mean value respectively.

Dataset	COS	COR	CAN	EUC	MIN	MAN	CHEB	MAH
Body	0.97	0.96	0.90	0.87	0.87	0.87	0.85	0.68
	0.97	0.96	0.90	0.84	0.87	0.87	0.85	0.68
Iris	0.97	0.96	0.94	0.90	0.88	0.86	0.88	0.79
	0.97	0.96	0.94	0.89	0.88	0.86	0.89	0.78
Wine	0.91	0.95	0.84	0.81	0.84	0.90	0.87	0.79
	0.88	0.90	0.85	0.89	0.85	0.89	0.89	0.64
Glass	0.62	0.61	0.61	0.61	0.60	0.61	0.59	0.60
	0.64	0.63	0.64	0.61	0.59	0.61	0.57	0.57
Ecoli	0.81	0.79	0.83	0.82	0.83	0.85	0.82	0.81
	0.81	0.80	0.80	0.86	0.85	0.86	0.85	0.82
SPECT	0.80	0.77	0.81	0.80	0.80	0.81	0.80	0.80
	0.79	0.77	0.82	0.81	0.79	0.81	0.81	0.80

the SPECT dataset. We also notice that the performances of EUC, MIN, MAN, CAN and CHEB distances are very similar, and that they often scored moderately, in comparison to the highest and the lowest scores. For example, based on the scores obtained for the Body dataset, the distance measures may be grouped into three groups: 1) the COS distance and COR distance measure in one group where the scores fall in the range [0.96 − 0.97], 2) the CAN, EUC, MIN, MAN and CHEB distance measures in the second group where the range is [0.85 − 0.90], and 3) the MAH distance measure which scores the lowest (0.68). As observed from the results, the EUC distance measure, which is often used in the spectral cluster analysis algorithms, may not always be a suitable choice.

We observe that, if the COS distance or the COR distance measure is used instead of the EUC distance, on average the performance improved by 7.42% (F-measure) and 8.17% (G-means), respectively, for our datasets. In Table 6 and Table 7, we provide the evaluation scores from the NJW (K-means) algorithm. In this case also, the results showed almost the same trend as the results from the SM (NCut) algorithm. For both the evaluation measures, the results indicate that the MAH distance measure often scored the lowest scores over a range of datasets. Among the six datasets, in five of the cases

Table 5. G-means scores for Numeric datasets. Algorithm: SM (NCut), Splitting points: Zero and Mean value respectively.

Dataset	COS	COR	CAN	EUC	MIN	MAN	CHEB	MAH
Body	0.97	0.96	0.90	0.87	0.87	0.87	0.85	0.71
	0.97	0.97	0.90	0.83	0.84	0.87	0.85	0.71
Iris	0.97	0.96	0.94	0.90	0.88	0.87	0.88	0.80
	0.97	0.96	0.94	0.90	0.89	0.87	0.89	0.80
Wine	0.92	0.95	0.86	0.83	0.86	0.91	0.87	0.80
	0.89	0.91	0.86	0.90	0.86	0.90	0.89	0.67
Glass	0.65	0.64	0.65	0.66	0.64	0.64	0.64	0.62
	0.66	0.65	0.67	0.64	0.62	0.64	0.61	0.60
Ecoli	0.82	0.80	0.83	0.82	0.83	0.86	0.83	0.81
	0.81	0.81	0.81	0.86	0.86	0.86	0.85	0.82
SPECT	0.81	0.80	0.82	0.82	0.81	0.81	0.80	0.82
	0.81	0.80	0.83	0.82	0.82	0.82	0.81	0.81

Table 6. F-measure scores of the NJW (K-means) algorithm (tested on numeric dataset)

Dataset	COS	COR	CAN	EUC	MIN	MAN	CHEB	MAH
Body	0.88	0.88	0.83	0.79	0.79	0.80	0.78	0.64
Iris	0.81	0.82	0.79	0.78	0.80	0.78	0.79	0.69
Wine	0.94	0.96	0.97	0.85	0.97	0.86	0.83	0.50
Glass	0.54	0.54	0.53	0.55	0.56	0.58	0.54	0.50
Ecoli	0.65	0.72	0.63	0.74	0.73	0.75	0.65	0.48
SPECT	0.77	0.77	0.77	0.77	0.77	0.77	0.76	0.77

Table 7. G-means scores of the NJW (K-means) algorithm (tested on numeric dataset)

Dataset	COS	COR	CAN	EUC	MIN	MAN	CHEB	MAH
Body	0.88	0.88	0.83	0.79	0.79	0.80	0.78	0.67
Iris	0.81	0.83	0.79	0.79	0.80	0.79	0.79	0.69
Wine	0.94	0.96	0.97	0.86	0.97	0.87	0.84	0.57
Glass	0.56	0.55	0.56	0.58	0.58	0.62	0.56	0.55
Ecoli	0.69	0.74	0.68	0.75	0.75	0.77	0.68	0.52
SPECT	0.80	0.80	0.80	0.80	0.80	0.80	0.79	0.80

the MAH distance scored the lowest scores. These datasets are: Body, Iris, Wine, Glass, and Ecoli. Furthermore, in none of the cases, the EUC distance measure scored the highest score. For two datasets (i.e. Body and Iris), the COR distance and the COS distance performed well, and for the Wine and Glass datasets, the scores were very close to the highest scores achieved. The MAN distance performed well for the Ecoli and Glass datasets.

The Friedman Test which was used to measure the statistical significance of our results, gives p-values of 0.0332 (SM (NCut)) and 0.0097 (NJW (K-means)), respectively. Since the p-values are less than 0.05, this indicates the results are statistically significant. Our results showed that the MAH distance often performed poorly when compared to the rest of the distance measures according to the cluster evaluation measures. We noticed that, when the MAH distance is used, the spectral clustering algorithms produced imbalanced clusters. Here the clusters are imbalanced when one partition contains relatively fewer objects than the other cluster. We also noticed that the objects that are placed in the smaller cluster are the objects that have the lowest degree. Recall from Section 2 that the degree is the total similarity value from one object to the rest of the objects in a dataset. In spectral clustering, the objects are considered as nodes in the graph, and a partition separates objects where the total within cluster similarity is high and the between cluster similarity is very low. Therefore, when the degree is low for an object, compared to the rest of the objects, it indicates that the object is less similar than most of the objects in the dataset. Now, the equation of the MAH distance defines an ellipsoid in n-dimensional space ([22], [23]). The distance considers the variance (how spread out the values are from the mean) of each attribute as well as the covariance (how much two variables change together) of the attributes in the datasets. It gives less weight to the dimensions with high variance and more weight to the dimensions with small variance. The covariance between the attributes allows the ellipsoid to rotate its axes and increase and decrease its size [23]. Therefore, the distance measure is very sensitive to the extreme points [24].

Figure 5 illustrates a scenario showing the MAH distance between the objects. In this figure, the distance between object 1 and 2 will be less than the distance between object 1 and 3, according to the MAH distance. This is because, object 2 lies very close to the main axes along with the other objects, whereas the object 3 lies further away from the main axes. Therefore, in such situations, the MAH distance will be large. For numeric data, the similarity will be very low when the distance is very large. The function in Equation 1, which is used to convert a distance value into a similarity value, will give a value close to zero when the distance is very large. Therefore, the degree from this object to the remainder of the objects becomes very low and the spectral methods separate these objects from the rest. This is one of the possible reasons for the MAH distance performing poorly. It either discovers imbalanced clusters or places similar objects wrongly into two different clusters. Consequently, one possible way to improve the performance of the MAH distance measure might be by changing the value of σ to a larger value. In this way, we may prevent the similarity to have a very small value.

Our results also indicate that the COR and COS distances performed best for four out of six datasets. Both of the distance measures calculate the relative distance from a fixed point (mean or zero, respectively). Therefore, two objects with a similar pattern will be more similar even if their sizes are different. The EUC, MAN, MIN, CAN and CHEB distance measures, however, calculate the absolute distance (i.e. straight line distance from one point to another). For instance, the Body dataset partitions the objects into two main clusters, one with larger body dimensions and another cluster with smaller body dimensions. According to the true cluster information, the larger body dimensions denote the Male population and the smaller body dimensions denote the Female population. When compared to the true clusters, we observed that several individuals, whose body dimensions are comparatively lower than the average body dimensions of the Male population, are placed with the individuals from the Female population by the distance measures that calculate the absolute distance. Conversely, Female individuals with larger body dimensions than the average body dimensions of the Female population are placed with the individuals from Male population. Therefore,

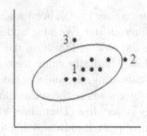

Fig. 5. A scenario depicting the Mahalanobis (MAH) distance between three points

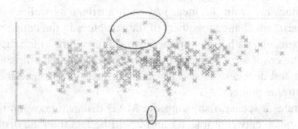

Fig. 6. Example of cluster assignments of the Body dataset. The circles are used to point to the several individual members that are placed differently.

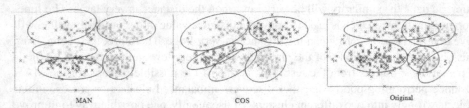

Fig. 7. Example of cluster assignments of the Ecoli dataset. (Left) The clusters obtained by using the MAN distance measure, (Middle) the clusters obtained by using the COS distance, and (Right) the original true clusters.

these individuals that fall very close to the boundary of the two true clusters, are placed differently by the distance measures that calculate the absolute distance (i.e. EUC distance, MAN distance, and MIN distance) than the distance measures that consider the relative distance (i.e. COR distance and COS distance). In such cases, the COS distance and the COR distance correctly identify these individuals. In Figure 6 we plot the first two attributes of the Body dataset when the EUC distance is used as the distance measure. The object marked with a smaller circle is an example of a Male individual with smaller body dimensions. When the EUC distance is used as the distance measure in the spectral clustering algorithm, this object is placed with the Female population. The objects marked with the larger circle illustrate the reversed situation, where Female individuals with larger body size are placed with the individuals from Male population. In both situations, the COR and COS distances placed the objects within their own groups. In Figure 7, we provide the clusters from the Ecoli dataset when the MAN (Left) and the COS (Middle) distances are used. The farthest right figure (with the title Original) depicts the true clusters. The objects, according to the true clusters, overlap between the clusters in a number of situations (e.g. the objects marked with circle 2 and 4, or the objects marked with circle 3 and 5). This indicates that there are several objects in the dataset that may be very similar, but are placed in two different true clusters. When the spectral clustering algorithms are applied to this dataset, both the COS and MAN distances divide the true cluster marked with circle 1 (in Figure 7) into two different clusters. However, the clusters produced by the COS distance contain members from true cluster 1 and 3, whereas the clusters produced by the MAN distance contain the members from true cluster 1. The figure indicates that the shape of the clusters produced by the COS distance are more elongated toward the origin, which is the reason why some of the members from true cluster 3 are included.

Discussion. The main conclusion drawn from our results thus indicate that the MAH distance needs special consideration. This measure tends to create imbalanced clusters and therefore results in poor performance. In addition, the distance measures based on the relative distances (i.e. COR and COS distance measure) outperformed the distance measures based on the absolute distance. We noticed that, in such cases, the objects that reside in the boundary area are correctly identified by the relative distance measures. These boundary objects are slightly different from the other members of their own group and may need special attention. This is due to the fact that the COR and COS measures consider the underlying patterns in between the objects from a fixed point (i.e. mean or zero), in contrast to the absolute distance approaches. Therefore, the Euclidian (EUC) distance which is a commonly used absolute distance measure in clustering domains, may not always be a good selection for the spectral clustering algorithm, especially in domains where we are sensitive to outliers and anomalies.

In summary, under certain conditions, the GOWER similarity coefficient and the LAFLIN coefficient perform similarly. The constraints are as follows: 1) the dataset does not include asymmetric binary variables, and 2) the distance and similarity measures for each of the variables are the same. Recall from Section 4.3 that our results for numeric variables indicate that the Euclidian distance may not be the best choice for numeric variables. This choice seems to impact the performance of the LAFLIN coefficient, which may be improved by using a different distance measure for the numeric variables.

4.4 Experimental Results for Mixed Datasets

In Table 8, we present the F-measure and G-means scores for the Mixed datasets, when tests are applied on the SM (NCut) algorithm. The results from NJW (K-means) algorithm are given in Table 9. Figure 8 presents a graphical representation of our results. Our results from the external evaluation scores show that, the GOWER coefficient performed well for the Automobile and Dermatology dataset. The LAFLIN's coefficient also performed well for two of the datasets. The datasets are CRX and Hepatitis. For Post Operative and Soybean datasets, both the coefficients scored the same scores. In contrast, when the tests are applied on the NJW (K-means) algorithm, our results indicate that the GOWER coefficient performed slightly better than the LAFLIN's. In four out of six datasets the GOWER scored slightly higher scores than the LAFLIN's

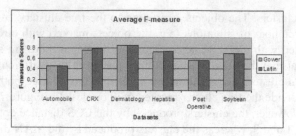

Fig. 8. Average F-measure scores for the Mixed Datasets when tested on the SM(NCut) algorithm

Table 8. F-measure and G-means scores for Mixed datasets. Algorithm: SM (NCut), Splitting points: zero and mean value, respectively.

	F-measure			G-means	
Dataset	GOWER	LAFLIN	Dataset	GOWER	LAFLIN
Automobile	0.46	0.44	Automobile	0.49	0.48
	0.46	0.46		0.50	0.49
CRX	0.76	0.79	CRX	0.76	0.79
	0.76	0.79		0.76	0.79
Dermatology	0.85	0.87	Dermatology	0.86	0.87
	0.84	0.82		0.85	0.83
Hepatitis	0.71	0.73	Hepatitis	0.74	0.75
	0.73	0.75		0.75	0.77
Post Operative	0.56	0.56	Post Operative	0.57	0.57
	0.56	0.56		0.58	0.57
Soybean	0.70	0.70	Soybean	0.72	0.72
	0.70	0.70		0.72	0.72

Table 9. F-measure and G-means scores from the NJW (K-means) algorithm (tested on mixed dataset)

	F-measure		G-means	
Dataset	GOWER	LAFLIN	GOWER	LAFLIN
Automobile	0.47	0.45	0.48	0.46
CRX	0.76	0.80	0.76	0.80
Dermatology	0.84	0.82	0.86	0.84
Hepatitis	0.71	0.74	0.74	0.76
Post Operative	0.52	0.47	0.53	0.49
Soybean	0.57	0.53	0.61	0.56

coefficient. The datasets are Automobile, Dermatology, Post Operative, and Soybean. For this case also, the LAFLIN's coefficient performed best for the same two datasets (i.e. CRX and Hepatitis) as our previous test on the SM (NCut) algorithm. However, we also noticed from the scores that the difference between the performances of the two coefficients is very low. The p-values from the Friedman test are 0.3173 and 0.4142, respectively. Since, both the values are greater than 0.05, the difference between the performance of the two coefficients is not statistically significant.

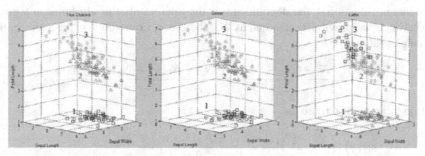

Fig. 9. Comparison of numeric functions on Iris dataset. (From left) the clusters obtained from the true clusters, the clusters obtained from the numeric function of the GOWER coefficient, and the clusters obtained from the numeric function of the LAFLIN coefficient.

In this part, we analyze the coefficients to determine the relationship between them. The equations for the two coefficients are given in Table 2. For the GOWER coefficient, the term $\delta_{ij}^{(f)}$ is an indicator variable associated with each of the variables present in the dataset and the term $d_{ij}^{(f)}$ is the distance or dissimilarity calculated for each variable for objects i and j. We also know from the description given in Table 2 that $\delta_{ij}^{(f)} = 0$ for the asymmetric binary variables and for all the other types $\delta_{ij}^{(f)} = 1$. In our datasets, all of the attributes are numeric, nominal, or symmetric binary. Therefore, the denominator of the GOWER's equation represents the total number of variables in the dataset. Also, recall that the GOWER coefficient is a dissimilarity measure, where the dissimilarity between the two objects, i and j, falls in between 0 and 1. This equation is converted into a similarity measure by subtracting from 1. Therefore, the equation for the GOWER similarity coefficient is:

$$s(i,j) = 1 - \frac{\sum_{f=1}^{p} \delta_{ij}^{(f)} d_{ij}^{(f)}}{\sum_{f=1}^{p} \delta_{ij}^{(f)}} \qquad (2)$$

Let $N = \sum_{f=1}^{p} \delta_{ij}^{(f)}$ be the total number of attributes and for each attribute $\delta_{ij} = 1$, then Equation 2 becomes,

$$s(i,j) = 1 - \frac{\sum_{f=1}^{p} 1 * d_{ij}^{(f)}}{N} = \frac{N - \sum_{f=1}^{p} d_{ij}^{(f)}}{N} \qquad (3)$$

Notice from the equation of LAFLIN's coefficient, s_i is the total similarity value of attribute type i, and N_i is the total number of variables of attribute type i. In our datasets, the attribute types are numeric (N_1 and s_1), nominal (N_2 and s_2), and symmetric binary (N_3 and s_3). Therefore, the equation becomes,

$$s(i,j) = \frac{N_1.s_1 + N_2.s_2 + N_3.s_3}{N_1 + N_2 + N_3} \qquad (4)$$

Notice that in Equation 4, the denominator is the total number of attributes in a given dataset, which we previously denoted as N. Therefore, Equation 4 is the same as the following equation:

$$s(i,j) = \frac{N - (N_1 - N_1.s_1 + N_2 - N_2.s_2 + N_3 - N_3.s_3)}{N} \tag{5}$$

Equation 5 can be re-written as:

$$s(i,j) = \frac{N - (N_1(1 - s_1) + N_2(1 - s_2) + N_3(1 - s_3))}{N} \tag{6}$$

At this point, the GOWER equation given in Equation 3 and the LAFLIN's coefficient given in Equation 6, both have similar patterns. They have the same denominator. However, they differ only in the terms in numerator. As mentioned previously, d_{ij} is the distance or dissimilarity between the two objects i and j, whereas, $(1 - s_1)$ is also a dissimilarity measure. Both functions handle nominal and binary variables in the same way. Therefore, this implies that the difference in the equations occurs due to the functions selected for the numeric attributes which are handled differently by the two coefficients. This is one of the reasons that the difference between the performances of both of the coefficients is very low. In our tests, we used the Euclidean distance for the LAFLIN's coefficient, whereas, the GOWER coefficient uses the distance measure given in Equation 7.

$$d_{ij}^{(f)} = \frac{|x_{if} - x_{jf}|}{max_h x_{hf} - min_h x_{hf}} \tag{7}$$

We use the two numeric functions with the spectral clustering algorithms and apply them on the Iris dataset from the UCI repository [3] to evaluate their performances. Figure 9 illustrates the clusters obtained from the true clusters (left), the clusters obtained from the numeric function of the GOWER coefficient (middle), and the clusters obtained from the numeric function of the LAFLIN coefficient (right). We notice that both of the measures correctly cluster the objects from true cluster 1. However, the difference between them is clear in true cluster 2 and cluster 3. Notice that these two true clusters have objects that overlap near the boundary of the clusters. The objects located at the boundary usually have attribute values slightly different from the other members of their own true clusters. The numeric function for the GOWER coefficient correctly distinguishes several objects near the boundary. However, the LAFLIN coefficient, which used the Euclidean distance to compute the distance between the objects, placed the objects which are located near the boundary, in two different clusters. We notice that the clusters formed from this measure have a shape similar to a sphere. This may be the reason for this measure performing slightly differently than the function of the GOWER coefficient.

Discussion. In summary, under certain conditions, the GOWER similarity coefficient and the LAFLIN coefficient perform similarly. The constraints are as follows: 1) the dataset does not include asymmetric binary variables, and 2) the distance and similarity measures for each of the variables are the same. Recall from Section 4.3 that our results for numeric variables indicate that the Euclidian distance may not be the best choice for numeric variables. This choice seems to impact the performance of the LAFLIN coefficient, which may be improved by using a different distance measure for the numeric variables.

5   Conclusions

The selection of proximity measures is a crucial step that has a significant impact on the quality and usability of the end results of the cluster analysis algorithms. This fact is frequently overlooked, leading to a degrading of the potential knowledge being discovered. This paper presented an explorative and comparative study of the performance of various proximity measures when applied to the spectral clustering algorithms. In particular, our study address the question when, and where, the choice of proximity measure becomes crucial in order to succeed. Our results indicate that proximity measures needs special care in domains where the data is highly imbalanced and where in is important to correctly cluster the boundary objects. These cases are of special interest in application areas such as rare disease diagnosis, financial market analysis and fraud detection.

Our future work will consider a diverse selection of datasets. We aim to evaluate if our conclusions hold for sparse datasets with noise and many missing values. We will also extend our research to very large datasets with high dimensionality. For such datasets, these proximity measures may not perform as per our expectation. That is, with high dimensions, the data may become sparse and the distance computed from these measures may not capture similarities properly. In such cases, a different set of proximity measures may be required to deal with the problem of high dimensionality.

The selection of the most suitable proximity measures when specifically aiming to detect outliers and anomalies is another topic of future research. In order to reach a conclusion with higher generality, we are interested to see whether the conclusions drawn from our paper persist for other clustering algorithms. The development of additional measures for mixed data types, especially ones that do not use the Euclidian distance for numeric data, are also a significant issue which will benefit from being further researched.

References

1. Luxburg, U.: A Tutorial on Spectral Clustering. Statistics and Computing 17(4), 395–416 (2007)
2. Shi, J., Malik, J.: Normalized Cuts and Image Segmentation. IEEE Transactions on Pattern Analysis and Machine Intelligence 22(8), 888–905 (2000)
3. Bach, F.R., Jordan, M.I.: Learning Spectral Clustering, with Application to Speech Separation. J. Mach. Learn. Res. 7, 1963–2001 (2006)
4. Paccanaro, A., Casbon, J.A., Saqi, M.A.: Spectral Clustering of Protein Sequences. Nucleic Acids Res. 34(5), 1571–1580 (2006)
5. Ng, A.Y., Jordan, M.I., Weiss, Y.: On Spectral Clustering: Analysis and an Algorithm. In: Dietterich, T.G., Ghahramani, S.B. (eds.) Advances in Neural Information Processing Systems, vol. 14, pp. 849–856 (2001)
6. Verma, D., Meila, M.: A Comparison of Spectral Clustering Algorithms (2001)

7. Jain, A.K., Murty, M.N., Flynn, P.J.: Data Clustering: a Review. ACM Computing Surveys 31(3), 264–323 (1999)
8. Bach, F.R., Jordan, M.I.: Learning Spectral Clustering. In: Advances in Neural Information Processing Systems 16: Proceedings of the 2003 Conference, pp. 305–312 (2003)
9. Everitt, B.S.: Cluster Analysis, 2nd edn. Edward Arnold and Halsted Press (1980)
10. Kaufman, L., Rousseeuw, P.: Finding Groups in Data: An Introduction to Cluster Analysis. Wiley-Interscience (2005)
11. Meila, M., Shi, J.: A Random Walks View of Spectral Segmentation. In: International Conference on Artificial Intelligence and Statistics (AISTAT), pp. 8–11 (2001)
12. Webb, A.R.: Statistical Pattern Recognition, 2nd edn. John Wiley & Sons (2002)
13. Larose, D.T.: Discovering Knowledge in Data: An Introduction to Data Mining. Wiley-Interscience (2004)
14. Han, J., Kamber, M.: Data Mining: Concepts and Techniques, 2nd edn. Morgan Kaufmann Publishers Inc., San Francisco (2006)
15. Costa, I.G., de Carvalho, F.A.T., de Souto, M.C.P.: Comparative Study on Proximity Indices for Cluster Analysis of Gene Expression Time Series. Journal of Intelligent and Fuzzy Systems: Applications in Engineering and Technology 13(2-4), 133–142 (2002)
16. Steinbach, M., Karypis, G., Kumar, V.: A Comparison of Document Clustering Techniques. In: KDD Workshop on Text Mining (2000)
17. Kubat, M., Holte, R.C., Matwin, S.: Machine Learning for the Detection of Oil Spills in Satellite Radar Images. Machine Learning 30(2-3), 195–215 (1998)
18. Witten, I.H., Frank, E.: Data Mining: Practical Machine Learning Tools and Techniques, 2nd edn. Morgan Kaufmann (2005)
19. Japkowicz, N., Shah, M.: Performance Evaluation for Classification A Machine Learning and Data Mining Perspective (in progress): Chapter 6: Statistical Significance Testing (2011)
20. Heinz, G., Peterson, L.J., Johnson, R.W., Kerk, C.J.: Exploring Relationships in Body Dimensions. Journal of Statistics Education 11(2) (2003)
21. Asuncion, A., Newman, D.: UCI Machine Learning Repository (2007)
22. Lee, S.-W., Verri, A. (eds.): SVM 2002. LNCS, vol. 2388. Springer, Heidelberg (2002)
23. Abou-Moustafa, K.T., Ferrie, F.P.: The Minimum Volume Ellipsoid Metric. In: Hamprecht, F.A., Schnörr, C., Jähne, B. (eds.) DAGM 2007. LNCS, vol. 4713, pp. 335–344. Springer, Heidelberg (2007)
24. Filzmoser, P., Garrett, R., Reimann, C.: Multivariate Outlier Detection in Exploration Geochemistry. Computers and Geosciences 31(5), 579–587 (2005)
25. Aiello, M., Andreozzi, F., Catanzariti, E., Isgro, F., Santoro, M.: Fast Convergence for Spectral Clustering. In: ICIAP 2007: Proceedings of the 14th International Conference on Image Analysis and Processing, pp. 641–646. IEEE Computer Society, Washington, DC (2007)
26. Fischer, I., Poland, J.: New Methods for Spectral Clustering. Technical Report IDSIA-12-04, IDSIA (2004)
27. Teknomo, K.: Similarity Measurement, http://people.revoledu.com/kardi/tutorial/Similarity/
28. Boslaugh, S., Watters, P.A.: Statistics in a Nutshell. O.Reilly & Associates, Inc., Sebastopol (2008)

Comparing the Macroeconomic Responses
of US and Japan through Time Series Segmentation

Jian Cheng Wong[1], Gladys Hui Ting Lee[1], Yiting Zhang[1], Woei Shyr Yim[1],
Robert Paulo Fornia[2], Danny Yuan Xu[3], Chong Eu Lee[4],
Jun Liang Kok[4], and Siew Ann Cheong[4]

[1] Division of Mathematical Sciences, School of Physical and Mathematical Sciences,
Nanyang Technological University, 21 Nanyang Link, Singapore 637371, Republic of Singapore
[2] University of Colorado at Boulder, Boulder, CO 80309, U.S.A.
[3] Bard College, PO Box 5000, Annandale-on-Hudson, NY 12504, U.S.A.
[4] Division of Physics and Applied Physics, School of Physical and Mathematical Sciences,
Nanyang Technological University, 21 Nanyang Link, Singapore 637371, Republic of Singapore

Abstract. In this paper we performed time series segmentation on the high-frequency time series data of various US and Japanese financial market indices, and found that for both economies, the time series segments can be very naturally grouped into four to six classes, corresponding roughly with economic growth, economic crisis, market correction, and market crash. With this classification of the time series segments, we discovered that the US economy recovered completely in one year six months, whereas the Japanese economy recovered incompletely in two years three months from the 2000 Technology Bubble Crisis. In contrast to the slow recovery, the US and Japanese economies succumbed to the 2007 Subprime Crisis in two months and 21 days respectively. Minimal spanning tree analysis of the time series segments points to signs of recovery as early as Sep 2009 for the US, but no signs for recovery as late as Jun 2010 for Japan.

Keywords: Time series segmentation, Coarse graining, Macroeconomic cycle, Financial markets.

1 Introduction

The dynamics of complex systems with very many microscopic variables frequently settle onto a small number of low-dimensional manifolds, because of the nonlinear interactions between the microscopic variables. Each of these low-dimensional manifolds can be described by a few collective variables. From the point of view of statistical thermodynamics, each low-dimensional manifold represents a distinct macroscopic phase. For example, a macroscopic collection of water molecules can be found in three distinct phases. Below the critical temperature and pressure, liquid water and water vapor can be distinguished by their densities. Liquid water and solid ice can also be easily distinguished by their pair distribution functions, whose Fourier transforms can be easily probed using experimental techniques like X-ray diffraction or neutron scattering. But what if we do not know all these beforehand, and only have time series data on the water molecule displacements. Can we still conclude that water has three distinct phases?

A. Fred et al. (Eds.): IC3K 2011, CCIS 348, pp. 79–93, 2013.

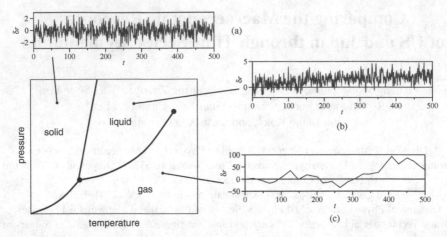

Fig. 1. A typical phase diagram showing where the solid, liquid, and gas phases of a substance occurs in the pressure-temperature (p-T) plane. Also shown in the figure are the equilibrium fluctuations δr in the displacement of a given atom in the (a) solid phase, with time-independent variance $\langle|\delta\mathbf{r}|^2\rangle \propto T$; (b) liquid phase, with a diffusive variance $\langle|\delta\mathbf{r}|^2\rangle \propto t$; and (c) gas phase, with long ballistic lifetimes.

From Figure 1, we see that the answer is affirmative. In solid ice, the displacement of a given water molecule fluctuates about an average point. This fluctuation becomes stronger with increasing temperature, but is time-independent. In liquid water at comparable temperatures, there are also strong displacement fluctuations. However, in addition to being temperature dependent, the fluctuations are also time dependent. This is because in liquid water, molecular trajectories are diffusive. Finally, in water vapor, molecular trajectories are ballistic, allowing us to distinguish it from liquid water. From this well known example, we understand that fast fluctuations of the microscopic variables are slaved to the low-dimensional manifold the system finds itself in.

In the case of water, we could from the above discussions discover its phases starting from only microscopic time series, since we know beforehand how these will be different statistically. But since it is simple statistics that differentiate phases, we can also discover them without any prior knowledge. If the system has gone through multiple phase transitions, we can detect these transitions by performing time series segmentation, which partitions the time series into a collection of segments statistically distinct from their predecessors and successors. If we then cluster these time series segments, we should be able to very naturally classify them into three clusters, each representing one phase of water. Alternatively, if we have many time series, some of which are in the solid phase, others in the liquid phase, and the rest in the gas phase, we can directly cluster the time series to find them falling naturally into three groups. The various methods for doing so are known as time series clustering.

These considerations are very general, and can be applied to diverse complex systems like the atmosphere, climate, earthquakes, and protein folding. In this paper, we focus on describing the time series segmentation method in Section 2, and how it can be applied to gain insights into the behavior of financial markets in Section 3. We then conclude in Section 4.

2   Methods

2.1   Optimized Recursive Segmentation

We start off with a time series $\mathbf{x} = (x_1, \ldots, x_N)$ which is statistically nonstationary. This means that statistical moments like the average and variance evaluated within a fixed window at different times are also fluctuating. However, we suspect that $\mathbf{x}$ might consist of an unknown number M of stationary segments from an unknown number P of segment classes. Since it is possible to arrive at reasonable estimates of M without knowing what P is, we will determine these two separately. The problem of finding M is equivalent to finding the positions of the $M - 1$ segment boundaries. This is the sequence segmentation problem [1,2].

There are various approaches to to time series and sequence segmentation [3,4,5,6,7,8]. In this study, we adopted the recursive entropic segmentation scheme proposed by Bernaola-Galván and coworkers [5,6] for biological sequence segmentation. For a time series of a continuous variable, we assume that all its segments are generated by Gaussian processes, i.e. within segment m, $x_i^{(m)}$ are normally distributed with mean μ_m and variance σ_m^2. Other distributions can be used, depending on what is already known about the time series statistics, how easy or hard parametrization is, and how easy or hard it is to calculate the probability distribution function. We chose Gaussian models for each segment because their parameters are easy to estimate, and their probability distribution functions are easy to calculate.

Given $\mathbf{x} = (x_1, \ldots, x_N)$, we first compute its one-segment likelihood

$$L_1 = \prod_{i=1}^{N} \frac{1}{\sqrt{2\pi\sigma^2}} \exp\left[-\frac{(x_i - \mu)^2}{2\sigma^2}\right] \tag{1}$$

assuming that the entire time series is sampled from a normal distribution with mean μ and variance σ^2. Next, we assume that $\mathbf{x} = (x_1, \ldots, x_t, x_{t+1}, \ldots, x_N)$ actually consists of two segments $\mathbf{x}_L = (x_1, \ldots, x_t)$, sampled from a normal distribution with mean μ_L and variance σ_L^2, and $\mathbf{x}_R = (x_{t+1}, \ldots, x_N)$ sampled from a normal distribution with mean μ_R and variance σ_R^2. The two-segment likelihood of $\mathbf{x}$ is thus

$$L_2(t) = \prod_{i=1}^{t} \frac{1}{\sqrt{2\pi\sigma_L^2}} \exp\left[-\frac{(x_i - \mu_L)^2}{2\sigma_L^2}\right] \times$$
$$\prod_{j=t+1}^{n} \frac{1}{\sqrt{2\pi\sigma_R^2}} \exp\left[-\frac{(x_j - \mu_R)^2}{2\sigma_R^2}\right]. \tag{2}$$

Taking the logarithm of the ratio of likelihoods, we obtain the Jensen-Shannon divergence [9]

$$\Delta(t) = \ln \frac{L_2(t)}{L_1}. \tag{3}$$

This is N times the more general definition $\Delta(P_L, P_R) = H(\pi_L P_L + \pi_R P_R) - \pi_L H(P_L) - \pi_R H(P_R)$ of the Jensen-Shannon divergence, with $\pi_L = N_L/N$,

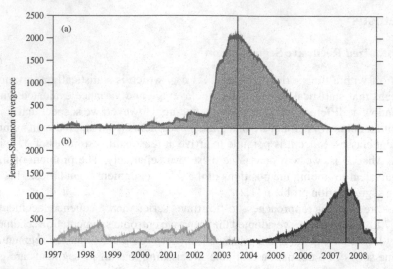

Fig. 2. Jensen-Shannon divergence spectrum of the Dow Jones Industrial Average index time series between January 1997 and August 2008 (red), based on which we identify (a) the first segment boundary to be around the middle of 2003 (marked by red vertical line). To further segment the left and right subsequences, we compute the Jensen-Shannon divergence spectra (green and blue respectively) entirely within the respective subsequences, and (b) find the locations of their divergence maxima.

$\pi_R = N_R/N$, and $H(P)$ is the Shannon entropy for the probability distribution P. The Jensen-Shannon divergence so defined measures how well the two-segment model fits the observed time series over the one-segment model. In practice, the Gaussian parameters $\mu, \mu_L, \mu_R, \sigma^2, \sigma_L^2, \sigma_R^2$ appearing in the likelihoods are replaced by their maximum likelihood estimates $\hat{\mu}, \hat{\mu}_L, \hat{\mu}_R, \hat{\sigma}^2, \hat{\sigma}_L^2$, and $\hat{\sigma}_R^2$.

To find the best segment boundary t^* to cut $\mathbf{x}$ into two segments, we run through all t, and pick $t = t^*$ such that

$$\Delta_{\max} = \Delta(t^*) = \max_t \Delta(t), \qquad (4)$$

as shown in Figure 2. At $t = t^*$, the left and right segments are the most distinct statistically. To find more segment boundaries, we repeat this one-to-two segmentation procedure for $\mathbf{x}_L$ and $\mathbf{x}_R$, and all subsequent segments. As the segments get shorter, the divergence maxima of new segment boundaries will also get smaller. When these divergence maxima become too small, the new segment boundaries will no longer be statistically significant. Further segmentation thus becomes meaningless.

There are two rigorous approaches to terminating the recursive segmentation in the literature, based on hypothesis testing [5,6] and model selection [11,10]. Based on the experience in our previous work [12], these most statistically significant segment boundaries are also discovered if we terminate the recursive segmentation when no new optimized segment boundaries with Jensen-Shannon divergence greater than a cutoff of $\Delta_0 = 10$ are found. This simple termination criterion sometimes result in long segments whose internal segment structures are masked by their contexts [13]. For these long segments, we progressively lower the cutoff Δ_0 until a segment boundary with strength $\Delta > 10$ appears. The final segmentation then consists of segment boundaries discovered through the automated recursive segmentation, as well as segment boundaries discovered through progressive refinement of overly long segments.

At each stage of the recursive segmentation, we also perform segmentation optimization, using the algorithm described in Ref. [14]. Given M segment boundaries $\{t_1, \ldots, t_M\}$, some of which are old, and some of which are new, we optimize the position of the mth segment boundary by computing the Jensen-Shannon divergence spectrum within the *supersegment* bounded by the segment boundaries t_{m-1} and t_{m+1}, and replace t_m by t_m^*, where the supersegment Jensen-Shannon divergence is maximized. We do this iteratively for all M segment boundaries, until all segment boundaries converge to their optimal positions. This optimization step is necessary, because of the *context sensitivity problem* [13]. Otherwise, statistically significant segment boundaries are likely to be masked by the context they are embedded within, and missed by the segmentation procedure.

2.2 Hierarchical Clustering

After the recursive segmentation terminates, we typically end up with a large number of segments. Neighboring segments are statistically distinct, but might be statistically similar to distant segments. We can group statistically similar segments together, to estimate the number P of time series segment classes. Various statistical clustering schemes can be used to achieve this (see for example, the review by Jain, Murty and Flynn [15], or texts on unsupervised machine learning). Since the number of clusters is not known beforehand, we chose to perform agglomerative hierarchical clustering, using the complete link algorithm. The statistical distances between segments are given by their Jensen-Shannon divergences.

In Figure 3, we show the hierarchical clustering tree for the Dow Jones Industrial Average index time series segments, which tells us the following. If we select a global threshold $\Delta > 739.1$, we end up with one cluster, whereas if we select a global threshold $249.3 < \Delta < 739.1$, we find two clusters. These clusters are statistically robust, because they are not sensitive to small variations of the global threshold Δ. However, they are not as informative as we would like them to be. Going to a lower global threshold of $\Delta = 30$, we find seven clusters. These seven clusters give us a more informative dynamical picture, but some of the clusters are not robust. If instead of a global threshold for all robust clusters, we allow local thresholds, i.e. $\Delta = 30$ to differentiate the deep blue and blue clusters, the green and yellow clusters, and $\Delta = 40$ to differentiate the orange and red clusters, we will find six natural and robust clusters.

These clusters are differentiated by their standard deviations, with deep blue being the lowest, and red being the highest. Based on the actual magnitudes of the standard deviations (also called *market volatilities* in the finance literature), we can also group the time series segments into four clusters: low (deep blue and blue), moderate (green), high (yellow and orange), and extremely high (red). As we will explain in Section 3, these four clusters have very natural interpretations as the growth (low-volatility), correction (moderate-volatility), crisis (high-volatility), and crash (extremely-high-volatility) macroeconomic phases.

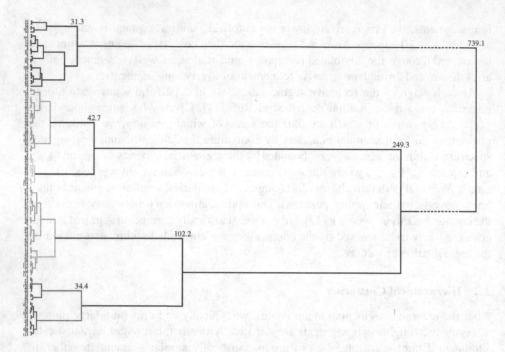

Fig. 3. The complete-link hierarchical clustering tree of the Dow Jones Industrial Average index time series segments assuming that the index movements within each segment are normally distributed. The differentiated clusters are coloured according to their standard deviations: low (deep blue and blue), moderate (green), high (yellow and orange), and extremely high (red). Also shown at the major branches are the Jensen-Shannon divergence values at which subclusters are merged.

2.3   Validation against Synthetic Data

To test the segmentation scheme, we perform several numerical experiments on artificial Gaussian time series. First, we set the standard deviations of the two 5,000-long segments to $\sigma_L = \sigma_R = 1.0$. We also fix the mean of the left segment at $\mu_L = 0$, and vary the mean μ_R of the right segment. As we can see from Table 1(a), the segmentation scheme found only the single segment boundary at $t^* = 5000$, for a difference in mean as small as $\Delta\mu = |\mu_L - \mu_R| = 0.1$. This is remarkable, because the standard deviations of both segments are $\sigma_L = \sigma_R = 1.0 > \Delta\mu$. As expected, the standard error for the boundary position decreases with increasing $\Delta\mu$.

Next, we set $\mu_L = \mu_R = 0$, fix $\sigma_L = 1.0$, and vary σ_R. Again, as we can see from Table 1(b), the single boundary at $t^* = 5000$ was found for ratio of standard deviations as close to one as $\sigma_R/\sigma_L = 0.9$. As expected, the standard error for the boundary position decreases with increasing disparity between σ_L and σ_R. Finally, we set $(\mu_L, \sigma_L) = (0, 1)$ and $(\mu_R, \sigma_R) = (0, 0.5)$, and vary the length N of the artificial time series, always keeping the segment boundary in the middle. From Table 1(c), we see that the boundary position is very accurately determined for time series as short as $N = 100$. We also see the standard error growing much slower than N.

Table 1. Positions and standard errors of the segment boundary discovered using the Jensen-Shannon divergence segmentation scheme, from 1,000 artificial Gaussian time series. In (a) and (b), we set $N_L = N_R = 5,000$, $(\mu_L, \sigma_L) = (0, 1)$, $\sigma_R = 1$, and vary μ_R. In (b), we set $N_L = N_R = 5,000$, $(\mu_L, \sigma_L) = (0, 1)$, $\mu_R = 0$, and vary σ_R. In (c), we set $(\mu_L, \sigma_L) = (0, 1)$, $(\mu_R = 0, \sigma_R) = (0, 0.5)$, and vary $N = N_L + N_R$.

(a)		(b)		(c)	
μ_R	$t^* \pm \Delta t^*$	σ_2	$t^* \pm \Delta t^*$	N	$t^* \pm \Delta t^*$
0.1	4990 ± 680	0.1	5000 ± 380	100	49 ± 7
0.2	4980 ± 500	0.2	5000 ± 350	200	98 ± 9
0.5	5000 ± 490	0.3	5010 ± 480	500	249 ± 15
1.0	4990 ± 260	0.4	5020 ± 420	1000	497 ± 41
2.0	5010 ± 330	0.5	5000 ± 280	2000	998 ± 28
5.0	5020 ± 410	0.7	5010 ± 320	5000	2500 ± 200
10.0	5000 ± 270	0.9	5000 ± 550	10000	5000 ± 290

Table 2. The ten segments of the $N = 10,000$ artificial Gaussian time series, and the segment boundaries obtained using the recursive Jensen-Shannon divergence segmentation scheme

m	start	end	μ_m	σ_m	$t^* \pm \Delta t^*$	m	start	end	μ_m	σ_m	$t^* \pm \Delta t^*$
1	1	1500	0.55	0.275	1497 ± 60	6	6501	7000	0.30	0.15	7002 ± 26
2	1501	2500	0.05	0.025	2500 ± 14	7	7001	8500	0.45	0.225	8500 ± 12
3	2501	3500	0.20	0.10	3504 ± 50	8	8501	9000	0.05	0.025	9001 ± 12
4	3501	5000	0.60	0.30	5010 ± 140	9	9001	9500	0.45	0.225	9500 ± 9
5	5001	6501	0.65	0.325	6497 ± 39	10	9501	10000	0.15	0.075	-

Following this, we recursively segmented 10,000 artificial Gaussian time series of length $N = 1000$, each consisting of the same 10 segments shown in Table 2. We also see in Table 2 that eight of the nine segment boundaries were accurately determined. The position of the remaining boundary, between segments $m = 4$ and $m = 5$, has a large standard error because it separates two segments that are very similar statistically.

Finally, we timed the MATLAB code that we used to implement the recursive segmentation. The spatial complexity of this scheme is $O(N)$, since we need to store the original time series and two other processed data arrays of the same length. The temporal complexity of the scheme, however, cannot be easily analyzed, because it depends on how many optimization iterations are needed, and how many segment boundaries are to be discovered. On a MacBook Pro with 2.4-GHz core-2 duo and 4-GB 1067-MHz DDR3 memory, the two-segment time series took 1 ± 1 ms to segment, for $N = 100$, and 63 ± 5 ms to segment, for $N = 10000$. The 10-segment time series with length $N = 10,000$ took 0.38 ± 0.03 s to segment, or 42 ± 3 ms for each boundary. We also segmented 30 50,000-point time series from a molecular dynamics simulation of penta-analine. 7084 boundaries were found after 114 s, which works out to 16 ms per boundary.

3 Case Study

3.1 Single US Time Series

For the single time series study, we chose the Dow Jones Industrial Average (DJI) index. This is a price-weighted average of the stock prices of the 30 most well

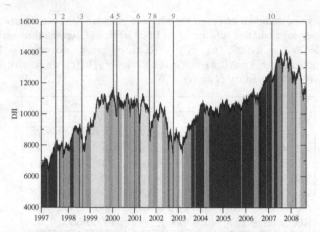

Fig. 4. Temporal distributions of the clustered segments superimposed onto the DJI time series. The red solid lines indicate the dates of important market events: (1) July 1997 Asian Financial Crisis; (2) October 1997 Mini Crash; (3) August 1998 Russian Financial Crisis; (4) DJI 2000 High; (5) NASDAQ Crash; (6) start of 2001 recession; (7) Sep 11 Attack; (8) end of 2001 recession; (9) DJI 2002 Low; (10) February 2007 Chinese Correction.

capitalized US companies. Tic-by-tic data between 1 January 1997 and 31 August 2008 was downloaded from the Thomson-Reuters Tickhistory database (https://tickhistory. thomsonreuters.com/TickHistory/login.jsp), and processed into a half-hourly index time series $\mathbf{X} = (X_1, \ldots, X_N)$. From $\mathbf{X}$, we then obtain the half-hourly index movement time series $\mathbf{x} = (x_1, \ldots, x_n)$, where $x_t = X_{t+1} - X_t$ and $n = N - 1$, which we assume consists of M statistically stationary segments. The half-hourly data frequency was chosen so that we can reliably identify segments as short as a single day. We do not go to higher data frequencies, because we are not interested in intraday segments.

As reported in Ref. [12], the clustered segments of the DJI tell very interesting stories when we plot how they are distributed over the January 1997 to August 2008 period. From Figure 4, we see that the DJI (as a proxy for the US economy as a whole) spends most of its time in the low-volatility phase (dark blue and blue) and the high-volatility phase (yellow and orange). Based on when it occurs, we can associate the low-volatility phase with economic expansion. We also see that the previous March to November 2001 economic contraction for the US is completely nested within the high-volatility phase. This suggests that the high-volatility phase has the natural interpretation as an economic crisis phase, which lasts longer than most official recessions. Interrupting both the low-volatility and high-volatility phases, we also find short-lived moderate-volatility phases (green), which we can therefore interpret as market corrections. In addition, even shorter-lived extreme-high-volatility phases (red) can be found almost exclusively within the high-volatility phase. These can be unambiguously associated with market crashes.

More importantly, the temporal distribution tells us that the US economy, as measured by the DJI, went into a five-year crisis period starting in mid-1998, before recovering in mid-2003. The US economy then enjoyed a remarkable four-year period of sustained growth, before succumbing to the Subprime Crisis in mid-2007. We also see in the temporal distribution the existence of year-long series of precursor shocks preceding each transition. These precursor shocks suggest on the surface that the July 1997 Asian Financial Crisis triggered the previous crisis, whereas market corrections in the Chinese markets, which started in May 2006, triggered the present crisis. Furthermore, the mid-2003 economic recovery is preceded by a year-long series of inverted shocks after the 2002 lows. Therefore, if the fundamental dynamics behind the US economy had not changed from 2002 to 2009, we expect from this single time series study that the US economy will emerge from the global financial crisis one year after the March 2009 lows, i.e. shortly after the first quarter of 2010. In contrast, the US National Bureau of Economic Research (NBER) announced in Sep 2010 that the US economic recession ended in Jun 2009.

3.2   Cross Section of US Time Series

The story of the US economy becomes even richer and more interesting, when we do a comparative segmentation and clustering analysis of the ten Dow Jones US (DJUS) economic sector indices [16]. Tic-by-tic data between 14 Feb 2000 and 31 Aug 2008 for these ten indices (see Table 3) were downloaded from the Thomson-Reuters TickHistory database. Since different indices have different magnitudes, we processed the raw data first into half-hourly time series $\mathbf{X}_i = (X_{i,1}, X_{i,2}, \ldots, X_{i,N})$ for each of the ten indices $i = 1, \ldots, 10$, before obtaining the half-hourly log-index movement time series $\mathbf{y}_i = (y_{i,1}, y_{i,2}, \ldots, y_{i,n})$, $i = 1, \ldots, 10$, $n = N - 1$, where $y_{i,t} = \ln X_{i,t+1} - \ln X_{i,t}$ for more meaningful comparison between the indices.

Table 3. The ten Dow Jones US economic sector indices

i	symbol	sector	i	symbol	sector
1	BM	Basic Materials	6	IN	Industrials
2	CY	Consumer Services	7	NC	Consumer Goods
3	EN	Energy	8	TC	Technology
4	FN	Financials	9	TL	Telecommunications
5	HC	Healthcare	10	UT	Utilities

From Figure 5, the first interesting observation we make is that it takes one and a half years for the US economy to recover from a financial crisis and just two months for it to completely enter a financial crisis. For the mid-2003 US economic recovery, the first two sectors to recover are EN and BM, and the last two sectors to recover are TL and TC. It is reasonable for EN and BM to recover first, since they are at the base of the economic supply chain. It is also reasonable that TC and TL were the last sectors to recover, since the previous financial crisis was the result of the technology bubble bursting. For the mid-2007 US economic decline, we find some surprises: instead of NC (which includes homebuilders and realties) and FN (which includes banks) being responsible for dragging the US economy down, fully half of the DJUS economic sectors entered the crisis phase before FN.

Guided by this coarse-grained picture of the US economy's slow time evolution, we can extract even more understanding from the high-frequency fluctuations [17]. We do this by computing the linear cross correlations.

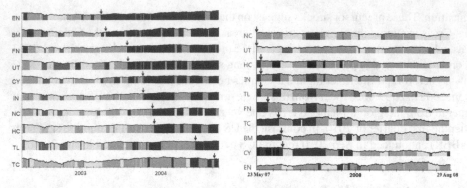

Fig. 5. Temporal distributions of clustered segments for the time series of all ten US economic sectors (left) between April 2002 and September 2004, showing the sequence of recovery from the mid-1998 to mid-2003 financial crisis, and (right) between 23 May 2007 and 29 August 2008, showing the sequence of descent into the present financial crisis

$$C_{ij} = \frac{\frac{1}{n}\sum_{t=1}^{n}(y_{i,t} - \bar{y}_i)(y_{j,t} - \bar{y}_j)}{\frac{1}{n-1}\sqrt{\sum_{t=1}^{n}(y_{i,t} - \bar{y}_i)^2}\sqrt{\sum_{t=1}^{n}(y_{j,t} - \bar{y}_j)^2}} \tag{5}$$

between the ten DJUS economic sector indices over different time intervals. We then look at the *minimal spanning tree* (MST) representation of these cross correlations [18,19,20]. The MST shows only the nine strongest links that do not incorporate cycles into the graph of the ten US economic sectors.

For the entire time series, from February 2000 to August 2008 (Figure 6(a), we see that IN, CY and NC, are at the centre of the MST, while the sectors HC, TC, TL, and UT lie on the fringe of the MST. This is consistent with the former group of sectors being of central importance, and the latter being of lesser importance to the US economy [21]. We also expect structural differences between MSTs constructed entirely within the previous crisis (2001–2002, Figure 6(b)), the previous growth (2004–2005, Figure 6(c)), and the present crisis (2008–2009, Figure 6(d)). Indeed, we see a chain-like MST which occurs for both crises, and a star-like MST which occurs for the growth phase. This association between MST topology and macroeconomic state is supported by the observations by Onnela et al., at the microscopic scale of individual stocks [22,23,24].

Speaking of 'green shoots' of economic revival that were evident in Mar 2009, Federal Reserve chairman Ben Bernanke predicted that "America's worst recession in decades will likely end in 2009 before a recovery gathers steam in 2010". We therefore looked out for a star-like MST in the time series data of 2009 and 2010. Star-like MSTs can also be found deep inside an economic crisis phase, but they very quickly unravel to become chain-like MSTs. A persistent star-like MST is therefore a statistical signature that the US economy is firmly on track to full recovery. More importantly, the closer the MST of a given period is to a star, the closer we are to the actual recovery. Indeed, the MST is already star-like for a moderate-volatility segment in Sep 2009 (see Figure 6(e)), and stayed robustly star-like throughout the Greek Debt Crisis of May/Jun 2010. The statistical evidence thus suggests that the US economy started recovering late 2009, and stayed the course through 2010. Bernanke was indeed prophetic.

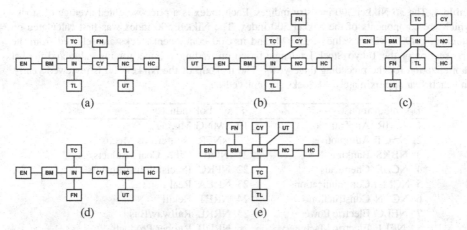

Fig. 6. The MSTs of the ten DJUS economic sectors, constructed using half-hourly time series from (a) February 2000 to August 2008, (b) 2001–2002, (c) 2004–2005, (d) 2008-2009, and (e) the moderate-volatility segment around Sep 2009. The first and the third two-year windows, (b) and (d), are entirely within an economic crisis, whereas the second two-year window, (c), is entirely within an economic growth period.

3.3   Cross Section of Japanese Time Series

As a comparison, we also segmented the 36 Nikkei 500 Japanese industry indices (see Table 4) between 1 Jan 1996 and 11 Jun 2010. Tic-by-tic data were downloaded from the Thomson-Reuters TickHistory database, and processed into half-hourly index time series $\mathbf{X}_i = \{X_{i,1}, X_{i,2}, \ldots, X_{i,N}\}$, $i = 1, \ldots, 36$. As with the US cross section study, we then obtain the half-hourly log-index movement time series $\mathbf{y}_i = (y_{i,1}, y_{i,2}, \ldots, y_{i,n})$, $i = 1, \ldots, 36$, $n = N - 1$, where $y_{i,t} = \ln X_{i,t+1} - \ln X_{i,t}$ for more meaningful comparison between the indices.

In this paper, we will focus on the 2005 near recovery of the Japanese economy, and the 2007 fall of the Japanese economy to the Subprime Crisis. From Figure 7, we see that NMNG started growing the earliest, NFIS started growing the latest, while NSPB did not seem to have grown at all between 2002 and 2005. We see also that the Japanese economy, led by NMNG and NELC, took two years and two months to completely recover from the back-to-back Asian Financial and Technology Bubble Crises. While the time scales of complete economic recovery appear to be different, very similar industries led the recovery processes of US and Japan.

Next, we look at how the Japanese economy succumbed to the Subprime Crisis. As we can see from Figure 7, the Japanese economy fell in five stages. The most important time scale in Japan's response to the Subprime Crisis is that associated with stage 2, which appears to be triggered by the start of the Subprime Crisis in US, and affected 21 out of 36 Nikkei 500 industries. Here, the Subprime Crisis swept through NISU to NTEQ in a mere 27 days. This is half the time it took for the US economy to fall from first to the last economic sector. As late as June 2010, most Japanese industries were still in the sustained crisis phase. Only NMNG, NWHO, NTRA, and NRET showed

Table 4. The 36 Nikkei 500 industry indices. Each index is a price-weighted average of stocks which are components of the Nikkei 500 index. The Nikkei 500 index was first calculated on January 4, 1972 with a value of 223.70, and its 500 component stocks are selected from the first section of the Tokyo Stock Exchange based on trading volume, trading value and market capitalization for the preceding three years. The makeup of the Nikkei 500 is reviewed yearly, and each year approximately 30 stocks are replaced.

i	symbol	industry	i	symbol	industry
1	NAIR	Air Transport	19	NMNG	Mining
2	NAUT	Automotive	20	NNFR	Nonferrous Metals
3	NBKS	Banking	21	NOIL	Oil & Coal Products
4	NCHE	Chemicals	22	NPRC	Precision Instruments
5	NCMU	Communications	23	NREA	Real Estate
6	NCON	Construction	24	NRET	Retail
7	NELC	Electric Power	25	NRRL	Railway/Bus
8	NELI	Electric Machinery	26	NRUB	Rubber Products
9	NFIN	Other Financial Services	27	NSEA	Marine Transport
10	NFIS	Fisheries	28	NSEC	Securities
11	NFOD	Foods	29	NSPB	Shipbuilding
12	NGAS	Gas	30	NSTL	Steel Products
13	NGLS	Glass & Ceramics	31	NSVC	Services
14	NISU	Insurance	32	NTEQ	Other Transport Equipment
15	NLAN	Other Land Transport	33	NTEX	Textiles & Apparel
16	NMAC	Machinery	34	NTIM	Pulp & Paper
17	NMED	Pharmaceuticals	35	NTRA	Trading Companies
18	NMIS	Other Manufacturing	36	NWHO	Warehousing

signs of early recovery from mid 2009 onwards. If the Japanese economy again takes two and a half years to completely recover, this will happen in the beginning of 2012.

Finally, we tracked how the MST change going from one segment to the next during the Subprime Crisis. In 21st century Japanese economy, NELI, NMAC, and NCHE are the growth industries, based on the fact that they are hubs consistently in all or most of the MSTs. NNFR and NRRL, which we consider quality industries, also become occasional hubs in the MSTs. In Figure 8, we see the NNFR and NRRL clusters of industries growing at the expense of the NCHE and NMAC clusters of industries, as we go from the Subprime3 period to the Subprime4 period. This tells us that the pheripheral industries went from being most strongly correlated with NCHE and NMAC to being most strongly correlated with NNFR and NRRL. We believe this is a signature of money leaving the NCHE and NMAC industries, and entering the NNFR and NRRL industries, i.e. a flight to quality [26,25] from NCHE/NMAC to NNFR/NRRL.

4   Conclusions

To conclude, we have performed time series segmentation on the Dow Jones Industrial Average index time series, the ten Dow Jones US Economic Sector indices, and the 36 Nikkei 500 Japanese Industry indices, as a concrete demonstration of its potential for knowledge discovery. From the single time series study, we found the time series segments very naturally fall into four to six clusters, which can be roughly associated

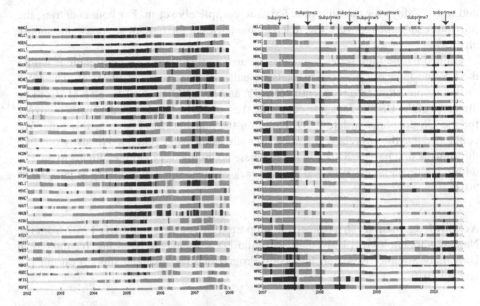

Fig. 7. Temporal distributions of the 36 Nikkei 500 Japanese industry indices from (left) January 2002 to December 2007, and (right) January 2007 to June 2010. In this figure, the growth segments are colored blue, correction segments are colored green, crisis segments are colored yellow or orange, and crash segments are colored red, whereas the eight corresponding segments within the Subprime Crisis are labeled 'Subprime1' to 'Subprime8'.

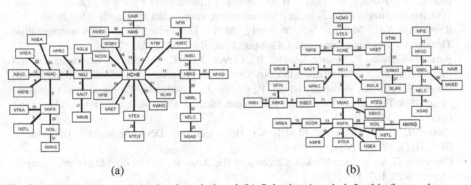

(a)   (b)

Fig. 8. MSTs for the (a) Subprime3 period and (b) Subprime4 period. In this figure, the number beside each link indicates the order in which the link was added to the MST, whereas the thicknesses of the links indicate how strong the correlations are between industries.

with the growth, crisis, correction, and crash macroeconomic phases. We also measured the lifetimes of the previous US crisis and growth phases to be about five years and four years respectively. From cross section studies, we found that the US economy took one-and-a-half years to completely recover from the Technology Bubble Crisis, but only two months to completely succumb to the Subprime Crisis. In contrast, the Japanese economy took two years and two months to completely recover from the previous crisis, and only 27 days for the Subprime Crisis to completely set in. For both countries, the previous economic recoveries were led by industries at the base of the economic supply chain.

Guided by the time series segments, we also analyzed the cross correlations within the US and Japanese financial markets, visualizing these in terms of MSTs. The MST visualizations allowed us to identify IN, CY, and NC, NCHE, NELI, and NMAC, to be the cores of the US and Japanese economies respectively. We detected an early recovery for the US economy in late 2009, based on the star-like MST seen at this time. We concluded that the US recovery gained strength, as the MST remained robustly star-like through the first half of 2010. For the Japanese economy, we identified flights to quality within the financial markets, and also the lack of clear signs of recovery as late as Jun 2010.

Acknowledgements. This research is supported by startup grant SUG 19/07 from the Nanyang Technological University. RPF and DYX thank the Nanyang Technological University Summer Research Internship Programme for financial support during Jun and Jul 2010.

References

1. Carlstein, E.G., Müller, H.-G., Siegmund, D.: Change-Point Problems. Lecture Notes-Monograph Series. Institute of Mathematical Statistics, vol. 23 (1994)
2. Chen, J., Gupta, A.K.: Parametric Statistical Change Point Analysis. Birkhäuser (2000)
3. Braun, J.V., Braun, R.K., Müller, H.-G.: Multiple Changepoint Fitting Via Quasilikelihood, With Application to DNA Sequence Segmentation. Biometrika 87(2), 301–314 (2000)
4. Ramensky, V.E., Makeev, V.J., Roytberg, M.A., Tumanyan, V.G.: DNA segmentation Through the Bayesian Approach. J. Comput. Biol. 7(1-2), 215–231 (2000)
5. Bernaola-Galván, P., Román-Roldán, R., Oliver, J.L.: Compositional Segmentation and Long-Range Fractal Correlations in DNA Sequences. Phys. Rev. E 53(5), 5181–5189 (1996)
6. Román-Roldán, R., Bernaola-Galván, P., Oliver, J.L.: Sequence Compositional Complexity of DNA Through an Entropic Segmentation Method. Phys. Rev. Lett. 80(6), 1344–1347 (1998)
7. Churchill, G.A.: Stochastic Models for Heterogeneous DNA Sequences. Bull. Math. Biol. 51(1), 79–94 (1989)
8. Churchill, G.A.: Hidden Markov Chains and the Analysis of Genome Structure. Comput. Chem. 16(2), 107–115 (1992)
9. Lin, J.: Divergence Measures Based on the Shannon Entropy. IEEE Trans. Infor. Theor. 37(1), 145–151 (1991)
10. Li, W.: DNA Segmentation as a Model Selection Process. In: Proceedings of the International Conference on Research in Computational Molecular Biology (RECOMB), pp. 204–210 (2001)
11. Li, W.: New Stopping Criteria for Segmenting DNA Sequences. Phys. Rev. Lett. 86(25), 5815–5818 (2001)
12. Wong, J.C., Lian, H., Cheong, S.A.: Detecting Macroeconomic Phases in the Dow Jones Industrial Average Time Series. Physica A 388(21), 4635–4645 (2009)
13. Cheong, S.-A., Stodghill, P., Schneider, D.J., Cartinhour, S.W., Myers, C.R.: The Context Sensitivity Problem in Biological Sequence Segmentation, arXiv:0904.2668 (2009)

14. Cheong, S.-A., Stodghill, P., Schneider, D.J., Cartinhour, S.W., Myers, C.R.: Extending the Recursive Jensen-Shannon Segmentation of Biological Sequences, arXiv:0904.2466 (2009)
15. Jain, A., Murty, M., Flynn, P.: Data Clustering: A Review. ACM Comput. Surv. 31(3), 264–323 (1999)
16. Lee, G.H.T., Zhang, Y., Wong, J.C., Prusty, M., Cheong, S.A.: Causal Links in US Economic Sectors, arXiv:0911.4763 (2009)
17. Zhang, Y., Lee, G.H.T., Wong, J.C., Kok, J.L., Prusty, M., Cheong, S.A.: Will the US Economy Recover in 2010? A Minimal Spanning Tree Study. Physica A 390(11), 2020–2050 (2011)
18. Kruskal, J.B.: On the Shortest Spanning Subtree of a Graph and the Traveling Salesman Problem. Proc. Am. Math. Soc. 7, 48–50 (1956)
19. Prim, R.C.: Shortest Connection Networks and Some Generalizations. Bell Labs Techn. J. 36, 1389–1401 (1957)
20. Mantegna, R.N.: Hierarchical Structure in Financial Markets. Eur. Phys. J. B 11, 193–197 (1999)
21. Heimo, T., Kaski, K., Saramäki, J.: Maximal Spanning Trees, Asset Graphs and Random Matrix Denoising in the Analysis of Dynamics of Financial Networks. Physica A 388, 145–156 (2009)
22. Onnela, J.-P., Chakraborti, A., Kaski, K., Kertész, J.: Dynamic Asset Trees and Black Monday. Physica A 324, 247–252 (2003)
23. Onnela, J.-P., Chakraborti, A., Kaski, K., Kertész, J., Kanto, A.: Asset Trees and Asset Graphs in Financial Markets. Phys. Scripta T106, 48–54 (2003)
24. Onnela, J.-P., Chakraborti, A., Kaski, K., Kertész, J., Kanto, A.: Dynamics of Market Correlations: Taxonomy and Portfolio Analysis. Phys. Rev. E 68(5), 056110 (2003c)
25. Baur, D.G., Lucey, B.M.: Flights and Contagion — An Empirical Analysis of Stock-Bond Correlations. J. Fin. Stability 5(4), 339–352 (2009)
26. Connolly, R., Stivers, C., Sun, L.: Stock Market Uncertainty and the Stock-Bond Return Relation. J. Fin. Quant. Anal. 40(1), 161–194 (2005)

A Block Coclustering Model for Pattern Discovering in Users' Preference Data

Nicola Barbieri, Gianni Costa, Giuseppe Manco, and Ettore Ritacco

High Performance Computing and Networking Institute of the Italian National Research Council
v. Pietro Bucci 41C, Arcavacata di Rende (CS), Italy
{barbieri,costa,manco,ritacco}@icar.cnr.it
http://icar.cnr.it/

Abstract. This paper provides a principled probabilistic co-clustering framework for missing value prediction and pattern discovery in users' preference data. We extend the original dyadic formulation of the *Block Mixture Model(BMM)* in order to take into account explicit users' preferences. *BMM* simultaneously identifies user communities and item categories: each user is modeled as a mixture over user communities, which is computed by taking into account users' preferences on similar items. Dually, item categories are detected by considering preferences given by similar minded users. This recursive formulation highlights the mutual relationships between items and user, which are then used to uncover the hidden block-structure of the data. We next show how to characterize and summarize each block cluster by exploiting additional meta data information and by analyzing the underlying topic distribution, proving the effectiveness of the approach in pattern discovery tasks.

Keywords: Collaborative Filtering, Recommender Systems, Block Clustering, Co-clustering.

1 Introduction

Collaborative Filtering (CF) is recently becoming the dominant approach in *Recommender Systems (RS)*. In literature, several *CF* recommendation techniques have been proposed, mainly focusing on the predictive skills of the system. Recent studies [10,2] have shown that the focus on prediction does not necessarily helps in devising good recommender systems. Under this perspective, CF models should be considered in a broader sense, for their capability to understand deeper and hidden relationships among users and products they like. Examples in this respect are user communities, item categories preference patterns within such groups. Besides their contribution to the minimization of the prediction error, these relationships are important as they can provide a faithful yet compact description of the data which can be exploited for better decision making.

In this paper we present a co-clustering approach to preference prediction and rating discovery, based on the *Block Mixture Model (BMM)* proposed in [6]. Unlike traditional CF approaches, which try to discover similarities between users or items using clustering techniques or matrix decomposition methods, the aim of the BMM is to partition

A. Fred et al. (Eds.): IC3K 2011, CCIS 348, pp. 94–108, 2013.

data into homogeneous block enforcing a simultaneous clustering which consider both the dimension of the preference data. This approach highlights the mutual relationship between users and items: similar users are detected by taking into account their ratings on similar items, which in turn are identified considering the ratings assigned by similar users. We extended the original BMM formulation to model each preference observation as the output of a gaussian mixture employing a maximum likelihood (ML) approach to estimate the parameter of the model. Unfortunately, the strict interdependency between user and item cluster makes difficult the application of traditional optimization approaches like EM. Thus, we perform approximated inference based on a variational approach and a two-step application of the EM algorithm which can be thought as a good compromise between the semantic of the original model and the computational complexity of the learning algorithm.

We reformulate standard pattern discovery tasks by showing how a probabilistic block model automatically allows to infer patterns and trends within each block. We show experimentally that the proposed model guarantees a competitive prediction accuracy with regards to standard state-of-the art approaches, and yet it allows to infer topics for each item category, as well as to learn characteristic items for each user community, or to model community interests and transitions among topics of interests. Experiments on both the Netflix and Movielens data show the effectiveness of the proposed model.

2 Preliminaries and Related Work

User's preferences can be represented by using a $M \times N$ rating matrix $\mathbf{R}$, where M is the cardinality of the *user-set* $\mathcal{U} = \{u_1, \cdots, u_M\}$ and N is the cardinality of the *item-set* $\mathcal{I} = \{i_1, \cdots, i_N\}$. The rating value associated to the pair $\langle u, i \rangle$ will be denoted as r_i^u. Typically the number of users and items can be very large, with $M >> N$, and preferences values fall within a fixed integer range $\mathcal{V} = \{1, \cdots, V\}$, where 1 denote the lower interest value. Users tend to express their interest only on a restricted number of items; thus, the rating matrix is characterized by an exceptional sparseness factor (e.g more than 95%). Let $\delta(u, i)$ be a rating-indicator function, which is equals to 1 if the user u has rated/purchased the item i, zero otherwise. Let $\mathcal{I}(u)$ denote the set of products rated by the user u: $\mathcal{I}(u) = \{i \in \mathcal{I} : \delta(u, i) = 1\}$; symmetrically, $\mathcal{U}(i)$ denotes the set of users who have expressed their preference on the item i.

Latent Factor models are the most representative and effective model-based approaches for CF. The underlying assumption is that preference value associated to the pair $\langle u, i \rangle$ can be decomposed considering a set of contributes which represent the interaction between the user and the target item on a set of features. Assuming that there are a set of K features which determine the user's interest on an given item. The assumption is that a rating is the result of the influence of these feature to users and items: $\hat{r}_i^u = \sum_{z=1}^{K} U_{u,z} V_{z,i}$, where $U_{u,z}$ is the response of the user u to the feature z and $V_{z,i}$ is the response on the same feature of the item i.

Several learning schema have been proposed to overcome the sparsity of the original rating matrix and to produce accurate models. The learning phase may be implemented in a deterministic way, via *gradient descent* [3] or, following a probabilistic approach, maximizing the log-likelihood of the model via the *Expectation Maximization* algorithm. The latter leads to the definition of the *Aspect Model*[7], known also as *pLSA*. According to the *user community variant*, the rating value r is conditionally independent of the user's identity given her respective community Z; thus, the probability of observing the rating value r for the pair $\langle u, i \rangle$ can be computed as $p(r|u, i) = \sum_{z=1}^{K} p(r|i, z)p(z|u)$, where $P(z|u)$ measures how much the preference values given by u fits with the behavior of the community z and $p(r|i, z)$ is the probability that a user belonging to the community z assigns a rating value r on i.

Only a few co-clustering approaches have been proposed for CF data. An application of the weighted *Bregman coclustering (Scalable CC)* to rating data is discussed in [4]. The *two-sided clustering model for CF* [7] is based on the strong assumption that each person belongs to exactly one user-community and each item belong to one groups of items, and finally the rating value is independent of the user and item identities given their respective cluster memberships. Let $C = \{c_1, \cdots, c_k\}$ be the user-clusters and let $c(u) : \mathcal{U} \to C$ be a function that maps each user to the respective cluster. Similarly, let $D = \{d_1, \cdots, d_L\}$ be a set of disjoint item-clusters, and $d(i) : \mathcal{I} \to D$ is the corresponding mapping function. According to the two-sided clustering model, the probability of observing the preference value r conditioned to the pair $\langle u, i \rangle$ is the following:

$$p(r|u, i, c(u) = c, d(i) = d) = p(r|c, d) \ .$$

where $p(r|c, d)$ are Bernoulli parameters and the cluster membership are estimated by employing a variational inference approach.

The *Flexible Mixture Model (FMM)* [8] extends the Aspect and the two sided model, by allowing each user/item to belong to multiple clusters, which are determined simultaneously, according to a coclustering approach. Assuming the existence of K user clusters indexed by c and L item clusters, indexed by d, and let $p(c_k)$ be the probability of observing the user-cluster k with $p(u|c_k)$ being the probability of observing the user profile u given the cluster k and using the same notations for the item-cluster, the joint probability $p(u, i, r)$ is defined as:

$$p(u, i, r) = \sum_{c=1}^{C} \sum_{d=1}^{D} p(c)p(d)p(u|c)p(i|d)p(r|c, d)$$

The predicted rating associated to the pair $\langle u, i \rangle$ is then computed as:

$$\hat{r}_i^u = \sum_{r=1}^{V} r \frac{p(u, i, r)}{\sum_{r'=1}^{V} p(u, i, r')}$$

The major drawback of the FMM relies on the complexity of the training procedure, which is connected with the computation of the probabilities $p(c, d|u, i, r)$ during the Expectation step.

A coclustering extension of the LDA[1] model for rating data have been proposed in [11]: the *Bi-LDA* employs two interacting LDA models which enforce the simultaneous clustering of users and items in homogeneous groups.

Other co-clustering approaches have been proposed in the current literature (see [12,14]), however their extension to explicit preference data, which requires a distribution over rating values, has not been provided yet.

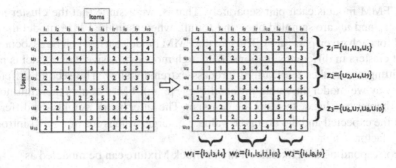

Fig. 1. Example Co-Clustering for Preference Data

3   A Block Mixture Model for Preference Data

In this section, we are interested in:devising how the available data fits into ad-hoc communities and groups, where groups can involve both users and items. Fig. 1 shows a toy example of preference data co-clustered into blocks. As we can see, a coclustering induces a natural ordering among rows and columns, and it defines blocks in the rating matrix with similar ratings. The discovery of such a structure is likely to induce information about the population, as well as to improve the personalized recommendations.

Formally, a *block mixture model (BMM)* can be defined by two partitions $(\mathbf{z}, \mathbf{w})$ which, in the case of preference data and considering known their respective dimensions, have the following characterizations:

- $\mathbf{z} = z_1, \cdots, z_M$ is a partition of the user set $\mathcal{U}$ into K clusters and $z_{uk} = 1$ if u belongs to the cluster k, zero otherwise;
- $\mathbf{w} = w_1, \cdots, w_N$ is a partition of the item set $\mathcal{I}$ into L clusters and $w_{il} = 1$ if the item i belongs to the cluster l, zero otherwise.

Given a rating matrix $\mathbf{R}$, the goal is to determine such partitions and the respective partition functions which specify, for all pairs $\langle u, i \rangle$ the probabilistic degrees of membership wrt. to each user and item cluster, in such a way to maximize the likelihood of the model given the observed data. According to the approach described [6,5], and assuming that the rating value r observed for the pair $\langle u, i \rangle$ is independent from the user and item identities, fixed z and w, the generative model can be described as follows:

1. For each u generate $z_u \sim Discrete(\pi_1; \ldots; \pi_K)$
2. for each i generate $w_i \sim Discrete(\psi_1; \ldots; \psi_L)$
3. for each pair (u, i):
    - detect k and l such that $z_{uk} = 1$ and $w_{il} = 1$
    - generate $r \sim N(\mu_k^l; \sigma_k^l)$

There are two main differences with respect to the FMM model introduced in the related work. First of all, in our model all cluster membership are assumed given a-priori, whereas FMM models each pair separately. That is, we assume that the cluster memberships z_u and w_i are sampled once and for all, whereas in the FMM model they are sampled for each given pair (u, i). Thus, in the FMM model, a use u can be associated to different clusters in different situations. Although more expressive, this model is prone to overfitting and makes the learning process extremely slow. The second difference is in the way we model the rating probability $p(r|z, w)$. FMM adopts the multinomial model, whereas we choose to adopt the gaussian. The latter better weights the difference between the expected and the observed value: i.e. larger values for $|\hat{r}_i^u - r_i^u|$ introduce a penalty factor.

The corresponding data likelihood in the Block Mixture can be modeled as

$$p(\mathbf{R}, \mathbf{z}, \mathbf{w}) = \prod_{u \in \mathcal{U}} p(z_u) \prod_{i \in \mathcal{I}} p(w_i) \prod_{(u,i,r) \in \mathbf{R}} p(r|z_u, w_i)$$

and consequently, the log-likelihood becomes:

$$L_c(\Theta; \mathbf{R}, \mathbf{z}, \mathbf{w}) = \sum_{k=1}^{K} \sum_{u \in \mathcal{U}} z_{uk} \log \pi_k + \sum_{l=1}^{L} \sum_{i \in \mathcal{I}} w_{il} \log \psi_l +$$
$$+ \sum_{\langle u,i,r \rangle \in \mathbf{R}} \sum_k \sum_l \left[z_{uk} w_{il} \log \varphi(r; \mu_k^l, \sigma_k^l) \right]$$

where Θ represents the whole set of parameters $\pi_1, \ldots, \pi_K, \psi_1, \ldots, \psi_L, \mu_1^1, \ldots, \mu_K^L,$ $\sigma_1^1, \ldots, \sigma_K^L$ and $\varphi(r; \mu, \sigma)$ is the gaussian density function on the rating value r with parameters μ and σ, i.e., $\varphi(r; \mu; \sigma) = (2\pi)^{-1/2} \sigma^{-1} exp \left(\frac{-1}{2\sigma^2} (r - \mu)^2 \right)$.

In the following we show how the model can be inferred and exploited both for prediction and for pattern discovery.

3.1   Inference and Parameter Estimation

Denoting $p(z_{uk} = 1|u, \Theta^{(t)}) = c_{uk}$, $p(w_{il} = 1|i, \Theta^{(t)}) = d_{il}$ and $p(z_{uk} w_{il} = 1|u, i, \Theta^{(t)}) = e_{ukil}$, the conditional expectation of the complete data log-likelihood becomes:

$$Q(\Theta; \Theta^{(t)}) = \sum_{k=1}^{K} \sum_u c_{uk} \log \pi_k + \sum_{l=1}^{L} \sum_i d_{il} \log \psi_l +$$
$$\sum_{\langle u,i,r \rangle \in \mathbf{R}} \sum_k \sum_l \left[e_{ukil} \log \varphi(r; \mu_k^l, \sigma_k^l) \right]$$

As pointed out in [5], the above function is not tractable analytically, due to the difficulties in determining e_{ukil}; nor the adoption of its variational approximation ($e_{ukil} = c_{uk} \cdot d_{il}$) allows us to derive an Expectation-Maximization procedure for $Q'(\Theta, \Theta^{(t)})$ where the M-step can be computed in closed form. In [5] the authors propose an optimization of the complete-data log-likelihood based on the *CEM* algorithm. We adapt the whole approach here. First of all, we consider that the joint probability of a a normal population x_i with $i = 1$ to n can be factored as: $\prod_{i=1}^{n} \varphi(x_i; \mu, \sigma) = h(x_1, \ldots, x_n) * \varphi(u_0, u_1, u_2; \mu, \sigma)$, where $h(x_1, \ldots, x_n) = (2\pi)^{-n/2}$, $\varphi(u_0, u_1, u_2; \mu, \sigma) = \sigma^{-u_0} exp\left(\frac{2u_1\mu - u_2 - u_0\mu^2}{2\sigma^2}\right)$ and u_0, u_1 and u_2 are the sufficient statistics.

Based on the above observation, we can define a two-way EM approximation based on the following decompositions of $\mathcal{Q}'$:

$$\mathcal{Q}'(\Theta, \Theta^{(t)}) = \mathcal{Q}'(\Theta, \Theta^{(t)}|\mathbf{d}) + \sum_{i \in \mathcal{I}} \sum_{l=1}^{L} d_{il} \log \psi_l - \sum_{u \in \mathcal{U}} \sum_{i \in \mathcal{I}(u)} d_{il}/2 \log(2\pi)$$

where

$$\mathcal{Q}'(\Theta, \Theta^{(t)}|\mathbf{d}) = \sum_{u=1}^{M} \sum_{k=1}^{K} c_{uk} (\log(\pi_k) + \tau_{uk})$$

$$\tau_{uk} = \sum_{l=1}^{L} \log\left(\varphi(u_0^{(u,l)}, u_1^{(u,l)}, u_2^{(u,l)}; \mu_k^l, \sigma_k^l)\right)$$

$$u_0^{(u,l)} = \sum_{i \in \mathcal{I}(u)} d_{il}; \quad u_1^{(u.l)} = \sum_{i \in \mathcal{I}(u)} d_{il} r_i^u; \quad u_2^{(u,l)} = \sum_{i \in \mathcal{I}(u)} d_{il} (r_i^u)^2$$

Analogously,

$$\mathcal{Q}'(\Theta, \Theta^{(t)}) = \mathcal{Q}'(\Theta, \Theta^{(t)}|\mathbf{c}) + \sum_{u \in \mathcal{U}} \sum_{k=1}^{K} c_{uk} \log \pi_k - \sum_{i \in \mathcal{I}} \sum_{u \in \mathcal{U}(i)} c_{uk}/2 \log(2\pi)$$

where

$$\mathcal{Q}'(\Theta, \Theta^{(t)}|\mathbf{c}) = \sum_{i=1}^{N} \sum_{l=1}^{L} d_{il} (\log(\psi_l) + \tau_{il})$$

$$\tau_{il} = \sum_{k=1}^{K} \log\left(\varphi(u_0^{(i,k)}, u_1^{(i.k)}, u_2^{(i,k)}; \mu_k^l, \sigma_k^l)\right)$$

$$u_0^{(i,k)} = \sum_{u \in \mathcal{I}(u)} c_{uk}; \quad u_1^{(i.k)} = \sum_{u \in \mathcal{I}(u)} c_{uk} r_i^u; \quad u_2^{(i,k)} = \sum_{u \in \mathcal{I}(u)} c_{uk} (r_i^u)^2$$

The advantage in the above formalization is that we can approach the single components separately and, moreover, for each component it is easier to estimate the parameters. In particular, we can obtain the following 2-step EM procedure:

E-Step (User Communities):
For each $u \in \mathcal{U}, i \in \mathcal{I}, k = \{1, \cdots, K\}$ compute:

$$c_{uk} = \frac{p(u|z_k) \cdot \pi_k}{\sum_{k'=1}^{K} p(u|z_{k'}) \cdot \pi_{k'}}$$

$$p(u|z_k) = \prod_{l=1}^{L} \varphi(u_0^{(u,l)}, u_1^{(u,l)}, u_2^{(u,l)}; \mu_k^l, \sigma_k^l)$$

M-Step (User Communities):
For each $k = \{1, \cdots, K\}$ compute:

$$\pi_k = \frac{\sum_{u \in \mathcal{U}} c_{uk}}{M}$$

For each $l = \{1, \cdots, L\}$ compute:

$$\mu_k^l = \frac{\sum_{u=1}^{M} \sum_{i \in \mathcal{I}(u)} c_{uk} d_{il} r_i^u}{\sum_{u=1}^{M} \sum_{i \in \mathcal{I}(u)} c_{uk} d_{il}}$$

$$(\sigma_k^l)^2 = \frac{\sum_{u=1}^{M} \sum_{i \in \mathcal{I}(u)} c_{uk} d_{il} (r_i^u - \mu_k^l)^2}{\sum_{u=1}^{M} \sum_{i \in \mathcal{I}(u)} c_{uk} d_{il}}$$

E-Step (Item clusters):
For each $i \in \mathcal{I}, l = \{1, \cdots, L\}$ compute:

$$d_{il} = \frac{p(i|w_l) \cdot \psi_l}{\sum_{l'=1}^{L} p(i|w_{l'}) \cdot \psi_{l'}}$$

$$p(i|w_l) = \prod_{k=1}^{K} \varphi(u_0^{(i,k)}, u_1^{(i,k)}, u_2^{(i,k)}; \mu_k^l, \sigma_k^l)$$

M-Step (Item clusters):
For each $l = \{1, \cdots, L\}$ compute:

$$\psi_l = \frac{\sum_{i \in \mathcal{I}} d_{il}}{N}$$

For each $k = \{1, \cdots, K\}$ compute:

$$\mu_k^l = \frac{\sum_{i=1}^{N} \sum_{u \in \mathcal{U}(i)} d_{il} c_{uk} r_i^u}{\sum_{i=1}^{N} \sum_{u \in \mathcal{U}(i)} d_{il} c_{uk}}$$

$$(\sigma_k^l)^2 = \frac{\sum_{i=1}^{N} \sum_{u \in \mathcal{U}(i)} c_{uk} d_{il} (r_i^u - \mu_k^l)^2}{\sum_{i=1}^{N} \sum_{u \in \mathcal{U}(i)} d_{il} c_{uk}}$$

3.2 Rating Prediction

The blocks resulting from a co-clustering can be directly used for prediction. Given a pair $\langle u, i \rangle$, the probability of observing a rating value r associated to the pair $\langle u, i \rangle$ can be computed according to one of the following schemes:

- *Hard-Clustering Prediction*: $p(r|i, u) = \varphi(r; \mu_k^l, \sigma_k^l)$, where $k = \text{argmax}_{j=1,\cdots,}$ $K c_{uj}$ and $l = \text{argmax}_{h=1,\cdots,L} d_{ih}$ are the clusters that better represent the observed ratings for the considered user and item respectively.
- *Soft-Clustering Prediction*: $p(r|i, u) = \sum_{k=1}^{K} \sum_{l=1}^{L} c_{uk} d_{il} \varphi(r; \mu_k^l, \sigma_k^l)$, which consists of a weighted mixture over user and item clusters.

The final rating prediction can be computed by using the expected value of $p(r|u, i)$.

In order to test the predictive accuracy of the BMM we performed a suite of tests on a sample of Netflix data. The training set contains $5, 714, 427$ ratings, given by $435, 656$ users on a set of $2, 961$ items (movies). Ratings on those items are within a range 1 to 5 (max preference value) and the sample is 99% sparse. The test set contains $3, 773, 781$ ratings given by a subset of the users ($389, 305$) in the training set over the same set of items. Over 60% of the users have less than 10 ratings and the average number of evaluations given by users is 13.

We evaluated the performance achieved by the BMM considering both the Hard and the Soft prediction rules and performed a suite of experiments varying the number of user and item clusters. Experiments on the three models have been performed by retaining the 10% of the training (user,item,rating) triplets as held-out data and 10 attempts have been executed to determine the best initial configurations. Performance results

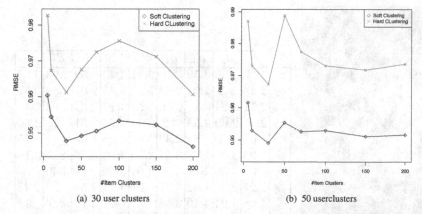

(a) 30 user clusters   (b) 50 userclusters

Fig. 2. Predictive Accuracy of BMM

measured using the RMSE for two BMM with 30 and 50 user clusters are showed in
Fig. 2(a) and Fig. 2(b), respectively. In both cases the soft clustering prediction rule
overcomes the hard one, and they show almost the same trend. The best result (0.9462)
is achieved by employing 30 user clusters and 200 item clusters. We can notice from
Tab. 1 that the results follow the same trend as other probabilistic models, like pLSA,
which on the same portion of the data achieves 0.9474 accuracy.

Table 1. RMSE of principal (co-)clustering approaches

Method	Best RMSE	K	H
BMM	**0.946**	**30**	**200**
PLSA	0.947	30	-
FMM	0.954	10	70
Scalable CC	1.008	10	10

4   Pattern Discovery Using BMM

The probabilistic formulation of the BMM provides a powerful framework for discovering hidden relationships between users and items. As exposed above, such relationships can have several uses in users segmentation, product catalog analysis, etc. Several
works have focused on the application of clustering techniques to discover patterns in
data by analyzing user communities or item categories. In [9] authors showed how the
pLSA model in its co-occurrence version can be used to infer the underlying task of
a web browsing session and to discover relationships between users and web pages.
Those approaches can be further refined by considering the co-clustering structure proposed so far, which increases the flexibility in modeling both user communities and
item categories patterns. Given two different user clusters which group users who have
showed a similar preference behavior, the BMM allows the identification of common
rated items and categories for which the preference values are different. For example,

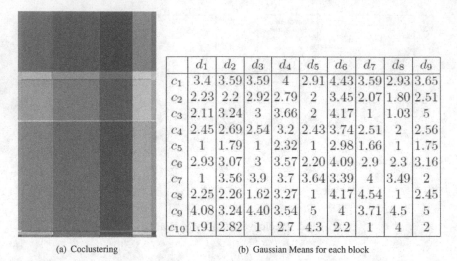

	d_1	d_2	d_3	d_4	d_5	d_6	d_7	d_8	d_9
c_1	3.4	3.59	3.59	4	2.91	4.43	3.59	2.93	3.65
c_2	2.23	2.2	2.92	2.79	2	3.45	2.07	1.80	2.51
c_3	2.11	3.24	3	3.66	2	4.17	1	1.03	5
c_4	2.45	2.69	2.54	3.2	2.43	3.74	2.51	2	2.56
c_5	1	1.79	1	2.32	1	2.98	1.66	1	1.75
c_6	2.93	3.07	3	3.57	2.20	4.09	2.9	2.3	3.16
c_7	1	3.56	3.9	3.7	3.64	3.39	4	3.49	2
c_8	2.25	2.26	1.62	3.27	1	4.17	4.54	1	2.45
c_9	4.08	3.24	4.40	3.54	5	4	3.71	4.5	5
c_{10}	1.91	2.82	1	2.7	4.3	2.2	1	4	2

(a) Coclustering   (b) Gaussian Means for each block

Fig. 3. Co-clustering Analysis: Rating Patterns

two user community might agree on action movies while completely disagree on one other. The identification of the topics of interest and their sequential patterns for each user community lead to an improvement of the quality of the recommendation list and provide the user with a more personalized view of the system. In the following we will discuss examples of pattern discovery and user/item profiling tasks.

The experiments in this section were performed considering the 1M MovieLens dataset[1], which contains $1,000,209$ ratings given by $6,040$ users on approximately $3,900$ movies. Each user in this dataset has at least 20 ratings and a list of genres is given for each movie. The latter information will be used to validate the the discovered block structure.

4.1   Co-clustering Analysis

The relationships between groups of users and items captured by the BMM can be easily recognized by analyzing the distribution of the preference values for each cocluster. Given a co-cluster $\langle k, l \rangle$, we can analyze the corresponding distribution of rating values to infer the preference/interest of the users belonging to the community k on item of the category l. Fig. 3(a) shows graphically a block mixture model with 10 users clusters and 9 item clusters built on the MovieLens dataset. A hard clustering assignment has been performed both on users and clusters: each user u has been assigned to the cluster c such that $c = \text{argmax}_{k=1,\cdots,K}\, c_{uk}$. Symmetrically, each item i has been assigned to the cluster d such that: $d = \text{argmax}_{l=1,\cdots,L}\, d_{il}$. The background color of each block $\langle k, l \rangle$ describes both the density of ratings and the average preference values given by the users (rows) belonging to the k-th group on items (columns) of the l-th category: the background intensity increases with the average rating values of the coclusters, which are given in Fig. 3(b). Each point within the coclusters represents a rating, and again

---

[1] http://www.grouplens.org/system/files/ml-data-10M100K.tar.gz

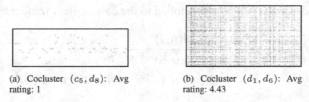

(a) Cocluster (c_5, d_8): Avg rating: 1

(b) Cocluster (d_1, d_6): Avg rating: 4.43

Fig. 4. Cocluster Analysis

an higher rating value corresponds to a more intense color. The analysis underlines interesting tendencies: for example, users belonging to the user community c_1 tend to assign higher rating values than the average, while items belonging to item category d_6 are the most appreciated. A zoom of portions of the block image is given in Fig. 4(a) and in Fig. 4(b). Here, two blocks are characterized by opposite preference behaviors: the first block contains few (low) ratings, whereas the second block exhibit a higher density of high value ratings.

4.2 Item-Topic Analysis

A structural property of of interest is the item-topic dependency. Given a set of F topics $\mathcal{G} = \{g_1, \cdots g_F\}$ and assuming that each item is tagged with at least one of those, we can estimate the relevance of each topic within item clusters through a variant of the *tf-idf* measure [15], namely *topic frequency - inverse category frequency* (*tf-icf*).

The *topic frequency* (similar to the *term frequency*) of a topic g in a cluster d_l can be defined as:

$$tf_{g,d_l} = \frac{\sum_{i \in d_l} \frac{\delta(g \in \mathcal{Q}_i)}{|\mathcal{Q}_i|} \sum_{u \in \mathcal{U}} \delta(u,i)}{\sum_{g'=1}^{F} \sum_{i \in d_l} \frac{\delta(g' \in \mathcal{Q}_i)}{|\mathcal{Q}_i|} \sum_{u \in \mathcal{U}} \delta(u,i)}$$

In a scenario, where items are associated with several topics (genres), and where the number of topics is much lower than size of the itemset, it is high likely that all topics appear at least one in each item category. According to this consideration, the standard definition of *idf* would be useless for our purposes. We, hence, provide an alternative formulation based on entropy [13], namely *inverse category frequency* (*icf*) for a topic g is:

$$icf_g = 1 + p(g) \log_2[p(g)] + [1 - p(g)] \log_2[(1 - p(g))]$$

Here, $p(g)$ represent the prior probability of observing a item-genre and is computed as $p(g) = \sum_{l=1}^{L} p(g|d_l) \cdot p(d_l)$, where $p(g|d_l) = tf_{g,d_l}$ and $p(d_l) = \psi_l$.

By combining the above definitions we can finally obtain the *tf-icf* measure for a topic g in a category d_l as:

$$tf\text{-}icf_{g,d_l} = tf_{g,d_l} \times icf_g$$

We can also exploit the fact that BMM provides a soft assignment to clusters, and provide an alternative version of *tf* as:

$$tf_{g,d_l} = \frac{\sum_{i \in d_l} \frac{\delta(g \in \mathcal{Q}_i)}{|\mathcal{Q}_i|} \cdot d_{il} \sum_{u \in \mathcal{U}} \delta(u,i)}{\sum_{g'=1}^{F} \sum_{i \in d_l} \frac{\delta(g' \in \mathcal{Q}_i)}{|\mathcal{Q}_i|} \cdot d_{il} \sum_{u \in \mathcal{U}} \delta(u,i)}$$

The above considerations can be also applied to the case of item frequency:

$$if_{i,d_l} = \frac{d_{il} \sum_{u \in \mathcal{U}} \delta(u,i)}{\sum_{i' \in d_l} d_{i'l} \sum_{u \in \mathcal{U}} \delta(u,i')}$$
$$icf_i = 1 + p(i) \log_2[p(i)] + [1 - p(i)] \log_2[(1 - p(i))]$$

where:

$$p(i) = \frac{|\mathcal{U}(i)|}{|\mathcal{U}|}$$

The topic and item relevance described so far can be directly employed to identify and measure the interest of each user community into topics and items. More specifically, we can measure the interest of a user community c_k for a topic g as:

$$CI_t(c_k, g) = \frac{\sum_{l=1}^{L} \mu_k^l \cdot tf\text{-}icf_{g,d_l}}{\sum_{g'=1}^{F} \sum_{l=1}^{L} \mu_k^l \cdot tf\text{-}icf_{g',d_l}}$$

The item-based counterpart follows straightforwardly:

$$CI_i(c_k, j) = \frac{\sum_{l=1}^{L} \mu_k^l \cdot if\text{-}icf_{j,d_l}}{\sum_{j'=1}^{F} \sum_{l=1}^{L} \mu_k^l \cdot if\text{-}icf_{j',d_l}}$$

where j is the item target.

Evaluation. The MovieLens dataset provides for each movie a list of genres. This information can be used to characterize each item category, by exploiting the within-cluster topic relevance discussed so far. The *tf-icf* measure of observing each genre within each item category is given in Tab. 2, where the dominant topic is in bold.

Table 2. *tf-icf* measures for each genre in each movie category

	d_1	d_2	d_3	d_4	d_5	d_6	d_7	d_8	d_9
Action	0.03640	0	0.07375	**0.06054**	0.05152	0	0.05624	0.06966	0
Adventure	0.01981	0	0.04237	0.04339	0.03813	0	0.03828	0	0
Animation	0.01591	0	0.00660	0.00926	0.01801	**0.24622**	0.00999	0	0
Children's	0.01581	0	0.03228	0.01643	0.02261	0	0.02855	0	0
Comedy	0.04137	0.03559	0.05403	0.05185	0.04730	0.06209	**0.05685**	0.10228	0
Crime	0.03319	0	0.01585	0.02217	0.01973	0	0.02515	0	0
Documentary	0.01423	0	0.00028	0.00053	0.00291	0	0.00341	0	**0.94466**
Drama	**0.09777**	0.00923	0.07720	0.05247	0.06209	0.04839	0.05099	0.06727	0
Fantasy	0.00553	0	0.01175	0.01579	0.01171	0	0.01559	0	0
Film-Noir	0.01485	0	0.00029	0.00123	0.00580	0	0.00113	0	0
Horror	0.01570	**0.53057**	0.08225	0.02691	0.01569	0	0.04014	0.03426	0
Musical	0.01739	0	0.00619	0.00914	0.02224	0	0.01088	0	0
Mystery	0.01697	0	0.00832	0.02757	0.00958	0	0.00952	0	0
Romance	0.03470	0	0.02395	0.05776	0.05092	0.09889	0.04625	0	0
Sci-Fi	0.02818	0	0.06247	0.04644	0.03843	0	0.04150	0	0
Thriller	0.04613	0	0.05851	0.05052	0.04771	0	0.05057	0	0
War	0.03902	0	0.01268	0.01041	0.01442	0.12291	0.00716	**0.11860**	0
Western	0.01653	0	0.00625	0.00704	0.00641	0	0.00875	0	0

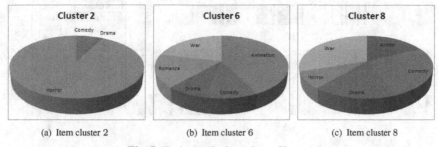

| (a) Item cluster 2 | (b) Item cluster 6 | (c) Item cluster 8 |

Fig. 5. Topic Analysis on Item Clusters

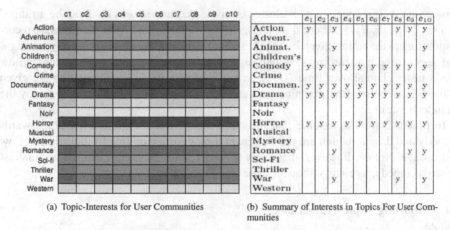

	c_1	c_2	c_3	c_4	c_5	c_6	c_7	c_8	c_9	c_{10}
Action	y		y					y	y	y
Advent.										
Animat.			y							y
Children's										
Comedy	y	y	y	y	y	y	y	y	y	y
Crime										
Documen.	y	y	y	y	y	y	y	y	y	y
Drama	y	y	y	y	y	y	y	y	y	y
Fantasy										
Noir										
Horror	y	y	y	y	y	y	y	y	y	y
Musical										
Mystery										
Romance			y						y	y
Sci-Fi										
Thriller										
War			y					y		y
Western										

| (a) Topic-Interests for User Communities | (b) Summary of Interests in Topics For User Communities |

Fig. 6. Topic relevance within the user communities

The pie charts in Fig. 5(a), Fig. 5(b) and Fig. 5(c) show the distribution on topics for different item clusters. We can observe different patterns: d_2 is characterized by a strong attitude for horror movies, animation is the dominant topic in cluster 6, and d_8 is summarized by the war genre. Finally, the cluster d_9 shows a predominance of drama movies.

The dominant genres in each item cluster, i.e., with higher *tf-icf*, are (in order): d_1, Drama; d_2, Horror; d_3 Horror; d_4, Action; d_5, Drama; d_6, Animation; d_7, Comedy; d_8, War; d_9, Documentary.

Fig. 6(a) shows the $CI_t(g, c_k)$ values (in gray scale). We can further analyze such values to infer the interest of a user community for a given topic. In particular, a community exhibits a high interest for a topic if the corresponding CI_t value is sufficiently higher than the average CI_t value of all the other topics. Figure 6(b) summarizes the associations among user communities and item topics. For example, users in c_8 exhibit preferences for the *Action* and *War* genres.

4.3 User Profile Segmentation

The topics of interest of a user may change within time and consecutive choices can influence each other. We can analyze such temporal dependencies by mapping each user's

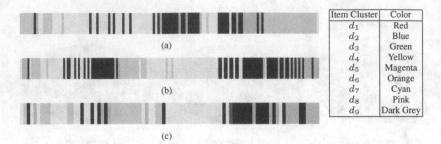

Item Cluster	Color
d_1	Red
d_2	Blue
d_3	Green
d_4	Yellow
d_5	Magenta
d_6	Orange
d_7	Cyan
d_8	Pink
d_9	Dark Grey

Fig. 7. User Profile Segmentation

choice into their respective item cluster. Assume that movieLens data can be arranged as a set $\{\bar{u}_1, \ldots, \bar{u}_M\}$, where $\bar{u} = \{\langle r_i^u, i, t_i^u \rangle \forall i \in \mathcal{I}(u)\}$ and t_i^u is the timestamp corresponding to the rating given by the user u on the item i. By chronologically sorting $\bar{u}$ and segmenting it according to item cluster membership, we can obtain a view of how user's tastes change over time. Three example of user profile segmentation are given in the figures below (the mapping between item categories and colors is given by the included table).

In practice, we can assume that the three users show a common attitude towards comedy and drama, which are the dominant topics corresponding to the colors yellow and orange. Notice, however, that users (b) and (c) are prone to change their interest towards comedy, as clearly shown by the change in color.

4.4 Modeling Topic Transitions

Based on the above observations, we aim at estimating the sequential connections among topics: In practice, we would like to analyze which item categories are likely to next capture the interests of a user. Those sequential patterns can be modeled by exploiting *Markov Models*. The latter are probabilistic models for discrete processes characterized by the Markov properties. We adopt a Markov Chain property here, i.e., a basic assumption which states that any future state only depends from the present state. This property limits the 'memory' of the chain which can be represented as a digraph where nodes represent the actual states and edges represent the possible transitions among them.

Assuming that the last observed item category for the considered user is d_i, the user could pick an item belonging to the another topic d_j with probability $p(d_j|d_i)$. Thus, we need to estimate all the *transition probabilities*, starting from a $|L + 1|$ x $|L + 1|$ *transition count matrix* $\mathcal{T}_c$, where $\mathcal{T}_c(i, j)$ stores the number of times that category j follows i in the rating profile of the users.[2]

The estimation we provide is rather simple, corresponding to a simple frequency count:

$$p(d_j|d_i) = \frac{\mathcal{T}_c(i, j)}{\sum_{j=1}^{L+1} \mathcal{T}_c(i, j')}$$

---

[2] We assume two further states ϵ, representing the initial choice, and β, representing the last choice.

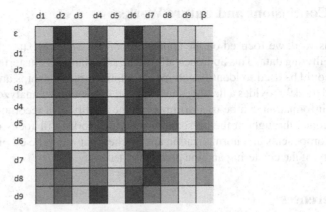

Fig. 8. Transition Probabilities Matrix

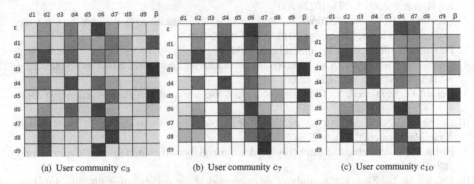

(a) User community c_3 (b) User community c_7 (c) User community c_{10}

Fig. 9. Transition Probabilities Matrix

Fig. 8 represents the overall transition probability matrix, which highlights some strong connection among given categories. As instance, the item categories having drama as dominant genre, d_4, d_6 and d_9 are highly correlated as well as d_2, d_7 and d_8 which correspond to comedy movies.

It is interesting to compare how the transition probabilities change within different user communities. Fig. 9 shows the transitions for three different communities. Notice that, besides common transition patterns, each community has some distinctive transitions that characterize their population. For all the considered user communities, the most likely initial item category is d_6; while the first and the last community reproduced in the example show a strong attitude corresponding to the transition $d_8 \rightarrow d_2$, this is instead a weak pattern within c_7. The same consideration can be done for the transition $d_9 \rightarrow d_7$, which is strong for c_7 and c_{10}, while users belonging to c_3 are more prone to the transition towards d_6.

The analysis of the transition probabilities can be hence exploited for generating new recommendations enforcing topic diversity [16] in the top-K lists of items by taking into account not exclusively the current topic of interest but the ones that more likely could be connected to it.

5   Conclusions and Future Works

In this work we focused on the application of the Block Mixture Model to Collaborative Filtering data. This approach allows the simultaneous clustering of users and items and could be used to identify and measure hidden relationships among them. The proposed model provides a flexible and powerful framework to analyze the users' behavior. This information can be used to improve the quality of a recommendation system, as mentioned throughout the presentation. Future works will focus on embedding baseline components and normalization approaches that might be employed to improve the quality of the clustering and the prediction accuracy.

References

1. Blei, D.M., Ng, A.Y., Jordan, M.I.: Latent dirichlet allocation. The Journal of Machine Learning Research 2003 3, 993–1022 (2003)
2. Cremonesi, P., Koren, Y., Turrin, R.: Performance of recommender algorithms on top-n recommendation tasks. In: RecSys 2010, pp. 39–46 (2010)
3. Funk, S.: Netflix update: Try this at home (2006)
4. George, T., Merugu, S.: A scalable collaborative filtering framework based on co-clustering. In: ICDM 2005, pp. 625–628 (2005)
5. Gerard, G., Mohamed, N.: Clustering with block mixture models. Pattern Recognition 36(2), 463–473 (2003)
6. Govaert, G., Nadif, M.: An em algorithm for the block mixture model. IEEE Trans. Pattern Anal. Mach. Intell. 27(4), 643–647 (2005)
7. Hofmann, T., Puzicha, J.: Latent class models for collaborative filtering. In: IJCAI 1999, pp. 688–693 (1999)
8. Jin, R., Si, L., Zhai, C.: A study of mixture models for collaborative filtering. Inf. Retr. 2006 9(3), 357–382 (2006)
9. Jin, X., Zhou, Y., Mobasher, B.: Web usage mining based on probabilistic latent semantic analysis. In: KDD 2004, pp. 197–205 (2004)
10. McNee, S.M., Riedl, J., Konstan, J.A.: Being accurate is not enough: How accuracy metrics have hurt recommender systems. In: ACM SIGCHI Conference on Human Factors in Computing Systems, pp. 1097–1101 (2006)
11. Porteous, I., Bart, E., Welling, M.: Multi-hdp: a non parametric bayesian model for tensor factorization. In: AAAI 2008, pp. 1487–1490 (2008)
12. Shan, H., Banerjee, A.: Bayesian co-clustering. In: ICML 2008 (2008)
13. Shannon, C.E.: Prediction and entropy of printed english. Bell Systems Technical Journal 30, 50–64 (1951)
14. Wang, P., Domeniconi, C., Laskey, K.B.: Latent Dirichlet Bayesian Co-Clustering. In: Buntine, W., Grobelnik, M., Mladenić, D., Shawe-Taylor, J. (eds.) ECML PKDD 2009, Part II. LNCS, vol. 5782, pp. 522–537. Springer, Heidelberg (2009)
15. Wu, H.C., Luk, R.W.P., Wong, K.F., Kwok, K.L.: Interpreting tf-idf term weights as making relevance decisions. ACM Trans. Inf. Syst. 26, 13:1–13:37 (2008)
16. Ziegler, C.-N., McNee, S.M., Konstan, J.A., Lausen, G.: Improving recommendation lists through topic diversification. In: WWW 2005, pp. 22–32 (2005)

Learning Attack Features from Static and Dynamic Analysis of Malware

Ravinder R. Ravula, Kathy J. Liszka, and Chien-Chung Chan

Department of Computer Science, University of Akron, Akron, OH, U.S.A.
{liszka,chan}@uakron.edu

Abstract. Malware detection is a major challenge in today's software security profession. Works exist for malware detection based on static analysis such as function length frequency, printable string information, byte sequences, API calls, etc. Some works also applied dynamic analysis using features such as function call arguments, returned values, dynamic API call sequences, etc. In this work, we applied a reverse engineering process to extract static and behavioral features from malware based on an assumption that behavior of a malware can be revealed by executing it and observing its effects on the operating environment. We captured all the activities including registry activity, file system activity, network activity, API Calls made, and DLLs accessed for each executable by running them in an isolated environment. Using the extracted features from the reverse engineering process and static analysis features, we prepared two datasets and applied data mining algorithms to generate classification rules. Essential features are identified by applying Weka's J48 decision tree classifier to 1103 software samples, 582 malware and 521 benign, collected from the Internet. The performance of all classifiers are evaluated by 5-fold cross validation with 80-20 splits of training sets. Experimental results show that Naïve Bayes classifier has better performance on the smaller data set with 15 reversed features, while J48 has better performance on the data set created from the API Call data set with 141 features. In addition, we applied a rough set based tool BLEM2 to generate and evaluate the identification of reverse engineered features in contrast to decision trees. Preliminary results indicate that BLEM2 rules may provide interesting insights for essential feature identification.

Keywords: Malware, Reverse Engineering, Data Mining, Decision Trees, Rough Sets.

1 Introduction

Malware, short for malicious software, is a sequence of instructions that perform malicious activity on a computer. The history of malicious programs started with "Computer Virus", a term first introduced by [6]. It is a piece of code that replicates by attaching itself to the other executables in the system. Today, malware includes viruses, worms, Trojans, root kits, backdoors, bots, spyware, adware, scareware and any other programs that exhibit malicious behaviour.

Malware is a fast growing threat to the modern computing world. The production of malware has become a multi-billion dollar industry. The growth of the Internet, the

A. Fred et al. (Eds.): IC3K 2011, CCIS 348, pp. 109–125, 2013.
© Springer-Verlag Berlin Heidelberg 2013

advent of social networks and rapid multiplication of botnets has caused an exponential increase in the amount of malware. In 2010, there was a large increase in the amount of malware spread through spam emails sent from machines that were part of botnets [12]. McAfee Labs have reported 6 million new botnet infections in each month of 2010. They also detected roughly 60,000 new malware for each day of 2010 [13]. Symantec discovered a daily average of 2,751 websites hosting malware in January 2011 [14]. Antivirus software, such as Norton, McAfee, Sophos, Kaspersky and Clam Antivirus, is the most common defense against malware. The vendors of these antivirus programs apply new technologies to their products frequently in an attempt to keep up with the massive assault. These programs use a signature database as the primary tool for detecting malware. Although signature based detection is very effective against previously discovered malware, it proves to be ineffective against new and previously unknown malware. Malware programmers bypass the known signatures with techniques like obfuscation, code displacement, compression and encryption. This is a very effective way to evade signature based detection. Antivirus companies are trying hard to develop more robust antivirus software. Some of the techniques include heuristics, integrity verification and sandboxing. However, in practice, they are not really very effective in detecting new malware. We are virtually unprotected until the signature of each new threat is extracted and deployed.

Signature detection is mostly accomplished using manual methods of reverse engineering. This is timely and work intensive. With the staggering number of malware generated each day, it is clear that automated analysis will be imperative in order to keep up. Hence, we cannot depend solely on traditional antivirus programs to combat malware. We need an alternative mechanism to detect unidentified threats.

In an effort to solve the problem of detecting new and unknown malware, we have proposed an approach in the present study. The proposed approach uses reverse engineering and data mining techniques to classify new malware. We have collected 582 malicious software samples and 521 benign software samples and reverse engineered each executable using both static and dynamic analysis techniques. By applying data mining techniques to the data obtained from the reverse engineering process, we have generated a classification model that would classify a new instance with the same set of features either as malware or a benign program.

The rest of the paper is organized as follows. Section 2 discusses previous work based on detection of malware using data mining techniques. Section 3 presents the reverse engineering techniques used in our work. Section 4 explains the data mining process and the machine learning tools we used for the experiments. Here we present and discuss the results and finally, section 5 concludes the study and suggests possible future work.

2   Literature Review

Significant research has been done in the field of computer security for the detection of known and unknown malware using different machine learning and data mining approaches.

A method for automated classification of malware using static feature selection was proposed by [8]. The authors used two static features extracted from malware and benign software, Function Length Frequency (FLF) [7] and Printable String Information

(PSI) [23]. This work was based on the hypothesis that "though function calls and strings are independent of each other they reinforce each other in classifying malware". Disassembly of all the samples was done using IDA Pro and FLF, PSI features were extracted using Ida2DB.

The authors used five classifiers; Naive Bayes, SVM, Random Forest, IB1 and Decision Table. The best results were obtained by AdaBoostM1 with Decision Table yielding an accuracy rate of 98.86%. It was also observed that the results obtained by combining both features were more satisfactory than using each kind of features individually.

[19] used different data mining techniques to detect unknown malware. They used three approaches for static analysis and feature identification; binary profiling, strings and byte sequences. Binary profiling was only applied to PE files. Other approaches were used for all programs.

Binary profiling was used to extract three types of features; 1) list of Dynamic Link Libraries (DLL) used by the PE, 2) function calls made from each DLL and 3) unique function calls in each DLL. The "GNU Strings" program was used to extract printable strings. Each string was used as a feature in the dataset. In the third method for features extraction, the hexdump [15] utility identified byte sequences, which were used as features.

The authors applied the rule based learning algorithm RIPPER [7] to the 3 datasets with binary profiling features, Naïve Bayes classifier to data with string and byte sequence features and finally six different Naïve Bayes classifiers to the data with byte sequence features. To compare the results from these approaches with traditional signature based method, the authors designed an automatic signature generator.

With RIPPER they achieved accuracies of 83.62%, 89.36%, and 89.07% respectively for datasets with features DLLs used, DLL function calls and Unique Calls in DLLs. The accuracies obtained with Naïve Bayes and Multi-Naïve Bayes were 97.11% and 96.88%. With the Signature method they achieved 49.28% accuracy. Multi-Naïve Bayes produced better results compared to the other methods.

In [23], the information in PE headers was used for the detection of malware, based on the assumption that there would be a difference in the characteristics of PE headers for malware and benign software as they were developed for different purposes. Every header (MS DOS header, file header, optional header and section headers) in the PE was considered as a potential attribute. For each malware and benign program, position and entry values of each attribute were calculated. In parallel, attribute selection was performed using Support Vector Machines. The dataset was tested with an SVM classifier using five-fold cross validation. Accuracies of 98.19%, 93.96%, 84.11% and 89.54% were obtained for virus, email worm, Trojans and backdoors respectively. Detection rates of viruses and email worms were high compared to the detection rates of Trojans and backdoors.

In [10], multiple byte sequences from the executables were extracted from PE files and combined to produce n-grams. Five hundred relevant features were selected by calculating the information gain for each feature. Several data mining techniques like IBk, TFIDF, Naïve Bayes, Support Vector Machine (SVM) and decision trees applied to generate rules for classifying malware. The authors also used "boosted" Naïve Bayes, SVM and decision tree learners. The boosted classifiers, SVM and IBk produced good results compared to the other methods. The performance of classifiers was improved by boosting and the overall performance of all the classifiers was better with the large dataset compared with the small dataset.

Komashinskiy and Kotenko [11] used position dependent features in the Original Entry Point (OEP) of a file for detecting unknown malware. Decision Table, C4.5, Random Forest, and Naïve Bayes were applied on the prepared dataset. Three assumptions were made for this work. 1) Studying the entry point of the program known as Original Entry Point (OEP) reveals more accurate information. 2) The location of the byte value of OEP address was set to zero. The offsets for all bytes in OEP were considered to be in the range [-127,127]. 3) Only a single byte can be read for each position value. The dataset contained three features; Feature ID, position and byte in position.

Feature selection was performed to extract more significant features. The resulting data was tested against all classifiers and the results were compared based on ROC-area. Random Forest outperformed all the other classifiers.

A specification language was derived in [5] based on the system calls made by the malware. These specifications are intended to describe the behaviour of malware. The authors also developed an algorithm called MINIMAL that mines the specifications of malicious behaviour from the dependency graphs. They applied this algorithm to the email worm Bagle.J, a variant of Bagle malware.

Clean and malicious files were executed in a controlled environment. Traces of system calls were extracted for each sample during execution. The dependencies between the system call arguments were obtained by observing the arguments and their type in sequence of calls. A dependency graph was constructed using system calls and their argument dependencies. A sub graph was then extracted by contrasting it with the benign software dependence graph such that it uniquely specifies the malware behaviour. A new file with these specifications would be classified as malware.

The Virus Prevention Model (VPM) to detect unknown malware using DLLs was implemented by [22]. Malicious and benign files were parsed by a program "dependency walker" which shows all the DLLs used in a tree structure. The list of APIs used by main program directly, the DLLs invoked by other DLLs other than main program and the relationships among DLLs which consists of dependency paths down the tree were collected. In total, 93,116 total attributes were obtained. After pre-processing there were 1,398 attributes. Of these, 429 important attributes were selected and tested. The detection rate with RBF-SVM classifier was 99.00% with True Positive rate of 98.35% and False Positive rate of 0.68%.

A similarity measure approach for the detection of malware was proposed by [21], based on the hypothesis that variants of a malware have the same core signature, which is a combination of features of the variants of malware. To generate variants for different strains of malware, traditional obfuscation techniques were used. The source code of each PE was parsed to produce an API calling sequence which was considered to be a signature for that file. The resulting sequence was compared with the original malware sequence to generate a similarity measure. Generated variants were tested against eight different antivirus products. The detection rate of SAVE was far better than antivirus scanners.

In [2], a strain of the Nugache worm was reverse engineered in order to study its underlying design, behaviour and to understand attacker's approach for finding vulnerabilities in a system. The authors also reverse engineered 49 other malware executables in an isolated environment. They created a dataset using features such as the MD5 hash, printable strings, number of API calls made, DLLs accessed and URL referenced. Due to the multi-dimensional nature of the dataset, a machine learning tool, BLEM2 [3], based on rough set theory was used to generate dynamic patterns which would help in classifying an unknown malware. As the size of the dataset was small, a very few number of decision rules were generated and the results were generally not satisfactory.

In another work by Ahmed et al. [1] based on dynamic analysis, spatio-temporal information in API calls was used to detect unknown malware. The proposed technique consists of two modules; an offline module that develops a training model using available data and an online module that generates a testing set by extracting spatio-temporal information during run time and compares them with the training model to classify run time process as either benign or malicious. In the dynamic analysis, spatial information was obtained from function call arguments, return values and were divided into seven subsets socket, memory management, processes and threads, file, DLLs, registry and network management based on their functionality. Temporal information was obtained from the sequence of calls. The authors observed that some sequences were present only in malware and were missing in benign programs.

Three datasets were created by combining benign program API traces with each malware type. The three datasets were combinations of benign-Trojan, benign-virus and benign-worm. They conducted two experiments. The first one studied the combined performance of spatio-temporal features compared to standalone spatial or temporal features. The second experiment was conducted to extract a minimal subset of API categories that gives same accuracy as from the first experiment. For this, the authors combined API call categories in all possible ways to find the minimal subset of categories that would give same classification rate as obtained in first experiment. For the first experiment, the authors obtained 98% accuracy with naive Bayes and 94.7% accuracy with J48 decision tree. They obtained better results with combined features as compared standalone features. The detection rate of Trojans was less compared to viruses and worms.

In the second experiment, combination of API calls related to memory management and file I/O produced best results with an accuracy of 96.6%.

In some of the above mentioned works, only static features such as byte sequences, printable strings and API call sequences were used. Though effective in detecting known malware, they would be ineffective if the attackers use obfuscation techniques to write malware. To solve this problem, some other works [1, 2] used dynamic detection methods. The work done in [1] used only dynamic API call sequences. Using only API calls may not be effective in detecting malware. In the work, malwares were reversed to find their behaviour and applied data mining techniques to the data obtained from reversing process. A very small number of rules were generated and the results were not effective as the experiments were conducted on very few numbers of samples.

Our work is different from all the above works as we combined static and behavioral features of all malware and benign software. It is an extension of the work done in [1] but it differs significantly, as we performed rigorous reverse engineering of each executable to find their inner workings in detail. We also used a large number (582 malicious and 521 benign) of samples which would facilitate determining more accurate behaviour of malicious executables.

3   Reverse Engineering

Reverse engineering malware can be defined as an analysis of a program in order to understand its design, components and its behavior to inflict damage on a computer system. The benefit of reverse engineering is that it allows us to see the hidden behavior of the file under consideration, which we can't see by merely executing it [18].

In the reverse engineering process we used both static and dynamic analysis techniques. There are many different tools available for each technique. All the tools used for our work are open source. In total, we reverse engineered 1103 PE (Portable Executable) files of which 582 were malicious executables and 521 were benign executables. All malicious executables were downloaded from Offensivecomputing.net and all benign executables were downloaded from Sourceforge.net and Brothersoft.com.

3.1   Controlled Environment

For static analysis of executables, we do not require a controlled environment. In this case, we do not run the executable to collect features. In the case of dynamic analysis, the code to run is malicious and dangerous. The environment for the reversing process must be isolated from the other hosts on the network. We apply the industry common standard, a virtual machine. Due to a strong isolation between the guest operating system in VM and host operating systems, even if the virtual machine is infected with a malware, there will be no effect of it on the host operating system.

For the analysis of malware we needed virtualization software that would allow quick backtrack to the previous system state after it has been infected by the malware. Each time a malware is executed in dynamic analysis process, it would infect the system. Analysis of subsequent malware had to be performed in a clean system. We chose VMware Workstation as virtualization software for our work.

3.2   Static Analysis

In general, it is a good idea to start analysis of any given program by observing the properties associated with it and predicting the behavior from visible features without actually executing it. This kind is known as static analysis. The advantage with static analysis is that it gives us an approximate idea of how it will affect the environment upon execution without actually being executed. However, most of the time, it is not possible to predict the absolute behavior of a program with just static analysis.

There are many different tools available that aid in static analysis of executables for example, decompilers, disassemblers, Source code analyzers and some other tools that help in extracting useful features from executables. The tools we used were Malcode Analyst Pack from idefense.com, PEiD from peid.has.it and IDA Pro Disassembler hex-rays.com.

3.2.1 Cryptographic Hash Function
A unique cryptographic hash value is associated with each executable file. This value differentiates each file from others. We started our reverse engineering process of each executable by calculating its hash value.

The reason for calculating the hash value is twofold. First, there is no unique standard for naming malware. There may be multiple names for a single piece of malware so by calculating hash value of each sample we know that all of them are indeed the same. This results in eliminating ambiguity in the reverse engineering process. The second reason is that if an executable is modified, its hash value will also be changed. That way we can identify that changes were made to the executable and thereby analyzing it to detect the changes made.

MD5, SHA1 and SHA256 are the widely used hash functions. We used Malcode Analyst Pack (MAP) tool to compute the MD5 (Message Digest 5) hash value of each PE file that we analysed.

3.2.2 Packer Detection

Malware authors employ various techniques to obfuscate the content of the malware they have written and making it unable to be reversed. Using packers is one of them. A packer is a program that helps in compressing another executable program, thereby hiding the content. Packers help malware authors hide actual program logic of the malware so that a reverse engineer cannot analyze it using static analysis techniques alone. Packers also help evade detection of the malware from antivirus programs.

In order to execute, a packed malware must unpack its code into memory. For this reason, the authors of the malware include an unpacker routine in the program itself. The unpacker routine is invoked at the time of execution of the malware and converts the packed code into original executable form. Sometimes they use multiple levels of packing to make the malware more sophisticated [9].

Detection of a packer with which a malware is packed is very important for the analysis of the malware. If we know the packer, we can unpack the code and then analyze the malware. We used the PEiD tool which is a free tool for the detection of packers. It has over 600 different signatures for the detection of different packers, cryptors and compilers. It displays the packer with which the malware is packed by simply opening the malware using PEiD. If the signature of the packer or compiler with which the malware is packed is not present in the PEiD database it will report that it didn't find any packers.

3.2.3 Code Analysis

The next step for better understanding the nature of malware is to analyze its source code. Although there are many decompilers that help in decompiling executables into high level languages, analyzing the malware by keeping the source code in low level language reveals more information. IDA Pro disassembler from DataRescue is a popular choice for the disassembly of executable program into Assembly Language.

We used the IDA Pro Disassembler for the code analysis of malware. In this step, we have gone through the assembly code of each PE file to find useful information and to understand the behavior of it. Following is the list of features that we were able to extract from the assembly code of PE files.

- Type of file from the PE header. If it was not a PE file, we discarded it.
- List of strings embedded in the code that would be useful for predicting the behavior of the PE.

- The programming language with which the PE was written.
- Compile date and time.

3.3   Dynamic Analysis

In static analysis of executables, we only analyze the static code of the executable and approximately predict its properties and behavior. We know that the authors of malware use techniques such as binary obfuscation and packing to evade static analysis techniques. To thoroughly understand the nature of the malware we cannot rely on static analysis techniques alone. If a program has to be run, the whole code must be unpacked and loaded into primary memory. Every detail of the executable program is revealed at run time regardless of how obfuscated the code is and what packer the executable is packed with [20]. In dynamic analysis, we observe the full functionality of the malware and its effect on the environment as it executes.

Tools that help in dynamic analysis of executables include debuggers, registry monitors, file system monitors, network sniffers and API call tracers. The tools we used in this step were Filemon, Regshot and Maltrap.

3.3.1   File System Monitor

When a program is executed it makes changes to the file system. The file system activity made by the program helps partly in determining its behavior. We used File Monitor (Filemon) from Microsoft Sysinternals to monitor file system activity of all the processes running on a windows system. It installs a device driver which keeps track of all the file system activity in the system. However, we only need the information related to a particular process under consideration, therefore, we can use a filter which lets us select a particular process for which we want to monitor file system activity by removing all the other processes from the list. Each file system activity made by the PE on the file system produces one line of output in the Filemon GUI window.

3.3.2   Registry Monitor

Windows Registry is a huge database hosting configuration details of the operating system and the programs installed on it. Registry entries are divided into hives and represented in a tree structure on Windows systems. Most applications use two hives frequently; HKLM and HKCU. HKLM stands for Hive Key Local Machine and contains settings of the operating system. HKCU stands for Hive Key Current User and contains configuration details for the user currently logged into the system.

Malware authors frequently use registries to infect systems. One usual technique employed   is   to   insert   an   entry   at   the   location HKEY_LOCAL_MACHINE\SOFTWARE\Microsoft\Windows\CurrentVersion\RUN so that each time system boots up the malware is executed. There is an extensive list of such keys in the Windows registry and they are used by attackers for their malicious purposes.

Regshot is a product of Microsoft Sysinternals that helps in the reverse engineering process by monitoring the Windows Registry. It lists the changes made in the windows registry upon installation of software. We used this tool to record the changes in the windows registry made by malware and benign software in the Windows Registry.

3.3.3 Api Call Tracer

Windows API (Application Program Interface) also known as Win API, is a long list of functions that provide system level services to the user programs. Every Windows application is implemented using the Win API functions.

Keeping track of the sequence of API Calls made by an application helps in the reverse engineering process. It allows us to go through each call and thereby predicting the behavior of that software. Maltrap is a software that lists the sequence of calls made by the software while execution.

4 Malware Detection Mining

In this section, we explain our implementation of the data mining process to find patterns that would help in classifying malware from benign software.

4.1 Feature Extraction

From static analysis of each sample in the reverse engineering process we have the MD5 hash of the file, file size in bytes, the packer used (if any, a determination of whether it contains unique strings, a time stamp and the programming language used to write the file.

From the dynamic analysis, for each file, we stored a log of file system activity, registry activity and the sequence of API calls made by the sample while running.

From the file system activity log we were able to extract three important features; the decisions of whether the sample under consideration attempted to write to another file, if the sample accessed another directory and all unique DLLs accessed during execution.

From the registry activity we extracted three features; registry keys added, registry keys deleted and registry values modified. The log contains all the registry keys modified by executables with their modified values. We removed the key values and recorded only the keys.

From the API call log we extracted the unique API calls made by each sample. We combined all the unique API calls made by each file and removed duplicates. Over the entire sample space, we identified 141 unique API calls which will be used as features to create a data set. We also noted if a sample contained any URL references or attempted to access the Internet, making. With this step we completed processing of raw data for feature selection.

4.2 Data Sets

We prepared two datasets: the first dataset with 15 features, 1103 instances (582 malicious and 521 benign) and the second dataset with 141 features, 1103 instances. We named the first dataset DRF (Dataset with Reversed Features) and the second dataset DAF (Dataset with API Call Features). Table 1 shows the list of 15 attributes and decision label in DRF. All the 141 attributes in DAF are of type binary except the decision label which is Boolean. The first three attributes File Name, File Size, and MD5 Hash do not provide useful information for classification purpose. We collected this information and retain it for tracking and other research purposes, however, they are removed from data sets used in the following experiments.

Table 1. Attributes in DRF

S.NO	Attribute Name	Type
1	FILE NAME	NOMINAL
2	FILE SIZE	NOMINAL
3	MD5 HASH	NOMINAL
4	PACKER	NOMINAL
5	FILE ACCESS	BINARY
6	DIRECTORY ACCESS	BINARY
7	DLLs	NOMINAL
8	API CALLS	NOMINAL
9	INTERNET ACCESS	BINARY
10	URL REFERENCES	BINARY
11	REGISTRY KEYS ADDED	NOMINAL
12	REGSITRY KEYS DELETED	NOMINAL
13	REGISTRY VALUES MODIFIED	NOMINAL
14	UNIQUE STRINGS	BINARY
15	PROGRAMMING LANGUAGE	NOMINAL
16	DECISION LABEL	BOOLEAN

In addition, values of attributes Registry Keys Added, Registry Keys Deleted, Registry Keys Modified, API CALLs and DLLs varies widely from very small to very large. They were discretized by using tools available in Weka. The ranges we obtained for each attribute after transforming the dataset are shown in Table 2.

Table 2. Discretized values

	ATTRIBUTE NAME	DISCRETIZED VALUES
1	KEYS ADDED	(-INF-1] (1-INF)
2	REGISTRY VALUES MODIFIED	(-INF-12.5] (12.5-INF)
3	API CALLS	(-INF -5.5] (5.5-22.5] (22.5-41.5] (41.5- INF)
4	DLLs	(-INF -16.5] (16.5- INF)

We prepared second dataset from DRF by replacing the discrete values with the discretized values shown in Table 2. We call it DDF (Dataset with Discretized Features).

4.3   Experimental Results

We conducted experiments on the datasets derived from reversed features: DRF and DDF, and data set derived from API call features: DAF. The J48 decision tree and Naïve Bayes algorithms in WEKA [24] were used to generate classifiers. We applied 5-fold cross validation with 80-20 splits for training and testing.

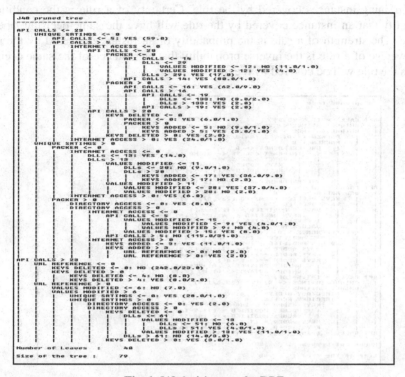

Fig. 1. J48 decision tree for DRF

Figure 1 shows that the number of API Calls made by a PE was selected as the root node of the decision tree. API Calls, Unique Strings, URL References, Internet Access, Packer, Registry Keys Deleted, Directory Access and Registry Keys Modified were the most used attributes in the classification model although we note that it used all the other attributes in the dataset.

The decision tree in Figure 2 shows that 9 attributes, API Calls, Unique Strings, URL References, Packer, Registry Keys Deleted, Registry Keys Modified, Directory Access, Registry Keys added and Internet Access were used in the classification model. In this case, the attributes DLLs and File Access were not used in the decision rules for classification.

Decision tree classifiers such as J48 are top-down learners that select an attribute to split a tree node in each step of the decision tree building process. In contrast, the machine learning algorithm BLEM2 [3] is a learner for generating if-then decision rules from bottom-up. In each step, an attribute-value pair is selected to form the condition part of a decision rule. Each rule learned by BLEM2 is minimal, and the entire set of rules learned is also minimal. It is a revision of the LEM2 algorithm [4], which is based on rough set theory introduced by Pawlak [16]. The major difference is that rules learned by BLEM2 are associated with four factors: support, certainty, strength, and coverage. These factors can be interpreted as Bayesian probabilities derived from the training data which was first introduced in [17]. We say that an instance is covered by a rule if it matches the condition of the rule. The support of a rule is the number of training instances covered by the rule. Certainty of a rule is the conditional probability that an instance covered by the rule will have the same decision label as the rule. The strength of a rule is the probability of an instance covered by the rule. The coverage of a rule is the inverse probability that an instance of the same decision label as the rule will be covered by the rule.

```
J48 pruned tree
------------------

API CALLS = (5.5-23.5]
|   UNIQUE SRTINGS <= 0: YES (230.0/37.0)
|   UNIQUE SRTINGS > 0
|   |   PACKER <= 0: YES (84.0/22.0)
|   |   PACKER > 0
|   |   |   URL REFERENCE <= 0
|   |   |   |   INTERNET ACCESS <= 0: NO (118.0/34.0)
|   |   |   |   INTERNET ACCESS > 0
|   |   |   |   |   KEYS ADDED = (-inf-1]: YES (5.0)
|   |   |   |   |   KEYS ADDED = (1-2.5]: YES (0.0)
|   |   |   |   |   KEYS ADDED = (2.5-inf): NO (2.0)
|   |   |   URL REFERENCE > 0: YES (8.0/1.0)
API CALLS = (23.5-41.5]
|   URL REFERENCE <= 0
|   |   KEYS DELETED <= 0: NO (239.0/23.0)
|   |   KEYS DELETED > 0
|   |   |   VALUES MODIFIED = (12.5-inf)
|   |   |   |   KEYS DELETED <= 4: NO (2.0)
|   |   |   |   KEYS DELETED > 4: YES (7.0/1.0)
|   |   |   VALUES MODIFIED = (-inf-12.5]: NO (7.0)
|   URL REFERENCE > 0
|   |   VALUES MODIFIED = (12.5-inf): YES (13.0/1.0)
|   |   VALUES MODIFIED = (-inf-12.5]
|   |   |   KEYS ADDED = (-inf-1]: NO (7.0)
|   |   |   KEYS ADDED = (1-2.5]: YES (2.0/1.0)
|   |   |   KEYS ADDED = (2.5-inf)
|   |   |   |   UNIQUE SRTINGS <= 0: YES (4.0)
|   |   |   |   UNIQUE SRTINGS > 0: NO (7.0/2.0)
API CALLS = (41.5-inf)
|   UNIQUE SRTINGS <= 0: YES (20.0/1.0)
|   UNIQUE SRTINGS > 0
|   |   DIRECTORY ACCESS <= 0: YES (2.0)
|   |   DIRECTORY ACCESS > 0
|   |   |   KEYS ADDED = (-inf-1]
|   |   |   |   VALUES MODIFIED = (12.5-inf): YES (3.0)
|   |   |   |   VALUES MODIFIED = (-inf-12.5]: NO (2.0)
|   |   |   KEYS ADDED = (1-2.5]: NO (2.0)
|   |   |   KEYS ADDED = (2.5-inf): NO (16.0/4.0)
API CALLS = (-inf-5.5]: YES (102.0/8.0)

Number of Leaves  :     23

Size of the tree :     40
```

Fig. 2. J48 decision tree for DDF

In our experiment, we applied BLEM2 to generate decision rules from the DDF dataset. Rules with a minimum support of 7 were selected for analysis. There are 11 attributes selected by BLEM2 to form decision rules. Tables 3 shows the attributes selected by BLEM2 together with the frequencies of attribute-value pairs appeared in the rules. Table 4 shows the 21 decision rules generated by BLEM2 for the decision label "YES", and Table 5 shows the 13 decision rules for the decision label "NO".

From Table 3, API CALLS and REGISTRY KEYS ADDED are identified as the most relevant attributes by BLEM2. In contrast, API CALLS is selected as the root of the decision tree in Figure 2. It is also clear that URL REFERENCES is not relevant in classifying benign software. It seems that both FILE ACCESS and URL REFERENCES may not be essential due to their low frequency; however, they cannot be ignored, since they have high support and certainty factors according to rules shown in Table 4 and 5. In general, we will need to consider the Bayesian factors of BLEM2 rules to derive a threshold for identifying essential attributes. This will be part of our future works.

Table 3. Attribute-value pair frequencies in BLEM2 rules with minimum support = 7

Attribute	Attribute Name	Freq. of YES (21 rules total)	Freq. of NO (13 rules total
C1	PACKER	17	12
C2	REGISTRY KEYS ADDED	20	13
C3	REGISTRY KEYS DELETED	9	10
C4	REGISTRY VALUES MODIFIED	19	12
C5	API CALLS	20	13
C6	DLLs	11	4
C7	FILE ACCESS	2	1
C8	DIRECTORY ACCESS	8	3
C9	INTERNET ACCESS	11	13
C10	URL REFERENCES	2	0
C11	UNIQUE STRINGS	16	12

Table 4. BLEM2 rules for DDF for the decision label "YES" with minimum support = 7

C1	C2	C3	C4	C5	C6	C7	C8	C9	C10	C 11	Supp	Cer	Stren	Cov
0	(-inf-1]	?	(-inf-12.5]	(5.5-23.5]	(16.5-inf)	?	1	0	?	0	43	0.811	0.039	0.074
0	(2.5-inf)	0	(-inf-12.5]	(5.5-23.5]	(16.5-inf)	?	?	?	?	0	30	0.909	0.027	0.052
?	?	?	?	?	?	0	?	?	?	?	26	1	0.024	0.045
1	(2.5-inf)	0	(12.5-inf)	(5.5-23.5]	(16.5-inf)	?	1	0	?	?	21	0.553	0.019	0.036
0	(-inf-1]	?	(12.5-inf)	(5.5-23.5]	(16.5-inf)	?	1	?	?	0	20	1	0.018	0.034
0	(-inf-1]	?	(-inf-12.5]	(5.5-23.5]	?	?	1	0	?	1	20	0.769	0.018	0.034
1	(2.5-inf)	0	(-inf-12.5]	(5.5-23.5]	(16.5-inf)	?	?	0	?	0	17	0.654	0.015	0.029
0	(-inf-1]	?	?	(5.5-23.5]	?	?	0	?	?	?	16	1	0.015	0.028
1	(-inf-1]	0	(-inf-12.5]	(5.5-23.5]	(16.5-inf)	1	1	0	?	0	15	0.455	0.014	0.026
?	(-inf-1]	?	(-inf-12.5]	(-inf-5.5]	(-inf-16.5]	?	1	?	?	0	14	1	0.013	0.024
0	(2.5-inf)	?	(-inf-12.5]	(5.5-23.5]	?	?	?	0	?	1	13	0.542	0.012	0.022
1	(2.5-inf)	0	(-inf-12.5]	(5.5-23.5]	(16.5-inf)	?	?	0	?	1	13	0.236	0.012	0.022
0	(2.5-inf)	?	(12.5-inf)	(5.5-23.5]	?	?	?	?	?	0	12	1	0.011	0.021
0	(2.5-inf)	0	(12.5-inf)	(5.5-23.5]	?	?	?	?	?	1	11	0.917	0.01	0.019
0	(-inf-1]	?	(12.5-inf)	(41.5-inf)	?	?	?	?	?	?	10	1	0.009	0.017
0	(-inf-1]	0	(12.5-inf)	(5.5-23.5]	(16.5-inf)	?	1	?	?	1	10	0.769	0.009	0.017
1	(-inf-1]	0	(-inf-12.5]	(5.5-23.5]	(16.5-inf)	?	?	0	?	1	10	0.2	0.009	0.017
1	(-inf-1]	0	(12.5-inf)	(5.5-23.5]	(16.5-inf)	?	?	0	?	0	9	0.9	0.008	0.016
1	(inf 1]	?	(12.5-inf)	(-inf-5.5]	?	?	?	?	?	1	7	1	0.006	0.012
?	(2.5-inf)	?	(-inf-12.5]	(5.5-23.5]	?	?	?	1	0	0	7	1	0.006	0.012
?	(-inf-1]	?	(-inf-12.5]	(5.5-23.5]	?	?	?	1	0	?	7	1	0.006	0.012

Table 5. BLEM2 rules for DDF for the decision label "NO" with minimum support = 7

C1	C2	C3	C4	C5	C6	C7	C8	C9	C10	C 11	Supp	Cer	Stren	Cov
1	(2.5-inf)	0	(-inf-12.5]	(23.5-41.5]	?	?	1	0	?	1	94	0.99	0.085	0.18
1	(1-2.5]	0	(-inf-12.5]	(23.5-41.5]	?	?	?	0	?	1	52	0.981	0.047	0.1
1	(2.5-inf)	0	(-inf-12.5]	(5.5-23.5]	(16.5-inf)	?	?	0	?	1	42	0.764	0.038	0.081
1	(-inf-1]	0	(-inf-12.5]	(5.5-23.5]	(16.5-inf)	?	?	0	?	1	40	0.8	0.036	0.077
?	(-inf-1]	0	(-inf-12.5]	(5.5-23.5]	(16.5-inf)	1	1	0	?	0	28	0.326	0.025	0.054
1	(2.5-inf)	0	(12.5-inf)	(23.5-41.5]	?	?	?	0	?	1	27	0.9	0.025	0.052
1	(2.5-inf)	0	?	(23.5-41.5]	?	?	1	0	?	0	27	0.9	0.025	0.052
1	(2.5-inf)	0	(12.5-inf)	(5.5-23.5]	?	?	1	0	?	1	14	0.667	0.013	0.027
0	(2.5-inf)	?	(-inf-12.5]	(5.5-23.5]	?	?	?	0	?	1	11	0.458	0.01	0.021
1	(-inf-1]	?	(-inf-12.5]	(23.5-41.5]	?	?	?	0	?	?	9	1	0.008	0.017
1	(2.5-inf)	0	(-inf-12.5]	(5.5-23.5]	(16.5-inf)	?	?	0	?	0	9	0.346	0.008	0.017
1	(1-2.5]	0	(-inf-12.5]	(23.5-41.5]	?	?	?	0	?	0	7	1	0.006	0.013
0	(2.5-inf)	?	(-inf-12.5]	(23.5-41.5]	?	?	?	0	?	1	7	1	0.006	0.013

In the experiment with the DAF data set, out of 141 attributes, 31 attributes were selected by J48 in the decision tree. The API Call "IsDebuggerPresent" is used as the root node in the tree. The most used attributes in the classification model are IsDebuggerPresent, WriteProcessMemory, RegSetValuesExW, GetVolumeInformationW, bind, CreateProcessW and Connect. While IsDebuggerPresent may seem surprising, it is actually very revealing. Malware writers often include code to check for the presence of a debugger during execution in order to detect reverse engineering attempts on their software. At that point they terminate the program in order to make analysis of their code more difficult.

The performance of the decision tree classifiers is shown in Table 6 where TP, TN, FP, FN denote True Positive, True Negative, False Positive, and False Negative rates respectively.

Table 6. Performance of J48 decision trees

Data	TP	TN	FP	FN	ROC Area	Overall Accuracy
DRF	0.793	0.809	0.191	0.207	0.843	80.09%
DDF	0.847	0.782	0.218	0.153	0.815	81.45%
DAF	0.832	0.947	0.053	0.168	0.917	89.14%

The Naïve Bayes algorithm of WEKA was applied to the same data, and the performance is shown in Table 7.

Table 7. Performance of Naïve Bayes

Data	TP	TN	FP	FN	ROC Area	Overall Accuracy
DRF	0.856	0.773	0.227	0.144	0.889	81.45%
DDF	0.865	0.836	0.164	0.135	0.912	85.07%
DAF	0.701	0.947	0.053	0.299	0.921	82.81%

From Table 6 and 7, it shows that Naïve Bayes performs slightly better than J48 tree classifier in experiments with DRF and DDF data sets, which has 15 attributes. In experiments with the DAF data, which has 141 attributes, the feature selection algorithm used in J48 provides an advantage. From our results, it is clear that discretization will improve performance for reverse engineered features as shown in both J48 and NB. In addition, feature reduction is essential in API CALLS data as indicated by the better performance of J48 over NB.

We did not evaluate the performance of BLEM2 rules in current experiments. Currently, an inference engine based on BLEM2 is under development. It will be interesting to see how the feature reduction of BLEM2 with Bayesian factors compares to J48 and NB.

5   Conclusions

In this work, the problem of detecting new and unknown malware is addressed. Present day technologies and our approach for the detection of malware are discussed. An isolated environment was set up for the process of reverse engineering and each executable was reversed rigorously to find its properties and behavior. On the data extracted from the reversing process, different data mining techniques were used to procure patterns of malicious executables and thereby classification models were generated. To test the models, new executables were supplied from the wild with the same set of features. The results thus obtained proved to be satisfactory. BLEM2 rules seem to provide insightful information for essential features identification and for developing inference engines based on Bayesian factors associated with the rules. One of our future works is to apply rough set feature reduction tools and BLEM2 inference engines to evaluate their performance in malware detection.

From analyzing the experimental results, we can conclude that finding static and behavioral features of each malware through reverse engineering and applying data mining techniques to the data helps in detecting new generation malware. Considering the rapidly increasing amount of malware appearing each day, this method of detection can be used along with current practice detection techniques.

We have reversed each strain of malware and benign executables to extract all the features we could with the help of the tools used by the computer security profession. However, we were not able to analyze the process address space of the executables in the physical memory as the memory analysis tools were released after we completed the reversing step. Analyzing the address space would reveal more interesting information about the processes and thereby analyzing their behavior more accurately.

Reversing each malware manually is a time consuming process and requires much effort with the thousands of new malware being generated. One way to cope up with this problem is to automate the whole reverse engineering process. Although there are some tools for automated reverse engineering, they do not record the full details of malware. A more specific tool that does rigorous reversing would help in combating large amounts of malware. We consider these two tasks as the future work that aid in detecting new malware more efficiently.

References

1. Ahmed, F., Hameed, H., Shafiq, M.Z., Farooq, M.: Using spatio-temporal information in API calls with machine learning algorithms for malware detection. In: AISec 2009: Proceedings of the 2nd ACM Workshop on Security and Artificial Intelligence, pp. 55–62. ACM, New York (2009)
2. Burji, S., Liszka, K.J., Chan, C.-C.: Malware Analysis Using Reverse Engineering and Data Mining Tools. In: The 2010 International Conference on System Science and Engineering (ICSSE 2010), pp. 619–624 (July 2010)
3. Chan, C.-C., Santhosh, S.: BLEM2: Learning Bayes' rules from examples using rough sets. In: Proc. NAFIPS 2003, 22nd Int. Conf. of the North American Fuzzy Information Processing Society, Chicago, Illinois, July 24-26, pp. 187–190 (2003)
4. Chan, C.-C., Grzymala-Busse, J.W.: On the two local inductive algorithms: PRISM, and LEM2. Foundations of Computing and Decision Sciences 19(3), 185–203 (1994)
5. Christodorescu, M., Jha, S., Kruegel, C.: Mining specifications of malicious behaviour. In: Proc. ESEC/FS 2007, pp. 5–14 (2007)
6. Cohen, F.: Computer Viruses. PhD thesis, University of Southern California (1985)
7. Cohen, W.: Learning Trees and Rules with Set-Valued Features. American Association for Artificial Intelligence, AMI (1996)
8. Islam, R., Tian, R., Batten, L., Versteeg, S.C.: Classification of Malware Based on String and Function Feature Selection. In: 2010 Second Cybercrime and Trustworthy Computing Workshop, Ballarat, Victoria Australia, July 19-July 20 (2010) ISBN: 978-0-7695-4186-0
9. Kang, M.G., Poosankam, P., Yin, H.: Renovo: A hidden code extractor for packed executables. In: Proc. Fifth ACM Workshop on Recurring Malcode, WORM 2007 (November 2007)
10. Kolter, J., Maloof, M.: Learning to detect malicious executables in the wild. In: Proc. KDD 2004, pp. 470–478 (2004)
11. Komashinskiy, D., Kotenko, I.V.: Malware Detection by Data Mining Techniques Based on Positionally Dependent Features. In: Proceedings of the 2010 18th Euromicro Conference on Parallel, Distributed and Network-based Processing, PDP 2010. IEEE Computer Society, Washington, DC (2010) ISBN: 978-0-7695-3939-3
12. Mcafee.com (2010a), http://www.mcafee.com/us/resources/reports/rp-quarterly-threat-q3-2010.pdf (retrieved)
13. Mcafee.com (2010b), http://www.mcafee.com/us/resources/reports/rp-good-decade-for-cybercrime.pdf (retrieved)
14. Messagelabs.com (2011), http://www.messagelabs.com/mlireport/MLI_2011_01_January_Final_en-us.pdf (retrieved)
15. Miller, P.: Hexdump. Online publication (2000), http://www.pcug.org.au/millerp/hexdump.html
16. Pawlak, Z.: Rough sets: basic notion. International Journal of Computer and Information Science 11(15), 344–356 (1982)
17. Pawlak, Z.: Flow graphs and intelligent data analysis. Fundamenta Informaticae 64, 369–377 (2005)
18. Rozinov, K.: Reverse Code Engineering: An In-Depth Analysis of the Bagle Virus. In: Information Assurance Workshop, IAW 2005. Proceedings from the Sixth Annual IEEE SMC, June 15-17, pp. 380–387 (2005)

19. Schultz, M.G., Eskin, E., Zadok, E., Stolfo, S.J.: Data Mining Methods for Detection of New Malicious Executables. In: Proceedings of the 2001 IEEE Symposium on Security and Privacy, pp. 38–49. IEEE Computer Society (2001)
20. Skoudis, E.: Malware: Fighting Malicious Code. Prentice Hall (2004)
21. Sung, A., Xu, J., Chavez, P., Mukkamala, S.: Static analyzer of vicious executables (save). In: Proc. 20th Annu. Comput. Security Appl. Conf., pp. 326–334 (2004)
22. Wang, T.-Y., Wu, C.-H., Hsieh, C.-C.: A Virus Prevention Model Based on Static Analysis and Data Mining Methods. In: Proceedings of the 2008 IEEE 8th International Conference on Computer and Information Technology Workshops, CITWORKSHOPS 2008, pp. 288–293 (2008)
23. Wang, T.-Y., Wu, C.-H., Hsieh, C.-C.: Detecting Unknown Malicious Executables Using Portable Executable Headers. In: Fifth International Joint Conference on INC, IMS and IDC, pp. 278–284 (2009)
24. Witten, I.H., Frank, E.: Data Mining: Practical Machine Learning Tools and Techniques, 2nd edn. (2005) ISBN: 0-12-088407-0

Improving Text Retrieval Accuracy by Using a Minimal Relevance Feedback

Francesco Colace[1], Massimo De Santo[1], Luca Greco[1], and Paolo Napoletano[2,*]

[1] Department of Electronic Engineering and Computer Engineering
University of Salerno, 84084 Fisciano, Italy
{fcolace,desanto,lgreco}@unisa.it
[2] DISCo (Department of Informatics, Systems and Communication)
University of Milan, Bicocca Viale Sarca 336
20126 Milan, Italy
napoletano@disco.unimib.it

Abstract. In this paper we have demonstrated that the accuracy of a text retrieval system can be improved if we employ a query expansion method based on explicit relevance feedback that expands the initial query with a structured representation instead of a simple list of words. This representation, named a mixed *Graph of Terms*, is composed of a directed and an a-directed subgraph and can be automatically extracted from a set of documents using a method for *term extraction* based on the *probabilistic Topic Model*. The evaluation of the method has been conducted on a web repository collected by crawling a huge number of web pages from the website ThomasNet.com. We have considered several topics and performed a comparison with a baseline and a less complex structure that is a simple list of words.

Keywords: Text Retrieval, Query Expansion, Probabilistic Topic Model.

1 Introduction

The widespread use of digital technologies in all aspects of daily life has improved knowledge about the behavior of the individual entities involved in a complex system. This has increased both conscious and unconscious collaborative modes of information/knowledge sharing/exchange: consider information systems like Amazon, e-bay, Twitter, Facebook, Wikis, e-marketplaces, Myspace, blogs and so on.

As a consequence, Intelligent Systems have been introduced to assist and augment this natural social process and so help people sift through available books, articles, web pages, movies, music, restaurants, jokes, grocery products, and so forth to find the most interesting and valuable information for them. All the existing intelligent systems are based on data mining methods which comprise also collaborative filtering and text mining techniques. These methods are either memory-based, model-based, content-based or hybrids. While the memory and model-based methods make use of the records contained in structured data (User X is quite interested inx product Y) to make predictions, the content-based methods analyze the content of textual information to match and find patterns. Leaving aside the memory and model-based methods, we focus only on the

---

* Corresponding author.

A. Fred et al. (Eds.): IC3K 2011, CCIS 348, pp. 126–140, 2013.
© Springer-Verlag Berlin Heidelberg 2013

content-based ones that, thanks to the massive use of the reviewing of the items activity by people, are becoming of great interest.

The content analysis is possible thanks to the findings obtained in the fields of text mining, text classification, text categorization as well as of sentiment analysis and detection, thus exploiting all the text retrieval theories. In the field of text retrieval the main problem is: "How can a computer tell which documents are relevant to the query, and, more importantly, which results are more relevant than others?"

There is of course no definitive answer, and all the existing approaches to solve this problem consider a different Information Retrieval model to represent a document in the document collection. We can divide all the existing methods into several categories: set-theoretic (including boolean) models, algebraic models and probabilistic models [9][1]. Although each method has its own properties, there is a common denominator: the *bag of words* approach to document representation.

The "bag of words" assumption claims that a document can be considered as a feature vector where each element in the vector indicates the presence (or absence) of a word, so that the information on the position of that word within the document is completely lost [9].

The elements of the vector can be weights and computed in different ways, for instance *BM25*, *tf-idf*, etc., thus a document can be considered as a list of weighted features (that are words). A query is considered as a document and so it is represented as a vector of weighted words.

The *term frequency-inverse document (tf-idf)* model is a weighting model used to give weights to the terms in a document collection by measuring how often a term is found within a document (*term frequency*), offset by how often the term is found within the entire collection (*inverse document frequency*).

In this paper we argue that a vector of weighted words, due to the inherent ambiguity of language (polysemy etc.), is not capable of discriminating between documents in the case of ad-hoc text retrieval tasks. Here the aim is to find the documents that best match the performed query (that is a topic). The ambiguity, in fact, can be reduced if we give more importance to words that convey concepts and that contribute to specify a topic, and if we assign less importance to those words that contribute to specify concepts and that, due to the fact that they can be more plausibly shared between concepts, can increase the ambiguity. This leads to a hierarchical structure that we call a mixed *Graph of Terms* and that can be automatically extracted from a set of documents using a global method for term extraction based on the Latent Dirichlet Allocation model implemented as the Probabilistic Topic Model.

We have employed the mixed *Graph of Terms* in a query expansion method based on explicit relevance feedback that expands the initial query with this new structured query representation. The evaluation of the method has been conducted on a web repository collected by crawling a huge number of web pages from the website ThomasNet.com. We have considered several topics and performed a comparison with a less complex structure that is a simple list of words. The results obtained, independently of the context, show that a more complex representation is capable of retrieving a greater number of relevant documents achieving a mean average precision of about 50%.

2   Query Expansion Techniques

It is well documented that the query length in typical information retrieval systems is rather short (usually two or three words [16], [15]) which may not be long enough to avoid the inherent ambiguity of language (polysemy etc.), and which makes text retrieval systems, that rely on a term-frequency based index, suffer generally from low precision, or low quality of document retrieval.

In turn, the idea of taking advantage of additional knowledge, by expanding the original query with other topic-related terms, to retrieve relevant documents has been largely discussed in the literature, where manual, interactive and automatic techniques have been proposed [12][9][1]. The idea behind these techniques is that, in order to avoid ambiguity, it may be sufficient to better specify "the meaning" of what the user has in mind when performing a search, or in other words "the main concept" (or a set of concepts) of the preferred topic in which the user is interested. A better specialization of the query can be obtained with additional knowledge, that can be extracted from *exogenous* (e.g. ontology, WordNet, data mining) or *endogenous* knowledge (i.e. extracted only from the documents contained in the repository) [2,22,9].

In this paper we focus on those techniques which make use of the "Relevance Feedback" (in the case of endogenous knowledge) which takes into account the results that are initially returned from a given query and so uses the information about the relevance of each result to perform a new expanded query. In the literature we can distinguish between three types of procedures for the assignment of the relevance: explicit feedback, implicit feedback, and pseudo feedback [1]. The feedback is obtained from assessors (or other users of a system) indicating the relevance of a document retrieved for a query. If the assessors know that the feedback provided is interpreted as relevance judgments then the feedback is considered as explicit; otherwise it is implicit. On the contrary, the pseudo relevance feedback automates the manual part of the relevance labeling by assuming that the top "n" ranked documents after the initial query are relevant and so finally performing relevance feedback as before under this assumption.

Most existing methods, due to the fact that the human labeling task is enormously annoying and time consuming [17,25], make use of pseudo relevance feedback. Nevertheless, fully automatic methods suffer from obvious errors when the initial query is intrinsically ambiguous. As a consequence, in recent years, some hybrid techniques have been developed which take into account a minimal explicit human feedback [21,11] and use it to automatically identify other topic related documents. The performance achieved by these methods is usually medium with a mean average precision of about 30% [21].

However, whatever the technique that selects the set of documents representing the feedback, the expanded terms are usually computed by making use of well known approaches for term selection such as Rocchio, Robertson, CHI-Square, Kullback-Lieber etc [23][7]. In this case the reformulated query consists in a simple (sometimes weighted) list of words.

Although such term selection methods have proven their effectiveness in terms of accuracy and computational cost, several more complex alternative methods have been proposed. In this case, they usually consider the extraction of a structured set of words so that the related expanded query is no longer a list of words, but a weighted set of clauses combined with suitable operators [5], [10], [18].

3   The Proposed Approach

The *vector of features* needed to expand the query is obtained as a result of an interactive process between the user and system. The user initially performs a retrieval on the dataset $\mathcal{D}$ by inputting a query to the system and later identifies a small set Ω_r of relevant documents from the hit list of documents returned by the system, that is considered as the training set $\Omega_r = \{\mathbf{d}_1, \ldots, \mathbf{d}_{|\Omega|}\} \subset \mathcal{D}$ (the relevance feedback).

Existing query expansion techniques mostly use the relevance feedback of both relevant and irrelevant documents. Usually they obtain the term selection through the scoring function proposed in [24], [7] which assigns a weight to each term depending on its occurrence in both relevant and irrelevant documents. Differently, in this paper we do not consider irrelevant documents.

Precisely, the *vector of features*, that we call the mixed *Graph of Terms*, can be automatically extracted from a set of documents Ω_r using a method for *term extraction* based on the *Latent Dirichlet Allocation* model [4] implemented as the *Probabilistic Topic Model* [13].

The general idea of this paper is supported by previous works [20] that have confirmed the potential of supervised clustering methods for term extraction, also in the case of query expansion [6,19].

3.1   Data Preparation

Texts can not be directly interpreted by a search engine and for this reason, an indexing procedure that maps a text into a compact representation of its content must be uniformly applied to the entire corpus and to the training set. Let us consider the case of Ω_r.

Each document can be represented, following the *Vector Space Model* [9], as a vector of term *weights*

$$\mathbf{d}_m = \{w_{1m}, \ldots, w_{|\mathcal{T}|m}\},$$

where $\mathcal{T}$ is the set of *terms* (also called *features*) that occur at least once in at least one document of Ω_r, and $0 \leq w_{nm} \leq 1$ represents how much a term t_n contributes to a semantics of document $\mathbf{d}_m$.

If we choose to identify terms with words, we have the *bag of words* assumption, that is $t_n = v_n$, where v_n is one of the words of a vocabulary. The *bag of words* assumption claims that each w_{nm} indicates the presence (or absence) of a word, so that the information on the position of that word within the document is completely lost [9].

To determine the weight w_{nm} of term t_n in a document $\mathbf{d}_m$, the standard tf-idf (*term frequency-inverse document frequency*) function can be used [26], defined as:

$$\text{tf-idf}(t_n, \mathbf{d}_m) = N(t_n, \mathbf{d}_m) \cdot \log \frac{|\Omega_r|}{N_{\Omega_r(t_n)}} \tag{1}$$

where $N(t_n, \mathbf{d}_m)$ denotes the number of times t_n occurs in $\mathbf{d}_m$, and $N_{\Omega_r(t_n)}$ denotes the document frequency of term t_n, i.e. the number of documents in Ω_r in which t_n occurs.

In order for the weights to fall within $[0, 1]$ interval and for the documents to be represented by vectors of equal length, the weights resulting from tf-idf are usually normalized by cosine normalization, given by:

$$w_{nm} = \frac{\text{tf-idf}(t_n, \mathbf{d}_m)}{\sqrt{\sum_{n=1}^{|\mathcal{T}|} (\text{tf-idf}(t_n, \mathbf{d}_m))^2}} \tag{2}$$

In this paper, before indexing, we have performed the removal of function words (i.e. topic-neutral words such as articles, prepositions, conjunctions, etc.) and we have performed the stemming procedure[1] (i.e. grouping words that share the same morphological root).

Once the indexing procedure has been performed, we have a matrix $|\mathcal{T}| \times |\Omega_r|$ of real values instead of the training set Ω_r. The same procedure is applied to the entire corpus $\mathcal{D}$.

3.2   A Mixed Graph of Terms

In this paper we have used a *global* method for *feature transformation* that considers pairs of words instead of single words as basic features thus obtaining a new space $\mathcal{T}_p$ of features. The dimensionality of such a new space is very high, much higher than $|\mathcal{T}|$, in fact: $|\mathcal{T}_p| \propto |\mathcal{T}|^2$. For this reason we need to reduce the transformed space in order to obtain a new space $\mathcal{T}_{sp}$ such that $|\mathcal{T}_{sp}| \ll |\mathcal{T}_p|$.

The method used to select the most representative pairs of words is based on the *Latent Dirichlet Allocation* model [4] implemented as the *Probabilistic Topic Model* [13] and this is the core of a new representation, named the mixed *Graph of Terms*, that consists of related pairs of words. The graph contains two kinds of relations between words, directed and undirected, and for this reason it is called *mixed*.

In the graph we can find several clusters of words and each cluster contains a set of words v_s that specifies, through a directed weighted edge, a special word, that we have named the *concept*, r_i, that is the centroid of such a cluster. The weight ρ_{is} can measure how far a word is related to a concept, or how much we need such a word to specify that concept, and it can be considered as a probability: $\rho_{is} = P(r_i|v_s)$. The resulting structure is a subgraph rooted on r_i (see fig. 1(a)).

Moreover, special words, namely *concepts*, can be linked together through undirected weighted edges, so forming a subgraph of pairs of centroids. The weight ψ_{ij} can be considered as the degree of semantic correlation between two concepts and it can be considered as a probability: $\psi_{ij} = P(r_i, r_j)$ (see fig. 1(a)).

Considering that each concept is a special word, we can say that the graph contains directed and undirected pairs of features that are all lexically denoted as words. Given the training set Ω_r of documents, the proposed method, through a learning procedure, selects a subset of pairs obtaining a number of pairs $|\mathcal{T}_{sp}| \ll |\mathcal{T}_p|$. In this way, the term extraction procedure is obtained by firstly computing all the semantic relatednesses

---

[1] Stemming has sometimes been reported to hurt effectiveness. However the recent tendency has been to adopt it, as it reduces both the dimensionality of the feature space and the stochastic dependence between terms.

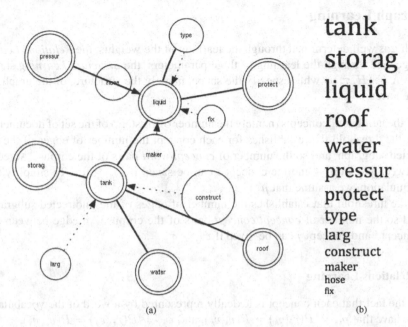

tank
storag
liquid
roof
water
pressur
protect
type
larg
construct
maker
hose
fix

(a)   (b)

Fig. 1. *Vector of features* for the topic *Storage Tanks*. 1(a) A mixed *Graph of Terms*. 1(b) A *List of Terms*.

between words and concepts, that is ρ_{is} and ψ_{ij}, and secondly selecting the right subset of pairs from all the possible ones. Before explaining in detail the learning procedure of a graph, we would like to highlight some aspects of this representation.

3.3   Graph and Document Representation in the Space $\mathcal{T}_{sp}$

A mixed *Graph of Terms* g can be viewed, following the *Vector Space Model* [9], as a vector of features t_n:

$$g = \{b_1, \ldots, b_{|\mathcal{T}_{sp}|}\},$$

where $|\mathcal{T}_{sp}|$ represents the number of pairs and each feature $t_n = (v_i, v_j)$ can be a *word/concept* or *concept/concept* pair. The weight b_n is named the *boost* factor and is equal to ψ_{ij} for both *word/concept* or *concept/concept* pairs.

Moreover, by following this approach, also each document of a corpus $\mathcal{D}$ can be represented in terms of pairs:

$$d_m = (w_{1m}, \ldots, w_{|\mathcal{T}_{sp}|m}),$$

where w_{nm} is such that $0 \leq w_{nm} \leq 1$ and represents how much term $t_n = (v_i, v_j)$ contributes to a semantics of document d_m. The weight is calculated thanks to the tf-idf model applied to the pairs represented through t_n:

$$w_{nm} = \frac{\text{tf-idf}(t_n, d_m)}{\sqrt{\sum_{n=1}^{|\mathcal{T}_{sp}|} (\text{tf-idf}(t_n, d_m))^2}} \tag{3}$$

4 Graph Learning

A graph $\mathbf{g}$ is well determined through the learning of the weights, the *Relations Learning* stage, and through the learning of three parameters, the *Structure Learning* stage, that are $\Lambda = (H, \tau, \mu)$ which specify the shape, namely the structure, of the graph. In fact, we have:

1. H: the number of concepts (namely the number of clusters) of the set of documents;
2. μ_i: the threshold that establishes for each concept the number of edges of the directed subgraph, and so the number of *concept/word* pairs of the corpus . An edge between the word s and the concept i can be saved if $\rho_{is} \geq \mu_i$. To simplify the formulation, we assume that $\mu_i = \mu, \forall i$;
3. τ: the threshold that establishes the number of edges of the undirected subgraph, and so the number of *concept/concept* pairs of the corpus. An edge between the concept i and concept j can be saved if $\psi_{ij} \geq \tau$.

4.1 Relations Learning

Due to the fact that each concept is lexically represented by a word of the vocabulary, then we have that $\rho_{is} = P(r_i|v_s) = P(v_i|v_s)$, and $\psi_{ij} = P(r_i, r_j) = P(v_i, v_j)$.

Considering that $P(v_i, v_j) = P(v_i|v_j)P(v_j)$, it is necessary, to learn all the relations between words, to compute the joint, or the conditional, probability $\forall i, j \in \{1, \cdots, |\mathcal{T}|\}$ and each $P(v_j) \forall j$.

We show here that the exact calculation of $P(v_j)$ and the approximation of the joint, or conditional, probability can be obtained through a smoothed version of the generative model introduced in [4] called Latent Dirichlet Allocation (LDA), which makes use of Gibbs sampling [13].

The original theory introduced in [13] mainly asserts a semantic representation in which documents are represented in terms of a set of probabilistic topics z. Formally, we consider a word u_m of the document $\mathbf{d}_m$ as a random variable on the vocabulary $\mathcal{T}$ and z as a random variable representing a topic between $\{1, \cdots, K\}$. The probability distribution of a word within a document $\mathbf{d}_m$ of the corpus can be obtained as:

$$P(u_m) = \sum_{k=1}^{K} P(u_m|z = k, \beta_k)P(z = k|\theta_m). \tag{4}$$

The generation of a document $\mathbf{d}_m$ can be obtained considering the generation of each word of the document. To obtain a word, the model considers three parameters assigned: α, η and the number of topics K. Given these parameters, the model chooses θ_m through $P(\theta|\alpha) \sim \text{Dirichlet}(\alpha)$, the topic k through $P(z|\theta_m) \sim \text{Multinomial}(\theta_m)$ and $\beta_k \sim \text{Dirichlet}(\eta)$. Finally, the distribution of each word given a topic is $P(u_m|z, \beta_z) \sim \text{Multinomial}(\beta_z)$.

As we have already discussed, we have used a smoothed version of Latent Dirichlet Allocation (LDA), which makes use of Gibbs sampling. The results obtained by performing this algorithm on a set of documents Ωr are two matrixes:

1. the *words-topics* matrix that contains $|\mathcal{T}| \times K$ elements representing the probability that a word v_i of the vocabulary is assigned to topic k: $P(u = v_i | z = k, \beta_k)$;
2. the *topics-documents* matrix that contains $K \times |\Omega r|$ elements representing the probability that a topic k is assigned to some word token within a document $\mathbf{d}_m$: $P(z = k | \theta_m)$.

In the same way, the joint probability between two words u_m and y_m of a document $\mathbf{d}_m$ of the corpus can be obtained by assuming that each pair of words is represented in terms of a set of topics z and then:

$$P(u_m, y_m) = \sum_{k=1}^{K} P(u_m, y_m | z = k, \beta_k) P(z = k | \theta_m) \qquad (5)$$

Note that the exact calculation of Eq. 5 depends on the exact calculation of $P(u_m, y_m | z = k, \beta_k)$ that can not be directly obtained through LDA. For this reason, we have introduced an approximation that considers words in a document as conditionally independent given a topic. In this way Eq. 5 can be written as:

$$P(u_m, y_m) \simeq \sum_{k=1}^{K} P(u_m | z = k, \beta_k) P(y_m | z = k, \beta_k) P(z = k | \theta_m). \qquad (6)$$

Note that Eq. 4 gives the probability distribution of a word u_m within a document $\mathbf{d}_m$ of the corpus. To obtain the probability distribution of a word u independently of the document we need to sum over the entire corpus:

$$P(u) = \sum_{m=1}^{M} P(u_m) \delta_m \qquad (7)$$

where δ_m is the prior probability for each document ($\sum_{m=1}^{|\Omega r|} \delta_m = 1$).

In the same way, if we consider the joint probability distribution of two words u and y, we obtain:

$$P(u, y) = \sum_{m=1}^{M} P(u_m, y_v) \delta_m \qquad (8)$$

Concluding, once we have $P(u)$ and $P(u, y)$ we can compute $P(v_i) = P(u = v_i)$ and $P(v_i, v_j) = P(u = v_i, y = v_j), \forall i, j \in \{1, \cdots, |\mathcal{T}|\}$ and so the relations learning can be totally accomplished.

4.2 Structure Learning

Given a set of documents, once each ψ_{ij} and ρ_{is} is known $\forall i, j, s$, letting the parameters $\Lambda_t = (H, \tau, \mu)_t$ assume a different set of values, we can observe a different structure of the graph $\mathbf{g}_t$ (here t is representative of different parameter values).

A way to learn the structure of the graph is to use an optimization based algorithm that searches for the best set of parameters Λ_t. In this case we need a scoring function and a searching strategy [3].

As we have previously seen, a g_t is a vector of features $g_t = \{b_{1t}, \ldots, b_{|\mathcal{T}_{sp}|t}\}$ in the space $\mathcal{T}_{sp}$ and each document of the training set Ω_r, as well as the documents of the corpus $\mathcal{D}$, can be represented as a vector $d_m = (w_{1m}, \ldots, w_{|\mathcal{T}_{sp}|m})$ in the space $\mathcal{T}_{sp}$. A possible scoring function is the cosine similarity between these two vectors:

$$S(g_t, d_m) = \frac{\sum_{n=1}^{|\mathcal{T}_{sp}|} b_{nt} \cdot w_{nm}}{\sqrt{\sum_{n=1}^{|\mathcal{T}_{sp}|} b_{nt}^2} \cdot \sqrt{\sum_{n=1}^{|\mathcal{T}_{sp}|} w_{nm}^2}} \tag{9}$$

and thus the optimization procedure would consist in searching for the best set of parameters Λ_t such that the cosine similarity is maximized $\forall d_m$.

By following this approach, the best g_t for the set of documents Ω_r is the one that produces the maximum score attainable for each of the documents when the same graph is used as a vector of features to measure the similarity of a set containing just those documents which have fed the graph builder. As a consequence, we obtain a score for each document d_m and then we have

$$S_t = \{S(g_t, d_1), \cdots, S(g_t, d_{|\Omega_r|})\},$$

where each score depends on the specific set $\Lambda_t = (H, \tau, \mu)_t$.

To compute the best value of Λ we can maximize the score value for each document, which means that we are looking for the graph which best describes each document of the repository from which it has been learned. It should be noted that such an optimization maximizes at the same time all $|\Omega_r|$ elements of S_t.

Alternatively, in order to reduce the number of the objectives being optimized, we can at the same time maximize the mean value of the scores and minimize their standard deviation, which turns a multi-objective problem into a two-objective one. Additionally, we can reformulate the latter problem by means of a linear combination of its objectives, thus obtaining a single objective function, i.e., *Fitness* ($\mathcal{F}$), which depends on Λ_t,

$$\mathcal{F}(\Lambda_t) = E[S_t] - \sigma[S_t],$$

where E is the mean value of all the elements of S_t and σ_m is the standard deviation. Summing up, the parameters learning procedure is represented as follows,

$$\Lambda^* = \underset{t}{\operatorname{argmax}}\{\mathcal{F}(\Lambda_t)\}.$$

We will see next how we have performed the searching strategy phase.

Since the space of possible solutions could grow exponentially, we have considered[2] $|\mathcal{T}_{sp}| \leq 100$. Furthermore, we have reduced the remaining space of possible solutions by applying a clustering method, that is the *K-means* algorithm, to all ψ_{ij} and ρ_{is} values, so that the optimum solution can be exactly obtained after the exploration of the entire space.

This reduction allows us to compute a graph from a repository composed of a few documents in a reasonable time (e.g. for 3 documents it takes about 3 seconds with a Mac OS X based computer, 2.66 GHz Intel Core i7 CPU and a 8GB RAM). Otherwise,

---

[2] This number is usually employed in the case of Support Vector Machines.

we would need an algorithm based on a random search procedure in big solution spaces. For instance, Evolutionary Algorithms would be suitable for this purpose, but would provide a slow performance. In fig. 1(a) we can see an example of a graph learned from a set of documents labeled as topic *Storage tanks*.

4.3   Extracting a Simpler Representation from the Graph

From the mixed *Graph of Terms* we can select different subsets of features so obtaining a simpler representation (see fig. 1(b)). Before discussing this in detail, we would recall that $\psi_{ij} = P(v_i, v_j)$ and $\rho_{is} = P(v_i|v_s)$ are computed through the topic model which also computes the probability for each word $\eta_s = P(v_s)$.

We can obtain the simplest representation by selecting from the graph all distinct terms and associating to each of them its weight $\eta_s = P(v_s)$. We name this representation the *List of Terms* (**w**), see fig. 1(b).

4.4   Consideration on the Method

It is important to make clear that the mixed Graph of Terms can not be considered as a co- occurrence matrix. In fact, the core of the graph is the probability $P(v_i, v_j)$, which we regard as a word association problem, that in the topic model is considered as a problem of prediction: given that a cue is presented, which new words might occur next in that context? It means that the model does not take into account the fact that two words occur in the same document, but that they occur in the same document when a specific topic (and so a context) is assigned to that document [13].

Furthermore, in the field of statistical learning, a similar structure has been introduced, named the Hierarchical Mixture of Experts [14]. Such a structure is employed as a method for supervised learning and it is considered as a variant of the well known tree-based methods. The similarity between such a structure and the proposed graph can be obtained by considering the "experts" as "concepts".

Notwithstanding this, the mixed Graph of terms is not a tree structure, and more importantly is not rigid but is dynamically built depending on the optimization stage. Moreover, the Hierarchical Mixture of Experts does not consider relations between experts which is, on the other hand, largely employed in the mixed Graph of Terms. Nevertheless, we will explore further connections between the two methods in future works.

5   Experiments

We have compared 2 different query expansion methodologies based on different *vector of features* with the baseline (b): the mixed *Graph of Terms* (g) and the *List of Terms* (**w**). The baseline (b) is the the *tf-idf* model without expansion of the query. We have embedded all the techniques in an open source text-based search engine, Lucene from the Apache project. Here the score function $S(q, d)$ is based on the standard vector cosine similarity[3] , used in a Vector Space Model combined with the Boolean Model [9]

---

[3] We have used the Lucene version 2.4 and you can find details on the similarity measure at
http://lucene.apache.org

Table 1. An example of a **g** for the topic *Storage Tank*

Conceptual Level		
Concept i	**Concept** j	**boost factor** (b)
tank	roof	1.0
tank	water	0.37
tank	liquid	0.14
...	...	...
liquid	type	0.44
liquid	pressur	0.21
...	...	...
Word Level		
Concept i	**Word** s	**boost factor** (b)
tank	larg	0.15
tank	construct	0.14
...	...	...
liquid	type	0.21
liquid	maker	0.12
liquid	hose	0.06
liquid	fix	0.01
...	...	...

which takes into account the boost factor b_k whose default value is 1, which is assigned to the words that compose the original query. Such a function permits the assignment of a rank to documents **w** that match a query **q** and permits the transforming of each *vector of features*, that is the **g** into a set of Boolean clauses. For instance, in the case of the **g**, since it is represented as pairs of related words, see Table 1, where the relationship strength is described by a real value (namely ψ_{ij} and ρ_{is}, the *Relation factors*), the expanded query is:

$$((tank \text{ AND } roof)^{1.0}) \text{ OR } ((tank \text{ AND } larg)^{0.15})...$$

As a consequence we search the pair of words *tank* AND *roof* with a boost factor of 1.0 OR the pair of words *tank* AND *larg* with a boost factor of 0.15 and so on.

5.1 Data Preparation

The evaluation of the method has been conducted on a web repository collected at the University of Salerno by crawling 154,243 web pages for a total of about 3.0 GB by using the website ThomasNet (http://www.thomasnet.com) as an index of URLs, the reference language being English[4]. ThomasNet, known as the "big green books" and "Thomas Registry", is a multi-volume directory of industrial product information covering 650,000 distributors, manufacturers and service companies within 67,000-plus industrial categories. We have downloaded webpages from the company websites related to 150 categories of products (considered as topics), randomly chosen from the ThomasNet directory.

---

[4] The repository will be public on our website to allow further investigations from other researchers.

Table 2. Number of words and pairs for each g

	♯ of words	♯ of pairs
Average Size	55	72

Table 3. Average values of performance

run	eMAP	eRprec	eP5	eP10	eP20	eP30	eP100
b	0.213	0.432	0.345	0.298	0.201	0.198	0.186
w	0.399	0.457	0.806	0.691	0.661	0.556	0.384
g	0.569	0.601	0.917	0.840	0.784	0.686	0.495

Note that even if the presence or absence of categories in the repository depends on the random choices made during the crawling stage, it could happen that webpages from some business companies cover categories that are different from those randomly chosen. This means that the repository is not to be considered as representative of a low number of categories (that is 150) but as a reasonable collection of hundreds of categories. In this work we have considered 50 test questions (queries) extracted from 50 out of the initial 150 categories (topics). Each original query corresponds to the name of the topic, for instance if we search for information about the topic "generator" therefore the query will be exactly "generator". Obviously, all the initial queries have been expanded through the methodologies explored in section 4.3. Here we show the summary results obtained on all the 50 topics.

5.2   Evaluation Measures

For each example the procedure that obtains the reformulation of the query, is explained as follows. A person, who is interested in the topic "generator", performs the initial query "generator" so interactively choosing 3 relevant documents for that topic, which represent the minimal positive feedback. From those documents the system automatically extracts the two *vectors of features*. In table 2 we show the average size of the list of words and the list of pairs, that is 55 and 72 respectively. The user has interactively assigned the relevance of the documents by following an *xml* based schema coding his intentions and represented as suggested by *TREC* [5].

The expanded queries have been performed again and for each context we have asked different humans to assign graded judgments of relevance to the first 100 pages returned by the system. Due to the fact that the number of evaluations for each topic, and so the number of topics itself, is small, the humans have judged, in contrast to the Minimum Test Collection method [8], all the results obtained. The assessment is based on three levels of relevance, *high relevant*, *relevant* and *not relevant*, assigned, to avoid cases of ambiguity, by following the *xml* based schema coding the user intentions.

The accuracy has been measured through standard indicators provided by [9] and based on *Precision* and *Recall*,

---

[5] The Text Retrieval Conference (TREC).

$$eAP = \frac{1}{ER} \sum_{i=1}^{k} \frac{x_i}{i} + \sum_{j>i} \frac{x_i x_j}{j} \qquad (10)$$

$$ePrec@k = eP@k = \frac{1}{k} \sum_{i=1}^{k} x_i \qquad (11)$$

$$ERprec = \frac{1}{ER} \sum_{i=1}^{ER} x_i \qquad (12)$$

$$ER = \sum_{i=1}^{n} x_i \qquad (13)$$

where eAP indicates the average precision on a topic, x_i and x_j are Boolean indicators of relevance, k is the cardinality of the considered result set (k=100) and ER is a subset of relevant documents[6]. The factor $ERprec$ is the precision at the level ER, while the measure $eMAP$ is the average of all eAPs over topics. The measure $eP@k$ is the precision at level k (for instance $eP5$ is the precision calculated by taking the top 5 results).

In table 3 we find summary results across topics for each *vector of features* and for the baseline (**b**). The overall behavior of the **g** method is better than the **w**.

6   Conclusions

In this work we have demonstrated that a mixed *Graph of Terms* based on a hierarchical representation is capable of retrieving a greater number of relevant documents than a less complex representation based on a list of words, even if the size of the training set is small and composed of only relevant documents.

These results suggest that our approach can be employed in all those text mining tasks that consider matching between patterns represented as textual information and in text categorization tasks as well as in sentiment analysis and detection tasks.

The proposed approach computes the expanded queries considering only endogenous knowledge. It is well known that the use of external knowledge, for instance WordNet, could clearly improve the accuracy of information retrieval systems, but we consider this as a future work.

References

1. Baeza-Yates, R., Ribeiro-Neto, B.: Modern Information Retrieval. ACM Press, New York (1999)
2. Bhogal, J., Macfarlane, A., Smith, P.: A review of ontology based query expansion. Information Processing & Management 43(4), 866–886 (2007)
3. Bishop, C.M.: Pattern Recognition and Machine Learning. Springer (2006)

---

[6] Note that, $ER = |R_{mGT} \cup R_{GT} \cup R_{LT} - R_{mGT} \cap R_{GT} \cap R_{LT}|$, where R_{vf} is the set of relevant and high relevant documents obtained for a given topic and vf=*vector of features*.

4. Blei, D.M., Ng, A.Y., Jordan, M.I.: Latent dirichlet allocation. Journal of Machine Learning Research 3, 993–1022 (2003)
5. Callan, J., Croft, W.B., Harding, S.M.: The inquery retrieval system. In: Proceedings of the Third International Conference on Database and Expert Systems Applications, pp. 78–83. Springer (1992)
6. Cao, G., Nie, J.Y., Gao, J., Robertson, S.: Selecting good expansion terms for pseudo-relevance feedback. In: Proceedings of the 31st Annual International ACM SIGIR Conference on Research and Development in Information Retrieval, SIGIR 2008, pp. 243–250. ACM, New York (2008)
7. Carpineto, C., de Mori, R., Romano, G., Bigi, B.: An information-theoretic approach to automatic query expansion. ACM Trans. Inf. Syst. 19, 1–27 (2001), http://doi.acm.org/10.1145/366836.366860
8. Carterette, B., Allan, J., Sitaraman, R.: Minimal test collections for retrieval evaluation. In: 29th International ACM SIGIR Conference on Research and Development in Information retrieval (2008)
9. Manning, C.D., Raghavan, P., Schütze, H.: Introduction to Information Retrieval. Cambridge University (2008)
10. Collins-Thompson, K., Callan, J.: Query expansion using random walk models. In: Proceedings of the 14th ACM International Conference on Information and Knowledge Management, CIKM 2005, pp. 704–711. ACM, New York (2005), http://doi.acm.org/10.1145/1099554.1099727
11. Dumais, S., Joachims, T., Bharat, K., Weigend, A.: SIGIR 2003 workshop report: implicit measures of user interests and preferences. SIGIR Forum 37(2), 50–54 (2003)
12. Efthimiadis, E.N.: Query expansion. In: Williams, M.E. (ed.) Annual Review of Information Systems and Technology, pp. 121–187 (1996)
13. Griffiths, T.L., Steyvers, M., Tenenbaum, J.B.: Topics in semantic representation. Psychological Review 114(2), 211–244 (2007)
14. Hastie, T., Tibshirani, R., Friedman, J.: The Elements of Statistical Learning. Springer (2009)
15. Jansen, B.J., Booth, D.L., Spink, A.: Determining the informational, navigational, and transactional intent of web queries. Information Processing & Management 44(3), 1251–1266 (2008)
16. Jansen, B.J., Spink, A., Saracevic, T.: Real life, real users, and real needs: a study and analysis of user queries on the web. Information Processing & Management 36(2), 207–227 (2000)
17. Ko, Y., Seo, J.: Text classification from unlabeled documents with bootstrapping and feature projection techniques. Inf. Process. Manage. 45, 70–83 (2009)
18. Lang, H., Metzler, D., Wang, B., Li, J.T.: Improved latent concept expansion using hierarchical markov random fields. In: Proceedings of the 19th ACM International Conference on Information and Knowledge Management, CIKM 2010, pp. 249–258. ACM, New York (2010), http://doi.acm.org/10.1145/1871437.1871473
19. Lee, C.-J., Lin, Y.-C., Chen, R.-C., Cheng, P.-J.: Selecting Effective Terms for Query Formulation. In: Lee, G.G., Song, D., Lin, C.-Y., Aizawa, A., Kuriyama, K., Yoshioka, M., Sakai, T. (eds.) AIRS 2009. LNCS, vol. 5839, pp. 168–180. Springer, Heidelberg (2009)
20. Noam, S., Naftali, T.: The power of word clusters for text classification. In: 23rd European Colloquium on Information Retrieval Research (2001)
21. Okabe, M., Yamada, S.: Semisupervised query expansion with minimal feedback. IEEE Transactions on Knowledge and Data Engineering 19, 1585–1589 (2007)
22. Piao, S., Rea, B., McNaught, J., Ananiadou, S.: Improving Full Text Search with Text Mining Tools. In: Horacek, H., Métais, E., Muñoz, R., Wolska, M. (eds.) NLDB 2009. LNCS, vol. 5723, pp. 301–302. Springer, Heidelberg (2010)

23. Robertson, S.E., Walker, S.: On relevance weights with little relevance information. In: Proceedings of the 20th Annual International ACM SIGIR Conference on Research and Development in Information Retrieval, SIGIR 1997, pp. 16–24. ACM, New York (1997)
24. Robertson, S.E.: On term selection for query expansion. J. Doc. 46, 359–364 (1991)
25. Ruthven, I.: Re-examining the potential effectiveness of interactive query expansion. In: Proceedings of the 26th Annual International ACM SIGIR Conference on Research and Development in Informaion Retrieval, SIGIR 2003, pp. 213–220. ACM, New York (2003), http://doi.acm.org/10.1145/860435.860475
26. Salton, G., McGill, M.J.: Introduction to modern information retrieval. McGraw-Hill (1983)

Using Distant Supervision for Extracting Relations on a Large Scale

Ludovic Jean-Louis, Romaric Besançon, Olivier Ferret, and Adrien Durand

CEA LIST, Vision and Content Engineering Laboratory,
Gif-sur-Yvette, F-91191, France

Abstract. Most of Information Extraction (IE) systems are designed for extracting a restricted number of relations in a specific domain. Recent work about Web-scale knowledge extraction has changed this perspective by introducing large-scale IE systems. Such systems are open-domain and characterized by a large number of relations, which makes traditional approaches such as handcrafting rules or annotating corpora for training statistical classifiers difficult to apply in such context. In this article, we present an IE system based on a weakly supervised method for learning relation patterns. This method extracts without supervision occurrences of relations from a corpus and uses them as examples for learning relation patterns. We also present the results of the application of this system to the data of the 2010 Knowledge Base Population evaluation campaign.

Keywords: Information extraction, Relation extraction, Distant supervision.

1 Introduction

In the domain of information extraction, the objective of relation extraction is to find whether two or more entities are semantically linked and when it is possible, to determine the nature of this link. In the work we present here, we are more specifically interested in extracting relations between named entities for building large-scale knowledge bases. Such building has been recently achieved in the context of Semantic Web by exploiting semi-structured data from open sources of information. One the most representative examples of this trend is the DBpedia project[1] [6], which built a large knowledge base from the semi-structured part of Wikipedia. The next step in this process is to supplement such knowledge bases by exploiting texts, which are a larger but more difficult to deal with source of information, and more particularly, extracting automatically relations between entities from them.

Work on relation extraction can be categorized according to the degree of supervision it requires. At a lower level of supervision, also called unsupervised extraction, the type of the relations to extract is not fixed *a priori*, neither by examples nor by a model. Only constraints about the linked entities, as their type, are set. The type of the extracted relations is defined *a posteriori*, by gathering similar relations. Such approach can be found in [28] or [3] for instance. The opposite approach, called supervised extraction, consists in fixing both the type of the target relations and the means for extracting them

---

[1] http://dbpedia.org/About

A. Fred et al. (Eds.): IC3K 2011, CCIS 348, pp. 141–155, 2013.
© Springer-Verlag Berlin Heidelberg 2013

from texts. It takes the form of either a handcrafted model, typically defined as a set of rules, or a model built by a machine learning algorithm from a set of contextualized relation examples coming from a manually annotated corpus. This second option is mostly represented by statistical machine learning models focusing on taking into account various kinds of features (lexical, syntactic, semantic ...) [33], for instance by finding kernel functions dealing with complex structures such as those produced by syntactic parsers [34].

Between these two extrema, weakly supervised approaches refer to cases where either examples or a model are provided but are not sufficient for developing a fully operational relation extraction system. As a consequence, this initial definition must be extended in an automatic way, generally by exploiting an unannotated corpus. Work in this area shows two main cases, that can be eventually combined, of underspecification of this initial definition:

- underspecification due to the extent of the definition. Only a small set of relation examples or an incomplete model are given;
- underspecification due to the nature of the definition, which occurs when the examples or the model have to be instantiated for being used.

The first case is classically tackled following the methodology of [15] by the means of a bootstrapping mechanism: starting from a model of the target relations made of a restricted set of initial examples or extraction rules, new examples are acquired from a corpus and used for completing the model. This two-step process is re-applied while the model is not developed enough for covering any new example. [1] is a typical application of such methodology in the case of relations between named entities. In a different context – the extraction of qualia structures – [12] is another example of the use of bootstrapping, by combining two types of systems, for learning linguistic patterns.

The second case is represented by the recent notion of "Distant supervision", introduced explicitly by [20] but already present in previous work about bootstrapping. Examples are given in this approach with an underspecified form as they are limited to pairs of entities: they are given both without context and without a linguistic form. The development of such approach is favored by the availability of large knowledge bases extracted from resources such as Wikipedia.

In this article, we present a large-scale information extraction method based on a weakly supervised learning of relation extraction patterns. Moreover, this learning starts from relation examples reduced to pairs of named entities. These pairs are then mapped onto a reference corpus for building the set of contextualized relation examples from which the extraction patterns are learned. This process comes under what we have called above the "Distant supervision" approach. We also present the results of the evaluation of our method in the framework defined by the KBP (Knowledge Based Population) track of the TAC 2010 (Text Analysis Conference) evaluation.

2 Overview

We focus in this work on a large-scale extraction of relations with the hypothesis that a specific knowledge base (KB) already exists. This KB is partially filled with relations

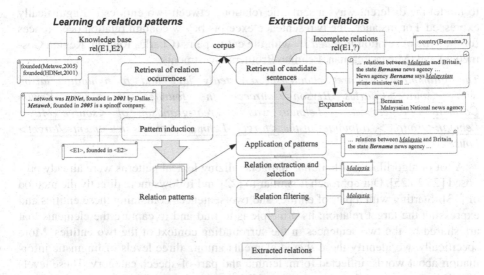

Fig. 1. Overview of the system

that are automatically acquired from semi-structured data. We limit our study to the relations between named entities because we do not focus our work on a specific domain where the entity recognition can be guided by a known terminology and we want to use entities that are usually well recognized. The idea of "large-scale" extraction actually covers several aspects. The first one corresponds to the large number of relation types that are considered, which implies that a rule-based approach based on handcrafted rules is hardly possible. A second aspect is the existence of a large number of existing relations (*i.e.* the association of two entity values with a relation type). These relations give a good starting point for machine learning techniques to learn a model for these types of relations. Finally, a third aspect is the large size of the collection of documents in which the new relations are searched, which implies the use of information retrieval techniques to retrieve good candidates on which a more sophisticated extraction is then performed (we cannot apply patterns for instance on all the sentences of such corpus).

This approach, as presented in Figure 1, is composed of two steps: a first step of *pattern learning* from occurrences of known relations and a step of *relation extraction* for the discovery of new relations. The first step starts with known instances of relations $R(E1,E2)$ to find occurrences of these relations in texts for covering as many different ways of expressing them as possible; then we use these occurrences to learn a set of patterns associated with the target type of relation. The second step starts with incomplete relations $R(E1,x)$, where the source entity $E1$ is known and the target entity x has to be discovered, and searches occurrences of relation R involving $E1$ in a collection of texts. The entity x is then extracted using the patterns learned in the first step. These two steps are described in more details in the following sections.

2.1 Relation Pattern Learning

Our procedure for learning relation patterns relies on the induction of lexical patterns from example sentences containing occurrences of the target relations. Its objective is to model the different ways a semantic relation between two entities is linguistically expressed. For instance, the two sentence excerpts below contain relation occurrences for the type of relation *founded_by* with the entity pairs (Charles Revson, Revlon Cosmetics) and (Mayer Lehman, Lehman Brothers investment):

The glamourous cabaret chanteuse reportedly had had a romantic liaison with <source>Charles Revson</source>, the founder of <target>Revlon Cosmetics</target> ... – Lehman was a great-grandson of <source>Mayer Lehman</source>, a founder of the <target>Lehman Brothers investment</target> house ...

A lot of algorithms for building and generalizing lexical patterns were already proposed [23,26,25]. Our approach is similar to [22] and follows more directly the method of [13]. Starting with a pair of entities and two sentences containing these entities and expressing the target relation, its principle is to find and to capture the elements that are shared by the two sentences in the surrounding context of the two entities. More specifically, we identify these shared elements among three levels of linguistic information about words: inflected form, lemma and part-of-speech category. These levels of information are produced by the OpenNLP[2] tools, also used for named entity recognition. Having these three levels enables the building of more expressive patterns that represent an interesting compromise in terms of generalization between the specificity of lexicalized elements and the more general nature of part-of-speech categories.

The induction of a pattern from two occurrences of relations relies more precisely on the three following steps:

- computation of the minimal edit distance between the two example sentences, that is to say, the minimal number of edit operation (insertion, deletion and substitution) that are necessary to turn one sentence into the other one. All the operations are given here the same weight;
- optimal alignment between the two example sentences from the matrix of distances between subsequences resulting from the computation of the edit distance. The classical algorithm for achieving such alignment is enhanced for enabling a match of two words at one of the three available levels of information when two words are tested for a substitution;
- building of patterns by completing alignments with two wildcard operators when it is necessary: (*s*) stands for 0 or 1 instance of any word while (*g*) represents exactly 1 instance of any word.

Table 1 shows the result of the induction of a pattern for the type of relation *founded_by* from the two sentence excerpts above.

This example illustrates our different levels of generalization: for a word such as *of*, only the inflected form is taken. In the case of a word such as *founder*, the inflected form is taken here but the lemma level would be selected for an excerpt such as *X, the founders of* ... At a higher level of generalization, the part-of-speech category *DET* (determiner) covers *a* and *the*, which makes the resulting pattern relevant for an excerpt such as *"Charles Kettering, another founder of DELCO ..."*. This example also illustrates the use of wildcards as a substitute for any word, that is to say the highest level

---

[2] http://opennlp.sourceforge.net/index.html

Table 1. Example of pattern induction

Charles Revson	, the	founder	of		Revlon Cosmetics
Mayer Lehman	, a	founder	of	the	Lehman Brothers investment
<source>	, *DET*	*founder*	*of*	*(*s*)*	*<target>*

of generalization. As it is always possible to generalize two sentences by a pattern only made of wildcards, fixing an upper limit to the number of wildcards that can used in the generalization process is necessary for having patterns that are specific enough to the target type of relation. Moreover, as our work is open domain and based on general named entities, we prefer to induce a large number of specific patterns rather than a small set of very general patterns to favor on precision. This argument also accounts for our choice of not generalizing patterns themselves by applying to them the generalization process described above. Thus, the maximal number of wildcards in a pattern is set to 1 in the evaluation of section 3.

In the context of distant supervision in which our work takes place, example sentences are not directly available but result from the mapping onto a corpus of relations given as pairs of entities (for instance, the pair (Ray Charles, Albany) for the type of relation *city_of_birth*). This process, which can also be viewed as the unsupervised annotation of a corpus with a set of known relations, is the key point of the distant supervision approach and can be expressed as the problem of combining examples of relations and generic means for detecting relations between entities in texts for extracting occurrences of relations. More concretely in our case, the example sentences are obtained by querying a search engine with pairs of entities corresponding to relations of the target type and by restricting its results to sentences that actually contain at least one of the pairs of entities used as queries. The nature of these restrictions has of course a direct impact on the quantity and the precision of final patterns: the more severe they are, the less example sentences we get but the better the induced patterns are. [2] adds for instance the constraint that the two entities of a relation pair must not be separated in a sentence by more than ten words.

Another important issue concerning the induction of patterns is its computational cost. This process is performed by considering each pair of example sentences, which can have a too high computational cost when the number of example sentences is significant: for 10,000 examples, around 50 millions of distinct pairs of sentences have to be compared ($n(n-1)/2$ exactly). The most straightforward way to solve this problem is to reduce drastically the number of example sentences before the induction of patterns. However, such solution implies having a smaller coverage of the different linguistic expressions of a type of relation if this reduction is performed blindly. Our solution to this problem exploits the fact that two sentences sharing a small number of words will not lead to an interesting pattern. The distance we use for inducing patterns – the edit distance – was chosen because of its ability to take into account the order of words but of course, it first depends on the number of words the two compared sentences share. As a consequence, the *a priori* filtering of example sentence pairs can be based on a similarity measure between sentences that only exploits a "bag of words" representation of them, as the *cosine* measure, and the application of a minimal threshold to these similarity values for discarding pairs that are not likely to lead to an interesting pattern. The *cosine* measure can be computed efficiently, either approximately, by using methods such as *Local Sensitive Hashing* [14], or without any approximation but the necessity to fix an *a priori* minimal similarity threshold, which corresponds to our case. We chose more precisely the *All Pairs Similarity Search* (APSS) algorithm, proposed in [4], which computes the *cosine* measure only for the pairs of objects – example sentences in our case – whose similarity is higher or equal to a fixed threshold. This algorithm relies on the incremental indexing of the objects whose similarity has to be evaluated and implements a set of optimizations of this indexing process based on both data gathered *a priori* about the features of these objects and their sorting according to these features.

Finally, learning patterns from a large number of example sentences often leads to several occurrences of the same pattern, either because an example sentence is found in several documents or because there are several occurrences of the same linguistic expression of a type of relation with different entity values (*Obama's height is 1.87m*; *Sarkozy's height is 1.65m*). As a consequence, we have two levels of filtering: first, a high similarity threshold is applied for identifying and discarding identical sentences; second, a lower threshold aims at checking that sentences are similar enough for inducing a meaningful pattern.

2.2   Relation Extraction

The extraction of new relations is done from the existing types of relations and given entities: we are searching to add knowledge to an existing knowledge base by adding missing attributes to entities already in the KB. The first step of relation extraction is the selection of candidate sentences that are likely to contain the expression of a relation. It starts from a query containing one entity associated with its type and the type of the target entity. The retrieval of the candidate sentences is performed, as in the pattern learning step, using a search engine in which the target corpus has been indexed. In our experiments, we used Lucene[3], with an indexing process taking into account the specific needs of our task: documents were segmented into excerpts of three sentences using a sliding window and the resulting segments were indexed by their plain words and their named entities with their type. Moreover, we also performed a kind of query expansion focusing on the source entity. Indeed, the source entity sometimes appears in the target base of documents under a slightly different form than in the query: for instance, *Bill Clinton* is often used instead of *William Jefferson Blythe III Clinton*, which is the normalized form of the entity in the KB. The expansion is based on an expansion database automatically built from Wikipedia[4]: each entity is expanded by all the formulations extracted from the redirection pages of Wikipedia for this entity. This expansion database contains alternative forms for 2.4 million entities and, starting from an entity such as *Barack Obama*, makes it possible to retrieve documents referring to {*B. Hussein Obama, Barack H. Obama Junior, Barack Obama Jr, Barack Hussein Obama Jr., etc.*}.

---

[3] http://lucene.apache.org/java/docs/index.html
[4] More precisely, we use the version of Wikipedia provided by the university of New York http://nlp.cs.nyu.edu/wikipedia-data

We then apply the patterns learned in the first step to all candidate sentences. The target entities extracted by these patterns are gathered and sorted. We only keep the most frequent entities: our hypothesis is that the more relevant the target entities are the more often they appear in documents together with the source entity. For relations with a unique target entity (*e.g. date_of_birth*), we choose the most frequent entity. For relations with several possible target values (*e.g. places_of_residence*), an arbitrary number of three values is taken since we do not have knowledge (either prior knowledge or learned from documents) about the correct number of values. Finally, a filter is applied to the target entities to check the compatibility of their value with constraints relative to the type of information we search. These constraints are defined by lists of values or regular expression. For instance, we check that the country of birth of a person is part of a list of known countries as the named entity type for the target entity – location – is not specific enough to guarantee the validity of the found information.

3 Evaluation

We present in this section our system' results on the data of the *Slot Filling* task of the TAC-KBP 2010 [31] evaluation. Our experiments have been carried out for English. The *Slot Filling* task matches the scope of our work as defined in section 2: the task aims at extracting from a large corpus the target entity of a relation, given that its source entity is part of a knowledge base that contains a large collection of examples of the target relation type. In this context, 42 relation types are considered, 16 relations for entity type ORGANIZATION (ORG) and 26 relations for entity type PERSON (PER). The list of these relation types is presented in Table 2. Note that all the experiments were conducted in parallel for all relation types on a 24 nodes (4 processors/node) cluster.

3.1 Evaluation Framework

The evaluation material from the TAC-KBP track is made of the following data:

- a 1.8 million documents text corpus (1,780,980 exactly) divided into 0.04% of transcripts (broadcast conversations, broadcast news, conversational telephone speech), 72.24% of newswire data and 27.72% Web texts;
- a knowledge base *(KB)* built from an October 2008 Wikipedia snapshot: each page containing an *infobox* was assigned a unique identifier together with an entity type among types *person*, *organization*, *geopolitical entity* and *unknown*, depending on the fields in the infobox. Typically, pages from infobox *Infobox_Actor* were associated with type *person*. Finally, 818,741 entries were selected to populate the KB, each entry being associated with a set of properties (the fields from the infobox) and a description text. As a consequence, relations in the KB are represented as tuples (identifier, infobox type, name, property, value), *e.g.*, (E0000437; Infobox_Actor; Julia Roberts; PER; birthplace; Atlanta);
- a mapping of Wikipedia properties to the relation types of the evaluation. For instance, *Infobox_Actor:birthplace* is mapped to *per:city_of_birth*. This mapping is a way of taking into account the heterogeneous nature of the labels of Wikipedia properties;

Table 2. Results for the different steps, for a subset of the 42 relation types

Relation type	Target type	Doc. Rec. (%)	Rel. Rec. (%)	# Learn.	# Test	# Induc.	# Pattern	Patt. Cov. (%)	# Ref.
*org*alternate_names	ORG	89.17	33.33	20,013	10,006	214	6,007	66.10	120
country_of_headquarter	LOC + list	91.04	55.22	18,401	9,200	2,110	185,158	69.56	67
dissolved	DATE	100	25	532	266	87	775	0	4
founded_by	ORG/PER	95.45	31.82	1,954	977	197	4,385	77.87	28
member_of	ORG	100	100	7,951	3,976	102	103	70	2
number_of_members	regexp + list	90.48	23.81	7,173	3,586	216	3,109	100	21
political_religious_affiliation	ORG	78.57	64.29	3,427	1,713	406	3,250	55.36	14
subsidiaries	ORG	82.69	28.85	5,588	2,794	498	3,764	56.48	52
top_members_employees	PER	91.48	37.22	40,929	20,464	108	1,010	70.57	223
website	regexp	78.26	30.43	30,813	15,407	32	28	0	23
per.age	regexp + list	85.32	32.14	157	79	3	1	0	109
alternate_names	PER	61.63	11.63	18,115	9,057	68	2,818	82.58	86
cause_of_death	list	100	0	1	1	0	0	0	2
country_of_birth	LOC + list	82.35	5.88	11,085	5,542	11,192	9,145,385	65.02	17
date_of_birth	DATE	90	20	11,689	5,845	30	22	0	20
employee_of	ORG	84.21	29.32	24,762	12,381	2,435	704,833	71.13	133
origin	list	81.58	42.11	37,626	18,813	2,710	276,653	74.41	76
parents	PER	78.13	9.38	1,314	657	37	604	77.78	64
religion	list	85.71	57.14	1,468	734	515	1,575	80	7
schools_attended	ORG + list	87.50	37.50	2,246	1,123	67	170	4.17	16
spouse	PER	80	35.56	5,385	2,693	3,094	314,329	80	45
title	list	84.55	52.77	31,574	15,787	8,797	1,573,512	49.07	343

Target type: final identification method of the target entity. *Doc. Rec.*: recall of sentence retrieval in terms of reference documents. *Rel. Rec.*: recall of candidate sentences in terms of reference documents. *# Learn.*: number of relations for pattern learning. *# Test*: number of relations for evaluating patterns. *# Induc.*: number of sentences containing relation occurrences used for pattern induction. *# Pattern*: number of patterns induced from occurrences of relations. *Patt. Cov.*: coverage of induced patterns. *# Ref.*: number of reference relations.

- a list of 100 source entities for which every possible target entities have to be extracted for all the target relation types. Among those entities, 15 are already present in the KB while 85 are new. Moreover, some of the relations for the source entities can not be found in the corpus. In this study, we only focus on the relations for which there is a target entity in the corpus[5], that is to say, a total of 2,069 relations. Their distribution according to their type is presented in column *# Ref.* of Table 2.

3.2   Evaluation of Pattern Learning

Patterns are used to confirm/deny the existence of a relation among two entities. As a consequence, it is important to ensure that the induced patterns have a high enough coverage to take into account as many variants as possible among the occurrences of relations. To assess the quality of these patterns, we divided the relations from the KB into a training set (2/3 of the relations) and a test set (1/3 of the relations) and we measured the coverage of the patterns by computing the percentage of relation occurrences of the test set that were found by applying the patterns learned from the relation occurrences of the training set. We used for this evaluation the previously described TAC-KBP 2010 corpus. It should be noted that using this corpus to evaluate the extraction of relations

---

[5] The list of target entities that are present in the corpus was built by the KBP organizers from the results of all participants.

does not invalidate its use for patterns learning since the relations are different for both tasks.

We provide in Table 2 the number of relations in the training and test sets in columns *# Learn.* and *# Test* respectively. The number of sentences that contain occurrences of relations used for pattern generalization is shown in the column *# Induc.* The number of patterns generated from these candidate sentences is shown in the column *# Pattern* of the same table.

For instance, if we consider relation type *org:alternate_names*, from 20,013 relations in the training set, only 214 candidate sentences demonstrating an evidence of the relation are selected. These 214 sentences are then used to generate 6,007 patterns that have a coverage of 66.10% (*i.e.* we find 66.10% of the sentences containing occurrences of the 10,006 test relations). The significant gap between the 20,013 relations and the 214 sentences is due to two main factors:

- a constraint applied during the selection of the candidate sentences: we only keep the sentences in which the named entities are fully recognized, whereas named entities can be partially (or improperly) recognized by linguistic processing;
- the nature of documents in the corpus: 72% of documents are news articles published between January 2008 and August 2009, which explains the lack of documents, if any, regarding some persons or organizations of the KB.

Detailed results regarding the pattern coverage for each relation type are presented in column *Patt. Cov.* of Table 2. As far as efficiency is concerned, the computation time for pattern generalization concerning for instance the relation type *per:country_of_birth* (11,192 sample sentences to compare) drops from 690mn and 5s without filtering to 0mn and 30s with filtering[6], which illustrates the benefit of this operation in terms of computation time.

3.3 Evaluation of Relation Extraction

The relation extraction process is composed of several steps, each of them influencing the overall result. Consequently, we have performed separate evaluations for the retrieval of candidate sentences and the core relation extraction process.

Retrieval of Candidate Sentences. A prerequisite for extracting relevant relations is ensuring that the search engine returns enough relevant documents so that we can identify the target entities. We measured the coverage based on the document search result, *i.e.* the percentage of documents retrieved by the index that are in the reference. We tried several strategies by testing different values for parameters such as the number of retrieved documents or whether to use query expansion or not. From this evaluation, the best configuration is to query the index using the source entities and their expansions together with considering the top 1,000 returned documents: this configuration allows retrieving 84.24% of reference documents. Detailed results by relation type are provided in the column *Doc. Rec.* of Table 2.

---

[6] The version with filtering being parallelized, the time given is a sum of the time recorded for each processor.

The candidate sentences for a given relation type are selected based on previously retrieved documents by ensuring that each sentence contains both the source entity and the entity type of the target entity. The quality and the number of candidate sentences are largely affected by the named entity recognition process. Since we do not have a reference for named entities in the corpus, we cannot evaluate the loss caused by entity recognition errors. However, we evaluated the proportion of reference documents in which we found candidate sentences. This information allows to set an upper bound for the percentage of relations that could be extracted if the following steps performed ideally. We obtained a total coverage of 37.55% of sentences belonging to documents of the reference. The breakdown by relation type is presented in the column *Rel. Rec.* of Table 2.

Relations Extraction. To evaluate the extracted relations, we used the metrics and the tools provided for the TAC-KBP campaign[7]. The judgment about the correctness of a relation is only based on the entity string with no restriction to the documents of the reference[8]. Table 3 summarizes our results regarding this evaluation – grouped for all relation types – and demonstrates the impact of the filtering of target entities in terms of recall, precision and f1-measure. Note that the filtering process ensures that target entities match some regular expressions and/or belong to a list of closed values. Column *Target type* in Table 2 presents the type of filtering applied for each relation type.

On one hand, results in Table 3 show that the filtering of target entities improves the performance of the system (average +2.74% f1-measure). On the other hand, they validate the assumption that patterns induced using the APSS are as relevant as those induced by considering every pair of relation examples (in this case, there is an improvement of +1.72% f1-measure on average).

Table 3. Impact of target entities filtering (measures given as %)

	Before filtering			After filtering		
	Recall	Precision	F1-measure	Recall	Precision	F1-measure
All relation pairs	16.28	11.20	13.26	18.07	13.66	15.56
APSS	16.90	12.76	14.54	18.67	16.87	17.72

Table 4 presents results from various systems on two similar corpus, KBP 2009 and KBP 2010 corpus, the latter adding to the first one Web documents and transcripts, *a priori* more difficult to process. These figures cover only the relations that are actually in the corpus. Hence, they integrate a constraint that the *Slot Filling* participants had to deal with and that is not taken into account in our system since it was developed outside the campaign, namely to decide whether the relation exists in the corpus. In this table, columns *2009* and *2010* denote the scores of the top three and last three systems for KBP 2009 and KBP 2010. [16] have shown that out of 492 reference relations, 60.4% were within the same sentence while the remaining 39.6% were cross-sentence, which

---

[7] http://nlp.cs.qc.cuny.edu/kbp/2010/scoring.html
[8] In fact, the reference is not complete as it was built using only TAC-KBP participants' outputs.

is very close to the findings of Stevenson in [29] about MUC data: such relations are handled by using coreference resolution (22.8%) or by applying inference mechanisms (16.8%). Therefore, we have distinguished in the column *2010 (a)* of Table 4 the scores of systems that are more directly comparable to ours because they only deal with relation extraction within the same sentence.

The top system of KBP 2010 [9] clearly outperforms others: +36.63% compared to the second and +4.68% compared to a human annotator. This performance is based both on the use of a manually annotated corpus – 3 million documents (a different corpus from the KBP corpus) – and the use of mechanisms for cross-sentence relation extraction: pronominal coreference resolution, metonymy between entities, resolution of semantic dependencies between words and entities, etc. Using an additional corpus seems to be a crucial factor compared to the other top systems while these ones differentiate themselves from the median results by taking into account cross-sentence relations. The worst results, lower in 2010, are mainly caused by systems under development.

Regarding our system, Table 4 situates our results in the average of KBP 2010 participants' results and in the top three systems based on within sentence relation extraction approaches. In the latter case, the most efficient approach (29.15% f1-measure) [8] uses a set of handcrafted rules that achieves a precision score (66.55%) equivalent to the best score of the campaign (66.80%) and a recall score (18.67%) lying in the average score (15.33%). This strong imbalance between precision and recall is rather symptomatic of manual approaches.

Table 4. Results on TAC-KBP data (f1-measure, as %)

TAC KBP systems	2009	2010	2010 (a)
Nb. submissions (N) / participants	N=16 / 8	N=31 / 15	N=18
Human annotator	58.99	61.10	61.10
1st score	34.35	65.78	29.15
2nd score	25.05	29.15	14.22
3th score	18	28.29	14.13
$(N-2)^{th}$ score	5.90	0.55	0.55
$(N-1)^{th}$ score	2.60	0.19	0.19
N^{th} score	1.75	0.08	0.08
Our system	–	17.72	17.72
Mean	13.43	17.49	9.71
Median	13.93	14.13	12.27

4 Related Work

Large scale relation extraction, within the meaning defined in section 2, is a recent issue. Nevertheless, by means of evaluations such as TAC-KBP, it has been the subject of several works suggesting different approaches. Concerning specifically the extraction of relations, three main trends appear: using statistical learning [2,18,11], using lexical patterns generalization [17,21,19,11] and finally, tuning already existing systems for relation detection [27,5]. Compared to the 2009 edition, rule-based approaches were introduced in KBP 2010, for instance by [8], as well as approaches based on *Distant supervision* and classifiers [30]. Our approach is based on lexical pattern generalization and makes the assumption, as in [20], that the mere presence of a pair of entities in a sentence is sufficient to indicate the effective presence of a relation between these entities. In fact this is not always the case and thus we believe it is important to filter the examples used for patterns generalization beforehand as suggested by [24].

Like our system, most systems developed for KBP 2009 do not exploit the dependencies among the relation types: for instance, there is an implicit link between the age and the birth date of a person. However, in [10], the authors show that the results obtained in [17] (31.96% f1-measure) can be improved (they get 34.81% f1-measure) by integrating dependencies between relations using inference rules based on a first order logic extension. In our work, we try to avoid integrating knowledge that is too dependent on the relation types in order to have a more generic approach, easily adaptable to other domains. Finally, [9] showed in KBP 2010 a very significant increase in terms of performance by integrating mechanisms for extracting relations beyond the sentence space: given the percentage of relations that occur between sentences, such mechanisms seem necessary and we plan to integrate them in our future work.

From a different angle, [17] distinguished itself in KBP 2009 by using a two-step relation extraction process: the first aimed at finding potential target entities within the documents of the evaluation corpus by using patterns of relations; the second aimed at finding additional potential target entities that had been missed by the first step, by applying the relation patterns on a recent Wikipedia snapshot. The potential target entities retrieved by the process were retained only if they can be found in a document from the corpus. Additional entity acquisition significantly increases their scores (they gain +9% f1-measure compared to [5]) but this process implies using an external corpus that can be viewed as closely related to the KB. In addition, results on KBP 2010 have shown that the overall performance could be improved without such complementary resource and that the effect of such process on final results were lower compared to KBP 2009 (we even observe a negative impact).

5   Conclusions and Perspectives

In this article, we present an information extraction system designed for the large-scale extraction of attribute relations between named entities. The "large-scale" qualification is meant for both the integration of a large number of types of relations and the search of these relations in a large corpus. This system is based on a weakly supervised approach in which the examples are limited to pairs of entities in relation. The extraction of relations is performed by the application of lexico-syntactic patterns that are learned from occurrences of relations automatically selected from the entity pairs of the examples and used to represent the relation types. We evaluate our approach using the evaluation framework from the *Slot Filling* task of the KBP evaluation campaign, concentrating on the problem of relation extraction itself (we did not consider the case where the relation is not present in the target corpus). The results obtained in this context are comparable to the results obtained by the participants of the 2010 campaign, which we consider promising for our system, since it is designed to be generic and is not tuned to deal with the specificities of the types of relations used in this campaign. We also show that specific techniques used to deal with the large-scale aspect of the task, such as the filtering of the examples with the APSS technique, do not decrease the performance and can even contribute to improve it.

We are currently working on the improvement of our system, trying to keep the idea of a generic system with respect to the type of relation considered. In particular, we focus on the pattern learning step: we are considering both the use of a more important number of examples to learn the patterns, typically by querying the Web, and the improvement of the quality of the examples. These two points are connected because getting more examples usually requires to relax constraints on the selection of examples, which generally increases the number of false examples. To avoid this drawback, we will explore the use of a relation filtering module [32] that is capable of determining if a sentence contains a relation between two entities or not without any consideration on the nature of the relation, similarly to [3]. Such module could also be used at the *Answer filtering* step of the relation extraction process to filter out occurrences of relations in which the source entity is not likely to be actually linked to the target entity. Futhermore, this kind of filtering, that validates the result of the application of one pattern, could be integrated in a multi-pattern strategy where the application of a pattern is considered as a feature and the extraction of a target entity relies on a configuration of such features, as in [7] in another context.

References

1. Agichtein, E., Gravano, L.: Snowball: Extracting relations from large plain-text collections. In: 5th ACM International Conference on Digital Libraries, San Antonio, Texas, USA, pp. 85–94 (2000)
2. Agirre, E., Chang, A., Jurafsky, D., Manning, C., Spitkovsky, V., Yeh, E.: Stanford-UBC at TAC-KBP. In: Second Text Analysis Conference (TAC 2009), Gaithersburg, Maryland, USA (2009)
3. Banko, M., Etzioni, O.: The Tradeoffs Between Open and Traditional Relation Extraction. In: ACL 2008: HLT, Columbus, Ohio, pp. 28–36 (2008)
4. Bayardo, R., Ma, Y., Srikant, R.: Scaling up all pairs similarity search. In: 16th International Conference on World Wide Web (WWW 2007), Banff, Alberta, Canada, pp. 131–140 (2007)
5. Bikel, D., Castelli, V., Radu, F., Jung Han, D.: Entity Linking and Slot Filling through Statistical Processing and Inference Rules. In: Second Text Analysis Conference (TAC 2009), Gaithersburg, Maryland, USA (2009)
6. Bizer, C., Lehmann, J., Kobilarov, G., Auer, S., Becker, C., Cyganiak, R., Hellmann, S.: DBpedia - A crystallization point for the Web of Data. Journal of Web Semantics 7, 154–165 (2009)
7. Bollegala, D., Matsuo, Y., Ishizuka, M.: WWW sits the SAT: Measuring Relational Similarity from the Web. In: 18th European Conference on Artificial Intelligence (ECAI 2008), Patras, Greece, pp. 333–337 (2008)
8. Byrne, L., Dunnion, J.: UCD IIRG at TAC 2010 KBP Slot Filling Task. In: Third Text Analysis Conference (TAC 2010), Gaithersburg, Maryland, USA (2010)
9. Chada, D., Aranha, C., Monte, C.: An Analysis of The Cortex Method at TAC 2010 KBP Slot-Filling. In: Third Text Analysis Conference (TAC 2010), Gaithersburg, Maryland, USA (2010)

10. Chen, Z., Tamang, S., Lee, A., Li, X., Passantino, M., Ji, H.: Top-Down and Bottom-Up: A Combined Approach to Slot Filling. In: Cheng, P.-J., Kan, M.-Y., Lam, W., Nakov, P. (eds.) AIRS 2010. LNCS, vol. 6458, pp. 300–309. Springer, Heidelberg (2010)

11. Chen, Z., Tamang, S., Lee, A., Li, X., Snover, M., Passantino, M., Lin, W.P., Ji, H.: CUNY-BLENDER TAC-KBP2010 Slot Filling System Description. In: Text Analysis Conference (TAC 2010), Gaithersburg, Maryland, USA (2010)

12. Claveau, V., Sébillot, P.: From efficiency to portability: acquisition of semantic relations by semi-supervised machine learning. In: 20th International Conference on Computational Linguistics (COLING 2004), Geneva, Switzerland, pp. 261–267 (2004)

13. Embarek, M., Ferret, O.: Learning patterns for building resources about semantic relations in the medical domain. In: 6th Conference on Language Resources and Evaluation (LREC 2008), Marrakech, Morocco (2008)

14. Gionis, A., Indyk, P., Motwani, R.: Similarity search in high dimensions via hashing. In: 25th International Conference on Very Large Data Bases (VLDB 1999), Edinburgh, Scotland, UK, pp. 518–529 (1999)

15. Hearst, M.: Automatic acquisition of hyponyms from large text corpora. In: 14th International Conference on Computational linguistics (COLING 1992), Nantes, France, pp. 539–545 (1992)

16. Ji, H., Grishman, R., Trang Dang, H.: Overview of the TAC 2010 Knowledge Base Population Track. In: Third Text Analysis Conference (TAC 2010), Gaithersburg, Maryland, USA (2010)

17. Li, F., Zheng, Z., Bu, F., Tang, Y., Zhu, X., Huang, M.: THU QUANTA at TAC 2009 KBP and RTE Track. In: Second Text Analysis Conference (TAC 2009), Gaithersburg, Maryland, USA (2009)

18. Li, S., Gao, S., Zhang, Z., Li, X., Guan, J., Xu, W., Guo, J.: PRIS at TAC 2009: Experiments in KBP Track. In: Second Text Analysis Conference (TAC 2009), Gaithersburg, Maryland, USA (2009)

19. McNamee, P., Dredze, M., Gerber, A., Garera, N., Finin, T., Mayfield, J., Piatko, C., Rao, D., Yarowsky, D., Dreyer, M.: HLTCOE Approaches to Knowledge Base Population at TAC 2009. In: Second Text Analysis Conference (TAC 2009), Gaithersburg, Maryland, USA (2009)

20. Mintz, M., Bills, S., Snow, R., Jurafsky, D.: Distant supervision for relation extraction without labeled data. In: ACL-IJCNLP 2009, Suntec, Singapore, pp. 1003–1011 (2009)

21. de Pablo-Sánchez, C., Perea, J., Segura-Bedmar, I., Martínez, P.: The UC3M team at the Knowledge Base Population task. In: Second Text Analysis Conference (TAC 2009), Gaithersburg, Maryland, USA (2009)

22. Pantel, P., Ravichandran, D., Hovy, E.: Towards terascale knowledge acquisition. In: 20th International Conference on Computational Linguistics (COLING 2004), Geneva, Switzerland, pp. 771–777 (2004)

23. Ravichandran, D.: Terascale knowledge acquisition. Ph.D. thesis, Faculty of the Graduate School University of Southern California, Los Angeles, CA, USA (2005)

24. Riedel, S., Yao, L., McCallum, A.: Modeling Relations and Their Mentions without Labeled Text. In: Balcázar, J.L., Bonchi, F., Gionis, A., Sebag, M. (eds.) ECML PKDD 2010, Part III. LNCS (LNAI), vol. 6323, pp. 148–163. Springer, Heidelberg (2010)

25. Ruiz-Casado, M., Alfonseca, E., Castells, P.: Automatising the learning of lexical patterns: An application to the enrichment of WordNet by extracting semantic relationships from Wikipedia. Data Knowledge Engineering 61, 484–499 (2007)

26. Schlaefer, N., Gieselmann, P., Schaaf, T., Waibel, A.: A Pattern Learning Approach to Question Answering Within the Ephyra Framework. In: Sojka, P., Kopeček, I., Pala, K. (eds.) TSD 2006. LNCS (LNAI), vol. 4188, pp. 687–694. Springer, Heidelberg (2006)

27. Schone, P., Goldschen, A., Langley, C., Lewis, S., Onyshkevych, B., Cutts, R., Dawson, B., MacBride, J., Matrangola, G., McDonough, C., Pfeifer, C., Ursiak, M.: TCAR at TAC-KBP 2009. In: Second Text Analysis Conference (TAC 2009), Gaithersburg, Maryland, USA (2009)
28. Shinyama, Y., Sekine, S.: Preemptive information extraction using unrestricted relation discovery. In: HLT-NAACL 2006, New York City, USA, pp. 304–311 (2006)
29. Stevenson, M.: Fact distribution in Information Extraction. Language Resources and Evaluation 40(2), 183–201 (2006)
30. Surdeanu, M., McClosky, D., Tibshirani, J., Bauer, J., Chang, A., Spitkovsky, V., Manning, C.: A Simple Distant Supervision Approach for the TAC-KBP Slot Filling Task. In: Text Analysis Conference (TAC 2010), Gaithersburg, Maryland, USA (2010)
31. TAC-KBP: Preliminary task description for knowledge-base population at TAC 2010 (2010)
32. Wang, W., Besançon, R., Ferret, O., Grau, B.: Filtering and clustering relations for unsupervised information extraction in open domain. In: 20th ACM International Conference on Information and Knowledge Management (CIKM 2011), pp. 1405–1414 (2011)
33. Zhou, G., Su, J., Zhang, J., Zhang, M.: Exploring various knowledge in relation extraction. In: 43rd Annual Meeting of the Association for Computational Linguistics (ACL 2005), Ann Arbor, USA, pp. 427–434 (2005)
34. Zhou, G., Zhang, M., Ji, D., Zhu, Q.: Tree kernel-based relation extraction with context-sensitive structured parse tree information. In: EMNLP - CoNLL 2007, Prague, Czech Republic, pp. 728–736 (2007)

Learning Effective XML Classifiers
Based on Discriminatory Structures and Nested Content

Gianni Costa, Riccardo Ortale, and Ettore Ritacco

ICAR-CNR, I87036 Rende (CS), Italy
{costa,ortale,ritacco}@icar.cnr.it

Abstract. Supervised classification aims to learn a model (or a classifier) from a collection of XML documents individually marked with one of a predefined set of class labels. The learnt classifier isolates each class by the content and structural regularities observed within the respective labeled XML documents and, thus, allows to predict the unknown class of unlabeled XML documents by looking at their content and structural features. The classification of unlabeled XML documents into the predefined classes is a valuable support for more effective and efficient XML search, retrieval and filtering.

We discuss an approach for learning intelligible XML classifiers. XML documents are represented as transactions in a space of boolean features, that are informative of their content and structure. Learning algorithms induce compact associative classifiers with outperforming effectiveness from the transactional XML representation. A preprocessing step contributes to the scalability of the approach with the size of XML corpora.

1 Introduction

XML is a standard model for data representation, that is well known for the flexibility and expressiveness with which it allows to organize textual content into hierarchical (and possibly irregular) logical structures.

The supervised classification of XML data into predefined classes consists in learning a model of the structural and content regularities (observed across a set of pre-classified XML documents), that discriminate each individual class. The resulting classifier can, hence, predict the class of a previously unseen XML document from the same applicative domain, by looking at its structure and content. This is useful to accomplish a wide variety of tasks in several applicative settings. Examples include: *(i)* the automatic organization of XML documents into hierarchical topic directories for content management; *(ii)* XML document filtering in information retrieval; *(iii)* devising suitable indexing techniques with which to improve the construction of query plans in the context of query processing. *(iv)* the extraction and incremental refinement of schemas or DTDs from structurally-homogeneous classes of schema-less XML documents,

Although a wide variety of approaches to XML classification can be found in the literature, such as [17,26,10,11,22], the induction of explicative classification models (i.e., concise and human-intelligible summarizations of the content and structural regularities that discriminate the individual classes) has not been a primary concern. Such

A. Fred et al. (Eds.): IC3K 2011, CCIS 348, pp. 156–171, 2013.

classification models are appealing because of their potential to offer an in-depth and actionable understanding of the relevant properties of very large corpora of XML data and, hence, are of great practical interest in all those settings in which XML classification is preliminarily performed to enable more effective and efficient XML search, retrieval and filtering. To elucidate, in the context of XML retrieval, these models provide an understanding of the spectrum of queries (i.e. the type of information as well as its organization) answerable by the different classes of XML documents, which is beneficial for both query formulation and optimization. Also, similarity search can substantially benefit from such classification models, that can be exploited to narrow the search space by discarding those classes, whose distinguishing regularities do not comply with the target properties specified by a user-supplied query.

In this paper, an approach to XML Classification by Content and Structure (XCCS) is discussed. XCCS relies on the solid and well-understood foundations of associative classification (i.e., classification based on suitable association rules) to induce an easily interpretable and highly expressive predictive model from the available XML data. The resulting predictive model is a compact set of rules, which discriminate the generic class from the others by means of content and structural regularities, that frequently occur in the class and are positively correlated with the class itself.

The discussion of XCCS involves four different aspects. First, tagging [23,26] is used to identify suitable features of the XML documents, that are actually informative of their content and structure, and allow to represent each XML document as a transaction in the resulting feature space. Second, algorithms are designed to perform associative classification on the transactional representation of the XML data. The proposed algorithms handle unbalanced class distributions, that are often encountered in the XML domain. One major advantage of XCCS is that it couples the benefits of associative classification, i.e., a high degree of both interpretability and expressiveness (that are well known in the literature) with a robust effectiveness. Third, the number of possible XML features to consider in the transactional representation of the XML documents is generally very high. This is generally a concern for the time efficiency and the scalability of rule learning and, thus, XCCS preliminarily performs a preprocessing phase in which an aggressive dimensionality reduction is achieved. Fourth, the classification performance of XCCS is comparatively evaluated over several XML corpora. Empirical evidence shows that XCCS scales to induce very compact classifiers with outperforming effectiveness from very large XML corpora.

The outline of this paper is as follows. Section 2 begins by introducing notation and preliminaries. Section 3 develops the XCCS approach. Section 4 presents and discusses the empirical results of a comparative evaluation of XCCS over several XML corpora. Section 5 provides a succinct overview of the state of the art in XML classification. Finally, section 6 concludes and highlights future research.

2 Preliminaries

We introduce the notation used throughout the paper as well as some basic concepts. The structure of XML documents without references can be modeled in terms of *rooted, labeled trees*, that represent the hierarchical relationships among the document elements (i.e., nodes).

Definition 1. XML Tree. *An XML tree is a rooted, labeled tree, represented as a tuple* $t = (r_t, V_t, E_t, \lambda_t)$, *whose individual components have the following meaning.* $V_t \subseteq \mathbb{N}$ *is a set of nodes and* $r_t \in V_t$ *is the root node of* t, *i.e. the only node with no entering edges.* $E_t \subseteq V_t \times V_t$ *is a set of edges, catching the parent-child relationships between nodes of* t. *Finally,* $\lambda_t : V_t \mapsto \Sigma$ *is a node labeling function and* Σ *is an alphabet of node tags (i.e., labels).*   □

In the above definition, the elements of XML documents and their attributes are not distinguished: both are mapped to nodes in the XML tree representation. Hereafter, the notions of XML document and XML tree are used interchangeably.

Let t be a generic XML tree. Nodes in V_t divide into two disjoint subsets: the set L_t of *leaves* and the set $V_t - L_t$ of *inner nodes*. An inner node has at least one child and contains no textual information. A leaf is instead a node with no children, that can contain only textual information.

A root-to-leaf path $p_l^{r_t}$ in t is a sequence of nodes encountered in t along the path from the root r_t to a leaf node l in L_t, i.e., $p_l^{r_t} = < r_t, \ldots, l >$. Notation $\lambda_t(p_l^{r_t})$ denotes the sequence of labels that are associated in the XML tree t to the nodes of path $p_l^{r_t}$, i.e., $\lambda_t(p_l^{r_t}) = < \lambda_t(r_t), \ldots, \lambda_t(l) >$. The set $paths(t) = \{p_l^{r_t} | l \in L_t\}$ groups all root-to-leaf paths in t.

Let l be a leaf in L_t. The set $terms(l) = \{\lambda_t(p_l^{r_t}).w_1, \ldots, \lambda_t(p_l^{r_t}).w_h, \lambda_t(p_l^{r_t}).\epsilon\}$ is a model of the information provided by l. Elements $\lambda_t(p_l^{r_t}).w_i$ (with $i = 1 \ldots h$) are as many as the distinct term stems in the context of l and seamlessly couple content information with its structural context. Therein, w_i is some term stem (obtained in the first step of the preprocessing in subsection 4.2) and $\lambda_t(p_l^{r_t})$ acts as a prefix specifying the location of l within the XML tree t, that allows to distinguish the occurrences of w_i in the context of distinct leaves. The unique element of the type $\lambda_t(p_l^{r_t}).\epsilon$ is instead informative only of the location of l within t: ϵ indicates the null string. $\lambda_t(p_l^{r_t}).\epsilon$ still provides some (purely-structural) information on l when the leaf does not contain textual information. Notice that prefixing a term (stem) with the labels of its locating root-to-leaf path is an instance of *tagging* [23,26].

Leaf terms and their prefixes are chosen as informative features of the XML data, with which to separate the classes of XML trees. Henceforth, for readability sake, we will write p instead of $\lambda(p)$ to mean the prefix of a term stem w.

Definition 2. XML Feature. *Let* t *be an XML tree. A prefixed term (stem) p.w is said to be a feature of* t *(or, equivalently, p.w occurs in* t), *denoted as* $p.w \preceq t$, *if the following two conditions hold. First, there exists a leaf* $l \in L_t$ *and, hence, a path* $p_l^{r_t} \in paths(t)$ *such that* $\lambda_t(p_l^{r_t}) = p$. *Second,* $p.w \in terms(l)$.   □

Assume that $S = \{p.w | \exists t \in \mathcal{D}$ such that $p.w \preceq t\}$ is a suitable selection of features from the XML trees in $\mathcal{D}$. S identifies an expressive feature space, in which to perform the induction of models for effective XML classification.

Let $\mathcal{D} = \{t_1, \ldots, t_N\}$ be a training database (or, equivalently, a forest) of N XML trees, each of which is associated with one label from the set $\mathcal{L} = \{c_1, \ldots, c_k\}$. The problem of XML classification can formalized as learning some suitable model $\mathcal{C} : 2^S \mapsto \mathcal{L}$ of the associations between the occurrence of the chosen features in the XML trees of $\mathcal{D}$ and the class labels of the same XML trees. The resulting classifier $\mathcal{C}$ is useful to predict the unknown class of a previously unseen XML tree t', on the basis of the features occurring in t'.

Ideally, the efficiency and scalability of XML classification should suffer neither from the dimensionality of S (i.e., the number $|S|$ of features), nor from the costs of the operations for the manipulation of tree-like structures. To meet such requirements, the dimensionality of the feature space corresponding to S is sensibly reduced in subsection 4.2. Additionally, XML data is represented in a convenient transactional form, that avoids the manipulation of XML trees.

The idea behind the transactional representation is that by looking at the elements in S as binary features, the available XML data can be projected into a feature space, wherein the occurrence of the individual features within each XML tree is explicitly represented. More precisely, if S denotes the selected collection of features, the XML trees from $\mathcal{D}$ can be modeled as transactions over a feature space $\mathcal{F} \triangleq \{\mathcal{F}_{p.w}|p.w \in S\}$. Here, the generic feature $\mathcal{F}_{p.w}$ is a boolean attribute, whose value indicates the presence/absence of the corresponding feature $p.w$ of S within the individual XML trees.

Let $\mathbf{x}^{(t)}$ be the transactional representation of an XML tree t. The value of each attribute $\mathcal{F}_{p.w}$ within $\mathbf{x}^{(t)}$ is true if $p.w$ is a feature of t, otherwise it is false. Hence, $\mathbf{x}^{(t)}$ can be modeled as a proper subset of $\mathcal{F}$, namely $\mathbf{x}^{(t)} \triangleq \{\mathcal{F}_{p.w} \in \mathcal{F}|p.w \preceq t\}$, with the meaning that the features explicitly present in $\mathbf{x}^{(t)}$ take value true, whereas the others assume value false. The original database $\mathcal{D}$ can hence be represented in transactional form as $\mathcal{D}^{(F)} = \{\mathbf{x}^{(t_1)}, \ldots, \mathbf{x}^{(t_N)}\}$, whereas the class associated with the generic transaction is denoted as $class(\mathbf{x}^{(t)})$. Hereafter, to keep notation uncluttered, the transactional database and the generic transaction will be denoted, respectively, as $\mathcal{D}$ and $\mathbf{x}$.

In this paper, XML classification is approached through associative classification [12], a powerful enhancement of conventional rule learning, that results from the integration of two fundamental tasks in data mining, namely, association rule mining and classification. Associative classifiers retain the advantages of traditional rule-based classification models (i.e., interpretability, expressiveness and high effectiveness) and, also, tend to achieve a better predictive performance [18].

The necessary concepts concerning associative classification in the domain of the transactional representation of the XML trees are formalized next.

The notion of class association rule is the starting point.

Definition 3. *Class Association Rule. Let $\mathcal{F}$ be a feature space, deriving from the selection of certain features of the XML data. Also, assume that $\mathcal{D}$ (the so-called training data) is a database of XML trees represented as transactions over $\mathcal{F}$ and that $\mathcal{L}$ is a set of class labels. A class association rule (or, equivalently, a CAR) $\mathbf{r} : \mathbf{I} \rightarrow c$ is a pattern that catches the association (i.e. the co-occurrence) in $\mathcal{D}$ of some subset $\mathbf{I}$ of $\mathcal{F}$ with a class label c belonging to $\mathcal{L}$. $\mathbf{I}$ and c are said to be, respectively, the antecedent and consequent of the CAR.* □

Essentially, a CAR relates the occurrence of a certain combination of features in a transaction corresponding to an XML tree with one particular class.

A rule $r : I \rightarrow c$ is said to *cover* a (labeled or unlabeled) transaction $x \in \mathcal{D}$ (and, dually, x is said to trigger or fire r) if the condition $I \subseteq x$ holds. The set $\mathcal{D}_r$ of transactions covered by r is defined as $\mathcal{D}_r = \{x \in \mathcal{D} | I \subseteq x\}$.

The notions of support and confidence are typically employed to define the interestingness of a rule r.

Definition 4. *CAR Support and Confidence*. *A transaction* $x \in \mathcal{D}$ *supports a CAR* $r : I \rightarrow c$ *if it holds that* $I \subseteq x$ *and* $c = class(x)$. *The support of* r, *denoted as* $supp(r)$, *is the fraction of transactions in* $\mathcal{D}$ *that support* r. *The confidence or predictive strength of* r, *denoted by* $conf(r)$, *is defined as* $conf(r) = \frac{supp(r)}{supp(I)}$, *where* $supp(I)$ *is the fraction of transactions in* $\mathcal{D}$ *including the subset* I. $\square$

Hereafter, a CAR r is actually interesting if it meets certain minimum requirements on its support and confidence and, additionally, if its antecedent and consequent are positively correlated [2]. This avoids misleading CARs (i.e., CARs with negative correlation despite a high confidence) when classes are skewed, which is often encountered in the XML domain.

Definition 5. *Associative Classifier*. *An associative classifier* $\mathcal{C}$ *is a disjunction* $\mathcal{C} = \{r_1 \vee \ldots \vee r_k\}$ *of interesting CARs learnt from a database* $\mathcal{D}$ *of labeled transactions (representing XML trees with known class labels).* $\square$

An associative classifier is a set of CARs that assign an unlabeled XML tree (in transactional form) to a class if certain features occur in the tree. An approach to induce associative classifiers for effective XML classification is proposed next.

3 The XCCS Approach

XCCS is a general framework for the associative classification of XML data, that relies on CARs to model the associations between subsets of co-occurring features and the discriminated classes.

XCCS exploits a selection of features of the available XML data for the discrimination of the individual classes. XML classification in XCCS divides into model learning and prediction. The former learns an associative classifier $\mathcal{C}$ from a database of labeled XML trees in transactional form. The latter exploits $\mathcal{C}$ to predict the class of unlabeled XML trees.

3.1 The Overall Model-Learning Process

The model learning process in XCCS, sketched in fig. 1(a), receives four input parameters: a database $\mathcal{D}$ of XML trees, a set $\mathcal{S}$ of discriminatory features, a set $\mathcal{L}$ of class labels in $\mathcal{D}$ and one global threshold τ, from which the minimum support thresholds for the individual classes in $\mathcal{L}$ are derived.

Model learning preliminarily involves the definition (at line 1) of the space $\mathcal{F}$ of features related to the elements of $\mathcal{S}$ as well as the mapping of the individual XML trees in $\mathcal{D}$ to as many corresponding transactions over $\mathcal{F}$ (lines 4-7).

MODEL-LEARNING($\mathcal{D},\mathcal{L},\mathcal{S},\tau$)	PRUNE($\mathbf{R},\mathcal{D},\mathcal{L}$)

Require: a training dataset $\mathcal{D}$;
  a set $\mathcal{S}$ of substructures of the XML trees in $\mathcal{D}$;
  a set $\mathcal{L}$ of class labels in $\mathcal{D}$;
  and a support threshold τ;
Ensure: An associative classifier $\mathcal{C} = \{r_1 \vee \ldots \vee r_k\}$;
1: let $\mathcal{F} \leftarrow \{\mathcal{F}_s \mid s \in \mathcal{S}\}$ be the feature space;
2: $\mathbf{R} \leftarrow \emptyset$;
3: $\mathcal{D}' \leftarrow \emptyset$;
4: **for each** $t \in \mathcal{D}$ **do**
5:   $\mathbf{x} \leftarrow \{\mathcal{F}_s \mid \mathcal{F}_s \in \mathcal{F}, s \preceq t\}$;
6:   $\mathcal{D}' \leftarrow \mathcal{D}' \cup \{\mathbf{x}\}$;
7: **end for**
8: $\mathbf{R} \leftarrow$ MINECARS $(\mathcal{F}, \mathcal{D}', \tau)$;
9: $\mathcal{C} \leftarrow$ PRUNE $(\mathbf{R})$;
10: RETURN $\mathcal{C}$
	Require: a set $\mathbf{R}$ of CARs;
	  a set $\mathcal{D}$ of transactions;
	  a set $\mathcal{L}$ of class labels in $\mathcal{D}$;
	Ensure: a classifier $\mathcal{C}$;
	  /* Rule ordering according to the devised total order $\ll$ */
	M1: $\mathbf{R} \leftarrow$ ORDER($\mathbf{R}$);
	M2: $\mathcal{C} \leftarrow \emptyset$;
	M3: $\mathcal{T} \leftarrow \mathcal{D}$;
	M4: **for each** $c \in \mathcal{L}$ in increasing order of occurrence frequency
	M5: **while** there are still CARs in $\mathbf{R}$ that target c **do**
	M6: choose the next rule $r : \mathbf{I} \rightarrow c$ from $\mathbf{R}$;
	M7: **if** $F^{(c)}(\mathcal{C} \cup \{r\}) > F^{(c)}(\mathcal{C})$ **then**
	M8: $cur\_length \leftarrow length(\mathcal{C} \cup \{r\})$;
	M9: $\mathbf{R} \leftarrow \mathbf{R} - \{r\}$;
	M10: $\mathcal{C} \leftarrow \mathcal{C} \cup \{r\}$;
	M11: $\mathcal{D}_r \leftarrow \{\mathbf{x} \in \mathcal{D} \mid \mathbf{I} \subseteq \mathbf{x}\}$;
	M12: $\mathcal{D} \leftarrow \mathcal{D} - \mathcal{D}_r$;
	M13: $min\_length \leftarrow cur\_length$;
	M14: **end if**
	M15: **if** $\|\mathcal{D}\| = 0$ **then**
	M16: continue at line M20;
	M17: **end if**
	M18: **end while**
	M19: **end for**
	M20: **if** $\|\mathcal{D}\| > 0$ **then**
	M21: $c^* \leftarrow argmax_{c \in \mathcal{L}}\, supp(c, \mathcal{D})$;
	M22: **else**
	M23: $c^* \leftarrow argmax_{c \in \mathcal{L}}\, supp(c)$;
	M24: **end if**
	M25: $\mathcal{C} \leftarrow \mathcal{C} \cup \{\emptyset \rightarrow c^*\}$;
	M26: RETURN $\mathcal{C}$;

The overall model learning process (a)	The PRUNE procedure (b)

Fig. 1. The scheme of the XCCS algorithm

The MINECARS procedure discovers a potentially large set $\mathbf{R}$ of CARs targeting the classes in $\mathcal{L}$. MINECARS is an Apriori-based procedure, that searches for meaningful CARs in the training data $\mathcal{D}$. MINECARS enhances the basic Apriori algorithm [1] by incorporating two effective mechanisms, i.e., multiple minimum class support [13] and complement class support [2], with which to distill, within each class in $\mathcal{D}$, an appropriate number of CARs with a positive correlation between their antecedents and consequents, even in the presence of class imbalance (that typically leads to unveil misleading CARs, i.e., rules with a negative correlation between their antecedents and consequents).

Due to space restrictions, the scheme of the MINECARS procedure is omitted.

Finally, the rule set $\mathbf{R}$ is distilled into a compact associative classifier $\mathcal{C}$ through the pruning method PRUNE, that is covered next in detail.

3.2 Distilling an Associative-Classification Model

Due to the inherently combinatorial nature of the associative patterns, MINECARS may yield a large number of CARs, which are likely to overfit the training data and provide contrasting predictions. To avoid such issues, a compact and accurate classifier is distilled from the rule set $\mathbf{R}$ through the covering method PRUNE, illustrated in fig. 1.(b).

PRUNE initially orders (at line M1) the available CARs according to the total order $\ll$, which is inspired to the one introduced in [12]. Precisely, given any two CARs $r_i, r_j \in R$, r_i precedes r_j, which is denoted by $r_i \ll r_j$, if (i) $conf(r_i)$ is greater than $conf(r_j)$, or (ii) $conf(r_i)$ is the same as $conf(r_j)$, but $supp(r_i)$ is greater than $supp(r_j)$, or (iii) $conf(r_i)$ is the same as $conf(r_j)$ and $supp(r_i)$ is identical to $supp(r_j)$, but $length(r_i)$ is less than $length(r_j)$. The length of a CAR $r : I \to c$ is the number of features in the antecedent of r, i.e., $length(r) = |I|$.

If two CARs r_i, r_j have equal confidence, support and length, then $r_i \ll r_j$ if r_i was generated before r_j.

A covering process (lines M4- M19) then seeks a compact classifier $\mathcal{C}$, consisting of a minimal number of CARs from R, that attain a high predictive accuracy over unlabeled transactions (representing unclassified XML trees).

The covering process attempts the maximization of the effectiveness $F(\mathcal{C})$ of the resulting classifier $\mathcal{C}$ across all classes. $F(\mathcal{C})$ is evaluated in terms of the macro-averaged F-measure [14] of $\mathcal{C}$, which is defined as follows

$$F(\mathcal{C}) = \frac{1}{|\mathcal{L}|} \sum_{c \in \mathcal{L}} F^{(c)}(\mathcal{C})$$

where $F^{(c)}(\mathcal{C})$ is the effectiveness (or, also, the predictive performance) of $\mathcal{C}$ over the generic class c, described below. $F(\mathcal{C})$ assigns a same relevance to the effectiveness of $\mathcal{C}$ over the different classes, regardless of the occurrence frequency of the individual classes in the training data. This is especially useful in the presence of class imbalance, since $F^{(c)}(\mathcal{C})$ is not dominated by the predictive performances of $\mathcal{C}$ over the most frequent classes across the transactions.

The covering process increases $F(\mathcal{C})$ by separately acting on each $F^{(c)}(\mathcal{C})$, through the selection (at line M6) of CARs from R that, when appended to $\mathcal{C}$ (at line M10), improve the predictive performance of the resulting classifier over c without a significant loss in compactness.

For each class c, covering scans (according the $\ll$ order) the different CARs of R that target c. A CAR $r : I \to c$ is appended to $\mathcal{C}$ if $F^{(c)}(\mathcal{C} \cup \{r\})$ is greater than $F^{(c)}(\mathcal{C})$ (at line M7). In this case, r is removed from R (at line M9) and all transactions covered by r are dropped from $\mathcal{D}$ (at line M12).

Notice that $F^{(c)}(\mathcal{C})$ is 0 while $\mathcal{C}$ does not include any CARs targeting c.

The total order established over R (at line M1) plays a key role while covering the generic class c: it ensures that the CARs predicting c with highest implicative strength are appended to $\mathcal{C}$ since the early stages of the covering process, which is beneficial for $F^{(c)}$. Moreover, as covering of class c proceeds, the consequent removal of transactions from $\mathcal{D}$ operates as a pruning method, which increasingly tends to prevent from retaining in $\mathcal{C}$ those CARs targeting c, that have not yet been considered. This positively acts on both the effectiveness and the compactness of $\mathcal{C}$ since, according to the $\ll$ order, such CARs predict c with either a lower implicative strength or a higher specificity.

Notice that the different classes are separately covered (at line M4) in increasing order of their occurrence frequency, to avoid that transactions belonging to less frequent classes are removed from $\mathcal{D}$ while covering other classes with higher occurrence frequency. This would have the undesirable effect of avoiding an appropriate evaluation of the gain in effectiveness due to the addition to $\mathcal{C}$ of CARs targeting the foresaid less frequent classes.

Covering halts when either there are no more CARs to consider (which is caught, for each class c, at line M5), or all training transactions have been covered (which is caught at line M15), or the predictive performance of $\mathcal{C}$ cannot be further increased (which is caught, for each class c, at line M5).

The generic $F^{(c)}$ summarizes two further measures of class-specific effectiveness, i.e., the degree of precision $P^{(c)}$ and recall $R^{(c)}$ of classifier $\mathcal{C}$ in class c:

$$F^{(c)}(\mathcal{C}) = 2 \left[\frac{1}{P^{(c)}(\mathcal{C})} + \frac{1}{R^{(c)}(\mathcal{C})} \right]^{-1}$$

An increase in $F^{(c)}$, due to the addition of a CAR $\mathbf{r}$ to the current classifier $\mathcal{C}$, means that $\mathbf{r}$ produces an acceptable improvement of the predictive performance of $\mathcal{C}$, ascribable to a sensible gain in at least one between $P^{(c)}$ and $R^{(c)}$.

Precision $P^{(c)}(\mathcal{C})$ is the exactness of $\mathcal{C}$ within class c, i.e., the proportion of transactions that are actually of class c among the ones assigned by $\mathcal{C}$ to class c. Recall $R^{(c)}(\mathcal{C})$ is instead the completeness of $\mathcal{C}$ within c, i.e., the fraction of transactions of class c that are correctly predicted by $\mathcal{C}$. Formally, let $\mathcal{D}_{\mathcal{C}}^{(c)} = \{\mathbf{x} \in \mathcal{D} | \exists \mathbf{r} \in \mathcal{C}, \mathbf{r} : \mathbf{I} \to c, \mathbf{I} \subseteq \mathbf{x}\}$ be the set of transactions covered by the CARs of $\mathcal{C}$ predicting class c and $p_{\mathcal{C}}^{(c)} = \{\mathbf{x} \in \mathcal{D} | \exists \mathbf{r} \in \mathcal{C}, \mathbf{r} : \mathbf{I} \to c, \mathbf{I} \subseteq \mathbf{x}, class(\mathbf{x}) = c\}$ be the set of transactions correctly assigned by $\mathcal{C}$ to class c. Also, assume that $\sigma(c)$ is the overall number of transactions of class c. Precision $P^{(c)}(\mathcal{C})$ and recall $R^{(c)}(\mathcal{C})$ are defined as reported below

$$P^{(c)}(\mathcal{C}) = \frac{|p_{\mathcal{C}}^{(c)}|}{|\mathcal{D}_{\mathcal{C}}^{(c)}|} \qquad\qquad R^{(c)}(\mathcal{C}) = \frac{|p_{\mathcal{C}}^{(c)}|}{\sigma(c)}$$

Precision $P^{(c)}(\mathcal{C})$ and recall $R^{(c)}(\mathcal{C})$ provide complementary information on the effectiveness of $\mathcal{C}$ over c. Indeed, an improvement in precision alone, achieved by appending $\mathbf{r}$ to $\mathcal{C}$, would not say anything on the corresponding variation in the recall of $\mathcal{C} \cup \{\mathbf{r}\}$. Dually, an improvement in recall alone would not say anything on the corresponding variation in the precision of $\mathcal{C} \cup \{\mathbf{r}\}$.

The simultaneous increase of both precision and recall is a challenging issue in the design of algorithms for learning classification models, since it often happens that a gain in recall corresponds to a loss in precision and vice-versa. $F^{(c)}(\mathcal{C})$ is the harmonic mean of $P^{(c)}(\mathcal{C})$ and $R^{(c)}(\mathcal{C})$ and, hence, it is always closer to the smaller between the two. Therefore, an improvement of $F^{(c)}(\mathcal{C} \cup \{\mathbf{r}\})$ with respect to $F^{(c)}(\mathcal{C})$ ensures that an increase in recall due to the addition of $\mathbf{r}$ to $\mathcal{C}$ is not vanished by a serious loss in precision.

Let $\mathcal{D}_{\mathbf{r}}$ be set of transactions left in $\mathcal{D}$, that are covered by $\mathbf{r} : \mathbf{I} \to c$ (at line M11). Also, assume that $p_{\mathbf{r}}^{(c)}$ is the subset of those transactions in $\mathcal{D}_{\mathbf{r}}$ correctly classified by $\mathbf{r}$ into class c, i.e., $p_{\mathbf{r}}^{(c)} = \{\mathbf{x} \in \mathcal{D}_{\mathbf{r}} | class(\mathbf{x}) = c\}$. The updated values of precision $P^{(c)}(\mathcal{C} \cup \{\mathbf{r}\})$ and recall $R^{(c)}(\mathcal{C} \cup \{\mathbf{r}\})$, resulting from the addition of $\mathbf{r}$ to $\mathcal{C}$, are incrementally computed from $P^{(c)}(\mathcal{C})$ and $R^{(c)}(\mathcal{C})$ as follows

$$P^{(c)}(\mathcal{C} \cup \{\mathbf{r}\}) = \frac{|p_{\mathcal{C}}^{(c)}| + |p_{\mathbf{r}}^{(c)}|}{|D_{\mathcal{C}}^{(c)}| + |\mathcal{D}_{\mathbf{r}}|} \qquad\qquad R^{(c)}(\mathcal{C} \cup \{\mathbf{r}\}) = R^{(c)}(\mathcal{C}) + \frac{|p_{\mathbf{r}}^{(c)}|}{\sigma(c)}$$

When covering ends (line M19), the resulting classifier $\mathcal{C}$ is a list of predictive CARs grouped by the targeted class. The individual groups of CARs appear in $\mathcal{C}$ in increasing order of the occurrence frequency of the targeted class. Moreover, within each group, the CARs reflect the total order $\ll$ established over $\mathbf{R}$.

To conclude the discussion on PRUNE, the mutual exclusiveness and the exhaustive coverage of the CARs of any resulting classifier $\mathcal{C}$ must be touched.

A rule-based classifier is mutually exclusive if each input triggers no more than one rule. Generally, such a property does not hold for $\mathcal{C}$. Indeed, it is actually possible that multiple CARs are triggered by a same transaction. This is clearly undesirable because (some of) the triggered CARs may provide contrasting predictions. This problem is overcome in sec. 3.3.

Instead, the addition to $\mathcal{C}$ (at line M25) of a default rule $\emptyset \rightarrow c^*$ ensures exhaustive coverage, i.e., that every transaction is covered by at least one CAR of $\mathcal{C} \cup \{\emptyset \rightarrow c^*\}$. In particular, the default rule covers all those transactions uncovered by the CARs of $\mathcal{C}$ and assigns them to a suitable class c^*. This guarantees a maximum recall with very poor precision in class c^*. To attenuate the loss in precision, c^* can be reasonably chosen (at line M23) to be the class with highest occurrence frequency in the training data, which ensures the highest precision for the default rule. Depending on the overall coverage of $\mathcal{C}$, there are two alternative possibilities for the choice of the default class c^*. If there are still uncovered transactions after the termination of the covering process (at line M20), c^* is selected (at line M21) as the class with maximum occurrence frequency $supp(c, \mathcal{D})$ in the residual training data $\mathcal{D}$. Otherwise, if all transactions have been covered, c^* is chosen (at line M23) to be the class with highest occurrence frequency in the whole training data. In both cases, ties can be arbitrarily broken.

3.3 Prediction

Let $\mathcal{C}$ be an associative classifier induced by XCCS at the end of model-learning phase of fig. 1(a). Also, assume that $\mathbf{t}$ is an unlabeled (i.e. an unclassified) XML tree, whose transactional representation over the underlying feature space $\mathcal{F}$ is $\mathbf{x}$. The class predicted by $\mathcal{C}$ for $\mathbf{t}$ is $\mathcal{C}(\mathbf{x})$. To avoid conflicting predictions from multiple triggered CARs, $\mathcal{C}(\mathbf{x})$ is provided by the first CAR of $\mathcal{C}$ that covers $\mathbf{x}$.

4   Evaluation

The empirical behavior of XCCS is studied with major two objectives: (i) comparatively evaluating the effectiveness of XCCS across different domains; (ii) understanding the scalability of XCCS. All tests are performed on a Linux machine, with an Intel Core 2 Duo cpu, 4Gb of memory and 2Ghz of clock speed.

4.1   Data Sets

The behavior of XCCS is tested over several real XML data sets. Synthetic XML corpora are not considered for experimental purposes, since these are generally unlikely to provide coherent textual information in natural language.

Macro-averaged effectiveness results are obtained by performing a stratified 10-fold cross validation on the transactional representation of each data set. We choose four real-world XML data sets, that include textual information and are characterized by skewed distributions of the classes of XML documents.

Wikipedia is an XML corpus proposed in the INEX contest 2007 [9] as a major benchmark for XML classification and clustering. The corpus groups $96,000$ XML documents representing very long articles from the digital encyclopedia. The XML documents are organized into 21 classes (or thematic categories), each corresponding to a distinct *Wikipedia Portal*. A challenging aspect of the corpus is the ambiguity of certain pairs of classes such as, e.g., Portal:Pornography and Portal:Sexuality or Portal:Chistianity and Portal:Spirituality [9].

IEEE is a reference text-rich corpus, presented in [8], that includes $12,107$ XML documents representing full articles. These are organized into 18 classes corresponding to as many IEEE Computer Society publications: 6 Transactions and 12 other journals. A same thematic can be treated into two distinct journals.

DBLP is a bibliographic archive of scientific publications on computer science (http://dblp.unitrier.de/xml/). The archive is available as one very large XML file with a diversified structure. The whole file is decomposed into $479,426$ XML documents corresponding to as many scientific publications. These individually belong to one of 8 classes: article ($173,630$ documents), proceedings ($4,764$ documents), mastersThesis (5 documents), incollection ($1,379$ documents), inproceedings ($298,413$ documents), book ($1,125$ documents), www (38 documents), phdthesis (72 documents). The individual classes exhibit differentiated structures, despite some overlap among certain document tags (such as *title*, *author*, *year* and *pages*), that occur in (nearly) all of the XML documents.

The *Sigmod* collection groups 988 XML documents (i.e., articles from Sigmod Record) complying to three different class DTDs: IndexTermsPage, OrdinaryIssue and Proceedings. These classes contain, respectively, 920, 51 and 17 XML documents. Such classes have diversified structures, despite the occurrence of some overlapping tags, such as *volume*, *number*, *authors*, *title* and *year*.

4.2   Preprocessing

The high dimensionality (i.e. cardinality) of the feature space S may be a concern for the time efficiency and the scalability of model induction. In particular, when the classes of XML trees cannot be discriminated through the structural information alone and, hence, the content information must necessarily be taken into account, the number of XML features likely becomes very large if the XML documents contain huge amounts of textual data.

To reduce the dimensionality of the feature space S, the available XML data is preprocessed into two steps.

The first step addresses the textual information of the XML data and sensibly reduces the overall number of distinct terms in the leaves of the XML trees through token extraction, stop-word removal and stemming.

Dimensionality reduction is performed at the second step both to reduce overfitting and to ensure a satisfactory behavior of XCCS in terms of efficiency, scalability and compactness of the induced classifiers. The idea is partitioning S into groups of XML features, that discriminate the individual classes in a similar fashion. For this purpose, we explicitly represent the discriminatory behavior of the features in S and then group the actually discriminatory features through distributional clustering [3]. In particular, the discriminatory behavior of each feature $p.w$ in S is represented as an array $\mathbf{v}_{p.w}$ with as many entries as the number of classes in $\mathcal{L}$. The generic entry of $\mathbf{v}_{p.w}$ is the probability $P(c|p.w)$ of class c given the XML feature $p.w$. Clearly, $P(c|p.w) = \frac{P(p.w|c)P(c)}{P(p.w)}$ by Bayes rule, where probabilities $P(p.w|c)$, $P(c)$ and $P(p.w)$ are estimated from the data. Before clustering features, noisy (i.e. non discriminatory) features are removed from S. Specifically, a feature $p.w$ is noisy if there is no class c in $\mathcal{L}$, that is positively correlated with $p.w$ according to chi-square testing at a significance level 0.05. The arrays relative to the remaining features of S are then grouped through distributional clustering [3] into a desired number of feature clusters with similar discriminatory behavior.

Eventually, an aggressive compression of the original feature space S is achieved by replacing each feature $p.w$ of S with a respective synthetic feature, i.e., the label of the cluster to which $\mathbf{v}_{p.w}$ belongs. Distributional clustering efficiently compresses the feature space by various orders of magnitude, while still enabling a significantly better classification performance than several other established techniques for dimensionality reduction [3].

4.3 Classification Effectiveness

We compare XCCS against several other established competitors in terms of classification effectiveness. Both direct and indirect comparisons are performed.

The direct comparisons involve three competing classification approaches, that produce rule-based classifiers, i.e., XRULES [22], CBA [12] as well as CPAR [18]. These competitors are publicly available and, thus, can be compared against XCCS on each XML data set.

XRULES is a state-of-the-art competitor, that admits multiple cost-models to evaluate classification effectiveness. For each data set, we repeatedly train XRULES to suitably tune the minimum support threshold for the frequent subtrees to find in the various classes and, then, report the results of the cost model allowing the best classification performance.

CBA and CPAR are two seminal techniques for learning associative classifiers. CBA and CPAR are included among the competitors of XCCS to compare the effectiveness of the three distinct approaches to associative classification at discriminating classes in (high-dimensional) transactional domains.

To evaluate CBA and CPAR, we use the implementations from [6]. In all tests, both CBA and CPAR are trained on the transactional representations of the XML data sets used to feed XCCS. Again, CBA and CPAR are repeatedly trained on the transactional representation of each XML data set, in order to suitably tune their input parameters. For every data set, we report the results of the most effective classifiers produced by CBA and CPAR.

Through preliminary tests we noticed that, in all tests, a satisfactory behavior of XCCS can be obtained by fixing the support threshold τ of fig. 1(a) to 0.1. This is essentially due to the adoption of the minimum class support [13] in the MINECARS procedure (touched in section 3.1).

Fig. 2(a) summarizes the effectiveness of the chosen competitors across the selected data sets. Columns **Size** and **#C** indicate, respectively, the number of XML documents and classes for each corpus. Column **Model** identifies the competitors. **Rules** is the rounding off of the average number of rules of a classifier in the stratified 10-fold cross validation.

The effectiveness of each classifier is measured in terms of average precision (**P**), average recall (**R**), average F-measure (**F**). More precisely, the values of **P**, **R** and **F** are averages of precision, recall and F-measure over the folds of the stratified 10-fold cross validation of classifiers on the individual data sets. The maximum values of **P**, **R** and **F** on each data set are highlighted in bold.

Notice that we tabulate only the (best) results achieved by the approaches [7,15,20,21,19] in the respective papers. Some results were not originally measured and, hence, are reported as *N.A.* (short for *not available*).

Rules has no sense for [7,15,20,21,19] and, thus, its entry in the corresponding rows is left blank.

The symbol − that appears in three rows of fig. 2(a) reveals that XRULES did not successfully complete the tests over *Wikipedia*, *IEEE* and *DBLP*. The enumeration of the frequent embedded subtrees within each class and the consequent generation of predictive structural rules (satisfying the specified level of minimum class-specific support) are very time-expensive steps of XRULES, especially when the underlying number of XML documents is (very) large. In all completed tests, XRULES is less effective than XCCS. In addition, the huge number of rules produced by XRULES makes the resulting classification models difficult to understand (and, hence, hardly actionable) in practice.

The classification performance of CBA is inferior than that of XCCS on the selected data sets. Moreover, as said in sec. 3.2, interpreting CBA classifiers may be cumbersome, since their rules are not ordered by the targeted class.

CPAR exhibits a satisfactory classification performance on the chosen XML corpora. Nonetheless, CPAR is still less effective and compact than XCCS.

The approaches [7,15,20] and [21,19] exhibit generally inferior classification performances than XCCS on the *Wikipedia* and *IEEE* corpora, respectively.

To conclude, XCCS consistently induces the most effective classifiers on the chosen corpora. As far as compactness is concerned, such classifiers are generally comparable to the ones induced by CBA and significantly more compact than the ones induced by XRULES and CPAR. The effectiveness of XCCS confirms its general capability at handling XML data with skewed class distributions.

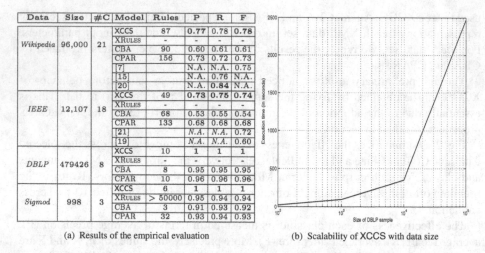

Data	Size	#C	Model	Rules	P	R	F
Wikipedia	96,000	21	XCCS	87	0.77	0.78	0.78
			XRULES	-	-	-	-
			CBA	90	0.60	0.61	0.61
			CPAR	156	0.73	0.72	0.73
			[7]		N.A.	N.A.	0.75
			[15]		N.A.	0.76	N.A.
			[20]		N.A.	0.84	N.A.
IEEE	12,107	18	XCCS	49	0.73	0.75	0.74
			XRULES	-	-	-	-
			CBA	68	0.53	0.55	0.54
			CPAR	133	0.68	0.68	0.68
			[21]		N.A.	N.A.	0.72
			[19]		N.A.	N.A.	0.60
DBLP	479426	8	XCCS	10	1	1	1
			XRULES	-	-	-	-
			CBA	8	0.95	0.95	0.95
			CPAR	10	0.96	0.96	0.96
Sigmod	998	3	XCCS	6	1	1	1
			XRULES	> 50000	0.95	0.94	0.94
			CBA	3	0.91	0.93	0.92
			CPAR	32	0.93	0.94	0.93

(a) Results of the empirical evaluation   (b) Scalability of XCCS with data size

Fig. 2. Empirical results and scalability of of XCCS

4.4 Scalability

To study the scalability of XCCS, we produce five random samples from the *DBLP* corpus including, respectively, 100, 1000, 10000 and 100000 XML documents and investigate how execution time varies with the size of the underlying data. Notably, the overall execution time is actually the sum of two contributions: the time to perform the preprocessing tasks described in subsection 4.2 and the time to execute the model-learning scheme of fig. 1(a). The global support threshold τ of fig. 1(a) is fixed to 0.1.

Fig. 2(b) shows the execution time of XCCS against the increasing number of XML documents. In practice, XCCS scales linearly with the size of the data. This property means that the induction of classifiers with XCCS can be scaled up to very large databases of XML documents.

5 Related Work

A broad literature exists on XML classification. A selection of some state-of-the-art approaches is reviewed next.

Such approaches can be divided into two major categories. One family of approaches focuses only on the structural information of XML data in classifier induction and class prediction, whereas the other family of approaches performs a more sophisticated separation of the classes, by considering both the content and structural information of XML data.

Among the structure-oriented approaches, we here review [10,22,11].

In [10] each XML document is represented as a sequence of node labels obtained through a dept-first traversal of its tree structure. Sequential pattern mining is then applied to the sequences within each class in order to discover representative subsequences corresponding to frequent sub-trees, that are viewed as distinctive characteristics of the available classes of XML documents. An unlabeled XML document is classified into the class whose characteristic sequential patterns collectively exhibit the highest average matching score. The latter is measured, with respect to the sequence of the unlabeled XML document, through the notion of longest common subsequence.

XRULES [22] is a state-of-the-art technique for the structural classification of XML documents. XRULES initially enumerates all frequent embedded subtrees related to the individual classes of the underlying XML documents. These subtrees are then exploited to form a set of predictive structural rules. The generic structural rule relates the presence of a frequent subtree (i.e., the rule antecedent) within a given XML document to the likelihood that the same XML document belongs to the class predicted by the rule consequent.

A technique presented in [11] classifies XML data by mining the emerging attribute trees within the individual classes of a collection of labeled XML documents. An attribute tree is essentially a subtree that also takes into account the attributes associated to the nodes of the tree structures of the original XML documents. An attribute tree is said emerging if it frequently occurs in the XML trees of one class and rarely within those in any other class. In order to induce a classifier within the frequent-pattern paradigm, the emerging attribute trees are used as binary features for decision tree learning.

A general criticism to the structure-oriented approaches is their inability at separating the classes of the available XML documents, when a same structure is shared across all classes for content representation.

The more sophisticated category of approaches to XML classification based on both structure and content includes [17,26,24,20].

The approach in [17] explores the construction of expressive feature spaces for the classification of schema-less XML documents. In particular, focus is on features capturing content, structure and ontological information. The structural features of XML documents included in the feature space are twigs (i.e. triples of the form `<left-child label, parent label, right-child label>`). The proportion of content and structural features in the resulting class-specific feature space is chosen on the basis of empirical considerations on the extent of structural diversity of the training XML documents. Classification of XML data in the resulting feature space is performed through support vector machines [5].

A text classifier that can cope with semi-structured XML documents is proposed in [26]. The idea is to leverage structure for grouping terms from the content of a same node: such terms are treated together and distinguished from terms appearing in the content of other nodes. A novel representation of XML documents, called *structured vector model*, allows to view an XML document as a structured vector, whose elements are either terms of leaf nodes or, recursively, other (nested) structured vectors modeling the substructures in the XML document. The classification of XML documents is performed by means of a Naïve Bayes classifier, obtained by plugging into the (conditional) probability of a class (given an unseen XML document) an adaptation of the traditional Bernoulli generative model to the structured vector model.

The two approaches in [24,20] classify XML data by resorting on the Structured Link Vector Model (SLVM) [25]. The latter is essentially an extension of the traditional space vector model [4] for representing XML documents. Intuitively, the SLVM model represents an XML document as a collection of traditional vector space models, each being associated to a leaf node of the XML document itself. Together, these vectors form a term-by-leaf matrix, whose generic (i, j)-th entry corresponds to the *tfidf* value [4] of the i-th term within the j-th leaf node of the underlying XML document. The SLVM representation is used with support vector machine for classification in [20]. This approach is refined in [24], where the closed frequent subtrees within the tree structures of the available XML documents are exploited to incorporate the textual information into the SLVM model instead of the inclusion of individual leaves (which does not consider their location and hierarchical relationships within the XML documents).

Although effective, the approaches to the induction of XML classifiers from both content and structure have not sofar focused on learning explicative classification models, i.e., compact, human-intelligible and actionable summarizations of the discriminatory properties of the individual classes of XML documents. We believe that this is a major limitation in all those settings in which XML classification is preliminarily performed to enable more effective and efficient XML search, retrieval and filtering.

6   Conclusions and Further Research

XCCS is an approach to XML classification that induces clearly interpretable predictive models, which are of great practical interest for more effective and efficient XML search, retrieval and filtering. XCCS induces very compact classifiers with outperforming effectiveness from very large corpora of XML data. Empirical evidence revealed that XCCS scales to induces very compact classifiers with outperforming effectiveness from very large XML corpora.

Ongoing research aims to increase the discriminatory power of XML features by incorporating the textual context of words in the leaves of the XML trees. Also, we are studying enhancements of model learning in XCCS, with which to induce classification rules that also consider the absence of XML features.

References

1. Agrawal, R., Srikant, R.: Fast Algorithms for Mining Association Rules. In: Proc. of Int. Conf. on Very Large Data Bases, pp. 487–499 (1994)
2. Arunasalam, B., Chawla, S.: CCCS: A Top-Down Association Classifier for Imbalanced Class Distribution. In: Proc. of ACM SIGKDD Int. Conf. on Knowledge Discovery and Data Mining, pp. 517–522 (2006)
3. Baker, L., McCallum, A.: Distributional Clustering of Words for Text Classification. In: Proc. of ACM Int. Conf. on Research and Development in Information Retrieval, pp. 96–103 (1998)
4. Baeza-Yates, R., Ribeiro-Neto, B.: Modern Information Retrieval. Addison-Wesley (1999)
5. Burges, C.J.C.: A Tutorial on Support Vector Machines for Pattern Recognition. Data Mining and Knowledge Discovery 2(2), 121–167 (1998)
6. Coenen, F.: LUCS KDD implementations of CBA and CMAR. Dpt. of Computer Science, University of Liverpool, http://www.csc.liv.ac.uk/frans/KDD/Software/
7. de Campos, L.M., Fernández-Luna, J.M., Huete, J.F., Romero, A.E.: Probabilistic Methods for Structured Document Classification at INEX'07. In: Fuhr, N., Kamps, J., Lalmas, M., Trotman, A. (eds.) INEX 2007. LNCS, vol. 4862, pp. 195–206. Springer, Heidelberg (2008)

8. Denoyer, L., Gallinari, P.: Report on the XML Mining Track at Inex 2005 and Inex 2006. ACM SIGIR Forum 41(1), 79–90 (2007)
9. Denoyer, L., Gallinari, P.: Report on the XML Mining Track at Inex 2007. ACM SIGIR Forum 42(1), 22–28 (2008)
10. Garboni, C., Masseglia, F., Trousse, B.: Sequential Pattern Mining for Structure-Based XML Document Classification. In: Fuhr, N., Lalmas, M., Malik, S., Kazai, G. (eds.) INEX 2005. LNCS, vol. 3977, pp. 458–468. Springer, Heidelberg (2006)
11. De Knijf, J.: FAT-CAT: Frequent Attributes Tree Based Classification. In: Fuhr, N., Lalmas, M., Trotman, A. (eds.) INEX 2006. LNCS, vol. 4518, pp. 485–496. Springer, Heidelberg (2007)
12. Liu, B., Hsu, W., Ma, Y.: Integrating Classification and Association Rule Mining. In: Proc. of Conf. on Knowledge Discovery and Data Mining, pp. 80–86 (1998)
13. Liu, B., Ma, Y., Wong, C.K.: Improving an Association Rule Based Classifier. In: Zighed, D.A., Komorowski, J., Żytkow, J.M. (eds.) PKDD 2000. LNCS (LNAI), vol. 1910, pp. 504–509. Springer, Heidelberg (2000)
14. Manning, C., Raghavan, P., Schutze, H.: Introduction to Information Retrieval. Cambridge University Press (2008)
15. Murugeshan, M., Lakshmi, K., Mukherjee, S.: A Categorization Approach for Wikipedia Collection based on Negative Category Information and Initial Descriptions. In: Proc. of the Initiative for the Evaluation of XML Retrieval (INEX 2007), pp. 212–214 (2007)
16. Ning, P., Steinbach, M., Kumar, V.: Introduction to Data Mining. Addison Wesley (2006)
17. Theobald, M., Schenkel, R., Weikum, G.: Exploiting Structure, Annotation, and Ontological Knowledge for Automatic Classification of XML Data. In: Proc. of WebDB Workshop, pp. 1–6 (2003)
18. Yin, X., Han, J.: CPAR: Classification based on Predictive Association Rules. In: Proc. of SIAM Int. Conf. on Data Mining, pp. 331–335 (2003)
19. Xing, G., Guo, J., Xia, Z.: Classifying XML Documents Based on Structure/Content Similarity. In: Fuhr, N., Lalmas, M., Trotman, A. (eds.) INEX 2006. LNCS, vol. 4518, pp. 444–457. Springer, Heidelberg (2007)
20. Yang, J., Zhang, F.: XML Document Classification Using Extended VSM. In: Fuhr, N., Kamps, J., Lalmas, M., Trotman, A. (eds.) INEX 2007. LNCS, vol. 4862, pp. 234–244. Springer, Heidelberg (2008)
21. Yong, S.L., Hagenbuchner, M., Tsoi, A.C., Scarselli, F., Gori, M.: Document Mining Using Graph Neural Network. In: Fuhr, N., Lalmas, M., Trotman, A. (eds.) INEX 2006. LNCS, vol. 4518, pp. 458–472. Springer, Heidelberg (2007)
22. Zaki, M., Aggarwal, C.: XRules: An Effective Algorithm for Structural Classification of XML Data. Machine Learning 62(1-2), 137–170 (2006)
23. Bratko, A., Filipic, B.: Exploiting Structural Information for Semi-structured Document Categorization. Information Processing and Management 42(3), 679–694 (2006)
24. Yang, J., Wang, S.: Extended VSM for XML Document Classification Using Frequent Subtrees. In: Geva, S., Kamps, J., Trotman, A. (eds.) INEX 2009. LNCS, vol. 6203, pp. 441–448. Springer, Heidelberg (2010)
25. Yang, J., Chen, X.: A Semi-structured Document Model for Text Mining. Journal of Computer Science and Technology 17(5), 603–610 (2002)
26. Yi, J., Sundaresan, N.: A Classifier for Semi-Structured Documents. In: Proc. of ACM SIGKDD Int. Conf. on Knowledge Discovery and Data Mining, pp. 340–344 (2000)

A Fast Method for Web Template Extraction via a Multi-sequence Alignment Approach

Filippo Geraci[1] and Marco Maggini[2]

[1] Istituto di Informatica e Telematica, Consiglio Nazionale delle Ricerche, Pisa, Italy
[2] Dipartimento di Ingegneria dell'informazione, Università di Siena, Siena, Italy
filippo.geraci@iit.cnr.it, maggini@dii.unisi.it

Abstract. The increased richness of the page contents and the diffusion of content management systems are responsible for the impressive changes happened in the last decade in a typical Web site layout. In fact, most of the Web sites are endowed with a template which gives them a uniform graphical and functional structure. Templates, by themselves, do not change the informative content of the pages, but they are typically designed to enhance the usability by uniformly organizing the contents following a standardized arrangement of functional blocks and by providing navigation tools, like menus or banners. However, the additional information provided by the template can worsen the performances of many algorithms for automatic Web processing. In fact, templates are designed for human users and provide redundant information that is marginally correlated with the main contents of a given page. These additional parts act as a noise source for many automated tasks such as web crawling, indexing, page classification and clustering. Hence, a preprocessing step to detect and strip the parts related to the template is needed to extract only the specific contents of each page. The critical part for the automation of this process is the accurate detection of the template, given a minimal set of pages from a given site.

The template consists in parts of the HTML tag structure that are shared by all the pages from the site, and its detection is made difficult by the variable parts intermixed with them. We propose an algorithm for template extraction that is based on the alignment of the HTML sequences of a set of pages. This approach is quite fast since it exploits efficient alignment algorithms proposed in bioinformatics and does not require complex tree matching or visual layout analysis. The algorithm aligns the HTML tag streams from pairs of pages and extracts a set of candidate templates that are merged following a binary tree consensus schema to increase the algorithm precision. The experimental evaluation shows that 16 sample pages are enough to extract the site template with good accuracy. The effects of the template stripping on a clustering task are also investigated, showing that the clustering quality can be effectively improved.

1 Introduction

Automatic processing of Web contents is nowadays an important tool to provide added value services to users. For instance, search engines have become an indispensable and powerful gateway to access information on the Web, information extraction techniques are the basis to design systems for data integration and collection from unstructured Web

A. Fred et al. (Eds.): IC3K 2011, CCIS 348, pp. 172–184, 2013.

documents, sentiment analysis and topic monitoring from Web posts are attracting an increasing interest from companies. Most of these applications need to focus only on the most informative part of a Web document and the quality of their results may be heavily hindered by the large amount of side information that is usually embedded in Web documents. In fact, most of Web pages are usually dynamically constructed by content management systems or edited with design tools that use a fixed template to format them. Basically a template provides a common structure of all the pages that is reflected both in the graphical layout and in the skeleton of HTML tags that are used to format the page.

Many different techniques for template detection and removal have been proposed in the literature. All the reported results show that automatic processing techniques, like indexing, Web page classification and clustering, may benefit from template removal. Some early works approached the task by a layout analysis based on a graphical rendering of the page but their practical use is limited by the high computational costs [7]. The most recent algorithms try to exploit the structure of HTML tags by searching for common parts of the pages or blocks [1,4,9] or by an analysis of the document tree as defined by HTML tags [12,2]. A related application that exploits similar techniques is wrapper induction (see e.g. [11]), that aims at the automatic generation of information extraction modules by exploiting the common structure of the Web pages containing the data we are interested to collect.

Desirable features of an automatic template extraction algorithms are low computational costs since they may employed in large scale systems, easy adaptability to any template structure using a minimum number of sample pages, the use of a completely unsupervised (or a very limited supervised) approach to avoid the need of human effort that would limit its applicability, and high precision in the detection of the template structure. In this paper we present a template extraction algorithm that meets most of these requirements. The proposed technique exploits the Needleman and Wunsch alignment algorithm originally devised for DNA sequences [8]. This algorithm is exploited to find the most likely alignment among two sequences of HTML tags extracted from two different pages from the same site. Since the common sequence structure that is extracted from a pair of pages may be strongly dependent on the specific selected pages, a hierarchical consensus schema is adopted: given a certain number of different pairs of pages, the extracted common sequences, that constitute the current set of hypotheses for the target template, are recursively compared to further distillate the common parts. The comparisons form a binary consensus tree and the output of the algorithm is the template sequence available in the tree root. The experimental results show that the proposed method is able to yield a good precision and recall in the detection of the HTML tags belonging to the template given only a very limited number of sample pages from the site (in the considered setting only 16 pages are able to provide satisfactory performances). The algorithm is also efficient and the running time for template extraction is quite low. Finally, the evaluation shows that the proposed extraction technique can provide significant benefits in a Web mining task, namely Web page clustering, confirming similar results as those reported in the literature.

The paper is organized as follows. The next section describes the template extraction algorithm in details. Then section 3 reports both the evaluation of the accuracy in the prediction of the real template and an analysis of the impact of the template removal in a Web mining application (Web page clustering). Finally in section 4 the conclusions are drawn and the future developments are sketched.

2   Template Detection Algorithm

The proposed template extraction algorithm is based on the assumption that a Web template is made up of a set of overrepresented contiguous subsequences of HTML tags shared by the pages from a given web site. According to this assumption, some local parts of the template can be missing in some particular web page.

The first processing step consists in splitting each page into a sequence of tokens. In the following, a token corresponds to a single HTML tag or to the text snippet contained between two consecutive tags. The tokens are then normalized to remove the potential differences due to the human editing of the pages or irrelevant features (i.e. extra white spaces are removed, capital letters are lowered). Moreover tags are normalized by sorting their internal attributes. The normalized tokens form an alphabet of symbols over which the page sequences are generated. Hence, the problem of finding the common parts in two Web pages can be cast as the computation of the global alignment of the two corresponding sequences of normalized tokens. The alignment can be obtained by exploiting a modified version of the DNA global alignment algorithm due to Needleman and Wunsch [8], which is based on a dynamic programming technique.

However, a single alignment of only two Web pages is likely to produce a poor approximation of the template since they can share the same tokens just by chance or a portion of the real template could be missing in one of the two pages. Hence, in order to compute a more reliable template, the alignment procedure is repeated exploiting a set t pages using a recursive procedure. At each step the k input sequences are paired and aligned to yield $k/2$ template profile candidates, starting from the initial t sequences representing the available samples. The template candidate originating from a given pair of sequences is obtained by pruning those tokens that have receive low evidence in the alignment steps performed so far. The procedure iterates until it remains only a single profile that is returned as final Web template (i.e. $\log_2(t)$ steps are required).

2.1   The Alignment Algorithm

Two strings s_1 and s_2 are aligned by inserting white spaces (gaps) into them, such that the probability of finding the same symbol in the same position of the two modified strings is maximized, and that only one string can contain a gap in a certain position. An alignment corresponds to the sequence of operations required to align the two strings. Consider the case in which we are comparing the i-th symbol of string s_1 and the j-th symbol of string s_2 (later referred to as $s_1[i]$ and $s_2[j]$). There are three options: skip both symbols and move to the next one for both the strings; insert a gap in $s_1[i]$ or in $s_2[j]$. We assign a score (or a penalty) to each operation so as to compute a global score for the whole alignment. In particular we reward the case in which $s_1[i] = s_2[j]$ giving score 1, we ignore the case in which we skip $s_1[i]$ and $s_2[j]$ because they are different, and we penalize the insertion of a gap giving score -1. The goal is to find an alignment with highest score among all the possible alignments.

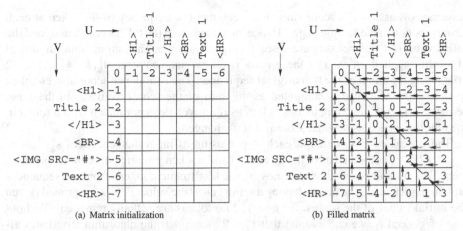

(a) Matrix initialization   (b) Filled matrix

Fig. 1. Similarity matrix for two small HTML snippets

In the considered application, the strings correspond to sequences of tokens extracted from the HTML pages, as described before. Let $U = \{u_1, \ldots, u_n\}$ and $V = \{v_1, \ldots, v_m\}$ be the sequences of tokens representing two Web pages. The alignment algorithm builds a $(n+1) \times (m+1)$ matrix S (called *similarity matrix*) such that each entry contains the score of the best alignment of a prefix of U and V. We initialize the first row and column of S as follows: $S[0][i] = S[i][0] = -i$. Figure 1(a) shows the matrix S initialized for a small portion of two HTML pages.

Once initialized, we fill each line of the similarity matrix S from left to right. Consider the case in which we are filling the cell in position i, j. If we insert a gap in position i of U, the partial score of the alignment becomes $S[j][i-1]-1$. Symmetrically, if we insert a gap in position j of the sequence V, the partial alignment score becomes $S[j-1][i] - 1$. If we align v_j and u_i there are two cases: the two tokens match, thus score becomes $S[j-1][i-1] + 1$, the tokens do not match, thus the score does not change. Since we are interested in the best possible alignment, the algorithm chooses the alternative that maximizes the score. Note that the value of $S[i][j]$ is always derived from one of its neighbours. In a separate matrix $\hat{S}$ we keep trace of the path from a cell of S to its predecessor. Figure 1(b) shows the final similarity matrix and a visual representation of the corresponding $\hat{S}$.

The final phase of the algorithm navigates $\hat{S}$ from the bottom right position to the top left so as to extract the aligned tokens. Given the element $\hat{S}[j][i]$, if its predecessor is $\hat{S}[j][i-1]$ the algorithm outputs v_i, symmetrically, if the predecessor is $\hat{S}[j-1][i]$, the algorithm outputs u_j. When the predecessor of $\hat{S}[j][i]$ is $\hat{S}[j-1][i-1]$ we have a match/missmatch. In the first case we output indifferently u_j or v_j, otherwise we output both tokens. Note that this sequence of tokens has to be inverted to match the original order.

2.2   The Template Distillation Algorithm

Given a set of t input pages represented by their normalized token sequences s_i^0, $i = 1, \ldots, t$, we can randomly pair these pages to obtain $t/2$ pairs (s_i^0, s_j^0). Each token in these sequences is assigned a score value that accounts the frequency of the token at each given position in the sample pages. Hence, at the first step the tokens are all assigned the initial value of $1/t$, since they are observed in just one case. The output of the alignment algorithm applied on each of these pairs is an aligned sequence $s_k^1, k = 1, \ldots, t/2$ containing the union of the two original sequences. The confidence score of each token in the output sequence is computed as the sum of the scores available in the input sequences for the matching tokens, whereas if a token does not match with a token in the other sequence its original score is left unchanged.

The procedure is repeated at each step r using as input the sequences $s_k^{r-1}, k = 1, \ldots, t/2^{r-1}$, computed after the alignment and a token pruning phase in the previous step. The pruning phase aims at removing the less frequent tokens from the sequences and, in particular, all the tokens having a score below the value $2^{r-2}/t$ are pruned (given the initial values of the scores, for $r = 1, 2$ no tokens are actually removed). Without loss of generality, we can assume that $t = 2^n$ such that the algorithm terminates after $n = \log_2(t)$ steps when the final sequence corresponding to the output template is computed. The distillation procedure can be described as a binary tree whose leaves correspond to the input pages whereas the internal nodes are the result of the recursive distillation process. Nodes in the higher levels represent more reliable template hypotheses since they collect the evidence extracted by combining the alignments originating from a larger set of pages. Finally the final result is the represented by the sequence available at the root of the tree. Hence, it is easy to verify that the procedure requires to perform a total of $2^n - 1 = t - 1$ alignments.

The sequence available at the root node is finally pruned by removing those tokens whose score is below $1/2$, that is that the evidence of their presence is supported by less than half of the pages used in the process (this is the value used in the experiments but a different threshold could be used).

3   Experimental Results

The computational cost and the impact of the template extraction algorithm in a typical Web mining task were evaluated in a set of experiments. In particular, the first evaluation was designed to assess the ability of the algorithm to predict exactly which tokens belong to the template for a given Web site, whereas a second benchmark has been performed to investigate the contribution of the algorithm on the final outcome of an information retrieval task (namely Web page clustering).

The experiments were performed on a Mac equipped with a 3.33 Ghz Intel core 2 Duo, 8 Gb of RAM and the Mac OS X v. 10.6.7 operating system. The algorithm was implemented in Python. Even if we could speed up the algorithm by exploiting the intrinsic parallelism of the CPU by performing many alignments of pages in parallel, we decided to implement the algorithm as a sequential program to avoid biases in the running time evaluation due to the particular CPU architecture. However we would like to remark that, using an architecture with $t/2$ processors, it is possible to speed up the algorithm from $O(t)$ to $O(\log_2(t))$.

Table 1. The Web sites collected in the benchmark dataset, with the related average number of tokens in a typical page and in the corresponding template

Web site	# tokens	# template tokens
PI: punto-informatico.it	876	686
Ansa: ansa.it	2245	1794
Wikipedia: en.wikipedia.org	2058	467
allMusic: allmusic.com	1559	282
zdNet: zdnet.com	2882	2820

3.1 Evaluation of Template Prediction

The main goal of the first set of experiments was to assess the ability of the algorithm to predict which tokens belong to the Web site template. To the best of our knowledge there is not a public available dataset of Web templates to be used for the comparison of template extraction algorithms. To perform an accurate evaluation the dataset should contain a set of pages extracted from Web sites that exploit different template structures and for each of them the actual template, that is the ground truth, should be available. Hence, we decided to select five Web sites with different size and characteristics in their templates (i.e. the ratio between the number of template tokens and the overall number of tokens in a typical page). For each of them, the "true" template was manually extracted. The sites we selected are reported in table 1, that shows for each of them the average number of tokens per page and the number of tokens in the associated template.

The task of extracting the template from a Web page (or a set of pages) is made complex because of two reasons:

- to discriminate whether a certain section of HTML is part or not of the template can result in an arbitrary choice;
- the template can contain some tags which require to be manually modified.

The first situation can be found in Web sites that are divided into sections, where each section shares with the others most of the template except some details. In these cases, we could define a more general template for the whole site or more specific templates for each section. Another interesting case is that of on–line newspapers in which often the home page has a slightly different structure than the other pages. An example of the second case is a template featuring a context menu such that each page contains a link pointing to itself or to the category to which it belongs. Another case is when the pages contain random generated banners/ads for which the template allocates some space, but the target link changes every time.

Table 2 reports the precision and recall of the computed template against the manually extracted ground truth, for different values of the parameter t. Even if there is not a theoretical limitation for the value of parameter t we narrowed the tests only to powers of 2.

It is not surprising that the recall tends to have a monotonically decreasing trend with respect to the value of the parameter t. In fact, when using more pages it is likely to evidence slight differences in the use of the template tags that prevent the algorithm to select them. In contrast the precision tends to increase. This is due to the fact that the

Table 2. Precision, recall and number of tags in the extracted template for different values of the parameter t

	PI: punto-informatico.it					Ansa: ansa.it				
# Pages	2	4	8	16	32	2	4	8	16	32
Precision	0.720	0.804	0.842	0.855	0.811	0.681	0.816	0.891	0.856	0.859
Recall	0.940	0.940	0.930	0.868	0.739	0.973	0.957	0.903	0.908	0.892
# tags	915	808	757	697	625	2651	2159	1822	1909	1869
	Wikipedia: en.wikipedia.org					allMusic: allmusic.com				
# Pages	2	4	8	16	32	2	4	8	16	32
Precision	0.155	0.226	0.569	0.906	0.911	0.119	0.196	0.395	0.702	0.725
Recall	0.946	0.905	0.899	0.931	0.907	0.950	0.946	0.939	0.936	0.936
# tags	3012	1864	738	480	466	2242	1362	671	377	365
	zdNet: zdnet.com									
# Pages	2	4	8	16	32					
Precision	0.674	0.874	0.890	0.934	0.959					
Recall	0.941	0.917	0.915	0.910	0.777					
# tags	4051	2961	2908	2748	2285					

higher is the number of pages exploited in the computation of the consensus, the lower is the probability for a token not belonging to the template to be shared enough to be kept in the final hypothesis. On the other hand, negative effects in the precision due to the local discrepancy of some pages are also mitigated by the higher number of pages.

With the purpose of better investigating the effects of the choice of t in the final template prediction quality, we report in figure 2 the f-measure for the experiments described above. F-measure is the harmonic mean of precision and recall, and it is commonly used in information retrieval to evaluate the trade-off between these two measures. The plot shows how a overestimated value for t tends to slightly reduce the overall F-measure (as highlighted from PI and zdnet). On the other hand, a underestimated value of t can result in a poor F-measure (as highlighted from wikipedia and allmusic) due to a low precision.

We observed that precision is strongly affected by the ratio between the size of the template and the number of tokens in the considered web pages. In particular, for a given assignment of t, the higher the ratio, the higher the precision. We observed also that, for increasing values of t, the number of tokens returned from the algorithm decreases (see table 2). This latter observation suggests that precision is not influenced by the number of retrieved tokens effectively belonging to the template, but from the number of false positives.

According to Figure 2 we can conclude that setting $t = 16$ seems to be a good choice in most situations, at least when an estimation of the web template size can not be automatically accomplished.

Figure 3 reports the running time of the algorithm using the hardware and software configuration described earlier in this section. Not surprisingly the algorithm running time depends linearly on t. Note that for the assignment of $t = 16$ the algorithm is always able to extract the template in few minutes.

We expected also to find a strong correlation between the running time and the characteristics of the Web sites (average number of tokens, template size), but our

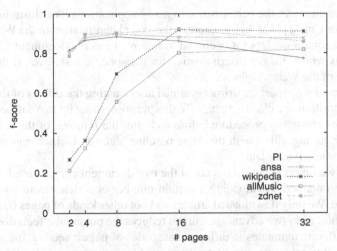

Fig. 2. F-measure of the template extraction algorithm with respect to the parameter t

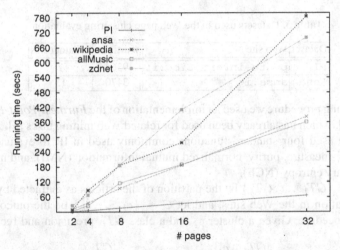

Fig. 3. Running time of the template extraction algorithm for increasing values of the parameter t

experiments confirm only partially this correlation. Consider the cases of allMusic and Ansa, the former has a lower number of tokens per page and the template size is smaller, however the running time is comparable to the latter. The explanation of this fact is that the running time is affected by the number of tokens in each alignment and not by either the page size or the template size. This feature depends from intrinsic characteristics of the Web pages, thus it is unpredictable in advance.

3.2 Web Page Clustering

The main application fields of Web template extraction algorithms are: information retrieval and Web mining. Experimental results in the literature have shown how these algorithms can improve the performance in many applications like clustering, classification and indexing [1] [12]. In this context, the exact identification of the Web template itself is not the main goal, but it is crucial to remove from the document corpus those repeated parts which are not discriminative for the specific task, even if they are not properly part of the Web template.

In this section we report experiments aimed at evaluating the effects of the template extraction algorithm in a clustering task. To this purpose we set up two datasets to which we applied the clustering procedure before and after the removal of the template. We also compare our algorithm with the naive baseline algorithm which removes the real (manually extracted) template.

In table 3 we report some details about the two document corpora used. Both these corpora were obtained downloading a certain number of articles from an Italian on-line magazine. We only downloaded articles and not other kinds of pages (i.e. the home page). This choice has two advantages: first it reduces the potential effects due to the use of slightly different templates in different categories of pages; second, the subdivision of the articles into categories can be easily derived from the manual classification made by the web site owners.

Table 3. Datasets used in the Web page clustering evaluation

Dataset	Web site	# articles	# Categories
PI	punto-informatico.it	9364	7
Ansa	ansa.it	3470	13

As clustering procedure we used an implementation of the *Furthest-Point-First* (FPF) algorithm [6], which has already been used for related Web mining tasks [5]. As quality measures we used four standard functions commonly used in IR to evaluate clustering quality: F-measure, purity, normalized mutual information (NMI), and normalized complementary entropy (NCE).

Let $GT = \{GT_1, \ldots, GT_k\}$ be the partition of the articles as extracted by the original classification in the Web sites, and let $C = \{c_1, \ldots, c_k\}$ be the outcome of the clustering procedure. Given a cluster c_j and a class GT_i, precision and recall can be defined as

$$p_{i,j} = \frac{|GT_i \cap c_j|}{|c_j|}, \qquad r_{i,j} = \frac{|GT_i \cap c_j|}{|GT_i|} .$$

Let $F(GT_i, c_j)$ be the F-measure for a given cluster c_j and a class GT_i defined as the harmonic mean of precision and recall. We can extend the concept of F-measure to the entire clustering by using the following formula:

$$F = \sum_i \frac{|GT_i|}{n} \max_j (F(GT_i, c_j)),$$

where n is the sum of the cardinality of all the classes. The value of F is in the range $[0, 1]$ and a higher value indicates better quality.

Purity is devised to evaluate how well each cluster fits in a class of the ground truth. Each cluster is assigned to the class which share the highest number of articles with it. The higher number of shared documents the higher cluster quality according to purity.

The overall purity for a clustering is the sum of the contributions provided by each cluster. In details purity is defined as:

$$A = \sum_{j} \frac{|c_j|}{n} \max_{i} |GT_i \cap c_j| \ .$$

The value of A is in the range $[0, 1]$ and a higher value indicates better quality.

The normalized mutual information (see e.g. [10, page 110]), comes from information theory and is defined as follows:

$$NMI(C, GT) = \frac{2}{\log |C||GT|} \sum_{c \in C} \sum_{c' \in GT} P(c, c') \cdot \log \frac{P(c, c')}{P(c) \cdot P(c')}$$

where $P(c)$ represents the probability that a randomly selected article o_j belongs to c, and $P(c, c')$ represents the probability that a randomly selected article o_j belongs to both c and c'. The normalization, achieved by the $\frac{2}{\log |C||GT|}$ factor, is necessary in order to account for the fact that the cardinalities of C and GT are in general different [3]. Higher values of NMI mean better clustering quality. NMI is designed for hard clustering.

The normalized complementary entropy [10, page 108] is a modified version of entropy which ranges in the interval $[0, 1]$. The entropy of a cluster $c_j \in C$ is

$$E_j = \sum_{k=1}^{|GT|} -r_{k,j} \log r_{k,j}$$

where $r_{k,j}$ is the recall of cluster c_k for class GT_j. To normalize the entropy we divide E_j for $\log |GT|$. To evaluate the entropy for the whole clustering we can now sum the contribution of each cluster. Then we compute the complement of the entropy to make it coherent with the above described measures so that higher values mean better quality. The final formula for the normalized complementary entropy is the following:

$$NCE(C, GT) = 1 - \frac{1}{\log |GT|} \sum_{j \in 1}^{|C|} E_j$$

Table 4 reports an evaluation of the clustering quality for the datasets PI and Ansa in different cases: when the template is not removed (column *Original* in the tables), when the template is identified and removed by the template extraction algorithm (with different values of the parameter t), and when we remove the manually extracted template. For each measure, we highlighted the case reporting the highest performance.

A first important observation is that, the experiments confirm how template removal is an effective tool to improve the quality of Web information retrieval tasks. Another important observation is that these experiments confirm how setting $t = 16$ in the extraction algorithm is, in general, a good choice independently of the document corpus characteristics.

Hence, even if it works quite well in practice, the removal of the manually extracted template is not the best possible choice for improving the clustering quality. In fact, for

Table 4. Clustering performance when using the original pages, the proposed template removal algorithm with $t = 8, 16$, and the removal of the manually extracted template

	PI: punto-informatico.it				Ansa: ansa.it			
	Original	$t = 8$	$t = 16$	Baseline	Original	$t = 8$	$t = 16$	Baseline
F-measure	0.234	0.260	**0.276**	0.274	0.387	0.445	**0.449**	0.422
Purity	0.249	0.261	0.272	**0.279**	0.397	0.466	**0.471**	0.452
NMI	0.040	0.044	**0.059**	0.058	0.346	**0.425**	0.384	0.346
NCE	0.064	0.068	**0.084**	0.083	0.383	0.388	**0.421**	0.383

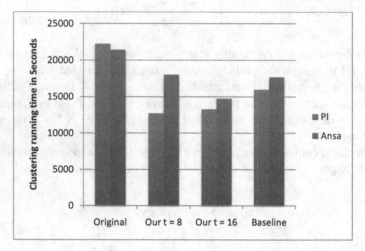

Fig. 4. Running time of the clustering algorithm for the datasets PI and Ansa before and after template removal

this task the perfect identification of the template is not the main issue. The experiments show that the automatic template extraction algorithm consistently provides a better result in the clustering task when compared to the removal of the manually extracted template.

Figures 4 and 5 show the benefits of Web template removal for clustering in terms of respectively clustering running time and reduction of the corpus size.

In particular, figure 4 reports the clustering running time for PI and Ansa. The experiments confirm that Web template removal has consistently positive effects on the clustering time. The extraction algorithm makes clustering faster than that obtained by the manual Web template removal. In the case of PI The proposed algorithm can contribute to save up to 50% of the clustering time (compared to the clustering of the original dataset). In the case of Ansa the amount of saved time depends on the choice of t. Figure 4 shows that, setting $t = 16$, the saved time is still about 40%.

Figure 5 shows that the extraction algorithm reduces the amount of used space to store the document corpus of a factor 3. Note that for the PI dataset the algorithm reduces the corpus size more than the removal of the manually extracted template. This is due to the fact that, in this case, the extraction algorithm has identified snippets of

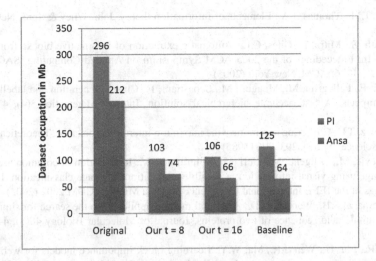

Fig. 5. Space requirements in MB for the datasets PI and Ansa before and after template removal

HTML, which are repeated in the most of pages, even if they are not part of the Web template (i.e. tags commonly used for article formatting).

4 Conclusions

The paper describes a template extraction technique that exploits a sequence alignment algorithm to detect the HTML structure that is shared by the pages from a target site. The algorithm is able to yield good retrieval performance given just a small number of pages from the site. It is also shown that the removal of the template extracted by the algorithm can significantly improve the performance of a Web page clustering application. Future work will concern the improvement of the alignment step by studying different solutions for assigning the similarity matrix on the basis of different features of the compared elements.

Acknowledgements. This work was partially supported by the Italian Registry of the ccTLD ".it" Registro.it.

References

1. Bar-Yossef, Z., Rajagopalan, S.: Template detection via data mining and its applications. In: Proceedings of the 11th International Conference on World Wide Web, WWW 2002, pp. 580–591. ACM, New York (2002)
2. Chakrabarti, D., Kumar, R., Punera, K.: Page-level template detection via isotonic smoothing. In: Proceedings of the 16th international conference on World Wide Web, WWW 2007, pp. 61–70. ACM, New York (2007)

3. Cover, T.M., Thomas, J.A.: Elements of information theory. John Wiley & Sons, New York (1991)
4. Debnath, S., Mitra, P., Giles, C.L.: Automatic extraction of informative blocks from webpages. In: Proceedings of the 2005 ACM Symposium on Applied Computing, SAC 2005, pp. 1722–1726. ACM, New York (2005)
5. Geraci, F., Pellegrini, M., Maggini, M., Sebastiani, F.: Cluster generation and labelling for web snippets: A fast, accurate hierarchical solution. Internet Matematics 3(4), 413–443 (2008)
6. Gonzalez, T.F.: Clustering to minimize the maximum intercluster distance. Theoretical Computer Science 38(2/3), 293–306 (1985)
7. Kovacevic, M., Diligenti, M., Gori, M., Milutinovic, V.: Recognition of common areas in a web page using visual information: a possible application in a page classification. In: Proceedings of the IEEE International Conference on Data Mining, ICDM 2002 (2002)
8. Needlemana, S.B., Wunsch, C.D.: A general method applicable to the search for similarities in the amino acid sequence of two proteins. Journal of Molecular Biology 48(3), 443–453 (1970)
9. Song, R., Liu, H., Wen, J.R., Ma, W.Y.: Learning block importance models for web pages. In: Proceedings of the 13th International Conference on World Wide Web, WWW 2004, pp. 203–211. ACM, New York (2004)
10. Strehl, A.: Relationship-based Clustering and Cluster Ensembles for High-dimensional Data Mining. Ph.D. thesis, University of Texas, Austin, US (2002)
11. Wong, T.L., Lam, W.: Adapting web information extraction knowledge via mining site-invariant and site-dependent features. ACM Trans. Internet Technol. 7 (February 2007)
12. Yi, L., Liu, B., Li, X.: Eliminating noisy information in web pages for data mining. In: Proceedings of the Ninth ACM SIGKDD International Conference on Knowledge Discovery and Data Mining, pp. 296–305 (2003)

TermWatch II: Unsupervised Terminology Graph Extraction and Decomposition

Eric SanJuan*

LIA & IUT STID, University of Avignon
339 chemin des Meinajaries, F-84000 Avignon Cedex 9, France
eric.sanjuan@univ-avignon.fr
http://termwatch.es

Abstract. We present a symbolic and graph-based approach for mapping knowledge domains. The symbolic component relies on shallow linguistic processing of texts to extract multi-word terms and cluster them based on lexico-syntactic relations. The clusters are subjected to graph decomposition based on inherent graph theoretic properties of association graphs of items (multi-word terms and authors). This includes the search for complete minimal separators that can decompose the graphs into central (core topics) and peripheral atoms. The methodology is implemented in the TermWatch system and can be used for several text mining tasks. In this paper, we apply our methodology to map the dynamics of terrorism research between 1990-2006. We also mined for frequent itemsets as a mean of revealing dependencies between formal concepts in the corpus. A comparison of the extracted frequent itemsets and the structure of the central atom shows an interesting overlap. The main features of our approach lie in the combination of state-of-the-art techniques from Natural Language Processing (NLP), Clustering and Graph Theory to develop a system and a methodology adapted to uncovering hidden sub-structures from texts.

Keywords: Knowledge domain maps, Natural language processing, Multi-word terms, Clustering, Graph decomposition, Formal concepts analysis, Frequent Itemsets.

1 Introduction

A timely awareness of recent trends in scientific domains is necessary to support several information intensive activities such as innovation, science and technology watch, business intelligence to name only a few. Such studies are usually conducted by analyzing the electronic literature available on line based on different approaches such as citation analysis, text and document clustering, pattern mining, novelty detection. Bibliometrics aims to elaborate indicators of the evolution of scientific activities using statistical and mathematical models. The two major bibliometric methods are co-citation and co-word analysis. Co-citation analysis has proved useful in highlighting major actors in a field

* Work supported by the French National Research Agency ANR via the project CAAS (Contextual Analysis and Adaptive Search - ANR 2010 CORD 001 02).

A. Fred et al. (Eds.): IC3K 2011, CCIS 348, pp. 185–199, 2013.

(the "who's who" of a field). Although some attempts have been made to work directly at the text level in bibliometrics, natural language processing (NLP) resources and capabilities have barely been tapped by this community. The most common NLP processing is limited to stemming [1] prior to clustering [2,3]. Text units have mainly been considered either as a bag-of-words or as a sequence of n-grams in the vast majority of topic mapping systems.

We take a different approach to text clustering and consider that a multi-disciplinary effort integrating surface linguistic techniques is necessary to elaborate indicators of topics trends at the level of texts. For this, we require a more fine-grained analysis, involving prior linguistic processing of the scientific literatures before applying statistical and mathematical models. The interesting features of our approach lie in the combination of state-of-the-art techniques from three disciplines: Natural Language Processing (NLP), Data Mining and Graph Theory. NLP enables us to extract meaningful textual units and identify relevant information between them, here multi-word terminological units. These text chunks correspond to domain concepts and the linguistic relations are lexical, syntactic and semantic variations. These variations are used in later stages of processing (clustering) to form topics through relations of synonymy and hyponymy/hypernymy and semantic relatedness. Prior grouping of term variants ensures that semantically close terms which reflect different aspects of the same topic are certain to end up in the same cluster at the end of the process. The linguistic theory behind the grouping of terms either by shared modifiers or by shared head is known as distributional analysis and was introduced by Harris [4]. It was later taken up by various studies in automatic thesaurus construction [5,6]. Ibekwe-Sanjuan [7] extended the types of identified relations and defined additional constraints like the position of added words and their number to avoid generating spurious variants. This approach has been implemented in the TermWatch system [8,9,10]. There co-occurrence (numerical) is optionally added during clustering as a means to capture the supplementary dimension of interactions between domain concepts. The end results are clusters of high semantic homogeneity which also capture the most salient association links.

TermWatch implements a hierarchical clustering algorithm to suit the characteristics of multi-word terms. This algorithm clusters the multi-word terms grouped into close semantic classes called components using optionally co-occurrence information. The clusters are represented as an undirected graph. The system has been applied successfully to text corpora from different domains and on several knowledge intensive tasks such as knowledge domain mapping in bio-technology [11], ontology population in the biomedical domain [12], opinion categorization of literature reviews [13].

Here we present an enhancement to the system by integrating a graph decomposition algorithm studied in [14] which enables the system to decompose complex graphs into more legible subgraphs representing coherent networks of research topics. This allows to split complex terminological networks of topics extracted by TermWatch based on their graph theoretic properties in order to identify sub-structures that represent highly connected sets of topics called central atom and distinct sets of topics called peripheral atoms.

In [15], we have applied previous versions of TermWatch II with an earlier implementation of the graph decomposition algorithm to scientific publications related to the

Sloan Digital Sky Survey[1]. Here, we apply the to mapping knowledge in terrorism research between 1990-2006. The datasets are publication records of peer-reviewed journal articles downloaded from the Web of Science (WoS). The input to our system are the titles and abstract, publication year and author fields of the records. We favored using the WoS database as it indexes high quality journals with high impact factor in their respective fields.

Since the sept 9/11 attack, a lot of attention has been focused on rapidly detecting indicators of potential terrorist threats. This corpus was built following a search on the WoS using the word "terrorism". $3,366$ bibliographic records were collected. Note that this corpus is not on individuals or groups involved in terrorist acts but rather on what researchers have been writing about terrorism: its effects on the victims and the general public, its forms, its means and ways to prepare for it. Previous studies have sought to map the terrorism domain either from this same perspective [16] or from that of groups actively involved in plotting and carrying out terrorist acts [17]. Of particular relevance to our study is the one done by C. Chen [16]. This author used the same database and the same query but on an earlier and shorter period (1990-2003). His results, validated by domain experts will serve as a "baseline" against which we compare our system's performance. Our analysis of the terrorism research dynamics is thus a follow-up of his study but using a different methodological approach.

The rest of the paper is structured as follows: Section 2 is a general description of TermWatch. Section 3 details the terminological graph extraction process. We then show in section 4 how an association graph can highlight a family of formal concepts and their relations based on the unique atom decomposition. Section 5 analyzes results obtained from the two case studies and section 6 draws some conclusions from this experiment.

2 Overview of TermWatch

TermWatch is designed to map research topics from unstructured texts and track their evolution in time. The system combines linguistic relations with co-occurrence information in order to capture all possible dimensions of the relations between domain concepts. It is currently run on-line on a LINUX server[2]. Standalone terminology graph construction and decomposition modules are available under the GNU public license (GPL). The processing of texts relies on surface linguistic relations between multi-word terms (MWTs) to build semantically tight clusters of topics. The processes leading from raw texts to the mapping of domain topics can be broken down into five major stages: multi-word term extraction, term variants identification, term clustering, graph decomposition and visualization. Figure 1 shows the overall process. As some components of the system have been described in previous publications [8,9,10], we will focus particularly on the graph decomposition algorithm of terminological graphs which aims to reveal a family of formal concepts and their relationships. A step-by-step procedure going from input texts to topic mapping consists in the following:

---

[1] http://www.sdss.org/
[2] http://system.termwatch.es

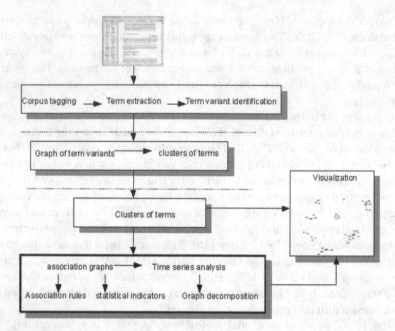

Fig. 1. Overview of the mapping knowledge domains process in TermWatch II

1. Build a scientific corpus reflecting a research question. The input corpus is composed of raw texts.
2. Terminological noun phrases (NPs) of maximal length are extracted using Tree-Tagger [18] or any POS tagger. A selection of NPs is done based on their syntactic structure and on our enhanced term weighting function in order to retain only domain terms.
3. Terms that are semantic variants of one another are detected and clustered in a hierarchical process. This results in a three level structuring of domain terms. The first level are the terms. The second level are components that group together terms semantically close terms or synonyms. Roughly, TermWatch's components generalize the notion of WordNet synsets [19] to multi-word terms. A clustering algorithm [20] is applied to this second level of term grouping based on a weighted graph of term variants. Components and clusters are labeled by their most active term and can be used as document features.
4. In the fourth stage, documents are indexed by cluster or component labels and the corresponding association graph is generated. The strength of the association is weighted based on different similarity measures and only those links that are above some threshold for all measures are considered.
5. Association graphs are decomposed into atoms [14]. An atom is a subgraph without clique separators. Each clique corresponds to a formal concept. Major atoms are detected and visualized using force directed placement algorithms. The periphery of big atoms is highlighted since it can reveal new concepts arising in a domain represented by a central more bigger atom.

If $< mod >^* < N >^+$ of $< mod >^* < N > + < prep1 >< verb >< mod >^* < N >^+$
then return: $< mod >^* < N >^+$ of $< mod >^* < N >^+$ **and** $< mod >^* < N >^+$ **where:**
$< mod >$ is a determiner or an adjective
$< N >$ is any of the noun tags
$< prep1 >$ is all the prepositions excluding "of"
* is the Kleene's operator (zero or n occurrences of an item)
$^+$ is at least one occurrence

Fig. 2. Example of contextual rules used to extract multi-word terms

3 Terminological Graph Extraction

3.1 Term Extraction

After the corpus has been tagged using TreeTagger [18], contextual rules are used to extract multi-word terms based on morphological and syntactic properties of terms. One such rule is shown in fig. 2. This rule favors the extraction of terminological noun phrases in a preposition structure where the preposition is "of". This preposition has been found to play an active role in the multi-word term formation process. More details of the rules can be found in [7]. The extracted terms can be simplex noun phrases (NPs) like "stress disorder" or complex ones like "posttraumatic stress disorder" which embeds simpler NPs. Also, terms are extracted in their two possible syntactic structures: NPs with prepositional attachment (execution of innocent victims) and compounds (innocent victims execution). This transformation operation, also known as permutation is useful for grouping together syntactic variants of the same concept that would otherwise be dispersed. No limit is imposed on the length of the extracted terms thus ensuring that new terms coined by authors of papers are extracted 'as is' and that existing domain concepts with multi-words are not altered or lost. By not resorting to the usual "bag-of-word" approach common in the IR and data mining communities, emergent domain terms can be identified in a timely manner because term extraction respects the structure of the domain terminology "in-the-making".

3.2 Generating a Graph of Semantic Term Variants

We studied linguistic operations between terms which are domain independent and can be used to build taxonomies, thesaurus or ontologies. These operations, called terminological variations, stem from two main linguistic operations: lexical inclusion and lexical substitution. By lexical inclusion, we refer to the case where a shorter term is embedded in a longer one through three specific operations: insertions (severe poisoning ↔ severe food poisoning), modifier or head word expansion ("disaster intervention" ↔ "disaster intervention call"). By lexical substitution, we refer to the case where terms of identical length share a subset of lexical items save one in the same position ("political violence threat" ↔ "political violence campaign"). Lexical inclusion often engenders hypernym/hyponym (generic/specific) relations between terms while the lexical

Table 1. Some synonyms acquired from the terrorism corpus using WordNet synsets

Term	Synonym identified using WordNet synsets
september 11 *wake*	september 11 *aftermath*
united states federal *agency*	united states federal *bureau*
risk society *conception*	risk society *concept*
Trauma type	*injury* type
Life-threatening problem	Serious problem
Cyber-terrorist attack	*hacker* attack

substitution tend to indicate a loose kind of semantic association between terms. Lexical substitutions between binary terms give rise to a highly connected graph of term variants (cliques) which may include some amount of noise (spurious relations). They are filtered using two criteria: we retain only those substitutions that involve terms of length ≤ 2 if the words in the same grammatical position are found in the same WordNet synset. Although there are many more types of linguistic relations, we restricted our choice to those that did not require heavy use of external semantic resources and were domain-independent, thus found in any well written text revolving around the same broad topic.

We also acquired explicit synonymy links between multi-word terms using Word-Net. To do this, we extended the single word-word relations in WordNet to multi-word terms by adding these restrictions: two multi-word terms are considered to be in a synonymy relation if two of their words are in the same WordNet synset, occupy the same grammatical role in the terms (both head words or modifier words) and are found in the same position. Table 1 shows some of the synonyms identified in this way. The italicized words were in the same WordNet synset.

Table 1 shows that the quality of the synonyms acquired through WordNet is indeed good. Explicit synonymy links ensure that concepts appearing under different names are not dispersed in different clusters at the end of the process. Table 2 gives examples of the different relations identified and the number of terms involved for the terrorism corpus.

Any relation between a set of documents and a set of features naturally induces a network of associations. Two features are associated if they index a substantial set of common documents. The association can therefore be weighted by a measure on the set of shared documents. The network of associations gives rise to a *feature × feature* symmetric matrix that can be analyzed using standard data mining approaches like clustering, factor analysis or latent semantic analysis. The output of these methods heavily depends on the choice of the association index. However, before applying any data mining process, the structure of the association network should be studied independently from the measure of associations.

The study of this structure becomes indispensable when features result from a complex text analysis process like multi-word terms (MWTs) extracted from abstracts in an automated procedure. Since these terms result from an unsupervised process, some amount of noise can be expected. The idea is then to use standard association measures

Table 2. Terminological variations identified between terms in the terrorism corpus

Variation type	example of Term	example of Variant	#Terms	#Links
Spelling	trauma *center*	trauma *centre*	93	138
Left exp.	food contamination	*pet* food contamination	1,799	2,709
Insertion	poisoning case	poisoning *medical intervention* case	41	60
Right exp.	disaster intervention	disaster intervention *call*	2,884	4,326
Modifier sub.	*acute* stress disorder	*posttraumatic* stress disorder	14,062	95,651
Head sub.	political violence *threat*	political violence *campaign*	13,810	125,385
Wordnet Mod. sub.	*Trauma* severity	*injury* severity	185	99
Wordnet Head sub.	terrorist *financing*	terrorist *funding*	396	217

to remove the most improbable associations. So, instead of working on a numeric matrix, we consider the binary matrix that indicates if an association between two multi-word terms is possible or not, without prejudice on its strength since it could result from some bias in the term selection procedure. Moreover, low frequency terms are essential when seeking for rare information like emerging new concepts and/or new relationships between concepts. This symmetric binary matrix gives rise to a non directed graph between multi-word terms. In the case of a corpus of documents constituted randomly, the structure of this graph corresponds to the usual small world frequently observed on co-word graphs [21]. In some cases, the extracted terminological network of possible associations shows an unexpected structure. TermWatch II aims to extract terminological graphs and to reveal this structure if it exists, based on advanced graph algorithm theory.

3.3   Term Clustering

The linguistic significance of each relation can be translated in terms of two possible roles: COMP and CLAS. Ideally, COMP relations are variations that induce near-semantic equivalence or synonymy links such as spelling variants, permutations, Word-Net synonyms, one-word modifier expansions and insertions. COMP relations are used to form a prior category of tight semantic clusters which serve as a first level of agglomeration. There is an edge between two nodes if one is a COMP variant of the other. By forming connected components, we group terms for which there is a sequence of variations in COMP. Since variations in COMP link only close semantically related terms, resulting connected components portray terms from the same concept family. Components are labeled by its most central term and can be used as document descriptors. CLAS relations are those that involve a topical shift between two terms, i.e., where the head word is different like head expansion and head substitution. For instance, the shift of focus from "criminal assault" to the victim in "criminal assault victim". This category of relations is used to aggregate the components formed by COMP relations in an agglomerative hierarchical process.

The strength of these links between components can be measured by the number of variations across them. In other to favor rare relations and eliminate noise, each variation is weighted by the inverse of its frequency in the corpus. Then the strength of the link between two components I, J is computed as follows:

Table 3. Main component of the cluster "terrorist attack" and related clusters

Terms in component "terrorist attack"
terrorist attack, presumed terrorist attack, limited terrorist attack, national terrorist attack, international terrorist attack, explosive terrorist attack, deliberate terrorist attack, deliberate smallpox terrorist attack, smallpox attack, covert smallpox attack, chemical terrorist attack, th terrorist attack, year terrorist attack
Some components in the clique around "terrorist attack"
anthrax infection, toxic chemical, medium representation, 9/11 event, september 11 attack, current PTSD, new york time, pharmaceutical industry, american history, united kingdom, potential terrorist, militant islam, safety sense, national terrorist attack impact, distress symptom, decontamination area, immigration policy

$$d(I, J) = \sum_{\theta \in CLAS} \frac{N_\theta(I, J)}{|\theta|} \qquad (1)$$

where $N_\theta(I, J)$ is the number of variations of type θ in a subset of relations not in COMP called CLAS ($CLAS \cap COMP = \emptyset$) that relate terms in I to terms in J. $|\theta|$ is the total number of variations in θ.

CLAS clusters can be then formed using any graph clustering algorithm based on this valued graph of components. TermWatch implemented CPCL (Clustering by Preferential Clustered Link) algorithm, first described in [20]. The principle of CPCL algorithm is to select at each iteration edges that are local maximums and merge iteratively together all nodes related by such edges. The advantage of this principle is that two nodes are merged not only based on the strength of their relation but also by considering all the relations in their neighborhood. The system then merges the components with the strongest relation at iteration t. We have shown in [10] that CPCL has a unique possible output and avoids part of the chain effect common to similar hierarchical clustering methods. CPCL is also different from usual hierarchical clustering (single, average, complete link) since more than one group of components can be clustered at different similarity values. We refer the reader to [10] for a more formal description as well as for a comparison with a larger family of clustering algorithms (variants of single-link, average link and variants of k-means). Table 3 shows as example, the content of the biggest component in the biggest cluster. Thez cluster has 78 terms and has been automatically labeled "terrorist attack" which is the label of its major component. The other terms in the cluster result from co-occurrence links. We also show in the lower part of this table, surrounding nodes around this cluster which form a clique.

4   Association Graph Analysis

Clustering a large corpus of terms can lead to several hundreds even if coherent clusters which are difficult to visualize (cluttered image). We also need to study the way in which these clusters are associated to documents.

4.1   Generating Association Graphs and Formal Concepts

In the context of Association mining as defined by Agrawal et al. [22] each document is related to the clusters that contain at least one term in the document. Clusters are then considered as items and each document defines an itemset. We shall call them document itemsets. The set of items can be extended to other fields (features) like authors. Given an integer threshold S, a frequent itemset is a set of items that are included in at least S document itemsets. There is no fixed size for frequent itemsets. Frequent itemset discovery in a data base allows to reveal hidden dependences in general. Frequent itemsets of size one are just frequent terms or authors. Frequent itemsets of size 2 induce an association graph where nodes are items and there is a link between two nodes i and j if the pair $\{i, j\}$ is a frequent itemset. Moreover, any frequent itemset defines a clique in the original association graph. Clearly, if $I = \{i_1, ..., i_n\}$ is a frequent itemset, then any pair i_k, i_p of elements in I is a frequent itemset of size two and defines an edge in the association graph but not necessarily on the graph of selected edges using a relevance measure. Thus all nodes $i_1, ..., i_n$ are related in the original association graph. However, not every clique in a graph induces a frequent itemset.

The resulting association graph being generally too dense to be visualized, it is usual to perform feature selection based on some measures like mutual information or log likelihood, to select most relevant edges. This approach has two drawbacks. First, the resulting graph structure depends on the selected measure1. Second, it is not adapted to highlight larger itemsets (triplets or more). Therefore, to visualize large frequent itemsets on the association graph, we need a decomposition approach that preserves cliques induced by frequent itemsets.

The theoretical framework of association discovery is Formal Concept Analysis (FCA) [23] based on Galois lattice theory. FCA offers a pragmatic way of formalizing the notion of concepts. It posits that to every real concept in a domain corresponds a formal concept in some database of specialized documents. In the present context, a formal concept consists of an extension made of a set D of documents, and an intension made of a set of items I such that a document d is related to all items in I if and only if d is in D. Thus a formal concept establishes an exact correspondence between a set of documents and a set of items. Frequent itemsets that are the intensions of some formal concept are called closed itemsets. We shall focus on graph decomposition methods that preserve the cliques induced by closed frequent itemsets.

4.2   Graph Decomposition

Algorithms to enumerate all closed frequent itemsets are exponential because the number of these frequent itemsets can be exponential. Moreover they are highly redundant [24]. Thus, available packages to mine them like state of the art arules from the R project [3] require the analyst to fix a maximal size for mined itemsets. Interestingness measures are then applied to rank them. However, the list of top ranked frequent itemsets heavily depends on the choice of this measure.

Our idea is to apply the results from recent research on graph theory [25,14] to extract sub-graphs that preserve special cliques that have a high probability to be closed

---

[3] http://cran.r-project.org/web/packages/arules/index.html

frequent itemsets. We focus on minimal clique separators, i.e. cliques whose removal from the original graph will result in several disjoint subgraphs. This leads to extracting maximal sub-graphs without minimal clique separators. These maximal sub-graphs are called atoms [14]. By revealing the atomic structure of a graph we also reveal: (i) special concepts that are interfaces between sub-domains or between domain kernels and external related objects; and (ii) aggregates of intrinsically related concepts at the heart of the domain. A key point of atom decomposition is that it is unique. It is an intrinsic graph property. It follows that the number of atoms and their size distribution can be considered as good indicators of their structure complexity. Moreover the atomic structure can be computed in quadratic time on the number of nodes: $O(\#vertex.\#edges)$.

In the case of mapping the structure of a domain based on a corpus of abstracts resulting from a multi-word query, it can be expected to find the concept corresponding to the query at the heart of the association graph in a central atom. This central atom should contain all concepts directly related to the domain as sub-cliques. Some of them should connect the domain with external concepts and thus should be at the intersection of the central atom with peripheral ones. The atom decomposition algorithm is implemented in a C++ program [14]. It computes the atomic graph structure and generates two images:

- the sub-graph that constitutes the central atom if it exists.
- the network of atoms to visualize those at the periphery and the way they are connected to the central atom.

We have experimentally checked that atoms do not break closed frequent itemsets at 98% [14]. In the result section, we shall focus on the central atom because we found out that in the corpus analyzed here and the one in [15], they have a surprisingly clear structure.

4.3 Graph Visualization

The atom graphs are generated in GDL format (Sander 1995) for AiSee[4]. GDL allows to define sub-graphs objects that can be displayed folded or wrapped in a colored background. We use this functionality to fold clique sub-graphs of nodes such that the probabilities $P(i/j)$ of finding one related to a document knowing that the other is related are equal for all pair of nodes in the clique. These cliques are then represented by a generic node to simplify the display of the graph without altering its structure. We use AiSee because this software implements optimized force direct graph display algorithms [26]. To analyze a complex graph structure. AiSee runs with maximal non crossing heuristics and a great number of iterations to approximate as far as possible a planar graph without crossing edges and separating non connected nodes clearly. The resulting images allow experts to quickly identify the main structural properties of the graph: maximal cycle length, connectivity, sub-cliques etc. Moreover, since nodes are labeled, domain specialists can also easily read these graphs using the browsing function of AiSee.

---

[4] http://www.aisee.com

5   Case Study

We present results on mapping the dynamics of research in terrorism research between 1990-2006. An association graph between cluster labels and authors was built and subjected to the graph decomposition algorithm. The analysis of results and evaluation of the graphs is done by comparing the structure of the central atom to the network obtained by Chen [16]. His study was on the same topic, using the same query on the same database (WoS) but on an earlier period (1990-2003). Given that he has already performed an evaluation of his results by sending questionnaires to domain experts, we highlight the similarities and differences in the map he obtained and more importantly show the evolution of research on terrorism since 2003.

5.1   Network of Atoms

The graph decomposition splits the association graph into a central and peripheral atoms. Owing to space limitations, we cannot show the images of the peripheral atoms[5]. We comment briefly on the most prominent ones. The map of atoms shows that indeed, it is a central atom on "biological terrorism" that makes the whole graph connected. Biological terrorism thus acts as a hub or a magnet for linking all the terrorism-related research. The most prominent peripheral atoms are somehow connected to this threat of bio-terrorism. The three biggest sub-graphs by number of atoms contained are "nuclear radiation" (37), "biological and chemical warfare" (25), "radiological dispersion device" (21).

5.2   Structure of the Central Atom

The central atom labelled "biological terrorism" can be unfolded to show its internal structure. We can clearly perceive three sub-graphs of clusters with some connections between them (fig. 3).

The topmost part reflects research on the psychological aftermath of september 11, 2001 attacks, namely posttraumatic stress disorders (PTSD). The middle part of the central atom corresponds roughly to two major clusters on "body injuries in terrorist bombing" and "health-care". The lower part of the graph reflects research on potential terrorists attacks using biological and nuclear weapons. The structure of these three sub-graphs echoes to a certain degree the network found in Chen [16] for the period 1990-2003. Mapping a hybrid network of cited documents and citing terms, he found three major groups of clusters reflecting three research threads: a first thread on "body injuries in terrorist bombing", a second bigger thread on "health care response to the threat of biological and chemical weapons", a third biggest and more recent thread on "psychological and psychiatric impacts of the september 11, 2001 terrorist attack" with terms like "United States" and "posttraumatic stress disorder" (PTSD) being very prominent. Globally, these three big threads of research are still present in 2006, albeit with significant changes. Since 2003, the first two threads on "body injuries" and

---

[5] Detailed views of all atoms and network can found on http://demo.termwatch.es

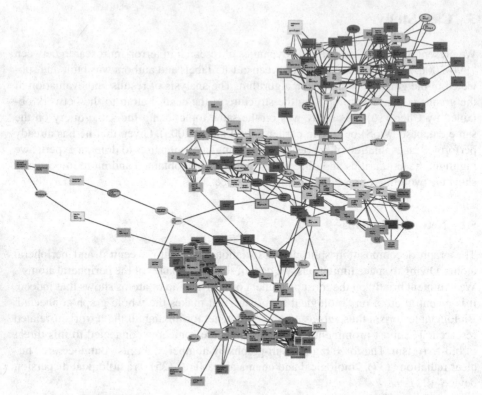

Fig. 3. Internal structure of the central atom on biological terrorism

"emergency medical care" have merged into one single thread while a new thread on bio-terrorism has emerged and become more prominent.

The system also computes statistical indicators from the Social Network Analysis [27] in order to characterize the relative position of nodes and their importance in the network. Nodes with high betweenness centrality values are possible transitions points from one research thread to another. "posttraumatic stress disorder" (PTSD) is the node with highest betweenness centrality. All other topmost nodes recall major terrorist threats ("traumatic event", "world health", "suicidal terrorist bombing", "biological terrorism", "mass destruction"). The three research threads portrayed by the three sub-graphs in the central atom are present in the first 20 nodes ranked by betweenness centrality: "posttraumatic stress disorder" (1st), "specific injury type" (8th), "primary injury blast" (18th), "biological terrorism" (6th).

The domination of red colour in the upper part of the central atom indicates that the majority of terms in these clusters appeared in the last period (2006). This sub-graph corresponds roughly to the most prominent thread found in [16] on "September 11" and "posttraumatic stress-disorder" (PTSD). This last term is still very much present three years later as shown by terminological variations found around this term, both in its developed form ("posttraumatic stress disorder symptom") and in abbreviated forms ("probable PTSD frequency", "PTSD symptom severity"," SCW-PTSD prevalence"). At the center of this sub-graph is the author node "Boscarino JA. His papers focused on psychological effects and PTSD caused by the 9/11, 2001 event. Among the pre-occupying health issues brought to light by this research thread is the increased use of drugs, alcohol and the increase in mental disorder among the population in the area surrounding the World Trade Center. This is evident in the surrounding cluster labels: physical health, psychological health, binge drinking, alcohol use, increased substance use, african-american, posttraumatic stress disorder symptom, psychotropic medication. Boscarino's studies were mostly carried out as sociological surveys by online questionnaire administration or telephone surveys (hence a cluster "random digital telephone surveys"). Researchers involved in this topic discovered an increased use of post-disaster medical services to combat PTSD predominantly among the white community, more prone to depression than the non white community. These findings were not yet visible in 2003.

Another difference or evolution observed in our graph and the network proposed by Chen [16] on the 1990-2003 data is the absence of the cluster "United States. This term has since been replaced by studies focusing on the precise area where the terrorist attack took place, hence the presence of the clusters labelled "new york resident, new york city, lower manhattan resident". It seems that with time, PTSD studies of the 9/11, 2001 terrorist attack have shifted from the nation-level crisis stance (The US was being attacked by terrorists) to a more localised and detailed level - the actual places where the attack took place and its effects on different segments of the population.

5.3 Mining Closed Frequent Itemsets on Terrorism Research

For complexity reasons, it is not possible to extract frequent itemsets whose extension has fewer than three documents, meanwhile we shall see that the atom graph allows us to identify interesting closed itemsets whose extension has only two documents. Using the apriori algorithm in R package, we found 1926 closed itemsets with a support of at least three documents of which 285 have more than three elements (three items). The largest closed frequent itemset without author names is: {*new york city, posttraumatic stress disorder, potential terrorist attack, same traumatic event, world trade center*}. The largest overall has 12 items: {*Parker G,Perl TM,Russell PK, biological terrorism, biological warfare, consensus-based recommendation, emergency management institution, MEDLINE database, nation civilian population, potential biological weapon, working group, world health*}. It appears that both itemsets can be clearly visualized on the central atom.

The graph layout moreover allows us to show how they these frequent itemsets are related and to point out distinct smaller concepts around them. When comparing the central atom structure with closed frequent itemsets, we find out that the upper part of the graph (9/11 & PTSD) is structured around the clique that corresponds to the longest closed frequent itemset without author name. The lower part (bioterrorism) is structurally organized around the clique that represents the longest frequent itemset containing authors items. It also contains the closed frequent itemset {*mass destruction, mass destruction weapon, nuclear weapon proliferation*}. The middle sub-graph focused on "physical injuries and emergency medical care" for victims of terrorist attacks point out formal concepts that connect the two parts of the graph. Apart the frequent item set {*blast lung injury,physical examination,primary blast injury*}, the extension of these formal concepts have only two documents and so cannot be directly computed by R arules library for complexity reasons (memory over stack). However they are essential to understand the relations between the upper and lower part of the graph that are clearly revealed by the visualisation of the graph structure. Finally, all extracted closed frequent itemsets correspond to the cliques in these two sub-graphs of the central atom.

6 Conclusions

We have presented a platform for mapping the dynamics of research in specialty fields. The distinctive features of this methodology resides in its clustering algorithm which is based primarily on linguistic (symbolic) relations and on its graph decomposition algorithm which renders complex terminological graph for comprehensible for domain analysts. The method has been able to identify the most salient topics in two different research domains and uncover the sub-structures formed by persistent and evolving research threads. More importantly, we have shown that it is possible, with limited linguistic resources, to perform a surface analysis of texts and use linguistic relation for clustering. To the best of our knowledge, this represents a unique and innovative approach to text clustering.

The graph decomposition algorithm offers a way of visualizing complex terminological graphs and revealing particular sub-structures contained therein. Mining frequent itemsets, in combination with evaluation by human experts, offer a joint and strong evidence of the significance of the maps produced for the domain.

References

1. Porter, M.F.: An algorithm for suffix stripping. Program: Electronic Library and Information Systems 40(3), 211–218 (2006)
2. Zitt, M., Bassecoulard, E.: Development of a method for detection and trend analysis of research fronts built by lexical or co-citation analysis. Scientometrics 30(1), 333–351 (1994)
3. Glenisson, P., Glänzel, W., Janssens, F., De Moor, B.: Combining full text and bibliometric information in mapping scientific disciplines. Information Processing and Management 41(6), 1548–1572 (2005)
4. Harris, Z.S.: Mathematical Structures of Language. Wiley, New York (1968)
5. Grefenstette, G.: Sqlet: Short query linguistic expansion techniques, palliating one-word queries by providing intermediate structure to text. In: Proceedings of Recherche d'Information Assiste par Ordinateur (RIAO), pp. 500–509 (1997)
6. Watcholder, N., Evans, D., Klavans, J.: Automatic identification of index terms for interactive browsing. In: Proceedings of the ACM IEEE Joint Conference on Digital Libraries, Roanoke, Virginia, pp. 116–124 (2001)
7. Ibekwe-SanJuan, F.: Terminological variation, a means of identifying research topics from texts. In: Proc. of Joint ACL-COLING 1998, Québec, Canada, August 10-14, pp. 564–570 (1998)
8. Ibekwe-SanJuan, F., SanJuan, E.: From term variants to research topics. Journal of Knowledge Organization (ISKO), Special Issue on Human Language Technology 29(3/4) (2003)

9. Ibekwe-SanJuan, F., SanJuan, E.: Mining textual data through term variant clustering: the termwatch system. In: Proc. of Recherche d'Information assistée par ordinateur (RIAO), Avignon, France, pp. 26–28 (April 2004)

10. SanJuan, E., Ibekwe-SanJuan, F.: Text mining without document context. Information Processing and Management 42, 1532–1552 (2006)

11. Ibekwe-SanJuan, F., Dubois, C.: Can syntactic variations highlight semantic links between domain topics? In: Proc. of the 6th International Conference on Terminology (TKE), Nancy, France, pp. 57–63 (August 2002)

12. Sanjuan, E., Dowdall, J., Ibekwe-Sanjuan, F., Rinaldi, F.: A symbolic approach to automatic multiword term structering. Computer Speech Language (CSL) 19(4), 524–542 (2005)

13. Chen, C., Ibekwe-SanJuan, F., SanJuan, E., Weaver, C.: Visual analysis of conflicting opinions. In: 1st International IEEE Symposium on Visual Analytics Science and Technology (VAST 2006), Baltimore - Maryland, USA, pp. 59–66 (2006)

14. Didi Biha, M., Kaba, B., Meurs, M.-J., SanJuan, E.: Graph Decomposition Approaches for Terminology Graphs. In: Gelbukh, A., Kuri Morales, Á.F. (eds.) MICAI 2007. LNCS (LNAI), vol. 4827, pp. 883–893. Springer, Heidelberg (2007)

15. Ibekwe-Sanjuan, F., SanJuan, E., Vogeley, M.S.E.: Decomposition of terminology graphs for domain knowledge acquisition. In: Shanahan, J.G., Amer-Yahia, S., Manolescu, I., Zhang, Y., Evans, D.A., Kolcz, A., Choi, K.S., Chowdhury, A. (eds.) CIKM, pp. 1463–1464. ACM (2008)

16. Chen, C.: Citespace ii: Detecting and visualizing emerging trends and transient patterns in scientific literature. JASIS 57(3), 359–377 (2006)

17. Chen, H., Wingyan, C., Qin, J., Reid, E., Sageman, M.: Uncovering the dark web: A case study of jihad on the web. Journal of the American Society for Information Science, JASIS 59(8), 1347–1359 (2008)

18. Schmid, H.: Probabilistic part-of-speech tagging using decision trees. In: Proceedings of International Conference on New Methods in Language Processing, Manchester, UK, vol. 12 (1994)

19. Miller, G.A.: Wordnet: A Lexical Database for English. In: HLT. Morgan Kaufmann (1994)

20. Ibekwe-SanJuan, F.: A linguistic and mathematical method for mapping thematic trends from texts. In: Proc. of the 13th European Conference on Artificial Intelligence (ECAI), Brighton, UK, pp. 170–174 (August 1998)

21. Ferrer i Cancho, R., Solé, R.V.: The small world of human language. Proceedings of The Royal Society of London. Series B, Biological Sciences 268, 2261–2266 (2001)

22. Agrawal, R., Imielińskivand, T., Swami, A.: Mining association rules between sets of items in large databases. SIGMOD Rec. 22, 207–216 (1993)

23. Ganter, B., Stumme, G., Wille, R. (eds.): Formal Concept Analysis. LNCS (LNAI), vol. 3626. Springer, Heidelberg (2005)

24. Zaki, M.J.: Closed itemset mining and non-redundant association rule mining. In: Liu, L., Özsu, M.T. (eds.) Encyclopedia of Database Systems, pp. 365–368. Springer US (2009)

25. Berry, A., Pogorelcnik, R., Simonet, G.: An introduction to clique minimal separator decomposition. Algorithms 3(2), 197–215 (2010)

26. Fruchterman, T.M.J., Reingold, E.M.: Graph drawing by force-directed placement. Software: Practice and Experience 21(11), 1129–1164 (1991)

27. Freeman, L.C.: A set of measures of centrality based on betweenness. Sociometry 40(1), 35–41 (1977)

Learning to Classify Text Using a Few Labeled Examples

Francesco Colace[1], Massimo De Santo[1], Luca Greco[1], and Paolo Napoletano[2,*]

[1] Department of Electronic Engineering and Computer Engineering,
University of Salerno, 84084 Fisciano, Italy
{fcolace,desanto,lgreco}@unisa.it
[2] DISCo (Department of Informatics, Systems and Communication)
University of Milan, Bicocca Viale Sarca 336
20126 Milan, Italy
napoletano@disco.unimib.it

Abstract. It is well known that supervised text classification methods need to learn from many labeled examples to achieve a high accuracy. However, in a real context, sufficient labeled examples are not always available. In this paper we demonstrate that a way to obtain a high accuracy, when the number of labeled examples is low, is to consider structured features instead of list of weighted words as observed features. The proposed vector of features considers a hierarchical structure, named a mixed Graph of Terms, composed of a directed and an undirected sub-graph of words, that can be automatically constructed from a set of documents through the probabilistic Topic Model.

Keywords: Text classification, Term extraction, Probabilistic topic model.

1 Introduction

The problem of supervised *text classification* has been extensively discussed in literature where metrics and measures of performance have been reported [4], [12], [9]. All the existing techniques have been demonstrated to achieve a high accuracy when employed in supervised classification tasks of large datasets.

Nevertheless, it has been found that only 100 documents can be hand-labeled in 90 minutes and in this case the accuracy of classifiers (amongst which we find Support Vector Machine based methods), learned from this reduced training set, could be around 30% [8].

This makes, most times, a classifier unfeasible in a real context. In fact, most users of a practical system do not want to carry out labeling tasks for a long time only to obtain a higher level of accuracy. They obviously prefer algorithms that have a high accuracy, but do not require a large amount of manual labeling tasks [10][8]. As a consequence, we can affirm that, in several application fields we need algorithms to be fast and with a good performance.

Although each existing method has its own properties, there is a common denominator: the "bag of words" assumption to create term weights.

The "bags of words" assumption claims that a document can be considered as a *vector of features* where each element in the vector indicates the presence (or absence)

---

* Corresponding author.

A. Fred et al. (Eds.): IC3K 2011, CCIS 348, pp. 200–214, 2013.

of a word, so that the information on the position of that word within the document is completely lost [4].

It is well known that the main purpose of text mining techniques is to identify common patterns through the observation of such *vectors of features* and then to use such patterns to make predictions. Unfortunately, the accuracy of classification methods based on the "bags of words" decreases as the number of labeled examples decreases. In this case classifiers identify common patterns that are insufficiently discriminative because, due to the inherent ambiguity of language (polysemy etc.), vectors of weighted words are not capable of discriminating between documents.

The ambiguity, in fact, can be reduced if we give more importance to words that convey concepts and that contribute to specify a topic, and if we assign less importance to those words that contribute to specify concepts and that, due to the fact that they can be more plausibly shared between concepts, can increase the ambiguity.

This leads to a hierarchical structure that we call a mixed *Graph of Terms* and that can be automatically extracted from a set of documents $\mathcal{D}$ using a global method for term extraction based on Latent Dirichlet Allocation [2] implemented as Probabilistic Topic Model [6].

Here we propose a linear single label supervised classifier that is capable, based on a *vector of features* represented through a mixed *Graph of Terms*, of achieving a better performance, in terms of accuracy, than existing methods when the size of the training set is about 1.4% of the original and composed of only positive examples.

To confirm the discriminative property of the graph we have evaluated the performance through a comparison between our methodology and a term selection methodology which considers the *vector of features* formed of only the list of concepts and words composing the graph and so where relations have not been considered. We have also compared our method with linear Support Vector Machines. The results, obtained on the top 10 classes of the ModApte split from the Reuters-21578 dataset, show that our method, independently of the topic, is capable of achieving a better performance.

2   Problem Definition

Following the definition introduced in [12], a supervised *Text Classifier* may be formalized as the task of approximating the unknown target function $\Phi : \mathcal{D} \times \mathcal{C} \to \{T, F\}$ (namely the expert) by means of a function $\hat{\Phi} : \mathcal{D} \times \mathcal{C} \to \{T, F\}$ called the *classifier*, where $\mathcal{C} = \{c_1, ..., c_{|\mathcal{C}|}\}$ is a predefined set of *categories* and $\mathcal{D}$ is a set of *documents*.

If $\Phi(d_m, c_i) = T$, then d_m is called a positive example (or a member) of c_i, while if $\Phi(d_m, c_i) = F$ it is called a negative example of c_i.

The categories are just symbolic labels: no additional knowledge (of a procedural or declarative nature) of their meaning is usually available, and it is often the case that no metadata (such as e.g. publication date, document type, publication source) is available either. In these cases, the classification must be accomplished only on the basis of knowledge extracted from the documents themselves, namely *endogenous knowledge*.

In practice, we consider an initial corpus $\Omega = \{d_1, ..., d_{|\Omega|}\} \subset \mathcal{D}$ of documents pre-classified under $\mathcal{C} = \{c_1, ..., c_{|\mathcal{C}|}\}$. The values of the total function Φ are known for every pair $(d_m, c_i) \in \Omega \times \mathcal{C}$.

We consider the initial corpus to be split into two sets, not necessarily of equal size:

1. the *training* set: $\Omega_r = \{d_1, \ldots, d_{|\Omega_r|}\}$. The classifier Φ for the categories is inductively built by observing the characteristics of these documents;
2. the *test* set: $\Omega_e = \{d_{|\Omega_r|+1}, \ldots, d_{|\Omega|}\}$, used for testing the effectiveness of the classifiers.

Here we consider the case of *single-label* classification, also called *binary*, in which, given a category c_i, each $d_m \in \mathcal{D}$ must be assigned either to c_i or to its complement $\overline{c}_i$. In fact, it has been demonstrated that, through transformation methods, it is always possible to transform the multi-label classification problem either into one or more single-label classification or regression problems [12,13].

It means that we consider the classification under $\mathcal{C} = \{c_1, \ldots, c_{|\mathcal{C}|}\}$ as consisting of $|\mathcal{C}|$ independent problems of classifying the documents in $\mathcal{D}$ under a given category c_i, and so we have $\hat{\phi}_i$, for $i = 1, \ldots, |\mathcal{C}|$, classifiers. As a consequence, the whole problem in this case is to approximate the set of function $\Phi = \{\phi_1, \ldots, \phi_{|\mathcal{C}|}\}$ with the set of $|\mathcal{C}|$ classifiers $\hat{\Phi} = \{\hat{\phi}_1, \ldots, \hat{\phi}_{|\mathcal{C}|}\}$.

Once the classification problem has been defined we can start the pre-processing of the data through the data preparation and reduction steps.

2.1 Data Preparation

Texts can not be directly interpreted by a classifier and for this reason, an indexing procedure that maps a text d_m into a compact representation of its content must be uniformly applied to the training and test documents. In the following we consider the case of the training set.

Each document can be represented, following the *Vector Space Model* [4], as a vector of term *weights*

$$d_m = \{w_{1m}, \ldots, w_{|\mathcal{T}|m}\},$$

where $\mathcal{T}$ is the set of *terms* (also called *features*) that occur at least once in at least one document of Ω_r, and $0 \leq w_{nm} \leq 1$ represents how much term t_n contributes to a semantics of document d_{mm}.

If we choose to identify terms with words, we have the *bags of words* assumption, that is $t_n = v_n$, where v_n is one of the words of a vocabulary. The *bags of words* assumption claims that each w_{nm} indicates the presence (or absence) of a word, so that the information on the position of that word within the document is completely lost [4].

To determine the weight w_{nm} of term t_n in a document d_m, the standard tf-idf (*term frequency-inverse document frequency*) function can be used [11], defined as:

$$\text{tf-idf}(t_n, d_m) = N(t_n, d_m) \cdot \log \frac{|\Omega_r|}{N_{\Omega_r(t_n)}} \quad (1)$$

where $N(t_n, d_m)$ denotes the number of times t_n occurs in d_m, and $N_{\Omega_r(t_n)}$ denotes the document frequency of term t_n, i.e. the number of documents in Ω_r in which t_n occurs.

In order for the weights to fall in the $[0, 1]$ interval and for the documents to be represented by vectors of equal length, the weights resulting from tf-idf are usually normalized by cosine normalization, given by:

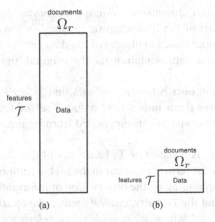

Fig. 1. Features-documents matrix. 1(a) In this case the number of features is much higher than the number of examples ($|\mathcal{T}| \gg |\Omega_r|$). 1(b). In this case $|\mathcal{T}| \ll |\Omega_r|$.

$$w_{nm} = \frac{\text{tf-idf}(t_n, \mathbf{d}_m)}{\sqrt{\sum_{n=1}^{|\mathcal{T}|}(\text{tf-idf}(t_n, \mathbf{d}_m))^2}} \qquad (2)$$

In this paper, before indexing, we have performed the removal of function words (i.e. topic-neutral words such as articles, prepositions, conjunctions, etc.) and we have performed the stemming procedure[1] (i.e. grouping words that share the same morphological root).

Once the indexing procedure has been performed, we have a matrix $|\mathcal{T}| \times |\Omega_r|$ of real values instead of the training set Ω_r, see Fig. 1(a). The same procedure is applied to the test set Ω_e.

2.2 Data Reduction

Usually, machine learning algorithms are susceptible to the problem named the *curse of dimensionality*, which refers to the degradation in the performance of a given learning algorithm as the number of features increases. In this case, the *computational cost* of the learning procedure and *overfitting* of the classifier are very common problems [3].

Moreover, from a statistical point of view, in the case of supervised learning, it is desirable that the number of labeled examples in the training set should significantly exceed the number of features used to describe the dataset itself.

In the case of text documents the number of features is usually high and particularly it is usually higher than the number of documents. In Fig. 1(a) we show the case of a training set composed of 100 documents and about 20000 features obtained following the data preparation procedure explained in the previous paragraph. As you can see, $|\mathcal{T}| \gg |\Omega_r|$ while it is desirable to have the opposite condition, that is $|\mathcal{T}| \ll |\Omega_r|$, as represented in Fig. 1(b).

---

[1] Stemming has sometimes been reported to hurt effectiveness, the recent tendency is to adopt it, as it reduces both the dimensionality of the feature space and the stochastic dependence between terms.

To deal with these issues, dimension *reduction techniques* are applied as a data pre-processing step or as part of the data analysis to simplify the whole data set (*global* methods) or each document (*local* methods) of the data set. As a result we can identify a suitable low-dimensional representation for the original high-dimensional data set, see Fig. 1(b).

In literature, we distinguish between methods that *select* a subset of the existing features or that *transform* them into a new reduced set of features. Both classes of methods can rely on a supervised or unsupervised learning procedure [3,12,4,5]:

1. *Feature Selection:* $\mathcal{T}_s$ is a subset of $\mathcal{T}$. Examples of this are methods that consider the selection of only the terms that occur in the highest number of documents, or the selection of terms depending on the observation of information-theoretic functions, among which we find the *DIA association factor, chi-square, NGL coefficient, information gain, mutual information, odds ratio, relevancy score, GSS coefficient* and others.

2. *Feature Transformation:* the terms in $\mathcal{T}_p$ are not of the same type as the terms in $\mathcal{T}$ (e.g. if the terms in $\mathcal{T}$ are words, the terms in $\mathcal{T}_p$ may not be words at all), but are obtained by combinations or transformations of the original ones. Examples of this are methods that consider generating, from the original, a set of "synthetic" terms that maximize effectiveness based on *term clustering, latent semantic analysis, latent dirichlet allocation, principal component analysis* and others. After a transformation we could need to reduce the number of the new features through a selection method thus obtaining a new set $\mathcal{T}_{sp}$ that is a subset of $\mathcal{T}_p$.

In this paper we have used a *global* method for *feature transformation* that considers pairs of words instead of single words as basic features thus obtaining a new space $\mathcal{T}_p$ of features. The dimensionality of such a new space is very high, much higher than $|\mathcal{T}|$, in fact: $|\mathcal{T}_p| \propto |\mathcal{T}|^2$. For this reason we need to reduce the transformed space in order to obtain a new space $\mathcal{T}_{sp}$ such that $|\mathcal{T}_{sp}| \ll |\mathcal{T}_p|$.

The method used to select the most representative pairs of words is based on the *Latent Dirichlet Allocation* [2] implemented as the *Probabilistic Topic Model* [6] and this is the core of the proposed classification method that we explain next.

3 Proposed Feature Selection Method

In this paper we propose a new method for feature selection that, based on the probabilistic topic model, finds the pairs among all the $|\mathcal{T}_p|$ that are the most discriminative. The method works on the initial data representation, that is the matrix $\mathcal{T} \times \Omega_r$, where the features are the single words, and extracts a new representation, named the mixed *Graph of Terms*, that consists of related pairs of words. The graph contains two kinds of relations between words, directed and undirected, and for this reason it is called *mixed*.

In the graph we can find several clusters of words and each cluster contains a set of words v_s that specify, through a directed weighted edge, a special word, that we have named the *concept*, r_i, that is the centroid of such a cluster. The weight ρ_{is} can measure how far a word is related to a concept, or how much we need such a word to specify that concept, and it can be considered as a probability: $\rho_{is} = P(r_i|v_s)$. The resulting structure is a subgraph rooted on r_i (see fig. 2(a)).

Moreover, special words, namely *concepts*, can be linked together through undirected weighted edges, so forming a subgraph of pairs of centroids. The weight ψ_{ij} can be considered as the degree of semantic correlation between two concepts and it can be considered as a probability: $\psi_{ij} = P(r_i, r_j)$ (see fig. 2(a)).

Considering that each concept is a special word, we can say that the graph contains directed and undirected pairs of features that are all lexically denoted as words. For this reason, the graph can be used to select the most important pairs from the space $\mathcal{T}_p$ in order to obtain a new reduced space $\mathcal{T}_{sp}$.

Given the training set Ω_r of documents, the proposed method, through a learning procedure, selects a subset of pairs obtaining a number of pairs $|\mathcal{T}_{sp}| \ll |\mathcal{T}_p|$. In this way, the term extraction procedure is obtained by firstly computing all the semantic relatednesses between words and concepts, that is ρ_{is} and ψ_{ij}, and secondly selecting the right subset of pairs from all the possible ones. Before explaining in detail the learning procedure of a graph, we would like to highlight some aspects of this representation.

3.1   Graph and Document Representation in the Space $\mathcal{T}_{sp}$

A graph **g** can be viewed, following the *Vector Space Model* [4], as a vector of features t_n:

$$\mathbf{g} = \{b_1, \ldots, b_{|\mathcal{T}_{sp}|}\},$$

where $|\mathcal{T}_{sp}|$ represents the number of pairs and each feature $t_n = (v_i, v_j)$ can be a *word/concept* or *concept/concept* pair. The weight b_n is named *boost* factor and is equal to ψ_{ij} for both *word/concept* or *concept/concept* pairs.

Moreover, by following this approach, also each document of a corpus can be represented in terms of pairs:

$$\mathbf{d}_m = (w_{1m}, \ldots, w_{|\mathcal{T}_{sp}|m}),$$

where w_{nm} is such that $0 \leq w_{nm} \leq 1$ and represents how much term $t_n = (v_i, v_j)$ contributes to a semantics of document $\mathbf{d}_m$. The weight is calculated thanks to the tf-idf model applied to the pairs represented through t_n:

$$w_{nm} = \frac{\text{tf-idf}(t_n, \mathbf{d}_m)}{\sqrt{\sum_{n=1}^{|\mathcal{T}_{sp}|} (\text{tf-idf}(t_n, \mathbf{d}_m))^2}} \tag{3}$$

3.2   Classifier Definition in the Space $\mathcal{T}_{sp}$

As we have seen before, the mixed *Graph of Terms* (**g**) learned from the training set Ω_r can be also represented as a vector of features in the $\mathcal{T}_{sp}$ space. If we learn a graph $\mathbf{g}_i$ from documents that are labeled as $\mathbf{c}_i$, then $\mathbf{g}_i$ it can be considered as representative of such labeled set of documents and considered as the expert $\hat{\phi}_i$ for the category $\mathbf{c}_i$ itself:

$$\mathbf{g}_i = \hat{\phi}_i = \{b_{1i}, \ldots, b_{|\mathcal{T}_{sp}|i}\}.$$

Using the expert we can perform a classification task by using a linear method that measures the similarity between the expert $\hat{\phi}_i$ and each document $\mathbf{d}_m$ represented in the space $\mathcal{T}_{sp}$.

Here we have considered as a measure of similarity the well known cosine similarity between vectors in a $\mathcal{T}_{sp}$ space and thus obtaining a ranking classifier $\forall i$:

$$CSV_i(\mathbf{d}_m) = \frac{\sum_{n=1}^{|\mathcal{T}_{sp}|} b_{ni} \cdot w_{nm}}{\sqrt{\sum_{n=1}^{|\mathcal{T}_{sp}|} b_{ni}^2} \cdot \sqrt{\sum_{n=1}^{|\mathcal{T}_{sp}|} w_{nm}^2}} \tag{4}$$

Such a ranking classifier for the category $\mathbf{c}_i \in \mathcal{C}$ consists in the definition of a function, the cosine similarity, that, given a document $\mathbf{d}_m$, returns a *categorization status value* ($CSV_i(\mathbf{d}_m)$) for it, i.e. a number between 0 and 1 that represents the evidence for the fact that $\mathbf{d}_m \in \mathbf{c}_i$, or in other words it is a measure of vector closeness in a $|\mathcal{T}_{sp}|$-dimensional space.

Following this criterion each document is then ranked according to its CSV_i value, and so the system works as a document-ranking text classifier, namely a "soft" decision based classifier. As we have discussed in previous sections we need a binary classifier, also known as a "hard" classifier, that is capable of assigning to each document a value T or F to measure the vector closeness.

A way to turn a soft classifier into a hard one is to define a threshold γ_i such that $CSV_i(\mathbf{d}_m) \geq \gamma_i$ is interpreted as T while $CSV_i(\mathbf{d}_m) \leq \gamma_i$ is interpreted as F. We have adopted an experimental method, that is the CSV *thresholding* [12], which consists in testing different values for γ_i on a subset of the training set (the *validation* set) and choosing the value which maximizes effectiveness. Next we show how such thresholds have been experimentally set.

4   Graph Learning

A graph g is well determined through the learning of the weights, the *Relations Learning* stage, and through the learning of three parameters, the *Structure Learning* stage, that are $\Lambda = (H, \tau, \mu)$ which specify the shape, namely the structure, of the graph. In fact, we have:

1. H: the number of concepts (namely the number of clusters) of the set of documents;
2. μ_i: the threshold that establishes for each concept the number of edges of the directed subgraph, and so the number of *concept/word* pairs of the corpus . An edge between the word s and the concept i can be saved if $\rho_{is} \geq \mu_i$. To simplify the formulation, we assume that $\mu_i = \mu, \forall i$;
3. τ: the threshold that establishes the number of edges of the undirected subgraph, and so the number of *concept/concept* pairs of the corpus. An edge between the concept i and concept j can be saved if $\psi_{ij} \geq \tau$.

4.1   Relations Learning

Due to the fact that each concept is lexically represented by a word of the vocabulary, then we have that $\rho_{is} = P(r_i|v_s) = P(v_i|v_s)$, and $\psi_{ij} = P(r_i, r_j) = P(v_i, v_j)$.

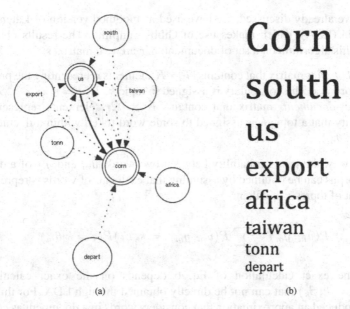

Corn
south
us
export
africa
taiwan
tonn
depart

(a)   (b)

Fig. 2. Part of the *Vector of features* for the topic *corn*. We have 2 concepts (double circles) and 6 words (single circles). Solid edges represent undirected relations (ψ_{ij}) while dotted edges represent directed relations (ρ_{is}). 2(a) A mixed *Graph of Terms*. 2(b) A *List of Terms*.

Considering that $P(v_i, v_j) = P(v_i|v_j)P(v_j)$, it is necessary, to learn all the relations between words, to compute the joint, or the conditional, probability $\forall i, j \in \{1, \cdots, |\mathcal{T}|\}$ and each $P(v_j)$ $\forall j$.

We show here that the exact calculation of $P(v_j)$ and the approximation of the joint, or conditional, probability can be obtained through a smoothed version of the generative model introduced in [2] called Latent Dirichlet Allocation (LDA), which makes use of Gibbs sampling [6].

The original theory introduced in [6] mainly asserts a semantic representation in which documents are represented in terms of a set of probabilistic topics z. Formally, we consider a word u_m of the document $\mathbf{d}_m$ as a random variable on the vocabulary $\mathcal{T}$ and z as a random variable representing a topic between $\{1, \cdots, K\}$. The probability distribution of a word within a document $\mathbf{d}_m$ of the corpus can be obtained as:

$$P(u_m) = \sum_{k=1}^{K} P(u_m|z = k, \beta_k)P(z = k|\theta_m). \tag{5}$$

The generation of a document $\mathbf{d}_m$ can be obtained considering the generation of each word of the document. To obtain a word, the model considers three parameters assigned: α, η and the number of topics K. Given these parameters, the model chooses θ_m through $P(\theta|\alpha) \sim \text{Dirichlet}(\alpha)$, the topic k through $P(z|\theta_m) \sim \text{Multinomial}(\theta_m)$ and $\beta_k \sim \text{Dirichlet}(\eta)$. Finally, the distribution of each word given a topic is $P(u_m|z, \beta_z) \sim \text{Multinomial}(\beta_z)$.

As we have already discussed, we have used a smoothed version of Latent Dirichlet Allocation (LDA), which makes use of Gibbs sampling. The results obtained by performing this algorithm on a set of documents Ωr are two matrixes:

1. the *words-topics* matrix that contains $|\mathcal{T}| \times K$ elements representing the probability that a word v_i of the vocabulary is assigned to topic k: $P(u = v_i|z = k, \beta_k)$;
2. the *topics-documents* matrix that contains $K \times |\Omega r|$ elements representing the probability that a topic k is assigned to some word token within a document $\mathbf{d}_m$: $P(z = k|\theta_m)$.

In the same way, the joint probability between two words u_m and y_m of a document $\mathbf{d}_m$ of the corpus can be obtained by assuming that each pair of words is represented in terms of a set of topics z and then:

$$P(u_m, y_m) = \sum_{k=1}^{K} P(u_m, y_m|z = k, \beta_k)P(z = k|\theta_m) \tag{6}$$

Note that the exact calculation of Eq. 6 depends on the exact calculation of $P(u_m, y_m|z = k, \beta_k)$ that can not be directly obtained through LDA. For this reason, we have introduced an approximation that considers words in a document as conditionally independent given a topic. In this way Eq. 6 can be written as:

$$P(u_m, y_m) \simeq \sum_{k=1}^{K} P(u_m|z = k, \beta_k)P(y_m|z = k, \beta_k)P(z = k|\theta_m). \tag{7}$$

Note that Eq. 5 gives the probability distribution of a word u_m within a document $\mathbf{d}_m$ of the corpus. To obtain the probability distribution of a word u independently of the document we need to sum over the entire corpus:

$$P(u) = \sum_{m=1}^{M} P(u_m)\delta_m \tag{8}$$

where δ_m is the prior probability for each document ($\sum_{m=1}^{|\Omega r|} \delta_m = 1$).

In the same way, if we consider the joint probability distribution of two words u and y, we obtain:

$$P(u, y) = \sum_{m=1}^{M} P(u_m, y_v)\delta_m \tag{9}$$

Concluding, once we have $P(u)$ and $P(u, y)$ we can compute $P(v_i) = P(u = v_i)$ and $P(v_i, v_j) = P(u = v_i, y = v_j), \forall i, j \in \{1, \cdots, |\mathcal{T}|\}$ and so the relations learning can be totally accomplished.

4.2   Structure Learning

Given a set of documents, once each ψ_{ij} and ρ_{is} is known $\forall i, j, s$, letting the parameters $\Lambda_t = (H, \tau, \mu)_t$ assume a different set of values, we can observe a different structure of the graph $\mathbf{g}_t$ (here t is representative of different parameter values).

A way to learn the structure of the graph is to use an optimization based algorithm that searches for the best set of parameters Λ_t. In this case we need a scoring function and a searching strategy [1].

As we have previously seen, a $\mathbf{g}_t$ is a vector of features $\mathbf{g}_t = \{b_{1t}, \ldots, b_{|\mathcal{T}_{sp}|t}\}$ in the space $\mathcal{T}_{sp}$ and each document of the training set Ω_r can be represented as a vector $\mathbf{d}_m = (w_{1m}, \ldots, w_{|\mathcal{T}_{sp}|m})$ in the space $\mathcal{T}_{sp}$. A possible scoring function is the cosine similarity between these two vectors:

$$S(\mathbf{g}_t, \mathbf{d}_m) = \frac{\sum_{n=1}^{|\mathcal{T}_{sp}|} b_{nt} \cdot w_{nm}}{\sqrt{\sum_{n=1}^{|\mathcal{T}_{sp}|} b_{nt}^2} \cdot \sqrt{\sum_{n=1}^{|\mathcal{T}_{sp}|} w_{nm}^2}} \tag{10}$$

and thus the optimization procedure would consist in searching for the best set of parameters Λ_t such that the cosine similarity is maximized $\forall \mathbf{d}_m$.

By following this approach, the best $\mathbf{g}_t$ for the set of documents Ω_r is the one that produces the maximum score attainable for each of the documents when the same graph is used as a vector of features to measure the similarity of a set containing just those documents which have fed the graph builder.

As a consequence, we obtain a score for each document $\mathbf{d}_m$ and then we have

$$\mathbf{S}_t = \{S(\mathbf{g}_t, \mathbf{d}_1), \cdots, S(\mathbf{g}_t, \mathbf{d}_{|\Omega_r|})\},$$

where each score depends on the specific set $\Lambda_t = (H, \tau, \mu)_t$.

To compute the best value of Λ we can maximize the score value for each document, which means that we are looking for the graph which best describes each document of the repository from which it has been learned. It should be noted that such an optimization maximizes at the same time all $|\Omega_r|$ elements of $\mathbf{S}_t$.

Alternatively, in order to reduce the number of the objectives being optimized, we can at the same time maximize the mean value of the scores and minimize their standard deviation, which turns a multi-objective problem into a two-objective one. Additionally, we can reformulate the latter problem by means of a linear combination of its objectives, thus obtaining a single objective function, i.e., *Fitness* ($\mathcal{F}$), which depends on Λ_t,

$$\mathcal{F}(\Lambda_t) = E[\mathbf{S}_t] - \sigma[\mathbf{S}_t],$$

where E is the mean value of all the elements of $\mathbf{S}_t$ and σ_m is the standard deviation. By summing up, the parameters learning procedure is represented as follows,

$$\Lambda^* = \operatorname*{argmax}_t \{\mathcal{F}(\Lambda_t)\}.$$

We will see next how we have performed the searching strategy phase.

Since the space of possible solutions could grow exponentially, we have considered[2] $|\mathcal{T}_{sp}| \leq 300$. Furthermore, we have reduced the remaining space of possible solutions by applying a clustering method, that is the *K-means* algorithm, to all ψ_{ij} and ρ_{is} values, so that the optimum solution can be exactly obtained after the exploration of the entire space.

---

[2] This number is usually employed in the case of Support Vector Machines.

This reduction allows us to compute a graph from a repository composed of a few documents in a reasonable time (e.g. for 3 documents it takes about 3 seconds with a Mac OS X based computer, 2.66 GHz Intel Core i7 CPU and a 8GB RAM). Otherwise, we would need an algorithm based on a random search procedure in big solution spaces. For instance, Evolutionary Algorithms would be suitable for this purpose, but would provide a slow performance. In fig. 2(a) we can see an example of a graph learned from a set of documents labeled as topic *corn*.

4.3   Extracting a Simpler Representation from the Graph

From the mixed *Graph of Terms* we can select different subsets of features so obtaining a simpler representation (see fig. 2(b)). Before discussing this in detail, we would recall that $\psi_{ij} = P(v_i, v_j)$ and $\rho_{is} = P(v_i|v_s)$ are computed through the topic model which also computes the probability for each word $\eta_s = P(v_s)$.

We can obtain the simplest representation by selecting from the graph all distinct terms and associating to each of them its weight $\eta_s = P(v_s)$. We name this representation the *List of Terms* (**w**), see fig. 2(b).

By using the list of terms we can perform a linear classification task considering both vectors of features and documents represented as vectors in the space $\mathcal{T}_s$ and by considering the cosine similarity in such a space.

4.4   Consideration on the Method

Here we wish to demonstrate that by using such a graph as a vector of features we are capable of achieving a good performance, in terms of accuracy, even if the size of the training set Υ_r is about 1.4% of the original Ω_r and is composed of only positive examples. We further wish to demonstrate that the performance of our approach is better than existing methods, such as support vector machines, based on feature selection instead of feature extraction.

How can we obtain a good performance when the training set is small? In this case, in fact, the number of documents is low while the number of features (for instance words), that occur in these documents, is higher, $|\mathcal{T}| \gg |\Upsilon r|$. Even if we perform data reduction through a selection method, we could still have $|\mathcal{T}_s| \gg |\Upsilon_r|$. If we follow the theory introduced at the beginning of this paragraph, in this case we have to say that the efficiency and accuracy of data analysis are low.

In this work we wish to demonstrate that a way to improve the performance when $|\mathcal{T}| \gg |\Upsilon_r|$ is to apply a method of feature extraction that discovers missing information between features in the original dataset and that maps the discovered information in a new augmented space $\mathcal{T}_p$ where such information can be emphasized.

The bags of words representation of a text introduces ambiguity when $|\mathcal{T}| \gg |\Upsilon_r|$ and we argue that the only way to reduce the ambiguity is to introduce another dimension of observation where it is possible to distinguish which word conveys which meaning. By using different pairs of words in the graph we are able to give more importance to words that convey concepts and that contribute to specify a topic and to assign less importance to those words that contribute to specify concepts and that, due to the fact that they can be more plausibly shared between concepts, can increase the ambiguity.

It is also important to make clear that the mixed *Graph of Terms* can not be considered as a co-occurrence matrix.

In fact, the core of the graph is the probability $P(v_i, v_j)$, which we regard as a word association problem, that in the topic model is considered as a problem of prediction: given that a cue is presented, which new words might occur next in that context? It means that the model does not take into account the fact that two words occur in the same document, but that they occur in the same document when a specific topic (and so a context) is assigned to that document [6].

Furthermore, in the field of statistical learning, a similar structure has been introduced, named the Hierarchical Mixture of Experts [7]. Such a structure is employed as a method for supervised learning and it is considered as a variant of the well known tree-based methods. The similarity between such a structure and the proposed graph can be obtained by considering the "experts" as "concepts".

Notwithstanding this, the mixed Graph of terms is not a tree structure, and more importantly is not rigid but is dynamically built depending on the optimization stage. Moreover, the Hierarchical Mixture of Experts does not consider relations between experts which is, on the other hand, largely employed in the mixed Graph of Terms. Nevertheless, we will explore further connections between the two methods in future works.

5 Evaluation

We have considered a classic text classification problem performed on the Reuters-21578 repository. This is a collection of 21,578 newswire articles, originally collected and labeled by Carnegie Group, Inc. and Reuters, Ltd.. The articles are assigned classes from a set of 118 topic categories. A document may be assigned to several classes or none, but the commonest case is a single assignment (documents with at least one class received an average of 1.24 classes).

For this task we have used the ModApte split which includes only documents that were viewed and assessed by a human indexer, and comprises 9,603 training documents and 3,299 test documents. The distribution of documents in classes is very uneven and therefore we have evaluated the system only on documents in the 10 largest classes [4][3].

Note that the graph is different from a simple list of key words because of the presence of two features: the relations between terms and the hierarchical differentiation between simple words and concepts. To demonstrate the discriminative property of such features we have to prove that the results obtained by performing the proposed approach are significantly better than the results obtained by performing the same classification task, through the cosine similarity, when the simple list of weighted words extracted from the graph is used as the vector of features.

In a single label, or binary, classifier we usually have a training set containing examples that are labeled as c_i or $\overline{c}_i$. The learned classifier is capable of assigning a new document to the category c_i or $\overline{c}_i$. The graph has been learned only from documents

---

[3] Note that considering the 10 largest classes means 75% of the training set and 68% of the test set.

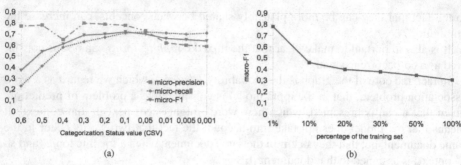

Fig. 3. 3(a) Tuning of the threshold for γ. 3(b) Different values of macro-F_1 for different percentages of the training set Υ_r.

labeled as c_i (positive examples) and documents belonging to the category $\bar{c}_i$ have not been used. For this reason, our method is not directly comparable with existing methods. Notwithstanding this, we have compared our approach with linear Support Vector Machines (SVM) learned on the same percentage of the training set but using both positive and negative examples. For SVM we have used a method for term selection based on mutual inference.

As a result, the aim of the evaluation phase is twofold:

1. To demonstrate the discriminative property of the graph compared with a method based only on the words from the graph without relations (named the Words List);
2. To demonstrate that the graph achieves a good performance when 1.4% of the training set is employed for each class. Here we report a comparison with SVM trained on the same percentage of the training set.

5.1   Measures

As discussed before, we have considered the *any-of problem* and so we have learned 10 two-class classifiers, one for each class, where the two-class classifier for class c_i is the classifier for the two classes c and its complement $\bar{c}_i$. For each of these classifiers, we have used several measures considering TP_i as true positive, TN_i as true negative, FP_i as false positive and FN_i as false negative for the category c_i ([12,4]):

- precision and recall for the category c_i: $P_i = \frac{TP_i}{TP_i+FP_i}$ and $R_i = \frac{TP_i}{TP_i+FN_i}$;
- micro-average precision and recall: $P_{micro} = \frac{\sum_{i=1}^{|C|} TP_i}{\sum_{i=1}^{|C|} TP_i+FP_i}$ and

  $R_{micro} = \frac{\sum_{i=1}^{|C|} TP_i}{\sum_{i=1}^{|C|} TP_i+FN_i}$
- F_1 measure for the category c_i: $F_{1i} = 2 \cdot \frac{P_i \cdot R_i}{P_i+R_i}$
- micro-average F_1: $F_{1micro} = 2 \cdot \frac{P_{micro} \cdot R_{micro}}{P_{micro}+R_{micro}}$
- macro-average F_1: $F_{1macro} = \frac{1}{|C|} \sum_{i=1}^{|C|} F_{1i}$

Table 1. Average dimension of the reduced training set Υ_r and original dimension of Ω_r. F_1 measure, F_{1micro} and F_{1macro} for the graph (**g**), word list (**w**) and support vector machines (SVM). The arrows column shows the increment of **g** performance compared with other methods.

Topic	Ω_r(KB)	Υ_r(KB)	g@max	g@av	w@max	w@av	SVM@max	SVM@av	
earn	957	14	91	76	82	69	95	66	⇓
acq	902	13	63	44	53	38	63	35	⇔
money-fx	476	7	46	30	39	23	37	09	⇑
grain	359	5	66	40	54	35	48	04	⇑
crude	356	5	70	42	60	40	47	10	⇑
trade	440	6	58	42	43	39	27	06	⇑
interest	267	4	50	34	38	24	09	01	⇑
ship	137	2	68	18	59	12	16	01	⇑
wheat	229	3	86	43	72	31	26	02	⇑
corn	153	2	65	23	54	16	10	02	⇑
		F_{1micro}	66	39	46	23	38	14	⇑
		F_{1macro}	74	53	56	33	61	28	⇑

5.2 Experiments

We have set the threshold γ for the categorization status value by evaluating aggregate measures: micro-precision, micro-recall and micro F1 (see Fig. 3(a)). We have chosen $\gamma = 0.1$ for all the topics.

After the classifier has been set, we have experimented with several dimensions of the reduced training set Υ_r and evaluated the performance through the macro-F_1. In Fig. 3(b) the behavior of the classifier shows a degradation of performance as the dimension of the training set increases. This suggests that the mixed graph of terms becomes less discriminative as the number of labeled examples increases. For this reason, we have considered Υ_r to be about 1.4% of Ω_r.

We have randomly selected the 1.4% from each training set (in table 1 is reported the comparison between the dimension of Υ_r and the original training set Ω_r) and moreover we have performed the selection 100 times in order to make the results independent of the particular document selection. As a result, we have 100 repositories and from each of them we have calculated 100 graphs by performing the parameters learning described above.

Due to the fact that each optimization procedure leads to a different structure of the graph, we have a different number of pairs for each structure. We have calculated the average number of pairs for each topic and the corresponding average number of terms. Note that the average size of $|\mathcal{T}_{op}|$ is 116, while the average size of $|\mathcal{T}_s|$ is 33. The overall number of features observed by our method is, independently of the topic, less than the number considered in the case of Support Vector Machines where we have employed a term selection process obtaining $|\mathcal{T}|_s = 300$.

In table 1 we have reported the F_1 measure, micro-F_1 and macro-F_1 obtained by the graph g, word list **w** and support vector machines (SVM). We have reported the best values and the average values obtained by performing the classification of all 100 examples of the reduced training set.

It is surprising how the proposed method, even if the training set is smaller than the original one, is capable of classifying in most cases with an accuracy sometimes comparable to and mostly better than Support Vector Machines. Note that the performance of the proposed method is, independently of the topic, better than the word list, so demonstrating that the graph representation possesses better discriminative properties than a simple list of words. Finally, it should be noticed that the good performance shown by the word list based method is due to the fact that the list of words is composed of the terms extracted from the graph demonstrating that the graph could be useful also to select the most discriminative words from the space $\mathcal{T}_s$.

6   Conclusions

In this work we have demonstrated that a *term extraction* procedure based on a mixed *Graph of Terms* representation is capable of achieving a better performance than a method based on a simple *term selection*, obtained considering only the words composing the graph, and a linear version of SVM. Moreover, we have demonstrated that the overall performance of the method is good even when only 1.4% of the training set has been employed.

References

1. Bishop, C.M.: Pattern Recognition and Machine Learning. Springer (2006)
2. Blei, D.M., Ng, A.Y., Jordan, M.I.: Latent dirichlet allocation. Journal of Machine Learning Research 3, 993–1022 (2003)
3. Blum, A.L., Langley, P.: Selection of relevant features and examples in machine learning. Artificial Intelligence 97, 245–271 (1997)
4. Manning, C.D., Raghavan, P., Schütze, H.: Introduction to Information Retrieval. Cambridge University (2008)
5. Fodor, I.: A survey of dimension reduction techniques. Tech. rep. (2002)
6. Griffiths, T.L., Steyvers, M., Tenenbaum, J.B.: Topics in semantic representation. Psychological Review 114(2), 211–244 (2007)
7. Hastie, T., Tibshirani, R., Friedman, J.: The Elements of Statistical Learning. Springer (2009)
8. Ko, Y., Seo, J.: Text classification from unlabeled documents with bootstrapping and feature projection techniques. Inf. Process. Manage. 45, 70–83 (2009)
9. Lewis, D.D., Yang, Y., Rose, T.G., Li, F.: Rcv1: A new benchmark collection for text categorization research. J. Mach. Learn. Res. 5, 361–397 (2004)
10. McCallum, A., Nigam, K., Rennie, J., Seymore, K.: A machine learning approach to building domain-specific search engines. In: Proceedings of the 16th International Joint Conference on Artificial Intelligence, vol. 2, pp. 662–667. Morgan Kaufmann (1999)
11. Salton, G., McGill, M.J.: Introduction to modern information retrieval. McGraw-Hill (1983)
12. Sebastiani, F.: Machine learning in automated text categorization. ACM Comput. Surv. 34, 1–47 (2002)
13. Tsoumakas, G., Katakis, I.: Multi-label classification: An overview. Int. J. Data Warehousing and Mining 2007, 1–13 (2007)

Part II

Knowledge Engineering
and Ontology Development

A System to Support Legal Case Building and Reasoning

Stefania Castellani[1], Nikolaos Lagos[1], Nicolas Hairon[1], Antonietta Grasso[1],
David Martin[1], and Frederique Segond[2]

[1] Xerox Research Centre Europe, 6, Chemin de Maupertuis, 38240, Meylan, France
{Stefania.Castellani,Nikolaos.Lagos,Antonietta.Grasso,
David.Martin}@xrce.xerox.com, nicolas.hairon@inria.fr
[2] Viseo, 34, Avenue Félix Viallet, 38000, Grenoble, France
fsegond@viseo.net

Abstract. We have designed a system to support collaborative legal case reasoning and building. The design is based on our understanding of the corporate litigation domain acquired through analysis of the literature, interviews of various parties involved in corporate litigation processes, and studies of the commercial tools already available. In this paper we illustrate the designed system and in particular the interaction modes that it supports that we believe address a number of the requirements that emerged through our analysis. We also describe its main components and their integration, including a knowledge model that represents the domain, and a natural language processing component for extracting semantic information. A description of a prototype system is also provided.

Keywords: Legal Case Reasoning and Building, Human Computer Interaction, Information Extraction, Sensemaking, Corporate Litigation Processes, Semantics-based Litigation.

1   Introduction

The work of corporate litigation lawyers is an interesting and challenging field to study and design for, both from an organizational and a technical point of view: it is a highly complex process, involving a variety of actors, who must manage and analyse huge corpora of information. The litigation process involves two main parts: 1) e-discovery [13] - the analysis of immense document sets to isolate only the documents relevant (i.e. responsive) to the case and 2) case construction – the finding of evidence and argument construction based on the contents of the set of relevant documents so determined. The case reasoning activity ultimately produces the defense or attack line to be used to settle or to go to court.

The primary goal of the searching and browsing facilities offered in current litigation tools is to find relevant documents, often using keyword/boolean based search techniques. Although this has proved to be relatively useful in the e-discovery phase, during case construction the emphasis shifts from finding documents to finding entities and actionable information derived from these entities [17], [20], [15]. This kind of search is an important part of the lawyers' work and tools currently on the market allow users to store information on relevant characters and events. However, there is

A. Fred et al. (Eds.): IC3K 2011, CCIS 348, pp. 217–231, 2013.
© Springer-Verlag Berlin Heidelberg 2013

little in the way of support to help users identify the relevant information and once the information is identified they must manually enter it in the tools database. Moreover little support is provided for collaborative work and information sharing among the members of a legal team working on a case.

On the basis of these observations and more specific requirements that we collected from interviews with lawyers and technology service providers, a review of litigation support tools currently available on the market, and an examination of the few available case studies [4-6] we are developing a work environment for lawyers. The system is meant to help lawyers search for information from the document collection associated with a legal case, build the case, reason about lines of inquiry, and share findings with colleagues working on the creation of an outline for the case. In particular, the system is designed to provide some forms of support for lawyers working to identify characters, e.g., people or organizations that have played a role in a case, events they have participated in, etc. Also the system aims at offering to the members of a legal team a shared representation of the legal case while being able to work individually on specific lines of inquiry.

Previous work on some components of the system has already been described in [12], [15]. In this paper we illustrate the overall design of the system and a first prototype for it based on semantic technologies that implement parts of the design.

The rest of the paper is organized as follows. Section 2 presents an analysis of current practices in litigation in relation to case building and reasoning. Section 3 illustrates the design of the system, including the overall architecture, its components, and the interaction modes that it supports. Section 4 presents our work for prototyping the system. Finally section 5 discusses related and future work.

2   Analysis of Current Litigation Practices

The litigation process usually involves two groups of lawyers: a key case team of senior lawyers (SLs) that starts the process, talks to the clients and generates the first documents, specifically letters of complaints, review protocols and the "issues" or main lines of inquiry. These are used by the responsiveness reviewers/issue-coders team, formed by junior lawyers (JLs) and paralegals (reaching up to 600 members), often organized in sub-teams, who read every single document from the usually very large set of documents that are potentially relevant to the case to determine the set of responsive documents. In this phase they also typically assign relevant documents to "issues", which are subtopics used later on to reason around and organize the case. After responsiveness review the still large set of remaining documents is further reviewed and filtered so that only the most important and relevant documents are seen by the key case team, which develops the case.

The role of technology in the above process is currently pretty limited, but is expanding in scope. Technology is mainly used to retrieve and store the document set for e-discovery, which may then be searched using keyword search. Even if more sophisticated technologies like conceptual searches have been suggested by technology providers, they have encountered resistance. Among the reasons for this is the need to be able to explain in court just how the documents have been filtered and why that method is valid. This situation however is changing. Technology that can semi-automatically categorize the documents and collaboratively assist in e-discovery is being developed and trialed [19]. Likewise, database-like tools [11] have appeared on the market to assist the phase of case construction by letting the teams to store relevant entities and construct an outline of case defense or attack.

Another important aspect to note is that the current largely manual legal work process implies a strict and procedural division of labour where the phases are distinct. There is little space for collaboration, and each refinement step weeds away documents following precise rules that provide material for the next phase and further sensemaking of the document set. We believe that these two aspects are interconnected and that the introduction of technology, while hopefully at first speeding up the simpler steps, will further dissipate the barriers between phases and enhance the phase of case construction. Procedurally, this is preferable as case building starts as soon as the legal case is issued but current logistical constraints enforce an unfortunate separation of discovery and case construction.

How do litigation lawyers search through the documents to construct the case? Attfield's study [4-6] provides insight on the reasoning surrounding manual case construction performed by a legal team. First of all, given the size of the dataset, and in order to appropriately distribute the labour, investigators usually need to gradually "decompose an investigation into meaningful and tractable chunks of enquiry" [5] taking into account any relevant information found during the investigation. This is important because of the need to separate and keep track of the "theories" eventually "eliminated when evidence found was contradictory or unsupportive" [5]. Additionally, a key point to understand is that the information that constitutes evidence for a fact can be contained across a set of documents. In these cases the risk is that something is not seen as relevant when first uncovered because the extra contextual information which will flesh it out as relevant has not (yet) been found and connected to it [4]. Thus as a potential area of support it would be useful for the lawyers to find, explore and manage the information within the entire document set (i.e. across documents) and to view low-level lines of enquiry in terms of a bigger picture [5], [12]. As reported in [6]: "The capability of iteratively selecting records and setting them aside, perhaps in addition to code filtering, would provide greater flexibility for exploring and discussing different possibilities." And also, "document references within event entries allowed the chronologies to act as indexes supporting the re-retrieval of raw evidence. [...] However, these links were not automated. More efficient access would be supported if the source documents could be accessed directly for the summary representation."

To complement these observations, we also found evidence from interviews we conducted that JLs have problems in managing consistency and getting up to date with the case. The current procedural practice is that they work under great time pressure and are involved just in assigning documents to issues rather than in creating "the case". However, if they had the support for this, they could communicate new knowledge in a timely fashion to the rest of the team (both JLs and SLs) working on the case. Therefore, as also reported in [5] methods to discuss findings, synchronise work and exchange information are vital.

Another aspect emerging from the literature and from analysis of the commercial databases for case construction is that legal reasoning evolves through manipulation of specific semantic entities. It is especially important for lawyers to identify key players and their relationships and to build chronologies of events relevant for a case [11], [5], [15]. As reported in [5-6]: lawyers can be searching for documents by focusing on particular time periods or on specific events, e.g. meetings, in support of their conjectures; "each team created one or more "issue" chronologies and, as these evolved important content was selected and consolidated into a single master chronology"; lawyers need to record information on events, times, participants and documents related.

On the basis of this analysis we have envisioned a future scenario of collaborative work for litigation and developed a design of a case building system, supporting it, which is currently under implementation. The next section illustrates the design of the system.

3   The Case Building System

The Case Building System (CBS) that we are developing aims at providing lawyers with tools to help them individually and collaboratively keep a record of their findings and lines of inquiry and seek new information from a document collection. CBS sits on top of a knowledge base containing case documents, such as letters of complaint (issued at the beginning of a litigation process) and responsive documents (as they emerge from the e-discovery phase). The knowledge base also contains information elements, (people, events, locations, etc.) automatically extracted by the system from those documents. Key features include:

• It is a unified system supporting case reasoning and building from the early stages of litigation.
• It supports recording multiple lines of inquiry and information discovery.
• It is strongly visual and interactive with several views based on key semantic dimensions (time, structure, etc.)
• It provides collaboration support, e.g. awareness of what colleagues have found, so that activities can be synchronised and findings shared.
• It is semi-automatic in its knowledge extraction and suggestions to let lawyers benefit from the power of content analysis whilst remaining in control.

Figure 1 shows the design of the user interface of the system. The user interacts with a visual environment organized in 4 interactive working areas that capture the current status of the lawyer's work, that is, current lines of thought and acquired and emerging information captured and displayed according to several dimensions.

• DocumentVisualiser (1 in Figure 1) allows the user to search for documents, in the whole collection or only within documents already included in the case, and navigate the contents of selected documents that display extracted information elements, e.g. events. The user can select information elements within documents to be explored and potentially included in the case by sending them to other views. In this way document evidence can be associated to information elements extracted in support of a line of inquiry.
• CastOfObjects (3 in Figure 1) provides a structured view of the information elements currently selected and saved by the users as salient information for the case, including characters, events, and the facts that constitute the various lines of inquiry. Characters represent the important "actors" of the story, e.g. a person that has played

a role in the case. Events are events that have happened, e.g. "John Doe met Jane Roe in Zurich in March 2000" or situations, e.g. "M. Jones is head of the human resources department". Facts are the units of case construction and collaboration and can be constructed from a number of information elements. This view allows the user to search for further elements to be included or for elements already considered as relevant, for example by other members of the team working on the case, to record additional selected elements, and to manually enter new elements. Also, the user can see elements that other members of the team have inserted.

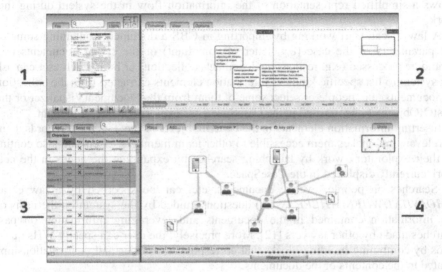

Fig. 1. The case building system's working area

- TimeLine (2 in Figure 1) shows a temporal dimension of the case in the form of a chronological visualisation of the events selected and saved by the users as relevant to the case. Events with fuzzy dates can be displayed with a special visual cue to differentiate them from the ones with precise dates. Users can select events to be inserted in the TimeLine either by selecting them from documents or by manually entering them. It will be visible if events have been manually added or extracted from documents. Events stored in the CastOfObjects will also be displayed. Moreover users will be able to express connections among the events. Also, the user should be able to see the elements that other members of the team have inserted.

- CaseSpace (4 in Figure 1) is an exploratory space where the lawyers can build and visualize networks of case elements. A network can be expanded starting by one of its nodes and searching for extracted connected elements or by manually adding new elements and connections. The displayed elements will have features that will represent diverse information including the "importance" of the element in the case and the degree of connection with other nodes which could be a combination of scores given by the system according to some properties and the score given by the users. The networks can be saved as clusters of evidence for a chunk of inquiry and shared with the other members of the legal team working on the case. Otherwise if the lawyer estimates that they do not support evidence for a chunk of inquiry they can be discarded.

This collaborative visual environment provides the user with tools to visualise and navigate documents and information, store important information that has then to be made persistent, and "play" with lines of inquiry. It should also provide the lawyers with a view of the current status of the team's work, and then a somewhat shared representation of the legal case that they can navigate and progressively enrich. According to our scenario of work, the lawyers of the team will work individually with the CBS conducting their investigations while the system supports the synchronization of their work and the collaborative construction of the global case. Figure 2 shows a simplified representation of the information flow in the system during this work.

A lawyer can start working by importing in CBS a document containing some of the parameters of the case (e.g. letter of compliant) or the set of documents that should be processed (e.g. responsive documents collection) or both. Each user can ask the system to list specific kinds of information elements extracted from the collection of documents, e.g. people or select some of them from the DocumentVisualizer or the CastOfObjects, to be inserted in the CaseSpace, for exploring a line of thought.

Inserting information elements in the CastOfObjects or the TimeLine records them as relevant and makes them accessible to other team members. Users can also continue their exploratory work by launching searches by expanding the nodes in the network currently displayed in the CaseSpace.

Searches on people, events, documents, etc. can be specified by a lawyer as WHO/WHAT/WHEN/WHERE (WHs) questions guided by the system on the basis of the information contained in the documents and leveraging information on past searches made by other lawyers [12]. More precisely, the user can specify WHs questions by combining building blocks that correspond to entities and their relationships related to the contents of the documents.

The user starts to build a question by choosing the type of answer he would like to get by selecting one of the building blocks among "WHO", "WHAT", "WHERE", and "WHEN". Then the system guides the user at each step of the question construction using a combination of structure, semantic and content-based mechanisms.

These mechanisms allows the system to show to the user the list of blocks that (s)he could use, that is, that would allow the definition of a request for information both syntactically valid and leading to some answers in the current corpus. The user can then select within this list the extension (s)he wants for the WH question. Figure 3 shows an example of question formulation (in order to search for "Who has worked for *Comp1* in 2002?") according to the designed interaction (for a more detailed description of the designed interaction see [12]). The CaseSpace is an exploratory space for building and visualizing networks of elements, extracted from the documents and/or built by the lawyers using their knowledge of the case. Selected networks can be saved and shared with other lawyers e.g. as a support to a line of inquiry or evidence of a fact.

The Timeline allows lawyers to capture and visualize the sequence of events selected during their work and annotate causal connections among them. Filters allow them to visualize their own chronologies or to see also the ones that other colleagues have reconstructed.

In order to support the interaction described so far we have designed the architecture of the CBS as shown in Figure 4. The Knowledge Model (KM) is used to represent the elements of information useful for the lawyers during the envisioned scenario of work described above. To automate (up to a certain degree) the process of finding information in a large corpus of documents a Natural Language Processing (NLP) system is used. The NLP system extracts the entities and relations defined in the KM.

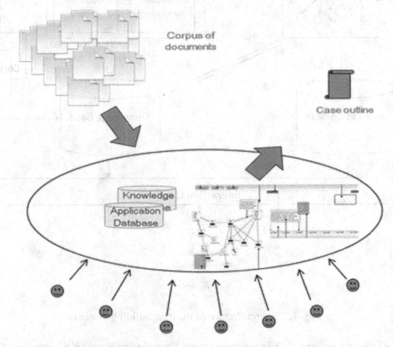

Fig. 2. A simplified representation of the information flow in the system

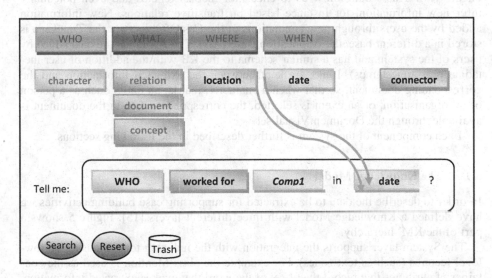

Fig. 3. Example of question formulation by building blocks composition

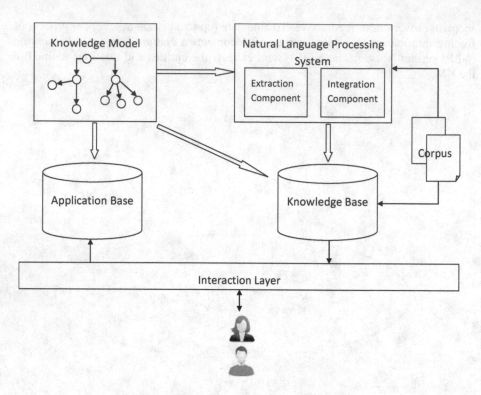

Fig. 4. The architecture of the case building system

The extracted information is inserted into a Knowledge Base (KB) structured based on the KM. This would allow us to check for inconsistencies and even potentially infer new information, for instance based on transitive relations. New information added by the users through new annotations in the CastOfObjects or the TimeLine is stored in a different base, the Application Base (AB). The AB holds the data added by users of the system and has a similar schema to the KB with the addition of user metadata (e.g. timestamps). Links are kept between the extracted information and the corresponding document, so that when a character, that is, an entity such as a person or an organisation, or an event is selected, the corresponding text in the document is available through the DocumentVisualiser.

Each component of the system is further described in the following sections.

3.1   The Knowledge Model

In order to describe the data to be extracted for supporting case building activities we have defined a Knowledge Model with three different layers [15]. Figure 5 shows a part of the KM' hierarchy.

The System layer supports the integration with the indexing tools focusing on low-level features (such as text zones). For example the class Mention is used as the container of attributes that record the offset of the word that evokes extracted information.

Part of this layer is also a class that represents the source of the information and serves the system in three different ways: record information provenance; point to the source in case manual verification of the results is required by the lawyers; and enable document searching according to document metadata.

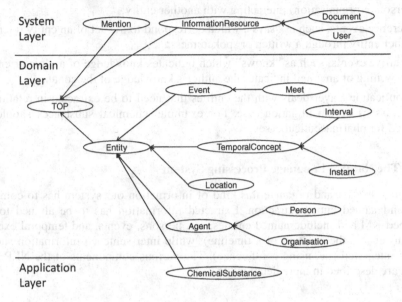

Fig. 5. Top concepts in the knowledge model's hierarchy

The Domain layer represents general concepts that we believe are useful to be considered in legal case building and reasoning activities. For instance, people and organisations are typical examples of characters that may have a role in a legal case. Special attention is given to the representation and analysis of events, as they serve as the core ingredient for supporting WHs questions. For instance, the role of the characters in a case is determined, among other factors, by the events in which they participate. Naturally that is a two way relation. The events that a key character participates in may be important for the case and the participants of a key event may be key characters.

One of the core requirements is therefore identifying the other factors, in addition to the participants, that make an event important. These include:

- The topic of an event, if any;
- The role of a character in the event;
- The relative time of an event in the chronology of the case;
- The location where the event took place.

Events are extracted from the collection of documents associated to a legal case. They may describe situations, e.g., meetings, actions, e.g., studying, or even statuses, e.g., belong to. The events identified will depend on the domain that the legal case covers. Additionally, we have identified a number of classes of relations among people and organisations that we believe to be of interest to lawyers, during case construction, independently from the litigation domain. Those classes correspond to events or event abstractions and include:

- Role-based events, such as "is employed by";
- Interaction-based events, such as "meets", which corresponds to the act of an entity (i.e. person or organisation) interacting with another entity;
- Reference events, such as "says", which correspond to the act of an entity referring to another entity through a written or spoken message;
- Cognitive events, such as "knows" which indicates knowledge of a topic or entity, e.g. the writing of an email indicates the author's knowledge of the contents.

The Application Layer deals with the entities that need to be extracted in relation to specific issues of the litigation case. For example, chemical substances should be extracted for pharmaceutical cases.

3.2   The Natural Language Processing System

In order to extract and organize this kind of information our system has to combine event and named entity extraction. Extracted information has to be aligned to the described KM and include named entities, e.g. persons, events, and temporal expressions (to enable the creation of a timeline), while inter-sentence information should also be integrated (i.e. using coreference). The various components of the NLP subsystem are described in more detail in [15].

3.3   The Knowledge and Application Databases

The information extracted from the NLP system is stored in a Knowledge Base (KB) that includes references to the initial information source (document ids). This allows the creation of links between the information that is presented to the system's users and the documents from which this information has been extracted, so that the user can verify the information in its context. The information extracted actually is a graph with links between entities and objects that describe those entities [15].

The Application Database (AD) holds information that is inserted in the system with a different means than the NLP system. This may include metadata (such as timestamps) but also and very importantly user created information. Suppose for example that the user has the ability to include another entity as the participant to an event, add information to an incomplete event representation, or even identify two events describing the same real world situation, then that information is stored in the AD and after validation it is propagated to the KB.

4   Prototype Implementation

We have designed and implemented a prototype of the CBS with the aim of providing the user with search and visualization facilities based on the interaction modes previously described. These facilities include a first version of the DocumentVisualiser, of the CaseSpace, and of the building blocks based search mechanism. The prototype follows a client-server architecture integrating four subsystems (Figure 6).

The **NLP subsystem** extracts semantic information from the document corpus according to the given set of ontologies. To this end, it takes as input the documents (plain text files) and ontologies (OWL files). The XIP Parser [1] has been used for this task, as described in [15]. The NLP system outputs a collection of RDF files containing the semantic information extracted from the documents.

The **Knowledge Base Management subsystem** manages the RDF files produced by the NLP system by storing them in a Knowledge Base (KB). Based on the assumption that a relational database (RDB), being a mature storage solution, would offer robustness, a RDB has been selected as the backend of the system. We could have chosen another storage target as well (e.g. native RDF store). A number of frameworks exist to enable this operation. Among them, we have used Jena [14], a Java library framework for developing semantic web applications based on W3C recommendations for RDF and OWL. Jena also supports SPARQL, the RDF query language, enabling us to pose queries on the KB. Jena API calls are encapsulated in the Query API. The Query API is a server-side service that:

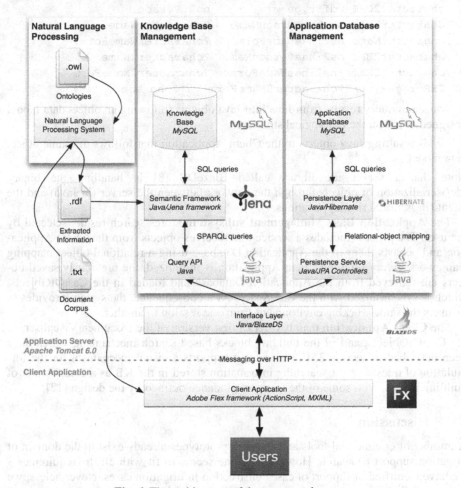

Fig. 6. The architecture of the prototyped system

• receives objects from the Client Application and translates them into SPARQL queries. For example the question "Who did John Doe meet in Zurich?" would generate the following SPARQL code:

```
SELECT DISTINCT ?characterNameForm
WHERE {
  ?eventURI rdf:type              ns1:Meet.
  ?eventURI ns1:hasParticipant    ?characterURI.
  ?eventURI ns1:hasParticipant    ?character2URI.
  ?eventURI ns1:hasLocation       ?locationURI.
  ?locationURI ns2:hasMention     ?locationMention.
  ?locationMention ns2:hasForm    ?locationForm.
FILTER regex(str(?locationForm), "Zurich", "i")
FILTER(?characterURI != ?character2URI)
  ?characterURI rdf:type          ns2:Person.
  ?character2URI rdf:type         ns2:Person.
  ?characterURI ns2:hasPersonName ?characterName.
  ?characterName ns2:hasNameForm  ?characterNameForm.
  ?character2URI ns2:hasPersonName ?character2Name .
  ?character2Name ns2:hasNameForm ?character2NameForm.
FILTER regex(str(?character2NameForm), "John Doe", "i")}
```

• translates query results from Jena into Java objects following an object data model designed to reflect the ontological structure.

• sends resulting Java objects to the Client Application that follows the same object data model.

Note that a server-side library called BlazeDS [8] is handling the binary (de/)serialization of objects through the network between the server (in Java) and the client (in ActionScript) applications.

The **Application Base Management subsystem** stores search results selected by the user. The process includes a service that receives objects from the Client Application and persists them in the Application Database using a relational-object mapping framework. Every time the Client Application is initialized, the previously saved objects are retrieved from the Application Database and loaded in the CastOfObjects, which is synchronized with the CaseSpace. As a consequence, the system provides to its users the same working environment from one session to another.

The **Client Application** mainly offers a first version of the DocumentVisualiser, of the CastOfObjects, and of the building blocks based search mechanism, which have been described in section 3. In particular it supports a visual mechanism for the formulation of questions for searching information stored in the KB as a composition of building blocks, with some of the forms of guidance defined by the design [12].

5   Discussion

A number of commercial tools and research prototypes already exist in the domain of litigation support or search. However, no one seems to fit with all the requirements we have identified in support of case construction in litigation cases. Nevertheless, we have identified a few interesting aspects from a few of them.

The Polestar system [18] provides built-in support for collecting textual facts from source documents and structured argumentation, plus awareness mechanisms which are based on activity on documents, but not on their contents. Entity Workspace [7] allows the organisation of extracted information in entity groups. However, there is no timeline and only a limited support for collaboration. Systems like Sandbox/TRIST [23] and JigSaw [21] have interesting visualisation capabilities, along the lines of CBS, but CBS is a more unified system with more collaboration support.

An interesting and quite widely used commercial database to support case analysis is the already mentioned CASEMAP and we have taken into account features provided by this tool in defining the requirements for our CBS, especially around the semantic structure of the information stored in the system (characters, etc.). On the other hand CASEMAP's system for creating entries is completely manual whereas a semi-automatic definition is supported in CBS.

The work described above is related to the provision of a complete work environment for lawyers to help them in the process of legal construction. While in the past the use of NLP in this domain was very limited, a more recent line of work combines NLP with reasoning procedures and representation models of a legal case, in order to find relevant case precedents (i.e. judgements related to previous cases). The central idea is that a case can be represented as a set of facts (or factors) that correspond to an abstraction level, appropriate for comparison based on legal norms and outcomes.

In initial works [2], [3], [9] factor assignment was done manually under the direction of experts. However, the high cost of development and maintenance pushed research towards the use of NLP techniques.

Bruninghaus and Ashley [10] present a framework called SMILE that utilises machine learning to assign automatically factors to different text passages based on a tagged collection. Although the results are promising, the tests have been performed on a small collection of documents because a similar larger annotated collection does not exist. A hierarchy or set of factors is developed manually.

Weber-Lee et al. [22] used a rule-based system for NLP to define the rhetorical structure of the texts and identify the parts in which the illocutionary expressions are present. They identify four main substructures: identification (i.e. surface features such as date, city, reporter and petition type), abstract (i.e. applicant and result), body (court decision and its foundations), closing (votes, date, place and names of participating attorneys). According to the different substructure, different rules are fired in the reasoning process.

As in our work, Maxwell et al. [16] move from factors and illocutionary expressions to the extraction of semantic events for legal case retrieval, where an event refers to any eventuality (event, state, and attribute) existing in the legal text. They report that although their results are promising, a larger evaluation of event-based extraction techniques as an enabler of understanding legal relevance should be carried out for reaching definite conclusions.

All these works have focused on case precedents and their retrieval. We rather argue that object retrieval can help in case construction activities, a different process.

While there were some very good reasons to select RDF and OWL related technologies to construct our architecture, such as model flexibility, explicit representation of semantics, out-of-the-box reasoners (for OWL or RDFS) and proliferation of freely available background knowledge (i.e. Linked Data), we have also found out that the combination we have selected is not highly scalable. This is a major issue in litigation where millions of pages are included in each case, which may mean hundreds of millions of entities and billions of triples. Possible approaches we are researching include: optimizing the query construction process (e.g. rather than using regular expressions within SPARQL FILTERs research the combination of free text search and RDF search), researching different store implementations (for example a native RDF store may have performed much better for SPARQL querying thanks to customised indexes), or developing native formats and schemas.

Another point we would like to address relates to the synchronization among different components of the system. For example, currently the schema of the AB is not automatically updated according to the KM while the number of building blocks in the search interface doesn't automatically reflect changes to the KM. Furthermore, while users can select and save search results in the AB we do not exploit user generated information in a more elaborate way (e.g. for incomplete information).

One of the main strengths of the approach we propose is related to the assumption that entities from different documents will create an interconnected graph that will enable the discovery of implicit information. However, we have found that merely annotating individual mentions of characters and events may enable a certain amount of new functionality, but there is more to be gained by recognizing that the same characters and events are mentioned multiple times in a single document and across multiple documents, and synthesizing richer representations that combine information from multiple sources. We have implemented simple coreference resolution mechanisms for mentions of persons, but this is only a start. The mechanisms could be enhanced to integrate encyclopedic knowledge from external sources (e.g. knowing that a referring expression "he" can't be coreferent with a name if the person with that name is known to be female), and need to be extended to other types of entities and to events. Reusing and integrating existing ontologies is also under investigation.

6   Conclusions

In this paper we have presented the overall design of the Case Building System that we are developing and the first prototype that we built for the system. This is ongoing work and testing the design of the system and complete its implementation will require time.

However, we believe that we have produced a novel and technically achievable design idea and that we have a good basis to evaluate, refine and evolve our concept with actual lawyers in realistic and then actual situations of use.

References

1. Ait-Mokhtar, S., Chanod, J.P., Roux, C.: Robustness beyond Shallowness: Incremental Deep Parsing. J. Nat. Lang. Eng. 8(2-3), 121–144 (2002)
2. Ashley, K.D., Rissland, E.L.: A Case-Based Approach to Modeling Legal Expertise. IEEE Intelligent Systems 3(3), 70–77 (1988)
3. Ashley, K.D., Aleven, V.: Reasoning Symbolically about Partially Matched Cases. In: Pollack, M.E. (ed.) 15th Int. Joint Conf. on Artificial Intelligence, vol. 1, pp. 335–341. Ed. Morgan Kaufmann Publishers, San Francisco (1997)
4. Attfield, S., Blandford, A.: E-discovery Viewed as Integrated Human-Computer Sensemaking: The Challenge of 'Frames'. In: 2nd Int. DESI Workshop (2008)
5. Attfield, S., Blandford, A., De Gabrielle, S.: Investigations within Investigations: a Recursive Framework for Scalable Sensemaking Support. In: CHI 2008 Workshop on Sensemaking (2008)

6. Attfield, S., Blandford, A.: Looking for Fraud in Digital Footprints: Sensemaking with Chronologies in a Large Corporate Investigation. Working paper, UCL Interaction Centre: London, UK (2009)
7. Bier, E.A., Ishak, E.W., Chi, E.: Entity Workspace: An Evidence File That Aids Memory, Inference, and Reading. In: Mehrotra, S., Zeng, D.D., Chen, H., Thuraisingham, B., Wang, F.-Y. (eds.) ISI 2006. LNCS, vol. 3975, pp. 466–472. Springer, Heidelberg (2006)
8. BLAZEDS,
   http://opensource.adobe.com/wiki/display/blazeds/BlazeDS
   (last accessed in July 2009)
9. Branting, L.K.: Representing and Reusing Explanations of Legal Precedents. In: 2nd Int. Conf. on Artificial intelligence and Law (ICAIL), pp. 103–110. ACM, NY (1989)
10. Brüninghaus, S., Ashley, K.D.: Progress in Textual Case-Based Reasoning: Predicting the Outcome of Legal Cases from Text. In: Cohn, A. (ed.) 21st National Conference on Artificial Intelligence, vol. 2, pp. 1577–1580. AAAI Press (2006)
11. CASEMAP, LexisNexis, articles,
    http://www.casesoft.com/training/articles.asp
    (last accessed in February 2011)
12. Castellani, S., Grasso, A., Benedetti, V., Lagos, N., Hairon, N.: A Semantics-Based Approach to Guide Formulation of Questions for Documentary Reconstruction Activities. Accepted and Presented at the 4th Int. Conf. on Advances in Semantic Processing, SEMAPRO 2010 (2010)
13. Electronic Discovery Reference Model, http://edrm.net (last accessed April 29, 2011)
14. JENA, http://jena.sourceforge.net/ (last accessed in July 2009)
15. Lagos, N., Segond, F., Castellani, S., O'Neill, J.: Event Extraction for Legal Case Building and Reasoning. In: Shi, Z., Vadera, S., Aamodt, A., Leake, D. (eds.) IIP 2010. IFIP AICT, vol. 340, pp. 92–101. Springer, Heidelberg (2010)
16. Maxwell, K.T., Oberlander, J., Lavrenko, V.: Evaluation of Semantic Events for Legal Case Retrieval. In: WSDM 2009 Workshop on Exploiting Semantic Annotations in Information Retrieval (ESAIR 2009), pp. 39–41. ACM, New York (2009)
17. Noel, L., Azemard, G.: From Semantic Web Data to Inform-Action: a Means to an End. In: Workshop SWUI (Semantic Web User Interaction), CHI 2008, Florence, Italie, April 5-10 (2008)
18. Pioch, N.J., Everett, J.O.: POLESTAR – Collaborative Knowledge Management and Sensemaking Tools for Intelligence Analysts. In: CIKM 2006, pp. 513–521. ACM (2006)
19. Privault, C., O'Neill, J., Renders, J.-M., Ciriza, V.: A New Tangible User Interface for Machine Learning Document Review. Artificial Intelligence and Law 18(4), 459–479 (2010); Special Issue on "E-Discovery", Ashley, K.D., Baron J.R., Conrad, J.G. (guest eds.)
20. Sheth, A., Arpinar, B., Kashyap, V.: Relationships at the Heart of Semantic Web: Modeling, Discovering, and Exploiting Complex Semantic Relationships. Technical Report, LSDIS Lab, Computer Science, Univ. of Georgia, Athens GA (2002)
21. Stasko, J., Gorg, C., Liu, Z.: Jigsaw: supporting investigative analysis through interactive visualisation. Information Visualisation 7, 118–132 (2008)
22. Weber-Lee, R., Barcia, R.M., da Costa, M.C., Rodrigues Filho, I.W., Hoeschl, H.C., D'Agostini Bueno, T.C., Martins, A., Pacheco, R.C.: A Large Case-Based Reasoner for Legal Cases. In: Leake, D.B., Plaza, E. (eds.) ICCBR 1997. LNCS, vol. 1266, pp. 190–199. Springer, Heidelberg (1997)
23. Wright, W., Schroh, D., Proulx, P., Skaburskis, A., Cort, B.: The Sandbox for Analysis - Concepts and Methods. In: CHI 2006. ACM (2006)

Different Approaches to Build Brief Ontologies

Julián Garrido[1], Stefano Mambretti[2], and Ignacio Requena[1]

[1] Dept. Computer Sciences and A.I., Granada University,
C/ Daniel Saucedo Aranda, 18071, Granada, Spain
[2] Wessex Institute of Technology (Ashurt) Southampton, U.K.
{jgarrido,requena}@decsai.ugr.es, stefano.mambretti@polimi.it

Abstract. This paper introduces the problem of building brief ontologies in order to use only a relevant portion of an ontology. It describes the procedure of generation that preserves partially the original taxonomy. Therefore, concepts definitions, ranges and properties domains are processed in order to be generalized. The paper also presents a methodology consisting of seven steps to build a brief ontology. This method is applied to an ontology for environmental assessment in order to build a specific brief ontology for floods.

Keywords: Knowledge Representation, Brief Ontology, Hazards, Floods.

1 Introduction

There have been different attempts of summarizing monolithic semantic networks and ontologies i.e. a methodology for partitioning a vocabulary hierarchy into trees [11]. This methodology refines pseudo-automatically a IS-A hierarchy of medical entities. In order to reach more simplicity, the methodology aims to create a set of very small trees where each concept has only one parent. However, this simplification makes the model unrealistic. Moreover, they use human evaluation to study how comprehensive are the resulting trees.

A comparison and description of pruning methods for bio-ontologies is done in [12]. It describes the suitability of the prunning method and if they should be avoided by showing their benefits and drawbacks. In general, these methods include two phases: i) the selection phase identifies relevant elements according to the user's goals. ii) The pruning phase uses the selection to remove irrelevant elements. They describe for each method: the knowledge base that the method uses, whether it supports integrity constraints, the level of automation, the type of selection strategy and the size of the final ontology. Although they plan to use metrics to assess the effectiveness, they only uses a group of experts.

A different approach divides ontologies into modules using partitioning based on structural properties [15]. Its criterion consists of building modules where the semantic connection between concepts in a module is maximized whereas the dependencies with concepts belonging to other modules are minimized. Firstly, it creates a weighted dependency graph, it does a partitioning and finally it optimizes the modules by isolating concepts, merging or duplicating concepts.

A. Fred et al. (Eds.): IC3K 2011, CCIS 348, pp. 232–246, 2013.

According to [13], an ontology view is a portion of an ontology that is specified as a query in some ontology-query language (analogously to databases). However, they extend this definition to ontologies that are defined by a traversal specification or by meta-information. They also present a tool able to accomplish management tasks such as comparing ontologies, merging them together, maintaining and comparing different versions. However, they declare as an open issue pruning the definitions of the concepts.

The concept of brief ontology was introduced in [5]. They define a brief ontology as the ontology which includes a small amount of knowledge referring to concepts existing in more generic ontologies. This concept provides relevant access to information in databases for web services-based and multi-agent architectures. Nonetheless, a formal definition of brief ontology is included in Section 2.

Methodologies like Methontology [7] point out the convenience of reusing other existing ontologies. However, the whole ontology has to be used even if only a small fraction is relevant for the problem. For this reason, the size of our ontology may grow with useless conceptualizations from other ontology.

In order to avoid this problem, brief ontologies may be used to obtain reduced versions of the ontology that the user wants to import. If these ontologies contain only the portion of knowledge that the user really needs then the size of the ontology will not increase unnecessarily. By contrast, the objective might be just isolating a portion of the knowledge in order to use only the brief ontology without unnecessary knowledge. Our approach presents a traversal method to build brief ontologies using not only concepts but also instances of concepts (individuals) as starter point. Moreover, the method is also fully compatible with pruning definitions of the concepts.

As an example, a case of study for floods management is presented because of the European normative [6] that encourages to the assessment of flood risks in order to do an adequate management of the problem. The starting point is an ontology for environmental assessment and a brief ontology for flood management is created as a base of a future knowledge-based system.

The paper is organized as follows, Section 2 introduces the concept of brief ontology, Section 3 describes the procedure for generation of brief ontologies taking into account two different scenarios i.e. the generation based on concepts and the generation based on individuals, Section 4 describe the case of study where a brief ontology for flood management is created. The final sections give the conclusions and list bibliography.

2 Brief Ontologies

According to [2], a typical DL (Description Logic) knowledge base comprises the TBox and the ABox. The TBox describes the intensional knowledge (terminology) and it is represented with declarations in order to describe general properties of the concepts. For instance, a concept may be defined as the intersection of other concepts. This type of simple definition allows defining a concept in terms of other previously defined concepts. However, the set of operators depends on the type of description logics that the language implements (OWL-DL). This sub-language implements the $\mathcal{SHOIN}(\mathcal{D})$ logic and it has less expressivity than OWL in order to reduce the computational complexity of reasoning and inferring [14]. It involves operators such as: union, intersection, complement, one of, existential restriction, universal restriction and cardinality restriction.

The ABox contains the extensional knowledge which is the knowledge that is specific to the individuals of the domain. It includes assertions about individuals i.e. using properties or roles to establish a relationship between individuals.

If an ontology is defined as the union of its TBox and ABox, a brief ontology is another ontology where the extensional and intensional knowledge have been restricted and modified in order to include only the relevant knowledge for a given context. This is formally described in the following definition where the TBox is represented as K_T and the ABox is represented as K_A. However, it is important to clarify before that if two concepts have the same name in O and O^B, then they are referred as equivalent with independence of their definitions. The concept of equivalence has the same consideration for individuals and roles.

Definition 1. For ontology $O = (K_T, K_A)$, a brief ontology is an ontology $O^B = (K_T^B, K_A^B)$ such that $K_T^B \sqsubseteq K_T \wedge K_A^B \sqsubseteq K_A$, and for every concept $C \in K_T$ and its equivalent $C^B \in K_T^B$, the definition of C exactly matches the definition of C^B or the definition of C^B is a generalization of the definition of C. Analogously, for every individual $v \in K_T$ and its equivalent $v^B \in K_T^B$, every assertion of v^B exactly matches the assertion of v or it is a generalization of the original assertion of v.

In other words, the brief ontology is a pseudo-copy of the original ontology that includes only a portion of the knowledge base (a subset of the intensional and extensional knowledge). It is referred as a portion because not all the concepts, individuals and roles of the original ontology are in the brief ontology. Moreover, it is considered a pseudo-copy because the definition of the concepts may be modified or generalized, and because some assertions of the individuals may be also ignored or generalized [10].

The exclusion of elements from the original ontology and the pseudo-copy are done in order to match with the restrictions for the brief ontology because the brief ontology pretends to be a simplified version of the original ontology.

3   Procedure of Generation

This section describes the extraction procedure of the brief ontologies. However, some considerations must be taken into account before describing the algorithm.

The brief ontology is built making a selective copy of the original ontology. The goal is to obtain a context-centered ontology where the context is considered the specification of the user for the relevant knowledge.

The extraction procedure is parameterized by this user specification because it is used as criteria to spread a traversal exploration on the original ontology and therefore it is used to decide whether an element of the original ontology is relevant and must be included in the brief ontology.

Traversal algorithms [1] are usually used in graphs theory to implement depth first or breadth first searches. These algorithms start at some node and then visit all the nodes that are reachable from the start node. If the graph is weighted then the strength of the relationships between nodes are usually defined with matrix. Therefore, traversal algorithms assist in the task of creating a sub-graph because they establish an order to visit nodes in the graph. Moreover, a threshold is useful to visit only nodes that are strongly connected and thus restricting the concept of reachable node.

Although an ontology is not a graph [3], a traversal exploration of an ontology implies analogously to consider two types of nodes i.e. concepts and individuals. Moreover, these elements are considered reachable whether there is some kind of relationship between them. It may be a parenthood relationship between concepts or a concept and its individuals, relationships of a concept with the concepts and individuals that are used in its definition and relationships between individuals that are represented in its assertions. Primitive values and datatype properties are not considered nodes and connections between nodes, therefore, those data are components of the node.

Whereas a threshold may be used to limit the nodes that are going to be visited in a weighted graph, other different mechanisms are used in ontologies. In particular, a set of properties are specified to restrict individuals and concepts that are visited during the traversal exploration of the ontology (this set of properties is named relevant properties). If two nodes are related with a property which is not a relevant property then the second node will not be reachable by this connection but it may be by another one.

If a property is relevant then the information that it gives is interpreted as significant for the purpose of the user. For the same reason, if a property is not in this set then all the information or semantic that it provides must be ignored and not included in the brief ontology.

The following subsections describe two different methods to build brief ontologies. The generation based on concepts should be considered if there is more interest on extracting the taxonomy of concepts rather than the individuals of the ontology. Depending on how the ontology is built, a brief ontology where all the individuals have been rejected is possible. However, if there is special interest in these individuals then the generation based on individuals is advisable.

3.1 Generation Based on Concepts

Algorithm 1 describes the generic procedure to build a brief ontology when the start set is compound of concepts. Its inputs are the original ontology, the set of main concepts (MC) where the traversal copy starts and the set of relevant properties (RR) to restrict the traversal exploration. The output is a new ontology that contains the relevant knowledge of the original ontology.

First of all, only the relevant properties (RR) will be created in the brief ontology and the rest of properties of the original ontology are ignored. After this, the traversal copy of concepts must be accomplished. This task is done for every concept that belongs to the set MC (first loop).

The traversal exploration of concepts involves spreading the algorithm to all the reachable nodes. In this case, it spreads to concepts and individuals by concept-concept, individual-individual, concept-individual and individual-concept connections. However, only concepts are labeled with positive evaluation to be created at this point. The reason is that the complete taxonomy of concepts for the brief ontology must be created before the creation of individuals, assertions or concept definitions.

In the second loop, a traversal exploration of concepts is started for each concept in MC. The next nested loop starts a traversal exploration of individuals for the ones that were reachable in the previous exploration. The traversal exploration of individuals is done following only individual-individual connections. All the individuals that are reached with the set of relevant properties RR are created in the brief ontology at this point.

All the concepts and individuals that have been included in the brief ontology are visited in the third loop. Firstly, a traversal exploration over the concepts is done in order to create the definition of concepts. Secondly, a traversal exploration over the individuals allows defining their assertions.

```
Algorithm 1
Input: Ontology O, main concepts MC,
       relevant properties RR
Outoput: brief ontology OB
Begin
  Create RR properties
  Foreach concept C in MC Begin
    Traversal exploration of concepts (RR,C)
    Create concepts with positive evaluation
  End For
  Foreach concept C in MC Begin
    Traversal exploration of concepts (RR,C)
    Foreach individual v in the exploration Begin
      Traversal exploration of individuals (RR,v)
      Create individuals with positive evaluation
    End For
  End For
  Foreach concept C in MC Begin
    Traversal exploration of concepts (RR,C)
    Foreach concept with positive evaluation Begin
      Create relevant definitions
    End For
    Foreach individual v in the exploration Begin
      Traversal exploration of individuals (RR,v)
      Create relevant assertions in individuals
      with positive evaluation
    End For
  End For
End
```

(Procedure to build brief using concepts as starting point)

Concepts and individuals may be reachable from different concepts and individuals and it may imply several traversal explorations in the same steps. Moreover, cycles may appear depending on the original ontology. In order to solve this problems, if a concept or individual has been computed in an iteration of the traversal algorithm then it does not have to be computed a second time in subsequent iterations.

For this reason, the complexity of a traversal exploration is lineal $O(n)$ if the number of relevant properties and main concepts are limited by constants. Hence, the efficiency of the algorithm corresponds to $O(n*m)$, being n, the number of concepts and m the number of individuals.

3.2 Generation Based on Individuals

Algorithm 2 describes the generic procedure to build a brief ontology when the starting set is compound of individuals. Its inputs are the original ontology, the set of individuals (MI) to start the traversal copy and the relevant properties (RR). The output is the brief ontology with the relevant knowledge.

This algorithm also starts creating the relevant properties in the brief ontology. Then, it continues with a traversal exploration of individuals for each individual in MI (first loop). The individuals cannot be created until the class they belong exists in the brief ontology. For this reason, the class of every individual found in the exploration is created in a nested loop immediately before its respective individual. At this point, all the direct connections between individuals that start in the MI are created. The next logical step is to spread the algorithm with a traversal exploration of all the concepts that were classes of the individuals. Hence, the complete taxonomy of concepts is in the brief ontology once finished this step.

Although the major part of individuals is already in the brief ontology, the individuals that are connected to concepts by its definition may not have been included. This requires a second loop where all the concepts that are classes of the individuals (which were found in the first loop of the algorithm) are again the starting point of a traversal exploration. Consequently, new individuals may be found in the concepts definition as a result of this exploration of concepts. These individuals are also starting point of a new traversal exploration of individuals and the new ones will be created in the brief ontology.

After finishing this second loop the complete taxonomy of concepts and individuals is in the brief ontology. However, individuals and classes are created empty at first attempt and it requires a second step to include its definitions and assertions. As a general rule, concepts must be created before individuals, these before definitions or assertions, and definitions before assertions.

In the third step, the algorithm consists of two nested traversal explorations of individuals (starting in the set MI) and its concepts. The definitions of the concepts are created at this moment. Nonetheless, it is important to remark that the original definition may be modified according to [10] due to some of the concepts or individuals of the definition may no longer exist in the brief ontology.

In the last step, a traversal exploration starts for every individual of MI and the assertions are created for every individual that is found during the exploration. Exploring the classes of these concepts, new individuals may be found and the algorithm ends creating the assertions for these individuals.

Although this algorithm increases the order of complexity compared to the case that is based on concepts, it has still polynomial efficiency.

```
Algorithm 2
Input: Ontology O, main individuals MI,
       relevant properties RR
```

```
Outoput: brief ontology OB
Begin
  Create RR properties
  Foreach individual v in MI Begin
    Traversal exploration of individuals (RR,v)
    Foreach individual p with positive evaluation Begin
      Create concept C that is class of p
      Create individual p
      Traversal exploration of concepts (RR,C)
      Foreach concept with positive evaluation Begin
        create concept
      End For
    End For
  End For

  Foreach individual v in MI Begin
    Traversal exploration of individuals (RR,v)
    Foreach individual p with positive evaluation Begin
      Select C that is class of p
      Traversal exploration of concepts (RR,C)
      Foreach individual u in the exploration Begin
       Traversal exploration of individual(RR,u)
       Create individuals with positive evaluation
      End For
    End For
  End For

  Foreach individual v in MI Begin
    Traversal exploration of individuals (RR,v)
    Foreach individual p with positive evaluation Begin
      Select C that is class of p
      Traversal exploration of concepts (RR,C) Begin
      Foreach concept with positive evaluation Begin
        Create definition
      End For
    End For
  End For

  Foreach individual v in MI Begin
    Traversal exploration of individuals (RR,v)
    Foreach individual p with positive evaluation Begin
      Create assertions of p
      Select C that is class of p
      Traversal exploration of concepts (RR,C) Begin
      Foreach individual v in the exploration Begin
        Create assertions of v
      End For
    End For
  .End For
End
```

(Procedure to build brief using individuals as starting point)

3.3   Range and Domain of Roles

There are some issues that may be found during the development of the brief ontologies because of all the concepts of the original ontology are not in the brief ontology. A restrictive property may define its range or domain using one or more concepts that were not included in the brief ontology. In order to solve this problem, three different approaches are considered.

In the first alternative, ranges and domains are not defined for any property. This is the easier and faster solution but also the one where more knowledge is lost. It might be useful whether only the taxonomy of concepts is relevant or whether the ontology will not be used for reasoning tasks.

The second one consists of defining ranges and domains of properties whose concepts belong to brief ontology. However, if the range or domain is defined using concepts that do not belong to the brief ontology, then it will not be defined.

In the third alternative, ranges and domains are defined in a first attempt if all the concepts in the concept expression belong to the brief ontology. The concept expressions are generalized in the remaining cases. The main idea of this alternative is that even though a concept is not in the brief ontology, an ancestor or ascendant might be in the brief ontology. Therefore, it is better to assign the ancestor rather than assigning nothing.

```
Algorithm 3
Input: Property R that belongs O, ontology O
       brief ontology Ob
Output: Property Rb that belongs to Ob.
Begin
  E = R.getDomainExpression()
  Eb = convertExpression(E, O, Ob)
  If Eb is not empty Then
     set Eb as domain of Rb
  End If
End
```

(Assignment of the domain of a property in the brief ontology)

Algorithm 3 represents the third alternative to assign the domain of a property. First, the concept expression that defines the domain for the property R is retrieved. The method *convertExpression* transform a concept expression defined in terms of concepts that belong to the ontology O into a concept expression in terms of the brief ontology Ob. If all terms in E are also in Ob then the conversion only consists of building an equivalent expression using concepts that belong to Ob. Otherwise, the concept expression Eb is built replacing concepts in E by its ancestor.

Let suppose the example of Figure 1 and a brief ontology O^B where the property *hasCar* and the concepts *Car* and *Person* are included. The domain and range may differ depending on the implemented alternative, there are three cases: i) using the first alternative, the property does not have defined domain and range. ii) in the second one, the domain is not defined because *Driver* is not in O^B whereas the range is the concept *Car*. iii) The third alternative defines *Car* as range for the property *hasCar* and *Person* as domain because it is the superClass of *Driver*.

```
<owl:ObjectProperty rdf:ID="hasCar">
  <rdfs:domain rdf:resource="#Driver"/>
  <rdfs:range rdf:resource="#Car"/>
</owl:ObjectProperty>

<owl:Class rdf:ID="Driver">
  <rdfs:subClassOf>
<owl:Class rdf:ID="Person"/>
  </rdfs:subClassOf>
</owl:Class>
```

Fig. 1. OWL fragment defining the *range* and *domain* for the property *hasCar*, and the class *Driver* as subclass of *Person*

It is noteworthy that the third alternative implies adding new knowledge to the brief ontology. The domain of the property was restricted to drivers in the original ontology. However, the domain is now more general and less strict or restrictive. The new statement makes, for example, that other subclasses of *Person* can use this property but it also changes the results of the inference processes.

4 Case of Study: Flood Management

First of all, building a brief ontology from a detailed one according to our needs is done by the generation process described in Section 3. Nonetheless, the complete semi-automatic procedure to build a brief model is detailed below.

1. Establish the aim and scope for the brief ontology because it allows setting the objectives and requirements for the ontology.
2. Selection of a detailed ontology where knowledge about the aim and scope is represented.
3. Analysis and study for the detailed ontology in order to be able to accomplish the next two steps.
4. Selection of the best type of extraction algorithm that is adapted to meet the requirements and characteristics of the ontology.
5. Selection of the starter point and restrictive properties.
6. Generation of the brief ontology.
7. Evaluation of the resulting model in order to see if it satisfies our requirements.

4.1 Aim and Scope

Flood is a body of water which overflows its usual boundaries over a land area with other land use, resulting in adverse impacts. The socioeconomic development in the floodplains and the reduction of the natural water retention by the land use increase the consequences of floods. For this reason, a European Directive [6] encourages the flood management and risk assessment. This management requires in general prevention, protection and mitigation actions [4].

The main goal for this case of study is to build a model for flood management.

4.2   Detailed Ontology

The detailed ontology will be the environmental impact assessment ontology that is originally described in [9]. This ontology was built with two purposes. Firstly, in order to provide and establish the conceptual framework of environmental assessment (EA) and secondly, to facilitate the development of methodologies and applications [8]. Indeed, the ontology was also born to be the knowledge base of an EA system.

4.3   Analysis of the Ontology

The EA ontology describes in essence the relationships between industrial activities and environmental impacts considering for instance the environmental indicators that should be controlled for every impact.

Although the ontology is focused on industrial activities and human actions, it is also taking under consideration natural processes and natural events as impacting actions. A natural process or event is considered an impacting action whether they interact with human activities and this interaction implies an increment of its environmental impact.

As a result of the inclusion of natural events, the EA ontology contains knowledge about floods and it allows using this detailed ontology as a base to extract the relevant knowledge about our case of study.

A deeper description of the EA ontology is found in [9]. It contains a description for the taxonomy of concepts, the properties and its justification.

4.4   Selection of the Algorithm

The selection of the algorithm may depend on the type of ontology and the portion of the ontology that the user is interested in. Some ontologies are built only as semantic models where no instances of concepts are stored. In this case, the application of the generation based on concepts is mandatory.

By contrast, other ontologies are used as knowledge base with a high number of instances. In this case, the user may be interested only in the semantic model represented by the taxonomy of concepts or the factual knowledge represented by the instances of concepts. The first case implies the utilization of the generation based on concepts whereas the second case requires the generation of brief ontologies based on individuals.

In our case of study, the generation based on concepts is chosen because of there is no special interest on individuals and we are interested only in the semantic model (taxonomy of concepts with their formal definitions).

4.5   Parameters of the Algorithm

Because of the traversal algorithm based on concepts has been chosen (Section 3.1), the parameters are a set of starter concepts and a set of properties to traverse. The selection of the starting point and the set of relevant properties of the EA ontology require the study and the analysis of the existing properties that are used in the concept definitions, i.e. studying how the concepts are related by these properties.

If the knowledge that the user has about the detailed ontology is not enough to choose the set of restrictive properties, a heuristic for the selection of relevant properties consist of the following steps:

1. The set of specific properties for the domain (floods) has to be identified.
2. Among them, a group of relevant properties is selected by studying its informal description, domain or range in order to understand its semantics.

   A property will be relevant in our domain depending on its semantics and its meaning. There are two different cases:
   - The property is specific for the targeted domain.
   - The property has general use in different domains but it is used in concept definitions of the targeted domain.
3. A temporary brief ontology is built by using this set of restrictive properties.
4. The resulting ontology is studied to identify new properties that are not specific in our domain but they are considered also relevant for the concepts of the brief ontology.
5. If new properties are selected, they are added to the set of restrictive properties and a new temporary brief ontology is built again until no more properties are selected.

Table 1. Set of relevant object properties

byMeansOf	characterizeRainfall	characterizeRainfall
dischargeAffectedBy	dischargeProducedBy	floodProducedBy
hasCharacterizingIndicator	hasDataSource	hasMitigatingAction
hasPreventiveAction	hasRecoveryAction	isCharacterizingIndicatorOf
isDataSourceOf	isObteinedWith	isParameterOf
isParametrizedBy	manageedBy	produce
produceDischarge	produceFlood	rainfallCharacterizedBy
use&Need		

For example, the property *floodProducedBy* has its domain in the concept *Flood* and it allows defining the causes of a flood. This property is considered relevant because it represents knowledge that we want in our brief ontology for floods. The table 1 includes the final set of relevant object properties for our case of study.

Regarding to the starting points or main concepts, the user should try to find some representative concepts in the targeted domain that are not connected by a traversal path with relevant properties. In our case study, the selection of the concept *FreshWaterFlood* is enough because of it is the best concept to represent our targeted domain.

4.6   Generation of the Brief Ontology

The construction of the brief ontology for floods is automatically carried out with the traversal algorithm once it has the starting point for the algorithm and the relevant properties.

As an example, the concept DDF[1] is subclass of the concept *RainfallStatistical-Analysis* and it has three existential restrictions over the property *isParametrizedBy*. When the concept *DDF* is added to the brief ontology, its relationship with other concepts and individuals is analyzed. These connections are represented in the definition below.

```
DDF
⊑ RainfallStatisticalAnalysis
∃ isParametrizedBy.RainfallDepth
∃ isParametrizedBy.RainfallDuration
∃ isParametrizedBy.RainfallFrequency
```

According to the traversal algorithm, because of the property *isParametrizedBy* is included in the set of relevant properties, the concepts *RainfallDepth*, *RainfallDuration* and *RainfallFrequency* will be added to the brief ontology and the traversal algorithm will continue through these concepts.

Figure 2 depicts a schema of the resulting model for flood assessment and management. The schema shows the main concepts and how they are related by the properties. The interpretation of the model is that floods are produced by a high level of water discharge or isolated events like bridge occlusion and embankment breaks. The discharge may be produced as well by rainfall, artificial water like canals, superficial waters like rivers or bad water regulation. In addition, water discharge is also affected by the catchment area.

The rainfall is frequently the main cause of high discharge and it is usually characterized by statistical analysis and design hyetograph. Finally, food management is the union of the preventive, mitigating and recovery actions that must be accomplished. However, the management also involves some processes like forecasting, economic evaluation, etc. (different agents like the municipality are in charge for each process).

4.7 Evaluation

The interpretation of the model is that floods are produced by a high level of water discharge or isolated events like bridge occlusion and embankment breaks. The discharge may be produced as well by rainfall, artificial water like canals, superficial waters like rivers or bad water regulation but it also is affected by the catchment area. The rainfall, which is usually the main cause of high discharge, is usually characterized by statistical analysis and design hyetograph. Finally, flood management is the union of the preventive, mitigating and recovery actions that must be accomplished. However, the management also involves some processes like forecasting, economic evaluation, etc. (different agents like the municipality are in charge for each process).

The detailed ontology contains 2054 named classes and this number has been reduced to 91 in the brief ontology. Therefore, the brief ontology for floods only includes the relevant knowledge for this case of study.

As [15] says, there is not golden standard to compare the results with and the goodness of the brief ontology depends on the application that will use the ontology.

---

[1] Rainfall statistical analysis whose acronym stands for Depth, Duration and Frequency.

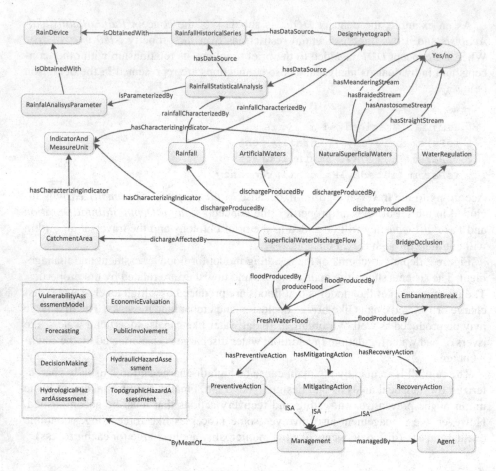

Fig. 2. Schema of the brief ontology

For this reason, the resulting brief ontology has been positively evaluated by several experts in the targeted domain (floods). Nonetheless, the quality of the brief ontology depends totally on the quality of the detailed ontology.

5   Conclusions

In general, brief ontologies have a wide range of advantages when, for some reason, the user or application does not wish to deal with the whole original ontology. Sometimes, the user is no interested in using all the information or the application is not capable of dealing with such a huge resources.

Moreover, reusing a large ontology when only a small portion is useful and relevant for our applications may involve unfavorable consequences i.e. the reasoning time increases with the size of the knowledge base and this issue may be essential in real-time applications. For this reason, the efficiency of our knowledge base is improved by isolating portions of knowledge from large ontologies in form of brief ontologies.

As an example, a case of study in flood management has been presented. A brief ontology is created specifying the initiator concept (flood) for the traversal algorithm and the set of relevant properties to decide which concepts on the ontology are relevant. The result has been an ontology where the number of concepts has been dramatically reduced and thus it contains only concepts related to flood.

As future work, it is planned to develop metrics to compare the detailed and brief ontologies such as the abstraction degree of equivalent concepts in both ontologies or the representativeness of the brief ontology. The paper describes a iterative approach to select the set of restrictive properties. For this reason, methods to automatize the selection of properties will be analyzed. For example, concepts and properties can be identified using a set of specific documents for the target domain. Therefore, these properties could be used as the set of relevant properties in the brief ontology.

Acknowledgements. This work has been partially supported by research projects (CICE) P07-TIC-02913 and P08-RNM-03584 funded by the Andalusian Regional Governments.

References

1. Aho, A.V., Ullman, J.D., Hopcroft, J.E.: Data Structures and Algorithms. Addison Wesley (1983)
2. Baader, F., Calvanese, D., McGuiness, D., Nardi, D., Patel- Schneider, P.: The Description Logic Handbook: Theory, Implementation and Applications. Cambridge University Press (2003)
3. Bizer, C., Seaborne, A.: D2rq - Treating Non-rdf Databases as Virtual Rdf Graphs (poster). In: The Semantic Web-ISWC (2004)
4. De Wrachien, D., Mambretti, S., Schultz, B.: Flood Management and Risk Assessment in Flood-prone Areas: Measures and Solutions. Irrigation and Drainage 60(2), 229–240 (2011)
5. Delgado, M., Pérez-Pérez, R., Requena, I.: Knowledge Mobilization Through Re-Addressable Ontologies. In: EUSFLAT Conf., pp. 154–158 (2005)
6. Directive 2007/60/CE of the European Parliament and of the Council on the Assessment and Management of Flood Risks (OJ L 288, 6.11.2007, p. 2734) (2007)
7. Fernández, M., Gómez, A., Pazos, J., Pazos, A.: Ontology of tasks and methods. IEEE Intelligent Systems and Their Applications 14(1), 37–46 (1999)
8. Garrido, J., Requena, I.: Knowledge mobilization to Support Environmental Impact Assessment. A Model and an Application. In: International Conference on Knowledge Engineering and Ontology Development, KEOD, pp. 193–199 (2010)
9. Garrido, J., Requena, I.: Proposal of Ontology for Environmental Impact Assessment. An Application with Knowledge Mobilization. Expert System with Applications 38(3), 2462–2472 (2011)
10. Garrido, J., Requena, I.: Towards Summarising Knowledge: Brief Ontologies. Expert System with Applications 39, 3213–3222 (2012)
11. Gu, H., Perl, Y., Geller, J., Halper, M., Singh, M.: A Methodology for Partitioning a Vocabulary Hierarchy into Trees. Artifcial Intelligence in Medicine 15(1), 77–98 (1999)

12. Kim, J., Caralt, J., Hilliard, J.: Pruning Bio-Ontologies. In: International Conference on System Sciences, Hawaii (2007)
13. Noy, N.F., Musen, M.A.: Traversing Ontologies to Extract Views. In: Stuckenschmidt, H., Parent, C., Spaccapietra, S. (eds.) Modular Ontologies. LNCS, vol. 5445, pp. 245–260. Springer, Heidelberg (2009)
14. Staab, S., Studer, R.: Handbook on Ontologies (International Handbooks on Information Systems). Springer (2009)
15. Stuckenschmidt, H., Schlicht, A.: Structure-Based Partitioning of Large Ontologies. In: Stuckenschmidt, H., Parent, C., Spaccapietra, S. (eds.) Modular Ontologies. LNCS, vol. 5445, pp. 187–210. Springer, Heidelberg (2009)

Graph-Based Semi-supervised Clustering for Semantic Classification of Unknown Words

Fumiyo Fukumoto and Yoshimi Suzuki

Interdisciplinary Graduate School of Medicine and Engineering, University of Yamanashi, Japan
{fukumoto,ysuzuki}@yamanashi.ac.jp

Abstract. This paper presents a method for semantic classification of unknown verbs including polysemies into Levin-style semantic classes. We propose a semi-supervised clustering, which is based on a graph-based unsupervised clustering technique. The algorithm detects the spin configuration that minimizes the energy of the spin glass. Comparing global and local minima of an energy function, called the Hamiltonian, allows for the detection of nodes with more than one cluster. We extended the algorithm so as to employ a small amount of labeled data to aid unsupervised learning, and applied the algorithm to cluster verbs including polysemies. The distributional similarity between verbs used to calculate the Hamiltonian is in the form of probability distributions over verb frames. The result obtained using 110 test polysemous verbs with labeled data of 10% showed 0.577 F-score.

1 Introduction

Semantic verb classification is not an end task in itself, but supports many NLP tasks, such as subcategorization acquisition ([18,15]), word sense disambiguation [29], and language generation [34]. Much of the previous work on verb classification has been to classify verbs into classes with semantically similar senses taken from an existing thesaurus or taxonomy. However, such a resource makes it nearly impossible to cover large, and fast-changing linguistic knowledge required for these NLP tasks, depending on text-type and subject domain. Let us take a look at the Levin-style semantic classes [23]. It consists of 3,024 verbs. Similarly, Japanese thesaurus dictionary called Bunrui-Goi-Hyo consists of 87,743 content words. Therefore, considering this resource scarcity problem, semantic classification of verbs which do not appear in the resource but appear in corpora has been an interest since the earliest days when a number of large scale corpora have become available.

A number of methodologies have been developed for verb classification. One such attempt is to apply clustering techniques to classification. However, two main difficulties arise in the use of clustering algorithms. The first is that we do not know how many classes there are in a given input. The usual drawback in many algorithms is that they cannot give a valid criterion for measuring class structure. The second is that the algorithm should allow each data point (verb) to belong to more than one cluster because of the existence of polysemous verbs. The aim of this work is to resolve these problems. We focus on unknown verbs including polysemies, and present a method for

A. Fred et al. (Eds.): IC3K 2011, CCIS 348, pp. 247–262, 2013.

classifying them into Levin-style semantic classes. We propose a graph-based semi-supervised clustering method which allows nodes (verbs) to belong to multiple clusters (senses). The essence of the approach is to define an energy function, called the Hamiltonian which achieves minimal energy when there is high within-cluster connectivity and low between-cluster connectivity. The energy minimum is obtained by simulated annealing. In this context, two verbs are "connected" if they share many of the same subcategorization frames. Comparing global and local minima of an energy function Hamiltonian, allows for the detection of overlapping nodes. We extended the algorithm so as to employ a small amount of labeled data to aid unsupervised learning, and clustered polysemous verbs. The distributional similarity between verbs used to calculate the Hamiltonian is in the form of probability distributions over verb frames. The results obtained using 110 test verbs including polysemies with labeled data of 10% showed 0.577 F-score, and it was comparable with previous work.

The rest of the paper is organized as follows. The next section provides an overview of existing techniques. Section 3 explains verb description, *i.e.*, verb frame patterns. Section 4 describes distributional similarity measures to compute semantic similarities between pairs of verbs. Section 5 explains our clustering algorithm. Finally, we report some experiments using 110 verbs including polysemies, and end with a discussion of evaluation.

2   Related Work

Much of the previous work on verb classification is to classify verbs into classes with semantically similar senses taken from an existing thesaurus or taxonomy. One attractive attempt is to use Levin-style semantic classes [23], as this classification includes the largest number of English verbs with fine-grained classes. Moreover, it is based on the assumption that the sense of a verb influences its syntactic behavior, particularly with respect to the choice of its arguments. Therefore, if we induce a verb classification on the basis of verb features, *i.e.*, syntactic information obtained from corpora, then the resulting classification should agree with a semantic classification to a certain extent.

Schulte [38] attempted to classify verbs using two algorithms: iterative clustering based on a definition by [13], and unsupervised latent class analysis as described by [36], based on the expectation maximization algorithm. Stevenson and Joanis compared their supervised method for verb classification with semi-supervised and unsupervised techniques [43]. Brew *et al.* focused on dimensionality reduction on the verb frame patterns, and applied a spectral clustering technique [30] to the unsupervised clustering of German verbs to Levin's English classes [4]. They reported that the results by a spectral clustering outperformed the standard k-means against all the evaluation measures including "F-measure" and all the distance measures including "skew divergence."

In the context of graph-based clustering of words, Widdows and Dorow used a graph model for unsupervised lexical acquisition [45]. The graph structure is built by linking pairs of words that participate in particular syntactic relationships. An incremental cluster-building algorithm using the graph structure achieved 82% accuracy at a lexical acquisition task, evaluated against WordNet 10 classes, and each class consists of 20 words. Matsuo *et al.* proposed a method of word clustering based on a word similarity measure by Web counts [25]. They used *Newman* clustering for the clustering algorithm, and reported that the results obtained with the algorithm were better than those obtained by average-link agglomerative clustering using 90 Japanese noun words. However, all these methods relied on hard-clustering models, and thus have largely ignored the issue of polysemy by assuming that words belong to only one cluster.

In contrast to hard-clustering algorithms, soft clustering allows that words to belong to more than one cluster. Much of the previous work on soft clustering is based on EM algorithm. The earliest work in this direction is that of Pereira *et al* [31], who described a hierarchical soft clustering method that clusters noun words. The clustering result was a hierarchy of noun clusters, where every noun belongs to every cluster with a membership probability. The initial data for the clustering process were frequencies of verb–noun pairs in a direct object relationship, as extracted from conditional verb–noun probabilities, the similarity of the distributions was determined by the KL divergence. The EM algorithm was used to learn the hidden cluster membership probabilities, and deterministic annealing performed the divisive hierarchical clustering. Schulte *et al.* [41] proposed a method for semantic verb classification that relies on selectional preferences as verb properties. The model was implemented as a soft clustering approach to capture the polysemy of the verbs. The training procedure used the EM algorithm to iteratively improve the probabilistic parameters of the model, and applied the MDL principle to induce WordNet-based selectional preferences for arguments within subcategorization frames. The results showed that after 10 training iterations the verb class model results were above the baseline results. Our work is similar to their method in the use of semi-supervised clustering, while they did not report in detail whether the clusters captured polysemic verbs. Moreover, the algorithm cannot assign unlabeled data to a new class other than known classes.

Korhonen *et al.* [19] used verb–frame pairs to cluster verbs relying on the information bottleneck. Our work is similar to their method in the use of 110 test verbs provided by [19], and focused especially on verbal polysemy. However, their method interpreted polysemy as represented by the soft clusters, *i.e.*, they used the Information Bottleneck, an iterative soft method with hardening of the output, while the method presented in this paper allows that verbs belong to more than one cluster.

3   Subcategorization Information

A typical word clustering task is to cluster words into classes based on their distributional similarity. Similarity measures based on distributional hypothesis compare a pair of weighted feature vectors that characterize two words [12,24,8].

Like much previous work on verb classification, we used subcategorization frame distributions to calculate similarity between verbs [38,4]. More precisely, [19] provided subcategorization frame data. They used the subcategorization acquisition systems of [5]. The system employs a robust statistical parser [6], which yields complete but shallow parses, and a comprehensive subcategorization frame classifier. It incorporates 163 subcategorization distinctions, a set of those found in the ANLT and COMLEX dictionaries [3,11]. A total of 6,433 verbs were first selected from COMLEX and British National Corpus [22]. Next, to obtain as comprehensive subcategorization frequency information as possible, up to 10,000 sentences containing an occurrence of each of these verbs were included in the input data for subcategorization acquisition. These sentences were extracted from five different corpora, including BNC [20]. We used these data to calculate similarity between verbs.

4   Distributional Similarity

There is a large body of work on distributional similarity measures. Here, we concentrate on eight more commonly used measures. In the following formulae, x and y refer to the verb vectors, their subscripts to the verb subcategorization frame values.

1. **The Binary Cosine Measure (bCos)**
   The cosine measures the similarity of the two vectors x and y by calculating the cosine of the angle between vectors, where each dimension of the vector corresponds to each of 163 subcategorization patterns and each value of the dimension is the frequency of each pattern. The binary cosine measure is a flattened version of the cosine measure in which all non-zero counts are replaced by 1.0.

2. **The Cosine Measure Based on Probability of Relative Frequencies (rfCos)**
   The differences between the cosine and the value based on relative frequencies of subcategorization frames are the values of each dimension, *i.e.*, the former are frequencies of each pattern and the latter are the probability of relative frequencies of each pattern.

3. **The Dice Coefficient (Dice)**
   The Dice Coefficient is a combinatorial similarity measure adopted from the field of Information Retrieval for use as a measure of lexical distributional similarity. It is computed as twice the ratio between size of the intersection of the two subcategorization patterns and the sum of the sizes of the individual subcategorization patterns:

$$Dice(x, y) = \frac{2 \cdot \mid F(x) \cap F(y) \mid}{\mid F(x) \mid + \mid F(y) \mid}.$$

4. **Jaccard's Coefficient (Jacc)**
   Jaccard's Coefficient can be defined as the ratio between the size of the intersection of the two subcategorization patterns and the size of the union of the subcategorization patterns:

$$Jacc(x, y) = \frac{\mid F(x) \cap F(y) \mid}{\mid F(x) \cup F(y) \mid}.$$

5. **L_1 Norm (L_1)**
   The L_1 Norm is a member of a family of measures known as the Minkowski Distance, for measuring the distance between two points in space. The L_1 distance between two verbs can be written as:

$$L_1(x,y) = \sum_{i=1}^{n} |x_i - y_i|.$$

6. **Kullback-Leibler (KL)**

Kullback-Leibler is a measure from information theory that determines the ineffi-
ciency of assuming a model probability distribution given the true distribution.

$$D(x||y) = \sum_{i=1}^{n} x_i * \log \frac{x_i}{y_i}.$$

KL is not defined in case $y_i = 0$. Thus, the probability distributions must be
smoothed. We used two smoothing methods, *i.e.*, Add-one smoothing and Witten
and Bell smoothing [46].[1] Moreover, two variants of KL, α-skew divergence and
the Jensen-Shannon, were used to perform smoothing.

7. **α-skew divergence (α div.)**

The α-skew divergence measure is a variant of KL, and is defined as:

$$\alpha div(x,y) = D(y \parallel \alpha \cdot x + (1 - \alpha) \cdot y).$$

Lee reported the best results with $\alpha = 0.9$ [21]. We used the same value.

8. **The Jensen-Shannon (JS)**

The Jensen-Shannon is a measure that relies on the assumption that if x and y are
similar, they are close to their average. It is defined as:

$$JS(x,y) = \frac{1}{2}[D(x \parallel \frac{x+y}{2}) + D(y \parallel \frac{x+y}{2})].$$

All measures except bCos, rfCos, Dice, and Jacc showed that smaller values indicate a
closer relation between two verbs. Thus, we used inverse of each value.

5 Clustering Method

We now proceed to a discussion of our modifications to the algorithm reported by
[33]; we call this semi-supervised RB algorithm. In this work, we focus on background
knowledge that can be expressed as a set of constraints on the clustering process. After
a discussion of the kind of constraints we are using, we describe semi-supervised RB
algorithm.

---

[1] We report Add-one smoothing results in the evaluation, as it was better than Witten and Bell
smoothing.

5.1   The Constraints

In semi-supervised clustering, a small amount of labeled data is available to aid the clustering process. Like much previous work on semi-supervised clustering [1,2], our work uses both must-link and cannot-link constraints between pairs of nodes [44]. Must-link constraints specify that two nodes (verbs) have to be in the same cluster. Cannot-link constraints, on the other hand, specify that two nodes must not be placed in the same cluster. These constraints are derived from a small amount of labeled data.

5.2   Clustering Algorithm

The clustering algorithm used in this study was based on a graph-based unsupervised clustering technique reported by [33]. This algorithm detects the spin configuration that minimizes the energy of the spin glass. The energy function Hamiltonian, for assignment of nodes into communities clusters together those that are linked, and keeps separate those that are not by rewarding internal edges between nodes and penalizing existing edges between different clusters. Here, "community" or "cluster" have in common that they are groups of densely interconnected nodes that are only sparsely connected with the rest of the network. Only local information is used to update the nodes which makes parallelization of the algorithm straightforward and allows the application to very large networks. Moreover, comparing global and local minima of the energy function allows the detection of overlapping nodes. Reichardt et $al.$ evaluated their method by applying several data, the college football network and a large protein folding network, and reported that the algorithm successfully detected overlapping nodes [32]. We extended the algorithm so as to employ a small amount of labeled data to aid unsupervised learning, and clustered polysemous verbs. Let v_i $(1 \leq i \leq n)$ be a verb in the input, and σ_i be a label assigned to the cluster in which v_i is placed. The Hamiltonian is defined as:

$$H(\{\sigma_i\}) = -\sum_{i<j}(A_{ij}(\theta) - \gamma p_{ij})\delta_{\sigma_i \sigma_j}. \tag{1}$$

Here, δ denotes the Kronecker delta. The function $A_{ij}(\theta)$ refers to the adjacency matrix of the graph. If both of the v_i and v_j are labeled data, it is defined by Eq. (2), otherwise it is defined by Eq. (3).

$$A_{ij}(\theta) = \begin{cases} 1 \text{ if } v_i \text{ and } v_j \text{ satisfy must-link} \\ 0 \text{ if } v_i \text{ and } v_j \text{ satisfy cannot-link.} \end{cases} \tag{2}$$

$$A_{ij}(\theta) = \begin{cases} 1 \text{ if } sim(v_i, v_j) \geq \theta \\ 0 \text{ otherwise.} \end{cases} \tag{3}$$

We calculated $sim(v_i, v_j)$, $i.e.$, similarity between v_i and v_j using one of the measures mentioned in Section 4. If θ is 0.9 for example, the value of the topmost 10% of the verb pairs are 1, and the remaining pairs are 0.

The matrix p_{ij} in Eq. (1) denotes the probability that a link exists between verb v_i and v_j, and is defined as:

$$p_{ij} = \sum_{i<j} \frac{A_{ij}(\theta)}{N(N-1)/2},\tag{4}$$

where N in Eq. (4) denotes the number of verbs and $N(N-1)/2$ is the total number of verb pairs. As the parameter γ in Eq. (1) increases, each verb is distributed into larger number of clusters. Eq. (1) thus shows comparison of the actual values of internal or external edges with its respective expectation value under the assumption of equally probable links and given data sizes. The minima of the Hamiltonian H are obtained by simulated annealing [17] as illustrated in Figure 1.

We applied the flow of the minima of the Hamiltonian shown in Figure 1 for M runs. We need to find a global minimum of the Hamiltonian. Each value of the $H_{\min}$ for M runs does not generally coincide with each other. Only the minimum among the values can be a global minimum and others are local minima. However, it often happens that one of the local minima is unexpectedly the minimum value, *i.e.*, a global minimum [2]. Thus, we regarded the minimum value which appears more than m times among the M results as the desired global minimum. We picked $H_{\min}$ and its corresponding all $\{\sigma_i\}_{\min}$. If a v_i belongs to more than two $\{\sigma_i\}_{\min}$, the v_i is regarded as a polysemous verb. The procedure of verb classification using the RB consists of four steps.

1. Input
   The input is a set of verbs $\{v_1, \cdots, v_n\}$, where n is the number of input verbs.
2. Calculation of similarity
   Similarities for each pair of v_i are calculated by using measures mentioned in Section 4.
3. Construction of adjacency matrix
   According to Eq. (2) and (3), adjacency matrix, $A_{ij}(\theta)$ is created.
4. Running RB algorithm
   The RB algorithm shown in Figure 1 is applied to the adjacency matrix, and clusters of verbs are obtained.

We note that the algorithm applies m times to find the minima of the Hamiltonian. Therefore, we parallelized the algorithm using the Message Passing Interface (MPI), as we applied simulated annealing for M runs. For implementation, we used a supercomputer, SPARC Enterprise M9000, 64 CPU, 1 TB memory.

6   Experiments

6.1   Experimental Setup

We used the data consisting of 110 test verbs constructed by Korhonen *et al.* [19]. There are two types: one is the monosemous gold standard, which lists only a single sense for

---

[2] The method to obtain the $H_{\min}$ does not warrant the value to be an actual global minimum, as it is based on the Monte-Carlo way.

Input {
$A_{ij}(\theta)$ // The adjacency matrix.
γ
}

Output {
$(H_{\min}, \{\sigma_i\}_{\min})$
// The minimum of the H and its corresponding clusters.
}

Initialization {
1. Assign an initial random value to each $\{\sigma_i\}$.
2. Calculate $H_{in} := H(\{\sigma_i\})$.
3. $H := H_{in}, H_{\min} := H_{in}$.
4. $T_{\max} := 10 \times |H_{in}|, T_{\min} := 0.1 \times |H_{in}|$.
// T is a parameter called the temperature.
// $T_{\max}$ and $T_{\min}$ are the maximum and minimum value of T, respectively
}

Simulated annealing {
For $t = 1, 2, \cdots, N$ {
// N is the number of iteration
$$T := T_{\max} - (T_{\max} - T_{\min}) \frac{(t-1)}{(N-1)}$$
// Let T be linearly decreased.
  For $i = 1, 2, \cdots, n$ {
    1. Calculate the change $\triangle H(\sigma_i \to s)$ from the current value of H
       when σ_i is changed to be s as:
$$\triangle H(\sigma_i \to s) = - \sum_{m\,(m \neq i)} (A_{im}(e) - \gamma p_{im}) \delta_{s\sigma_m}.$$
    2. Choose a random cluster label q as a candidate of σ_i, such that $q \neq \sigma_i$.
    3. Calculate $p(\sigma_i = q)$ as:
$$p(\sigma_i = q) = \frac{\exp(-\frac{1}{T}\triangle H(\sigma_i \to q))}{\sum_s \exp(-\frac{1}{T}\triangle H(\sigma_i \to s))}.$$
// $p(\sigma_i = q)$ is a probability that a cluster label of σ_i is q.
    4. If $r < p(\sigma_i = q)$, then σ_i and H are updated as follows:
       $H := H + \triangle H(\sigma_i = q), \sigma_i := q$
// r is a random value and $0 \leq r \leq 1$
    5. If $H < H_{\min}$, then
       $H_{\min} := H, \{\sigma_i\}_{\min} := \{\sigma_i\}$
  }
}
}

Fig. 1. Minima of the Hamiltonian by Simulated Annealing

each test verb corresponding to the most frequent sense in WordNet, and the other is the polysemous standard, which provides all senses for each verb. We used polysemous data in the experiments. We randomly selected 10% of verbs from 110 test verbs, and used

Table 1. Clustering results

	θ	γ	Sim	C	Prec	Rec	F
RB	0.2	1.0	Dice	48	0.536	0.626	0.577
EM	–	–	–	59	0.301	0.512	0.387

them as a labeled data. The remaining verbs are used as unknown verbs. The selection of labeled data was repeated 10 times. All results are averaged performance against 10 trials. The similarity between verbs are calculated by using subcategorization frame data provided by [19]. In the experiments, we experimentally set m and M in semi-supervised RB to 3 and 1,000, respectively.

For evaluation of verb classification, we used the precision, recall, and F-score, which were defined by [38], especially to capture how many verbs does the algorithm actually detect more than just the predominant sense. Precision was defined by the percentage of verb senses appearing in the correct clusters compared to the number of verb senses appearing in any cluster, and recall was defined by the percentage of verb senses within the correct clusters compared to the total number of verb senses to be clustered.

For comparison, we utilized the EM algorithm which is widely used as a soft clustering and semi-supervised clustering technique [41]. We followed the method presented in [37]. We used a probability distribution over verb frames with selectional preferences, and used up to 30 iterations to learn the model probabilities.

6.2 Basic Results

The results are shown in Table 1. "γ" and "θ" refer to the parameters used by semi-supervised RB algorithm. "Sim" indicates similarity measure reported in Section 4. We performed experiments by varying these values. Table 1 denotes the value that maximized F-score. "C" refers to the number of clusters obtained by the method, and "EM" shows the results obtained by EM algorithm. Table 1 shows that the results obtained by semi-supervised RB algorithm were comparable to those obtained using the EM algorithm. We note that the number of clusters obtained by EM algorithm is the number of different labeled verbs in the test data, as the algorithm cannot assign unlabeled data to a new class other than known classes (labeled verbs). The result obtained by semi-supervised RB shows that three new classes on an average are correctly obtained in 10 trials, while the number of clusters is smaller than that obtained by EM and the correct clusters, 62 clusters.

Table 2 shows the results by using each similarity measure. We can see from Table 2 that our core finding, that unknown words including polysemy actually aids verb classification, was robust across a wide variety of distributional similarity measures, although the Dice coefficient was decidedly the best such measure for this particular problem. The observation indicates that the RB algorithm, especially the minima of the Hamiltonian, demonstrates our basic assumption: a verb which belongs to multiple clusters will in general reduce the energy.

We recall that semi-supervised RB algorithm uses two parameters: γ and θ. We examined how these parameters affect overall clustering results. Figure 2 shows F-score of 110 verbs plotted against γ value for the top approach, *i.e.*, by running RB with Dice

Table 2. Results against each measure

Sim	θ	γ	C	Prec	Rec	F
Cos	0.1	1.1	39	0.402	0.583	0.476
rfCos	0.1	1.0	39	0.396	0.565	0.466
Dice	0.2	1.0	48	0.536	0.626	0.577
Jacc	0.1	0.7	36	0.314	0.785	0.449
L_1	0.2	0.7	46	0.378	0.724	0.497
KL	0.1	0.9	38	0.411	0.630	0.497
αdiv.	0.1	1.2	39	0.421	0.634	0.506
JS	0.4	1.0	37	0.380	0.539	0.446
EM	–	–	59	0.301	0.512	0.387

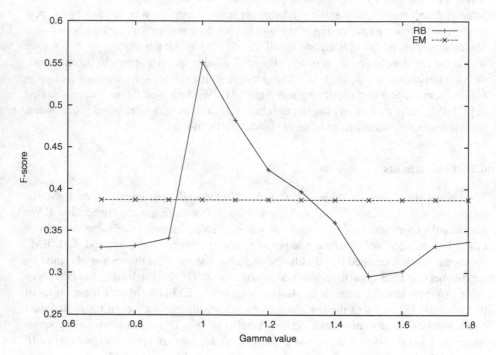

Fig. 2. F-score against γ values

coefficient as similarity measure. Similarly, Figure 3 shows F-score of 110 verbs plotted against θ value by running RB with Dice coefficient.

As can be seen clearly from Figure 2, the overall performance is extremely worse when the γ value is smaller than 0.9. This is because most of the clustering results show that verbs are classified into one of the clusters , while they are polysemous verbs. Similarly, the performance is worse when the γ value is larger than 1.3 because many of the clustering results show that verbs are classified into more than one clusters, while they are not polysemous verbs. This indicates that the RB with an extremely large or small value of γ is not effective for verb classification.

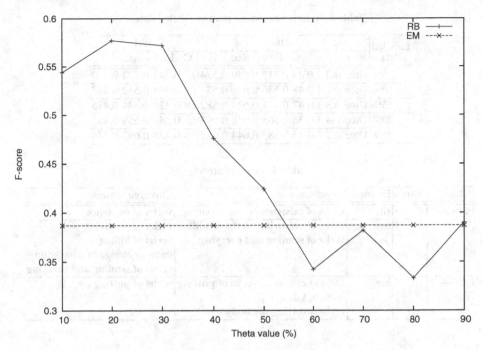

Fig. 3. F-score against θ values

Figure 3 also shows the impact against the θ values on the effectiveness for clustering verbs. As shown in Figure 3, the larger value of θ affects overall performance. This is because the number of selected pairs of verbs is few. As a result, pairs are not chose, even if they are semantically related. The result supports the usefulness of Dice coefficient, as the best performance was obtained when the θ value was around 20%, then the performance is decreased when the value of θ is large.

6.3   Efficacy against Unknown Words

It is interesting to note that how semi-supervised RB algorithm affects the ratio of labeled data. Table 3 shows the results. In table 3, "Labeled data" shows a rate of labeled data against the number of input data. Each value shown in Table 3 denotes the average accuracy, namely we randomly selected labeled data. The process is repeated 10 times for each ratio. The average accuracy is a ratio that the average number of unknown words which is assigned correctly divided by the total number of unknown words over 10 trials.

As can be seen clearly from Table 3, more labeled verbs improves overall performance in both methods, while the ratio equals to 5% and 10%, there is no statistical significance between F-scores, as the result was P-value $\geq$ .005. using micro sign test. Table 3 shows that the results obtained by semi-supervised RB are always better than those of EM at all of the ratios, especially precision. As we have shown in [19], each sense of the class is very delicate. Therefore, these results show the effectiveness of the method.

Table 3. Results against the ratio of labeled data

Labeled data	RB						EM			
	$Sim\ \theta$	γ	C	Prec	Rec	F	C	Prec	Rec	F
5%	Dice 0.2	1.0	44	0.513	0.650	0.574	43	0.241	0.632	0.350
10%	Dice 0.2	1.0	48	0.536	0.626	0.577	59	0.301	0.512	0.387
15%	Dice 0.3	1.0	47	0.541	0.629	0.582	58	0.320	0.634	0.425
20%	Dice 0.4	1.0	58	0.563	0.638	0.593	60	0.381	0.659	0.483
25%	Dice 0.2	1.0	43	0.583	0.644	0.612	58	0.488	0.697	0.574

Table 4. Types of errors

Pattern	#times	Example	Target sense(s)	Clustered sense(s)
Partial	23	sit	**verbs of existence**, verbs of putting	**verbs of existence**
	31	remove	**verbs of removing, verbs of killing** **verbs of sending and carrying**	**verbs of removing** **verbs of killing** verbs of change of possession **verbs of sending and carrying**
Poly → 1	4	hang	verbs of existence, **verbs of putting** verbs of killing **verbs involving the body**	**verbs of putting**

6.4 Error Analysis against Polysemy

We examined whether polysemous verbs were correctly classified into classes. Then, we manually analyzed clustering results obtained by running semi-supervised RB algorithm, with Dice coefficient as a similarity measure, which was the best quality F-score against the polysemic gold standard. 13 out of 71 polysemies (unlabeled data) were perfect classification: each polysemous verb was correctly classified into multiple clusters. For example, the polysemous verb, "drop" was correctly classified into four clusters. The words in italics denotes the majority sense, which corresponds to the sense according to [23].

Putting	{**drop** fill}
Change of State	{**drop** dry build}
Existence	{**drop** hang hit}
Motion	{**drop** walk travel}

Others were errors and classified into two patterns shown in Table 4. In Table 4, "Pattern" and "#times" refer to a type of an error and the numbers of errors, respectively. "Example" indicates example verbs, and "Target" sense(s)" denotes sense(s) that polysemies should be assigned. "Clustered sense(s)" refers to the sense(s) assigned by the system. "Partial" refers to partially correct: some senses of a polysemous verb were correctly identified, but others were not. The first example of this pattern is that "sit" has two senses, "verbs of existence" and "verbs of putting". However, only one sense:

"verbs of existence" was identified correctly. The second example is that three senses of the verb "remove" were correctly into the classes, while it was classified incorrectly into the class "verbs of change of possession".

"poly → 1" of "Pattern" refers to polysemous verb classified into only one cluster consisting of multiple senses. "hang" was classified into one cluster. However, the cluster consisted of four senses. There was no error type that the target senses and clustered senses did not completely match.

Error analysis against polysemies provided some interesting insights for further improvement. First, we should be able to obtain further advantages in efficiency and efficacy of the method by using hierarchical splits in the clusters, as the number of clusters obtained by semi-supervised RB algorithm was smaller than the number of correct clusters. One solution is to hierarchically apply RB algorithm, i.e., in the hierarchical approach, the classification problem can be decomposed into a set of smaller problems corresponding to hierarchical splits in the tree [28]. Roughly speaking, one first classifies to distinguish among classes at the top level, then lower level classification is performed only within the appropriate top level of the tree [31]. Each of these sub-problems can be solved much more efficiently, and hopefully more accurately as well. This is definitely worth trying with our method. Second, it is important to use other types of features, such as selectional preferences using semantic concepts from thesaurus like WordNet [41]. Third, we plan to apply the method to other thesaurus such as WordNet semantic classes, and other languages to evaluate the robustness of the method.

7 Conclusions

We have developed an approach for classifying unknown verbs including polysemies into Levin-style semantic classes. We proposed a graph-based semi-supervised clustering method which employs a small amount of labeled data to aid unsupervised learning. Moreover, the method allows verbs to belong to multiple senses. The results using the data consisting 110 test verbs was better than the EM algorithm, as the F-score obtained by the RB was 0.577 and that of the EM was 0.387. Moreover, we found that unknown words including polysemy actually aids verb classification, was robust across a wide variety of distributional similarity measures. To examine the effects of unknown words classification, we applied semi-supervised RB to the different ratio of labeled data against the number of input data. The results showed that RB is always better than the EM, even for a small number of labeled verbs, while more labeled verbs improves the overall performance in both methods. Future work includes (i) extending the method to deal with hierarchical splits in the clusters, (ii) incorporating other semantic concepts into the method, and (iii) applying the method to other dictionaries and languages.

Acknowledgements. The authors would like to thank the referees for their comments on the earlier version of this paper. This work was partially supported by the Telecommunications Advancement Foundation.

References

1. Bar-Hillel, A., Hertz, T., Shental, N., Weinshall, D.: Learning Distance Functions using Equivalence Relations. In: Proc. of the 20th International Conference on Machine Learning, pp. 11–18 (2003)
2. Bilenko, M., Basu, S., Mooney, R.J.: Integrating Constraints and Metric Learning in Semi-Supervised Clustering. In: Proc. of the 21st International Conference on Machine Learning, pp. 81–88 (2004)
3. Bouraev, B., Briscoe, E.J., Carroll, J., Carter, D., Grover, C.: The Derivation of a Grammatically-Indexed Lexicon from the Longman Dictionary of Contemporary English. In: Proc. of the 25th Annual Meeting of the Association for Computational Linguistics, pp. 193–200 (1987)
4. Brew, C., Walde, S.S.: Spectral Clustering for German Verbs. In: Proc. of 2002 Conference on Empirical Methods in Natural Language Processing, pp. 117–123 (2002)
5. Briscoe, E.J., Carroll, J.: Automatic Extraction of Subcategorization from Corpora. In: Proc. of 5th ACL Conference on Applied Natural Language Processing, pp. 356-363 (1997)
6. Briscoe, E.J., Carroll, J.: Robust Accurate Statistical Annotaion of General Text. In: Proc. of 3rd International Conference on Language Resources and Evaluation, pp. 1499-1504 (2002)
7. Chen, K.J., Chen, C.J.: Automatic Semantic Classification for Chinese Unknown Compound Nouns. In: Proc. of 38th Annual Meeting of the Association for Computational Linguistics, pp. 125–130 (2000)
8. Dagan, I., Lee, L., Pereira, F.C.N.: Similarity-based Models of Word Cooccurrence Probabilities. Machine Learning 34(1-3), 43–69 (1999)
9. Dorr, B.: Large-scale Dictionary Construction for Foreign Language Tutoring and Interlingual Machine Translation. Machine Translation 128(4), 271–325 (1997)
10. Galley, M., McKeown, K.: Improving Word Sense Disambiguation in Lexical Chaining. In: Proc. of 19th International Joint Conference on Artificial Intelligence, pp. 1486–1488 (2003)
11. Grishman, R., Macleod, C., Meyers, A.: Complex Syntax: Building a Computational Lexicon. In: Proc. of International Conference on Computational Linguistics, pp. 268–272 (1994)
12. Hindle, D.: Noun Classification from Predicate-Argument Structures. In: Proc. of 28th Annual Meeting of the Association for Computational Linguistics, pp. 268–275 (1990)
13. Hughes, J.: Automatically Acquiring Classification of Words, Ph.D. thesis University of Leeds (1994)
14. Jannink, J., Wiederhold, G.: Thesaurus Entry Extraction from an Online Dictionary. In: Proc. of Fusion 1999 (1999)
15. Kermanidis, K., Maragoudakis, M., Fakotakis, N., Kokkinakis, G.K.: Natural Language Engineering. Learning Verb Complements for Modern Greek: Balancing the Noisy Dataset 14(1), 71–100 (2008)
16. Kudo, T., Matsumoto, Y.: Fast Methods for Kernel-based Text Analysis. In: Proc. of 41st Annual Meeting of the Association for Computational Linguistics, pp. 24–31 (2003)
17. Kirkpatrick, S., Gelatt Jr., C.D., Vecchi, M.P.: Optimization by Simulated Annealing. Science 220(4598), 671–680 (1983)
18. Korhonen, A.: Subcategorization Acquisition, Ph.D. thesis University of Cambridge (2002)
19. Korhonen, A., Krymolowski, Y., Marx, Z.: Clustering Polysemic Subcategorization Frame Distributions Semantically. In: Proc. of the 41st Annual Meeting of the Association for Computational Linguistics, pp. 64–71 (2003)
20. Korhonen, A., Krymolowski, Y., Briscoe, T.: A Large Subcategorization Lexicon for Natural Language Processing Applications. In: Proc. of the 5th International Conference on Language Resources and Evaluation (2006)

21. Lee, L.: Measures of Distributional Similarity. In: Proc. of the 37th Annual Meeting of the Association for Computational Linguistics, pp. 25–32 (1999)
22. Leech, G.: 100 Million Words of English: The British National Corpus. Language Research 28(1), 1–13 (1992)
23. Levin, B.: English Verb Classes and Alternations. Chicago University Press (1993)
24. Lin, D.: Automatic Retrieval and Clustering of Similar Words. In: Proc. of 36th Annual Meeting of the Association for Computational Linguistics and 17th International Conference on Computational Linguistics, pp. 768–773 (1998)
25. Matsuo, Y., Sakaki, T., Uchiyama, K., Ishizuka, M.: Graph-based Word Clustering using a Web Search Engine. In: Proc. of 2006 Conference on Empirical Methods in Natural Language Processing (EMNLP 2006), pp. 542–550 (2006)
26. Mihalcea, R.: Unsupervised Large Vocabulary Word Sense Disambiguation with Graph-based Algorithms for Sequence Data Labeling. In: Proc. of the Human Language Technology/Empirical Methods in Natural Language Processing Conference, pp. 411–418 (2005)
27. Muller, P., Hathout, N., Gaume, B.: Synonym Extraction Using a Semantic Distance on a Dictionary. In: Proc. of the Workshop on TextGraphs, pp. 65–72 (2006)
28. Navigli, R.: A Structural Approach to the Automatic Adjudication of Word Sense Disagreements. Natural Language Engineering 14(4), 547–573 (2008)
29. Navigli, R.: Word Sense Disambiguation: A Survey. ACM Computing Surveys 41(2), 1–69 (2009)
30. Ng, A.Y., Jordan, M.I., Weiss, Y.: On Spectral Clustering: Analysis and an Algorithm. In: Advances in Neural Information Processing Systems 14. MIT Press (2002)
31. Pereira, F., Tishby, N., Lee, L.: Distributional Clustering of English Words. In: Proc. of the 31st Annual Meeting of the Association for Computational Linguistics, pp. 183–190 (1993)
32. Reichardt, J., Bornholdt, S.: Detecting Fuzzy Community Structure in Complex Networks with a Potts Model. Physical Review Letters 93(21) (2004)
33. Reichardt, J., Bornholdt, S.: Statistical Mechanics of Community Detection. Physical Review E 74 (2006)
34. Reiter, E., Dale, R.: Building Natural Language Generation Systems. Cambridge University Press (2000)
35. Resnik, P.: Using Information Content to Evaluate Semantic Similarity in a Taxonomy. In: Proc. of 14th International Joint Conference on Artificial Intelligence, pp. 448–453 (1995)
36. Rooth, M.: Two-Dimensional Clusters in Grammatical Relations. In: Inducing Lexicons with the EM Algorithm, AIMS Report 4(3) (1998)
37. Rooth, M., Riezler, S., Prescher, D., Carroll, G., Beil, F.: Inducing a Semantically Annotated Lexicon via EM-Based Clustering. In: Proc. of the 37th Annual Meeting of the Association for Computational Linguistics (1999)
38. Schulte im Walde, S.: Clustering Verbs Semantically according to their Alternation Behaviour. In: Proc. of the 18th International Conference on Computational Linguistics, pp. 747–753 (2000)
39. Schulte im Walde, S., Brew, C.: Inducing German Semantic Verb Classes from Purely Syntactic Subcategorisation Information. In: Proc. of the 40th Annual Meeting of the Association for Computational Linguistics (2002)
40. Schulte im Walde, S.: Experiments on the Automatic Induction of German Semantic Verb Classes. Computational Linguistics 32(2), 159–194 (2006)
41. Schulte im Walde, S., Hying, C., Scheible, C., Schmid, H.: Combining EM Training and the MDL Principle for an Automatic Verb Classification Incorporating Selectional Preferences. In: Proc. of the 46th Annual Meeting of the Association for Computational Linguistics, pp. 496–504 (2008)

42. Sinha, R., Mihalcea, R.: Unsupervised Graph-based Word Sense Disambiguation Using Measures of Word Semantic Similarity. In: Proc. of the IEEE International Conference on Semantic Computing, pp. 46–54 (2007)
43. Stevenson, S., Joanis, E.: Semi-Supervised Verb-Class Discovery using Noisy Features. In: Proc. of the 7th Conference on Natural Language Learning at HLT-NAACL 2003, pp. 71–78 (2003)
44. Wagstaff, K., Cardie, C., Rogers, S., Schroedl, S.: Constrained K-Means Clustering with Background Knowledge. In: Proc. of 18th International Conference on Machine Learning, pp. 577–584 (2001)
45. D. Widdows and B. Dorow, A Graph Model for Unsupervised Lexical Acquisition. In: Proc. of 19th International Conference on Computational Linguistics (COLING 2002), pp. 1093-1099, (2002)
46. Witten, I.H., Bell, T.C.: The Zero-Frequency Problem: Estimating the Probabilities of Novel Events in Adaptive Text Compression. IEEE Transactions on Information Theory 37(4), 1085–1094 (1991)

Ontology Engineering for the Autonomous Systems Domain

Julita Bermejo–Alonso, Ricardo Sanz, Manuel Rodríguez, and Carlos Hernández

Autonomous Systems Laboratory (ASLab), Universidad Politécnica de Madrid, Spain
jbermejo@etsii.upm.es,
{ricardo.sanz,manuel.rodriguezh,carlos.hernadez}@upm.es
http://www.aslab.org

Abstract. Ontologies provide a common conceptualisation that can be shared by all stakeholders in an engineering development process. They provide a good means to analyse the domain, allowing to separate descriptive from problem–solving knowledge. Our research programme on autonomous systems considered an ontology as the adequate mechanism to conceptualise the autonomous systems domain, and the software engineering techniques applied to such systems. This paper describes the ontological engineering process of such an ontology: OASys (Ontology for Autonomous Systems). Its development considered different stages: the specification of the requirements to be fulfilled by the ontology; the extraction of the actual features needed to implement the desired requirements; the conceptualisation phase with the design decisions to integrate the different domains, theories and techniques addressed by the ontological elements; and finally, the implementation of the ontology, which integrates both ontology engineering and software engineering approaches by using UML as the implementation language.

Keywords: Ontological engineering, Knowledge-based engineering, Autonomous systems.

1 Introduction

Knowledge Engineering research has addressed the use and the development of ontologies as a means to improve knowledge processes. An ontology as a conceptualisation of a specification [25], provides a solid basis to build knowledge bases for a greater functionality among users. Ontologies allow defining an abstract and simplified view of the concepts, their properties and their relationships within a domain of knowledge. Ontologies organise this knowledge in an appropriate structure, providing a representation vocabulary specialised for a domain. Ontologies formalise, structure and express the semantic content in the form of entities, their properties and their relationships, paying attention to the granularity of the ontological elements. On the other hand, ontologies are developed with a pragmatic focus, having in mind a context and an intended use for a particular domain, generally being developed following a design method or methodology.

Ontological Engineering refers to the different activities in the development process, the methodologies to support it, and the languages and tools used for the deployment of

A. Fred et al. (Eds.): IC3K 2011, CCIS 348, pp. 263–277, 2013.

an ontology [23]. This paper describes how we carried out the ontological engineering of an ontology for autonomous systems. We have developed this ontology, OASys, as a conceptual framework and software support for the domain of autonomous systems. Our approach has been to develop an ontology to consider not only the description but also the engineering process of this kind of systems, as part of a long-time research programme on a universal technology for autonomous systems. Our goal is to include both generic knowledge on systems, as well as the domain-specific one on autonomous systems, providing a common vocabulary for all the stakeholders. The underlying idea is that the ontology should express the concepts and consider the constraints or relationships in an explicit way under some ontological commitments, as the ontology will be readable by computers. This way the ontology would become an engineering artefact within a software process developed to define and to implement autonomous systems, with the ontological concepts being used at run-time by the autonomous system. Ontological domain models can drive typical development phases, such as requirements, design and implementation. The ontology so understood, is a mapping of the philosophical meaning of ontology into knowledge-based systems epistemology.

The paper is organised as follows. Section 2 reviews current research on engineering ontologies for the domains of autonomous systems, and software engineering. Section 3 summarises both the requirements necessary in our ontology for the domain of autonomous systems, as well as the way they were deployed. Section 4 explains the design decisions made whilst developing the ontology. Next, Section 5 describes the actual ontology obtained, formalised using software engineering techniques. Finally, Section 6 draws some concluding remarks on the ontology development, and additional tasks to carry out to improve and to refine it.

2   Related Work

Our ontology for autonomous systems (OASys) addressed two different but interrelated domains. Firstly, the domain of autonomous systems. Secondly, the domain of software engineering.

Related to the domain of autonomous systems, ontologies have addressed different kinds of autonomous systems: mobile robots, agent-based applications, and autonomic systems. For mobile robots, the ontologies have been used as a knowledge-representation mechanism to conceptualise their domain, their tasks or the environment where the mobile robots act. The research generally focuses on the description of the ontologies [76], on their use for a particular mobile robot or application [4], [63], [62], [28], and on the benefits achieved [67], [53]. For agent-based systems, the research on ontologies emphasises the necessity to share and to exchange knowledge among the agents in the system, and the problems of interoperatibility [42], [62]. In autonomic systems, ontologies support information exchange and integration [40], as part of the autonomous system [69], and as an explicit representation of data semantic and rules [66]. In general, the research on ontologies for autonomous systems have focused on their usage, rather than providing a detailed account of the ontological engineering process that obtained the ontology.

When it comes to the other domain of interest for our research, ontologies have been developed to act as domain ontologies to describe software engineering processes or technologies [31], [54], [1], [32], [18]. Additionally, ontologies have been used as software elements within the system's architecture to support the software process [16], [79]. The ontologies description has once again paid more attention to their benefits and use than to the specification, conceptualisation and formalisation of the ontological elements in the ontologies.

Our review pointed out the increasing use of ontologies for autonomous systems and for software engineering [7], as they provide a common understanding of the concepts, allow sharing and transfering knowledge, and manage knowledge scalability. Nevertheless, the existing research did not provide enough elements to infer how the ontologies were engineered, in terms of their specification, conceptualisation and formalisation. These aspects are more commonly addressed as part of ontological engineering efforts [75], [46], [13], [26], without a specific domain such as the autonomous system's one under consideration.

Our approach to develop OASys combined the detailed description of the ontological engineering process as well as the analysis of the specific features to fulfil the requirements of the ontology to be used for the description and engineering of autonomous systems.

3   OASys Specification

A key aspect whilst developing an ontology is to state its purpose, which drives the development and its ontological contents. Knowing what the ontology is to be developed for, allows focusing on the essential elements to be included. Additionally, it is necessary to define the type of ontology based on the subject of conceptualisation to consider. The level of abstraction, generality, and reusability of the ontological terms to be gathered in the ontology changes when considering an upper-level ontology from a domain one. Different design criteria can serve as guideline to support the ontological engineering [25]. Not all criteria can be met when designing an ontology. It is necessary to establish trade-offs between them and to compromise between the ontology design and its intended use.

3.1   OASys Requirements

- *Purpose:* the ontology would need to conceptualise the ontological elements to be used in the description of the autonomous system. Moreover, it aims at capturing the concepts required to define its generic engineering process. Our aim is to provide the system's developers with the ontological elements necessary both to describe and to engineer the autonomous system.
- *Type of ontology:* it would be a domain ontology to describe the autonomous system domain. Being a domain ontology allows a high level of usability as it captures the domain knowledge in a problem-solving independent manner, being its reusability constrained to autonomous systems related aspects.
- *Design criteria:* to assure its coherence and quality, the development would respect design criteria such as: clarity, extendibility, minimal encoding bias, and minimal ontological commitment.

- *Knowledge acquisition:* it would be made by considering different sources such as documents, existing ontologies, and experts. Documents serve as an input source for the ontological elements. Existing ontologies should also be reviewed, as the domain might have already been conceptualised, however with a different viewpoint or purpose. These existing ontologies should be selected, evaluated, and finally fully or partially reused, paying attention to the level of granularity (if the existing ontology covers the same level of detail as in the ontology under development). Domain experts also act as a source, since they provide their terminology in a domain they are familiar with.
- *Methodology:* the election of the methodology to follow during the ontology building is also an important factor. There is a wide range of methodologies to support and guide this process, as reviewed in [22], [48]. It would be necessary to assess them, to be reused in the ontology development.
- *Formalisation:* the ontology can be formalised using either traditional ontological languages or software engineering techniques. An analysis of the benefits and drawbacks of each option should be made to select the most convenient.

3.2 OASys Features

Once the ontology requirements were established, we considered the actual ontology features and additional elements to fulfil each one of them. This section describes how the requirements were finally deployed in the ontological engineering process of OASys.

- *Structure:* the ontology needed to address two different aspects in its structure, the knowledge contents and the intended use. The knowledge contents refer to the type of ontology, considering different levels of abstraction to separate generic knowledge from domain-specific one. The intended use relates to the purpose of the ontology, as the distinction between the knowledge on autonomous system description and the knowledge about its engineering process. To address the different levels of abstraction, the ontology has adopted a layered structure to address both generic and domain-specific knowledge. The upper layer contains the more abstract level knowledge. A lower layer gathers the ontological elements to charaterise an autonomous' system structure, function and behaviour. To tackle the intended purpose for both the autonomous system's description and its engineering, we found a sensible idea to consider two ontologies as part of OASys: the ASys Ontology and the ASys Engineering Ontology.
- *Design Criteria:* the design criteria were followed throughout the development of the ontology. To address the *clarity* criteria, existing ontologies and glossaries were reviewed to extract the ontological elements. Those concepts would be later discussed with the group members to commit to the desired meaning for our research, and defined in natural language. To cater for extensions in the future for the *extendibility* criteria, subontologies and packages organised the concepts. Subontologies group ontological elements at the different abstraction levels. Packages classify the concepts within a subontology according to a concrete aspect. These organising elements allow the extension or modification of the ontology without major

changes to its structure and composition. To consider the *minimal encoding bias*, intermediate tabular representations and graphs were used to define the different ontological constructs. *Minimal ontological commitment* was achieved considering only the fundamental concepts as agreed by the ontology users both at a generic knowledge, and at a domain–specific level.

- *Inputs and sources:* documents and existing ontologies were considered. Documents were analysed to come up with existing terminology and definitions for the different domains, subdomains, applications and aspects considered in the ontology's structure. They included articles in related journals, body of knowledge documents, and books. As underlying focus, the ideas developed in our research programme. The different sources were analysed to extract the ontological elements, checking for commonalities, mismatches and level of granularity. Experts were also questioned for their domain knowledge, as an additional input for ontological concepts.

- *Methodology:* from the available methodologies and methods, METHONTOLOGY [19] was chosen as a starting point for several reasons. First, the stages for the development process are well and clearly defined in an ontology life cycle. It also comprises different and further tasks to be considered, such as the ontology maintenance. Moreover, the conceptualisation activity is decomposed in different detailed tasks, with a proposed order. As a key element, intermediate representations, such as tables and graphs, can be easily understood both by domain experts and ontology developers. Finally, the methodology allows for flexibility in the process, the representation and the existence of evolving prototypes. Some additional guidelines described in [46], [43] were also considered.

- *Formalisation:* a software engineering general- and specific-purpose language, such as UML [51], was chosen to specify the ontology. We realised the limitations of UML for ontology development [3], [23]. Our decision was based on our review of ontologies for autonomous systems and software engineering where UML has been widely used [21], [68], [54]. Additionally, the Ontology Definition Metamodel (ODM) [52] opened the possibility of a later formalisation of OASys using traditional ontological languages such as OWL and RDF, by using the metamodels, mapping and profiles defined in it.

4   OASys Conceptualisation

Our ontological engineering process addressed the requirements and the features of the ontology described in Section 3. The specified characteristics were actually conceptualised at this stage of the development of the ontology. Even with the guidance of a methodology, some design decisions and trade-offs had to be considered whilst developing our ontology:

- **The Modular Structure.** For the ontology containing the elements for autonomous system's description, the ASys Ontology, it was clear the need to consider two different levels in the knowledge content. One level provided concepts for a generic system, without paying attention to autonomous properties (the System Subontology). A second level gathered ontological elements specific for the the capabilities

Table 1. ASys Ontology: Packages and Sources

SUBONTOLOGY	PACKAGE	PURPOSE	SOURCES
System	General Systems	To characterise system's structure, function and behaviour	[36], [39], [37], [38],
	Mereology	To represent whole-part relationships	[8], [10], [27], [35], [45], [44]
	Topology	To describe topological connections	[9], [45], [44]
ASys	Perception	To conceptualise the perceptive and sensing processes	[74], [71], [41]
	Knowledge	To consider the different kinds of knowledge an autonomous system uses	[61], [41], [30], [57], [2]
	Thought	To describe the reconfiguration and adaptation of goals	[73], [72], [80], [77], [70]
	Action	To specify the actions different actors will carry out	[41], [30]
	Device	To define the devices features	[55], [29], [14]

we consider key elements of autonomy in our research (the ASys Subontology, where ASys stands for autonomous system). To define the packages in each subontology, the underlying idea was to organise the ontological constructs in a way easy to change and to update, especially for the domain-specific knowledge that would evolve as our research would do. The packages, their purpose and the sources are shown in Table 1.

A similar process was followed to establish the modular structure of the ASys Engineering Ontology. Once again, two levels of knowledge were considered. The higher level concepts can be re-used for different engineering processes (the System Engineering Subontology), whereas the lower level ones specifically addressed the engineering process under development as part of our research on autonomous systems (the ASys Engineering Subontology). The final considered packages are shown in Table 2.

– **The Packages' Contents:** to define the ontological elements to be considered in each one of the subontologies, we followed a combination of top–down and a bottom–up approaches. The top–down approach allowed starting the ontology development with an intuitive analysis of the basic concepts and specifying them in detail afterwards. This approach was used to define the different packages to be contained in a particular subontology as described before, as well as a first overall description of the contents to be included in each one. For example, for the Thought package that conceptualises the goal-oriented process in the autonomous system, we considered at a first stage the necessity to include general goal-oriented terms such as goal, subgoal, goal structure, etc. Next, we followed a bottom-up approach to elicit the concepts finally contained in each package, by analysing the terms actually used in a given field of knowledge and trying to interpret them and their structural relations. Continuing with the Thought package as example, we analysed goal-oriented theories and terminology on this field as described in [80],

Table 2. ASys Engineering Ontology: Packages and Sources

SUBONTOLOGY	PACKAGE	PURPOSE	SOURCES
System Engineering	Requirement	To define stakeholders need and requirements	[33], [49], [51], [61]
	Perspective	To specify stakeholders concerns	[33], [34], [20]
	Engineering Process	To describe the engineering process as phases, tasks and products obtained	[50]
	Model–driven	To include model–driven theories	[47], [65]
ASys Engineering	ASys Requirement	To specialise stakeholders requirements for an autonomous system	[59], [58], [30]
	ASys Perspective	To describe and autonomous system from different perspectives	[55], [45], [78], [20]
	ASys Engineering Process	To describe the engineering process of an autonomous system	[60], [15], [64], [61], [17]

[77], [70]. Additional techniques described in [15] were used to identify the objects domain, such as underlying the nouns in the analysed texts, identifying causal objects (sources of actions or events), identifying real-world entities, physical devices, key concepts, or control elements.

– **The Concepts' Integration:** this process posed a twofold approach depending on the sources considered as input for a package. Some packages were based upon a concrete theory that provided the ontological elements, however not being expressed from an ontological viewpoint. Key concepts were identified following the bottom-up approach, establishing the fundamental concepts and relationships considering minimal ontological commitments. For other packages, the knowledge was covered by several sources, being necessary to assess the granularity of the terms, the existence of synonyms, and the suitability of the concepts for our research. This assessment process was especially relevant for the domain-specific packages, where not only our research ideas but also existing sources with a similar approach had to be considered. For example, the ASys Engineering Process package contents were obtained by analysing, mapping and manually merging the concepts described in [60], [64], combined with a review of existing model–based engineering methodologies in [17].

– **The Ontologies Intradependencies:** the original design idea was to develop self-contained subontologies, i.e., grouping concepts without depending on the ontological elements of any other subontology or package. However, this was not possible for two reasons. Firstly, the definition of some concepts in one package was based upon concepts belonging to another package. Secondly, the layered structure into generic and domain-specific knowledge, made necessary to assure the conceptualisation of the generic concept prior to the domain specialised ones.

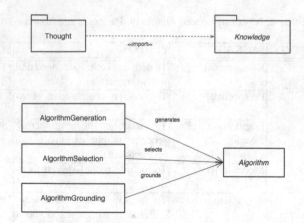

Fig. 1. Thought package usage of Knowledge package

For example, the Thought package defines the concepts of AlgorithmGeneration, AlgorithmSelection and AlgorithmGrounding as phases in the thinking process of an autonomous system. Their definition is based upon the concept Algorithm, that had to be previously define in the Knowledge package as a kind of knowledge in the autonomous system (Figure 1).

– **The ontologies interdependencies**: a second kind of dependency between the ontologies had to be considered, not so much as part of the conceptualisation of the ontologies content but to accomplish the intended use of the ontologies. The ASys Ontology conceptualises the elements to describe an autonomous system. The ASys Engineering Ontology does similarly with the terms of the autonomous system's engineering process. This process was conceptualised as different phases, tasks, and workproducts in the form of conceptual models to describe the stakeholders' needs, the autonomous system's structure, behaviour and function. These conceptual models use the ontological constructs of the ASys Ontology, thus their conceptualisation in terms of definition, attributes, relationships and axioms had to be previously made. These interdependencies were addressed and described in an ontology-based methodology [5], which describes and guides the conceptual modelling of an autonomous system based on the ontological constructs provided by the ASys Engineering Ontology, which in turn uses the elements in the ASys Ontology (Figure 2).

5   OASys Formalisation

Considering the requirements, their fulfilment and the design decisions described in former sections, the final ontology for autonomous systems (OASys) was formalised as two main ontologies (Figure 3): the ASys Ontology for the ontological elements related to the system's description, and the ASys Engineering Ontology to provide system's engineering ontological constructs. Each one of them was conceptualised and formalised as a standalone ontology, using the chosen methodology. Hence, OASys is in fact two ontologies grouped under the same name. However, they were conceived to be used in conjunction, with the ASys Engineering Ontology contents constructing and

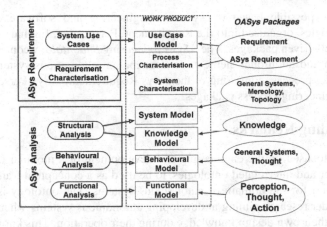

Fig. 2. OASys-based Methodology: phases, workproducts and OASys related packages

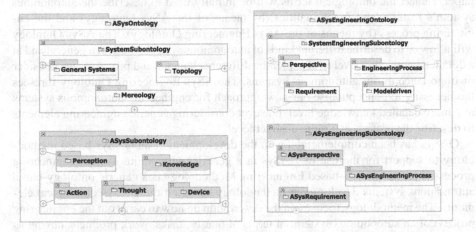

Fig. 3. OASys formalisation: ontologies, subontologies and packages

guiding the use of the ASys Ontology contents during an autonomous system's conceptual modelling.

1. **ASys Ontology:** as part of it, two subontologies were developed to cover from generic knowledge to domain–specific one regarding autonomous system's description. The *System Subontology* contains the generic knowledge on systems, organised into the General Systems, Mereology, and Topology packages. The *ASys Subontology* specialises and refines the previous concepts, adding autonomous systems specific ones, consisting of the Perception, Knowledge, Thought, Action, and Device packages.

2. **ASys Engineering Ontology:** two different subontologies were developed as part of this ontology to conceptualise the engineering process of autonomous systems, from a more abstract to domain–specific knowledge. The *System Engineering Subontology* gathers ontological elements for any system engineering process as

general as possible based on system's engineering and software engineering methodologies, organised into the Requirement, Perspective, Engineering Process and Model–driven packages. The *ASys Engineering Subontology* contains the specialisation and additional elements to describe an autonomous system's generic engineering process, consisting of the ASys Requirement, ASys Perspective, and ASys Engineering Process packages.

6 Concluding Remarks

Our research focused on the engineering and development of a modular ontology, as a set of smaller and interrelated ontologies, to be used as a conceptual framework and software support for the domain of autonomous systems. The ontology is the initial step in a broader research aiming at developing autonomous systems where such systems will use their own design knowledge during their operation. This knowledge will be represented in the form of conceptual models based on the ontology described in this paper. Hence, the ontological terms will be initially used to describe the autonomous system's features and functionalities (by means of the ASys Ontology) and the engineering process (by means of the ASys Engineering Ontology). The ASys Ontology will allow us to describe different kinds of autonomous systems, both at a general and at a detailed knowledge level to consider the different elements and processes we consider of importance in our autonomous systems. The ASys Engineering Ontology foresees the necessity to conceptualise our new approach for engineering autonomous systems at a more detailed knowledge level, however considering generic engineering elements to describe the process at a more abstract level.

OASys has been complemented with the development of an OASys-based methodology to support for the use of OASys in a generic autonomous system engineering process [7]. The OASys–based Engineering Methodology is a generic ontology-based autonomous systems development methodology based on the OASys ontological elements. The methodology focuses on the description on how to carry out the engineering process of an autonomous system, in terms of phases, tasks, work products, having as guideline the ontological elements in the System Engineering and ASys Engineering subontologies. Being OASys–based, the methodology considers the ontological elements required in the different tasks, by specifying the OASys packages to be used. The outcome of applying the methodology is a set of conceptual models that show the structural, behavioural and functional features of the autonomous system under study. Later on, these conceptual models will be used by the autonomous system itself as knowledge to perform their operation, following a model-based control paradigm.

To assess the suitability and shortcomings of the ontology and the related methodology, two testbeds have been considered to obtain these conceptual models [5], [6]. The first one, the Robot Control Testbed (RCT), is a collection of mobile robot systems, with a wide range of implementations and capabilities (from conventional SLAM based mobile robots to virtual ones inspired in rat brain neuroscience). A second testbed, the Process Control Testbed (PCT), involves the development of a robust control architecture for a chemical reaction system (with multiple steady states), providing the system with cognitive capabilities to carry out complex tasks such as fault diagnosis, alarm management, and control system reconfiguration from a single theoretical standpoint.

The application of the framework composed of the ontology (OASys) and the methodology (OASys–based Engienering Methodology) has allowed us to evaluate and to assess the ontological contents from the users' point of view. From the different ontology evaluation techniques [24], [12], [11] we adopted an application–based evaluation that consists in using the ontology in an application and evaluating the results. The goal is to determine what has been correctly defined by the ontology, what has not been defined in the ontology, and what has been incorrectly defined. By applying the ontology, we checked the consistency, completeness, and conciseness of the ontologies, subontologies, and packages.

The first conclusion from evaluating the ontology was that it has conceptualised domain knowledge both at a general level and at a more specific level, without being application-oriented. This approach has allowed us to model the testbeds at the level of detail required for their software development. However, the particular features of the testbeds have hinted a possible necessity to complement our ontology with subdomain or application specific knowledge. This will lead to additional analysis, mapping and integration aspects to be addressed as part of further research. For example the different data types corresponding to the different sensors in mobile robots had to be conceptualised and integrated in subpackages under the OASys Knowledge package. Secondly, the ontology structure was chosen to cater for different levels in the contents as well as different domains in use. The modelling of the testbeds using OASys showed the suitability of this multilevel modular approach, although pinpointing the complexity of considering in detail the dependencies among the packages and the two inner ontologies. The packaged structure allowed us to add new packages as our research evolved. As example, the Epistemic Control Loop package has been developed to conceptualise the operation of control loops based on models according to the ECL model [56], with dependencies from the Though package and others from the ASys Ontology.

References

1. Abran, A., Cuadrado, J., García-Barriocanala, E., Mendes, O., Sánchez-Alonso, S., Sicilia, M.: Engineering the ontology for the SWEBOK: Issues and techniques. In: Calero, C., Ruiz, F., Piattini, M. (eds.) Ontologies for Software Engineering and Software Technology, pp. 103–121. Springer, Heidelberg (2006)
2. ASLab Team: Core mental terminology: from an autonomous system perspective. Technical Report R-2006-XXX, Autonomous Systems Laboratory (ASLab) (2006)
3. Baclawski, K., Kokar, M., Kogut, P., Hart, L., Smith, J., Letkowski, J., Emery, P.: Extending the unified modeling language for ontology development. Software Systems Modeling 1, 142–156 (2002)
4. Barbera, T., Albus, J., Messina, E., Schlenoff, C., Horst, J.: How task analysis can be used to derive and organize the knowledge for the control of autonomous vehicles. Robotics and Autonomous Systems 49, 67–78 (2004)
5. Bermejo-Alonso, J., Sanz, R., Rodríguez, M., Hernández, C.: An Ontological Framework for Autonomous Systems Modelling. International Journal on Advances in Intelligent Systems 3, 4, 211–225 (2010)
6. Bermejo-Alonso, J., Sanz, R., Rodríguez, M., Hernández, C.: An Ontology–Based Approach for Autonomous Systems' Description and Engineering: The OASys Framework. In: Setchi, R., Jordanov, I., Howlett, R.J., Jain, L.C. (eds.) KES 2010, Part I. LNCS, vol. 6276, pp. 522–531. Springer, Heidelberg (2010)

7. Bermejo-Alonso, J.: OASys: ontology for Autonomous Systems. PhD thesis, Universidad Politécnica de Madrid (2010)
8. Borst, P., Akkermans, H., Top, J.: The PhySys ontology for physical systems. In: Ninth International Workshop Ninth International Workshop on Qualitative Reasoning. Department of Social Science Informatics (S.W.I.) University of Amsterdam, Amsterdam, The Netherlands, pp. 11–21 (1995)
9. Borst, P., Akkermans, H., Top, J.: Engineering ontologies. International Journal of Human-Computer Studies 46, 365–406 (1997)
10. Borst, W.N.: Construction of Engineering Ontologies for Knowledge Sharing and Reuse. PhD thesis, Centre for Telematics and Information Technology, University of Tweenty, Enschede, The Netherlands (1997)
11. Obrst, L., Ashpole, B., Ceusters, W., Mani, I., Ray, S.R., Smith, B.: The Evaluation of Ontologies. In: Baker, C.J.O., Cheung, K.-H. (eds.) Semantic Web: Revolutionizing Knowledge Discovery in the Life Sciences, 1st edn. Springer (2006)
12. Brank, J., Grobelnik, M., Mladenic, D.: A survey of ontology evaluation techniques. In: Conference on Data Mining and Data Warehouses, SiKDD (2005)
13. Corcho, O., Fernández-López, M., Gómez-Pérez, A.: Methodologies, tools and languages for building ontologies. where is their meeting point? Data and Knowledge Engineering 46, 41–64 (2003)
14. de la Mata, J.L.: CSTR overall specification: The main PCT testbed. Technical Report R-2009-001, Autonomous Systems Laboratory, ASLab (2009)
15. Douglass, B.P.: Real Time UML: advances in the UML for real–time systems, 3rd edn. The Addison–Wesley object technological. Addison-Wesley (2004)
16. Eberhart, A.: Ontology-Based Infrastructure for Intelligent Applications. Phd thesis, University of Saarbrücken (2003)
17. Estefan, J.A.: Survey of model–based systems engineering (MBSE) methodologies. Technical Report INCOSE-TD-2007-003-01 (Rev. B), Model Based Systems Engineering (MBSE) Initiative, International Council on Systems Engineering (INCOSE) (2008)
18. Falbo, R., Ruy, F., Moro, R.: Using ontologies to add semantics to a software engineering environment. In: 17th International Conference on Software Engineering and Knowledge Engineering (SEKE 2005), Taipei, China, pp. 151–156 (2005)
19. Fernández-López, M., Gómez-Pérez, A., Juristo, N.: METHONTOLOGY: from ontological art towards ontological engineering. In: Farquhar, A., Grüninger, M., Gómez-Pérez, A., Uschold, M., van der Vet, P. (eds.) AAAI 1997 Spring Symposium on Ontological Engineering, Stanford University, CA, U.S.A, pp. 33–40 (1997)
20. Friedenthal, S., Moore, A., Steiner, R.: A practical guide to SysML: The Systems Modeling Language. Morgan Kaufmann and OMG Press (2008)
21. Gasevic, D., Djuric, D., Devedzic, V.: Model Driven Architecture and Ontology Development. Springer, Heidelberg (2006)
22. Gómez Pérez, A., Fernández López, M., Corcho, M.: Methodologies and methods for building ontologies. In: Ontological Engineering: with examples from the Areas of Knowledge Management, e-Commerce and the Semantic Web. Advanced Information and Knowledge Processing, pp. 107–197. Springer (2004)
23. Gómez Pérez, A., Fernández López, M., Corcho, M.: Ontological Engineering: with examples from the areas of Knowledge Management, e-Commerce and the Semantic Web. Advanced Information and Knowledge Processing. Springer (2004)
24. Gómez Pérez, A.: Ontology Evaluation. In: Staab, S., Studer, R. (eds.) Handbook on Ontologies, pp. 251–274. Springer (2004)
25. Gruber, T.: Toward principles for the design of ontologies used for knowledge sharing. In: Guarino, N., Poli, R. (eds.) International Workshop on Formal Ontology in Conceptual Analysis and Knowledge Representation, Padova, Italy. Kluwer Academic Publishers (1993)

26. Grüninger, M., Fox, M.: Methodology for the design and evaluation of ontologies. In: Skuce, D. (ed.) IJCAI 1995 Workshop on Basic Ontological Issues in Knowledge Sharing, Montreal, Canada, pp. 6.1–6.10 (1995)
27. Guizzardi, G.: Ontological Foundations for Structural Conceptual Models. Phd thesis, University of Twente, The Netherlands (2005)
28. Hallam, J., Bruynickx, H.: An ontology of robotics science. In: Christensen, H.I. (ed.) European Robotics Symposium 2006 (STAR 22), pp. 1–14. Springer, Heidelberg (2006)
29. Hernández, C., Hernando, A.: RCT overall specification: Higgs platform. Technical Report R-2008-XXX, Autonomous Systems Laboratory, ASLab (2008)
30. Hernández, C., Sanz, R., López, I.: Consciousness in cognitive architectures: a principled analysis of RCS, Soar and ACT–R. Technical Report R–2008–004, Autonomous Systems Laboratory, ASLab (2008)
31. Hesse, W.: Ontologies in the software engineering process. In: Lenz, R. (ed.) Tagungsband Workshop on Enterprise Application Integration (EAI 2005), GITO–Verlag, Berlin (2005)
32. Hruby, P.: Ontology–based domain–driven design. In: Object-Oriented Programming, Systems, Languages and Applications (OOPSLA 2005), San Diego, California, U.S.A (2005)
33. IEEE: IEEE Standard Glossary of Software Engineering Terminology. IEEE Computer Society, New York, IEEE std 610.12 1990 edition (1990)
34. IEEE: IEEE Recommended Practice for Architectural Description for Software- Intensive Systems. Institute for Electrical and Electronics Engineering, New York, IEEE std 1471-2000 edition (2000)
35. Keet, C., Artale, A.: Representing and reasoning over a taxonomy of part–whole relations. Applied Ontology (1), 1–17 (2007)
36. Klir, G.J.: Approach to General Systems Theory. Van Norstrand Reinhold, New York (1969)
37. Klir, G.J.: Facets of Systems Science. Plenum Press (1991)
38. Klir, G.J., Elias, D.: Architecture of Systems Problem Solving. IFSR International Series on Systems Science and Engineering, vol. 21. Kluwer Academic Publishers (2003)
39. Klir, G.J.K.: The emergence of two–dimensional science in the information society. Systems Research 2(1), 33–41 (1985)
40. Lehtihet, E., Strassner, J., Agoulmine, N., Foghlú, M.Ó.: Ontology-Based Knowledge Representation for Self-governing Systems. In: State, R., van der Meer, S., O'Sullivan, D., Pfeifer, T. (eds.) DSOM 2006. LNCS, vol. 4269, pp. 74–85. Springer, Heidelberg (2006)
41. López, I.: A Foundation for Perception in Autonomous Systems. Phd thesis, Universidad Politécnica de Madrid (2007)
42. Malucelli, A., Palzer, D., Oliveira, E.: Combining ontologies and agents to help in solving the heterogeneous problem in e-commerce negotiations. In: International Workshop on Data Engineering Issues in E-Commerce (DEEC 2005), Tokyo, Japan, pp. 26–35. IEEE Computer Society (2005)
43. Mizoguchi, R.: Tutorial on ontological engineering - part 2: ontology development, tools and languages. New Generation Computing 22(1), 61–96 (2004)
44. Morbach, J., Bayer, B., Wiesner, A., Yang, A., Marquardt, W.: OntoCAPE 2.0: the upper level. Technical Report LPT–2008–25, RWTH Aachen University (2008)
45. Morbach, J., Wiesner, A., Marquardt, W.: A meta model for the design of domain ontologies. Technical Report LPT-2008-24, RWTH Aachen University (2007)
46. Noy, N., McGuinness, D.: Ontology development 101: A guide to creating your first ontology. Technical Report KSL–01–05, Stanford Knowledge Systems Laboratory (2001)
47. OMG: MDA Guide Version 1.0.1. Object Management Group (2003)
48. Pinto, H.S., Martins, J.P.: Ontologies: How can they be built. Knowledge and Information Systems 6(4), 441–464 (2004)
49. OMG: OMG SysML Specification. Object Management Group, v 1.1 edition (2008)

50. OMG: Software and systems process engineering meta–model specification version 2.0. OMG Formal Specification 2008–04–01, Object Management Group, Inc. (2008)
51. OMG: OMG Unified Modeling Language (OMG UML) Infrastructure and Superstructure Version 2.2 (2009)
52. OMG: Ontology Definition Metamodel Version 1.0. Object Management Group (2009)
53. Provine, R., Uschold, M., Smith, S.: Observations on the use of ontologies for autonomous vehicle navigation planning. In: 2004 AAAI Spring Symposium on Knowledge Representation and Ontologies for Autonomous Systems, Stanford, California (2004)
54. Ruiz, F., Hilera, J.: Ontologies for Software Engineering and Software Technology. Springer, Heidelberg (2006)
55. Rumbaugh, J., Jacobson, I., Booch, G.: The Unified Modeling Language Reference Manual, 2nd edn. Object Technology. Addison-Wesley (2004)
56. Sanz, R., Hernández, C., Gómez, J., Bermejo-Alonso, J., Rodríguez, M., Hernando, A., Sánchez, G.: Systems, models and self–awareness: towards architectural models of consciousness. International Journal of Machine Consciousness 1(2), 255–279 (2009)
57. Sanz, R., Hernández, C., Rodríguez, M.: ASys models: Model–driven engineering in ASys. Technical Report R–2007–016, Autonomous Systems Laboratory, ASLab (2007)
58. Sanz, R., López, I., Bermejo, J.: A rationale and vision for machine consciousness in complex controllers. In: Artificial Consciousness, pp. 141–155. Imprint Academic, Exeter (2007)
59. Sanz, R., López, I., Bermejo, J., Chinchilla, R., Conde, R.: Self-X: The control within. In: 16th IFAC World Congress. IFAC, Praga (2005)
60. Sanz, R., Matia, F., Puente, E.A.: The ICa approach to intelligent autonomous systems. In: Microprocessor–based and Intelligent Systems Engineering. Kluwer Academic Publishers (1999)
61. Sanz, R., Rodríguez, M.: The ASys vision: Engineering any-x autonomous system. Technical Report R-2007-001, Autonomous Systems Laboratory, ASLab (2008)
62. Schlenoff, C., Messina, E.: A robot ontology for urban search and rescue. In: 2005 ACM Workshop on Research in Knowledge Representation for Autonomous Systems, pp. 27–34. ACM Press, Budapest (2005)
63. Scrapper, C., Balakirsky, S.: Knowledge representation for on–road driving. In: 2004 AAAI Spring Symposium on Knowledge Representation and Ontologies for Autonomous Systems, Stanford, California (2004)
64. Segarra, M.J.: CORBA control systems. PhD thesis, Universidad Politécnica de Madrid (2005)
65. Stahl, T., Völter, M.: Model-Driven Software Development: technology, engineering, management. John Wiley and Sons, Ltd. (2006)
66. Stojanovic, L., Abecker, A., Stojanovic, N., Studer, R.: Ontology–based correlation engines. In: International Conference on Autonomic Computing (ICAC 2004), pp. 304–305 (2004a)
67. Stojanovic, L., Schneider, J., Maedche, A., Libischer, S., Studer, R., Lumpp, T., Abecker, A., Breiter, G., Dinger, J.: The role of ontologies in autonomic computing systems. IBM Systems Journal 43(3), 598–616 (2004b)
68. Tamma, V., Cranefield, S., Finin, T., Willmott, S.: Ontologies for Agents: Theory and Experiences. Whitestein Series in Software Agent Technologies and Autonomic Computing. Birkhäuser (2005)
69. Tziallas, G., Theodoulidis, B.: Building autonomic computing systems based on ontological component models and a controller synthesis algorithm. In: 14th International Workshop on Database and Expert Systems Applications (DEXA 2003), pp. 674–680, Prague, Czech Republic (2003)
70. University of Toronto: GRL ontology (2004)
71. UPM-ICEA-Team: Case studies of perception and system analysis. Technical Report ASLab-ICEA-R-2006-015, 1.0 Final, Autonomous Systems Laboratory (ASLab) (2006)

72. UPM-ICEA-Team: ICEA glossary: integration, cognition, emotion, autonomy. Technical Report ASLab-ICEA-R-2006-014, Autonomous Systems Laboratory (ASLab) (2006)
73. UPM-ICEA-Team: A vision of general autonomous systems. Technical Report ASLab-ICEA-R-2006-018, 1.0 Final, Autonomous Systems Laboratory (ASLab) (2006)
74. UPM-ICEA-Team: A vision of perception in autonomous systems. Technical Report ASLab-ICEA-R-2006-017, Autonomous Systems Laboratory (ASLab) (2006)
75. Uschold, M., King, M.: Towards a methodology for builiding ontologies. In: Skuce, D. (ed.) IJCAI 1995 Workshop on Basic Ontological Issues in Knowledge Sharing, Montreal, Canada, pp. 6.1–6.10 (1995)
76. Uschold, M., Provine, R., Smith, S., Schlenoff, C., Balikirsky, S.: Ontologies for world modeling in autonomous vehicles. In: 18th International Joint Conference on Artificial Intelligence, IJCAI 2003 (2003)
77. van Lamsweerde, A.: From System Goals to Software Architecture. In: Bernardo, M., Inverardi, P. (eds.) SFM 2003. LNCS, vol. 2804, pp. 25–43. Springer, Heidelberg (2003)
78. van Lamsweerde, A.: Requirements engineering: From craft to discipline. In: FSE 2008: 16th ACM Sigsoft International Symposium on the Foundations of Software Engineering, Atlanta, U.S.A (2008)
79. Wongthongtham, P., Chang, E., Dillon, T.: Towards ontology-based software engineering for multi-site software development. In: 3rd IEEE International Conference on Industrial Informatics (INDIN), Perth, Australia, pp. 362–365 (2005)
80. Yu, E.: Towards modelling and reasoning support for early–phase requirements engineering. In: 3rd IEEE International Symposium on Requirements Engineering (RE 1997), Washington, D.C., USA, pp. 226–235 (1997)

Cloud Services Composition Support by Using Semantic Annotation and Linked Data

Martín Serrano[*], Lei Shi, Mícheál Ó Foghlú, and William Donnelly

National University of Ireland Galway – NUIG,
Digital Enterprise Research Institute – DERI, Co., Galway, Ireland
`martin.serrano@deri.org`
Waterford Institute of Technology – WIT,
Telecommunications Software and Systems Group – TSSG, Co., Waterford, Ireland
`martin.serrano@ieee.org, {lshi,mofoghlu,wdonnelly}@tssg.org`

Abstract. Cloud computing is not only referred as synonym of on-demand usage of computing resources and services, but as the most promising paradigm to provide infinite scalability by using virtual infrastructures. In the other hand mobile technologies are scaling up to encompass every day a growing number of real and virtual objects in order to provide large-scale data applications, e.g. sensor-based intelligent communications networks, smart grid computing applications, etc. In those complex scenarios, cloud-based computing systems need to cope with diverse service demands in order to enable dynamic composition based on particular user's demands, variations in collected data broadband, fluctuation of data quality and to satisfy ad-hoc usage for personalized applications. Thus essential characteristics from cloud-native systems i.e. elasticity and multi-tenancy are fundamental requirements into large-scale data processing systems. In this paper we have investigated common practices on information sharing and domain ontological modelling to enable service composition of cloud computing service provisioning. This approach exploits the potential of semantic models in supporting service and application linkage by studying links between the complementary services. By using semantic modelling and knowledge engineering we can enable the composition of services. We discuss what implications this approach imposes on architectural design terms and also how virtual infrastructures and cloud-based systems can benefit from this ontological modelling approach. Research results about information sharing and information modelling by using semantic annotations are discussed. An introductory application scenario is depicted.

Keywords: Information Modelling, Semantic Annotation, Linked Data, Interoperability, Knowledge Management, Ontology Engineering, Cloud Computing, Service Composition, Elasticity, Multi-tenancy.

1   Introduction

Cloud computing is not only referred as synonym of on-demand usage of computing resources and services, but as the most promising paradigm to provide infinite

---

[*] IEEE member and Corresponding author.

A. Fred et al. (Eds.): IC3K 2011, CCIS 348, pp. 278–293, 2013.

scala-bility by means of using virtual infrastructures. However, far is the day this paradigm is to be fully implemented, for example, in the race for deploying cloud computing services solutions (SaaS - Software as a Service) an important challenge left aside is enabling the information to interoperate between the different service applications or service stacks (information sharing). Unfortunately, likewise, cloud service infrastructure implementations (IaaS - Infrastructure as a Service) has not fully run a coordinated course in terms of design and deployed solutions either. In middleware terms, middleware approaches (PaaS - Platforms as a Service) where the information exchange is crucial happens the same problem, even if this gap is smaller, there is no equal level development between design and implementation of software applications. As result of this uncoordinated development, diversity of approaches and multiple non-interoperable cloud solutions are in place.

In the other hand, today's communication and computing systems have less technological restrictions, for example mobile technologies are being developed with more powerful computing capacities. So, additional to the computing systems generating vast amount of information to be processed, it is not difficult to imagine every day mobile devices with increasing number of real sensors and virtual applications serving as providers in large-scale data applications and stored in cloud infrastructures. Following this premise the information interoperability looks like as a bottleneck where infrastructure and software encounter.

Linked data for information sharing is becoming an accepted best practice to exchange information in an interoperable and reusable fashion way [13], for example different communities over the Internet use the semantic web standards to enable interoperability and exchange information. This practice is actually being well accepted by other ICT's communities, when building enterprise solution(s) traditionally a series of combinations to use existing enterprise services (sub-services) is a very common practice, first by economic interest and second by technology restrictions, this practice has become so popular and today is know as service composition. As an important feature service composition can define in other services definition (recursively). Recursive service composition of business services is one of the most important features of Software Oriented Architectures (SOA), however which advantages it offers in cloud services? How SOA architectures influence cloud solutions allowing rapidly build new solutions based on the existing business services? Is this same recursive methodology applicable to cloud environments? As a fact we assume the amount of individual business services and their composed services are a growing tendency (at least in SOA design), and this practice allows a much easier implementation for new enterprise solutions.

Currently cloud computing architectures, as a design conception, enables capabilities for interoperability and information exchange between data and service levels, this feature facilitates service composition processes, a common practice in software oriented architectures (SOA). However aware to this requirement this feature is far to be fully implemented, and information exchange can't be done transparently, this fact promotes a race between academic and industry communities to investigate for designing the Cloud Computing architectures and service solutions enabling or facilitating this feature [23]. It is a fact, currently design approaches concentrate on defining individual business services to be implemented for stand alone applications and ad-hoc particular infrastructures. Design principles in Cloud Computing aligned with composed services practices contribute for transforming from isolated services (some times considered agnostics) to a more awareness services and integrated solutions. In this designs process many active academic and Information and Communications Technology (ICT) industry communities have participated with approaches to enable information interoperability. Mainly proposing the design conception in the area of Future Internet [2]; [3]; [7]; [12] where virtual infrastructure support design ideas.

Convergence towards Internet technologies for communications networks and application services has been a clear trend in the ICT domain in the past few years. Although widely discussed this exponentially increasing trend involve many issues of non-interoperable aspects where social, economic and political dimensions take place, all these issues a matter of end-user demands and service requirement.

The intention in this paper is not to define what the Knowledge Engineering means, but rather to view study and define a service-oriented design philosophy; coming through a revision about the role linked data and semantic modelling [13] can play to satisfy part of the mentioned shared information challenges. In Cloud-based systems services and infrastructure follow a common guideline; provide solutions in form of implemented interoperable mechanisms. Communications networks have undergone a radical shift from a traditional physical expansion environment with heavy expensive devises focused on applications-oriented perspective, towards converged service-oriented distributed software applications alike more powerful (shared to increase processing capacity) data centres architecture. In this radical shift Internet applications are the interaction interface between customer as end-user and network operators and service providers.

This paper focuses on information interoperability and Linked Data for controlling communication systems in the Future Internet of network and services. The extensible, reusable, common and manageable information Linked-Data layer is critical for this deployment. The novelty aspect of this approach relies on the fact that high level infrastructure representations do not use resources when they are not being required to support or deploy services. We optimize resources using this approach by classifying and identifying, by semantic descriptions in a knowledge-based fashion way what resources need to be used. Thus dynamically the service composition is executed and service deployed by result of knowledge–based analysis.

Organization of this paper is as follows: Section 2 presents the summary of challenges for an architecture-infrastructure interoperable, where information exchange (linked data processes) occurs to support application and network services. Section 3 discusses the analysis about composing services in the Cloud era and the role Software Oriented Architectures and Linked data plays in this ongoing transformation. Section 4 introduces our data link approach in form of meta-ontologies facilitating information interoperability and a demonstrator in form of functional architecture to support the inference approach. Section 5 presents some previous work and projects working in this topic and the state of the art outlook and finally some relevant references used in this paper are listed.

2   Information Sharing in Cloud Systems

Taking a broad view of state of the art, current development of data link interactions and converging communications, many of the problems present in current Internet about data and information management are generated by interoperability problems; we have identified three persistent problems:

1. Users are offered relatively small numbers of composed services, which they can not *personalise* to meet their *evolving needs*; communities of users can not tailor *services* to help *create, improve and sustain* their *social interactions*.

2. The *services* offered are typically *technology-driven* and static, designed to maximise usage of capabilities of underlying technologies and not to satisfy user requirements *per-se*, and thus cannot be *readily adapted* to their changing *operational context*.

3. Service providers cannot *configure* their infrastructure to *operate effectively* in the face of changing *service usage patterns* and technology deployment; infrastructure can only be *optimised*, on an *individual basis*, to meet specific low-level objectives, often resulting in sub-optimal operation in comparison to the more important *business and service user objectives*.

As the move towards convergence of communications systems and a more extended service-oriented architecture design and cloud computing gains momentum (VoIP is a clear example of this convergence) the academic research community is increasingly focussing on how to evolve technologies to enable dynamic service composition. In this sense we believe that addressing evolution of networking technologies in isolation is not enough; instead, it is necessary to take a holistic view of the evolution of communications services and the requirements they will place on the heterogeneous physical or virtual infrastructure over which they are delivered [23]; [15]; [21], However, communications technology is not part of this research, we concentrate in accessible information sharing features in this convergence of systems.

By addressing sharing information issues, composed systems must be able to exchange information and customize their services. So cloud environments can reflect individual and shared preferences in network and services and can be effectively managed to ensure delivery of critical services in a services-aware design view with general infrastructure challenges.

3   Service Composition in CLOUD

The business benefits of the cloud systems significantly reflects cost reduction and increase systems flexibility to react to user demands efficiently and by replacing, in a best practice manner, a plethora of proprietary hardware and software platforms with generic solutions supporting standardised development and scalable stacks over the Internet. Research initiatives addressing this cloud-based design trend and inspired mainly by software oriented architectures (SOA) requirements argue that the future rely in application layers above virtual infrastructures that can meet various requirements whilst keeping a very simplistic, almost unmanaged network. IP for the underlying Internet for example, GENI NSF-funded initiative to rebuild the Internet [18] is an example of this. Others argue that the importance of wireless access networks requires a more fundamental re-design of the core Internet Protocols themselves [4]; [1]. Whilst this debate races nothing is a clear outcome in terms of information interoperability or data models sharing.

We follow the idea of service agnostic designs (ad-hoc solutions) are not anymore a way to achieve interactive solutions in terms of information sharing capabilities for heterogeneous infrastructure support either to facilitate service composition in complex environments such cloud environments/applications.

A narrow focus on designing optimal networking protocols in isolation is too limited, instead a more abstracted view is required. This offers the advantage of non-dependency on physical infrastructures offering limited amount of services. In this view multiple services are now result of subservices, this method is commonly called composition. When meaning of various distributed protocols and delivering sub-services orchestrate multiple sub services, the operations (e.g. applications, computing processing, distribution of services, networking) can be done more efficiently. In other terms, a more realistic way of offering services is following mechanisms to organise operations according to changes in the parameters and based on users needs. However, realistically this new holistic view increasingly stops to become a matter of critical infrastructure, in this sense cloud computing infrastructures with virtualisation, as main driver is a promising alternative of solution to this stopping problem. Not only IT operators are today coming to realise the problem IP networks face up, also service-web providers recognize where new integrated Internet services are easier to design, deploy and manage when they are a result of composed sub-services in different virtual infrastructures providers systems.

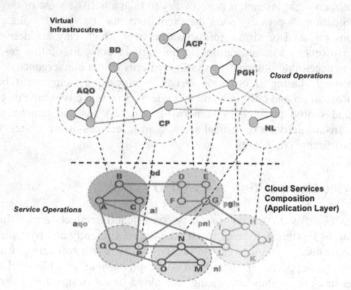

Fig. 1. Cloud service composition on virtual infrastructures

The figure 1 depicts the mentioned cloud service composition, its implementation relies on the inference plane [24], or knowledge layer [30] where the exchange of information (Linked-Data structures) [5] facilitates knowledge-driven support and generation of cloud composed services with operations by enabling interoperable information on networked connected objects [10]. From down to top and having cloud infrastructures representation as example, isolated components representations are depicted with no capacities of sharing information, linked data mechanisms are missing and "X" represented. In an upper Layer linked mechanism are represented and used to define virtual infrastructure operations and expose them externally. So the migrations towards composed services and networks increases providing solutions to a number of significant technical issues by using more standard information exchange and promoting sharing information. At the upper part of the linked data mechanisms are supported by ontology representations and ontology-based mapping allowing at the same time original services (e.g. ABC) can be managed effectively and most important offering open opportunities for a knowledge-based service-oriented support having a fundamental impact on cloud composition of services (e.g. BD, AQO, PGH, etc.) by a complete information sharing and sub-services representation (e.g. bd, cl, pnl, nl).

As a research work outcome, the composition of services in cloud can be represented by functions that identify variations in the values of its temporal content regarding with infrastructure dynamics (those functions are temporal marked based on its analogy on the fact cloud services depend on temporal variables to be deployed or removed). Those functions can be operated on using logic or mathematical expressions. Thus in this approach logic and mathematical operations are used for both cloud service representation $f(Xs_n)_m$ as well as infrastructure cloud operations $f(Ot_n)_m$. Such functions are representing possible cloud systems that can be distributed (for example, from different data clouds) as follow.

$$f(Xs_n)_m \rightarrow \text{Cloud Service Functions, n,m where } n>1 \text{ and } m > 0 \qquad (1)$$

$$f(Ot_n)_m \rightarrow \text{Operation Functions, t, n, m where } n>0 \text{ and } m > 0 \qquad (2)$$

The functions are independent of each other. The way to create a relation and compose services is by generating inclusive functions "F" which contains sets of cloud C service and cloud Operations functions. In these kind of functions, the constraints "$n > 1$" and "$m > 0$" are forced to be rewritten as "$n \geq 1$" and "$m \geq 1$", since the service and at least 1 operation must be valid. The expression representing this condition is then composed as follow:

$$F[f\{(Ot_n)_m\}\, f\{(Xs_n)_m\}] \rightarrow \text{Composed functions related with Operations} \qquad (3)$$

where $n \geq 1$ and $m \geq 1$

Integrating the expression by set theory arguments and using summation:

$$\sum F[\{(Ot_n)_m\}\{(Xs_n)_m\}] \rightarrow \text{Summation of functions} \qquad (4)$$

where $n \geq 1$ and $m \geq 1$

The proposed service infrastructure operations do not assume a "static" service model language (i.e., a particular, well-defined vocabulary that does not change) for expressing operations. This enables multiple operations to be processed in parallel, since service cloud conditions can be changed rapidly (e.g., new systems performance values can be defined at run-time, so those variables acting as context information are dynamic).

The variations of infrastructure when multiple service operations are being executed is represented by "p", and the expression representing such dynamic behaviour is as follows:

$$\sum_{Xs=1} F[\{(Ot_n)_m\}\{(Xs_n)_m\}] \, p \rightarrow \text{Service Composed Function referring to } "p"$$

(5)

infrastructure parameters

where $n \geq 1$ and $m \geq 1$ and $p \geq 1$ (when at least one service is active)

The essence of this function is to associate service cloud functions with service functions. This is accomplished using a set of policies.

A set of parameters is represented by the notation "pn", also the initial condition of cloud systems requires an initial set of parameters represented as "ps". Specifically, the functions can either match pre-defined functions that manipulate existing schema elements, or can be used to extend these schema elements to represent new information that is required by the service.

The final representation of the function is:

$$\sum_{Xs=1}^{ps+pn} F[\{(Ot_n)_m\}\{(Xs_n)_m\}] p^{ps+pn} \rightarrow \text{Composed Function referring to multiple}$$

(6)

"pn" infrastructure parameters

where $n \geq 1$, $m \geq 1$, $p \geq 1$, Cs is valid for $Cs \geq 1$ to "$ps+pn$"
 (including the total number of cloud service parameters)

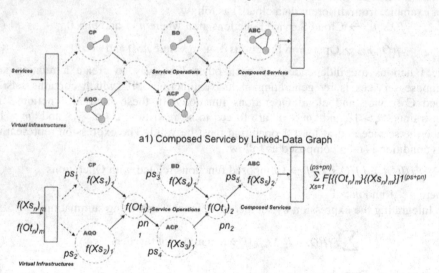

a1) Composed Service by Linked-Data Graph

a2) Composed Service by Linked-Data Representation

Fig. 2. Cloud service composition on virtual infrastructures

The total number of parameters, plus the number of non-repeated relationships between the context data and the service management operations data, is required in order to solve this function. The work to formalize the information by using first order logic-models is studied in [14]. For example, assume that one composed service has five parameters "$pn = 5$" and as initial condition "$ps = 1$", thus the parameters are used to represent its cloud service operations $Xs \geq 1$ to $Xs = ps+pn$, and further more assume that no context variations exist.

By using this modelling mechanism and as a manner of example we demonstrates that it is feasible for representing the capabilities of composing services based on constructed functions by dependency on cloud infrastructure parameters associated with cloud services $f(Xs_n)_m$ and cloud infrastructure operations $f(Ot_n)_m$ A graphical representation of this service composition is shown in the figure 2. a1) From left to tight there are service functions which can be combined by using an operator generating a linked data process which generates a composed function. a1) represent the service composition by using linked data nomenclature.

4   Linked Data and Semantic Annotation

A current activity, attracting the attention of many research and industrial communities is the formalization of data models (ontology engineering). Enabling information for management of services and control of operations is an example where this formalization is used [30]; [26]. This process focuses in the semantic enrichment task where descriptive references about simple data entries are used to extend data meaning (semantic aggregation), to for example, provide an extensible, reusable, common and manageable linked data plane, also referenced as inference plane [24]. Thus management information described in both enterprise and infrastructure data models (physical or virtual) with ontological data can be used inherently in both domains.

The semantic aggregation can be seen as a tool to integrate user data with the management service operations, to offers a more complete understanding of user's contents based on their operational relationships and hence, a more inclusive governance of the management of components in the infrastructure (resources, devices, networks, systems) and or services inclusive. The objective is sharing the integrated management information within different management systems (liked data). This approach is to use ontologies as the mechanism to generate a formal description, which represents the collection and formal representation for network management data models and endow such models with the necessary semantic richness and formalisms to represent different types of information needed to be integrated in network management operations. Using a formal methodology the user's contents represent values used in various service management operations, thus the knowledge-based approach over the inference plane [30] aims to be a solution that uses ontologies to support interoperability and extensibility required in the systems handling end-user contents for pervasive applications [24].

4.1   Service and Infrastructure Management by Using Ontological Modelling

The meta-ontology approach introduced in this section integrates concepts from the IETF policy standards [34]; [16] as well as the TM Forum SID model [32]; [28]. In this section important classes originally defined in the IETF, SIM and DEN-ng models, in telecommunications, are cited and implemented as ontologies, some other extended or adapted for communication services adaptability [25]. The meta-model defines a set of interactions between the information models, pervasive management service lifecycle models, and communications systems operations in order to define relationships and interactions between the classes from cited models and from the different knowledge domains. The Ontology mapping and Ontology construction process, which is a four-phase methodology is result of formal study to build up ontologies contained and studied from [11]; [8].

The formal language used to build the set of ontologies is the web ontology language (OWL) [19]; [20], which has been studied in order to be applied to cloud computing environments; additional formal definitions act as complementary parts of the modelling process. Formal descriptions about the terminology related within infrastructure management domain has been specified to build and enrich the proposal for integrating infrastructure and other service and operation management data within the information models. The objective is to create a more complete information model definition with the appropriate management operations using formal descriptions in OWL. The proposed meta-Ontology model uses concepts from policy-based management systems [29]; [31] to represent a system ruled by infrastructure policies in cloud environment (virtual infrastructures), which is an innovative aspect of this research work. Figure 3 shows the Ontology representation. The image represents the linked data representation process as result in the integration of classes related to the management operation class through the event class. The InfoEntity class interacts with other classes from different domains in order to represent information.

This representation simplifies the identification of interactions between the information models. These entity concepts, and their mutual relationships, are represented as lines in the figure. The InfoEntity class forms part of an Event class, and then the Event govern the policy functionality of a Managed Entity by taking into account context information contained in Events. This functionality enables exchange information as part of operations requested from a cloud service, and is represented as interaction between Event and InfoEntity.

The meta-Ontology model is driven by a set of service management operations each as part of the cloud-based service lifecycle. The service composition and its model representation contain the service lifecycle operations, as depicted in figure 3. In figure 3, service management operations, as well as the relationships involved in the management service lifecycle process, are represented as classes. These classes then are used, in conjunction with ontologies, to build the language that allows a *formal* form of English to be used to describe its actions that has effect into events. To do so information (InfoEntity) is underlayed in such relationships, which a correspondence with activities called "*events*" and related to Info Class.

The meta-model is founded on information models principles for sharing information and policy management promoting an integrated management, which is required by both cloud-bases systems as well as service management applications. The combination of data models and virtual infrastructures motivates the definition of extensible and scalable component frameworks to enable the generation of these liked diverse in nature sources of knowledge to realize a scalable management platform. The use of virtual infrastructures facilitates the treatment of them as service providers, this enable handling components as information units and process as operations. In this sense services description in form of composed services has to be created. Services will act as information units and the linked data connectors as operations to generate composed services. The composed services can be seen as the operations.

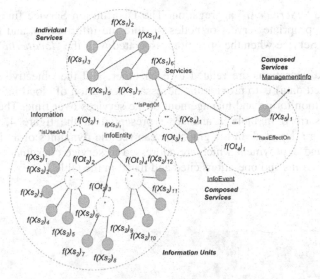

Fig. 3. InfoEntity with linked data for service-oriented control operations

The Table 1 contains the operations that are controlled by InfoEntity, the Service Editor is the Service Interface acting as the application that creates the new service and the Service Creator is the Service Interface acting as application that defines services.

Table 1. Control and representation of services interactions - UML model

Service (Offereed as InfoEntity)	Operation $f(Ot_2)_3$	Activty $f(Xs_1)_2$	Linked Target (Composed Services)
Application	whenEventArrives	Distribution	Service Distributor
	aServiceOn	Maintenance	Service Maintenace
	asCodeRelease	Editor	Service Editor
	startOf	Invocation	Service Invocation
	sServiceChange	Creator	Service Creator
Invocation	startOf	Application	Service Application
	aServiceStart	Editor	Service Editor
	aServiceAt	Execution	Service Execution
Distribution	eventFor	Maintenance	Service Maintenance
	aServiceNewr	Editor	Service Editor
	whenEventArrives	Application	Service Application
	aServiceOn	Consumer	Service Consumer
	whenServiceChange	Assurance	Service Assurance
Execution	locatedIn	Consumer	Service Consumer
	startOf	Assurance	Service Assurance
	aServiceAt	Invocation	Service Invocation
Assurance	whenServiceChange	Distribution	Service Distribution
	startOf	Execution	Service Execution
	whenServiceOff	Distribution	Service Consumer
Maintenance	aServiceOn	Application	Service Application
	eventFor	Distribution	Service Distributor
	aServiceChange	Editor	Service Editor

As an Example to understand the content of this table, assume that the service for deploying and updating the service code in certain cloud-based Application $f(Xs_1)_1$ has been created. The deployment activity for this code result in the generation of an event named *"aServiceOn"*, which instantiates a service with a relationship u operation $f(Ot_n)_m$ between the Application $f(Xs_1)_2$ and Maintenance $f(Xs_1)_3$. This in turn causes the data stream to be distributed via the Invocation Service $f(Xs_1)_4$ as defined by the *"aServiceAt"* aggregation. The Distribution Service finds the nearest and/or most appropriate servers or nodes to store the information, and then deploys the Execution Service when the operation associated with the *"locatedIn"* aggregation is instantiated.

The service operations are related to each other, and the objective to build this relations (linked data) is to provide the necessary activation of cloud infrastructure to guarantee the monitoring and management of the services over time. The service and operations described in table 1 and are represented in the figure 4, where these relationships, are depicted. In the graph we can see the dynamics of a virtual service provisioning and deployment which are addressed to service consumers by using the services descriptions and operations cited and listed in the table 1.

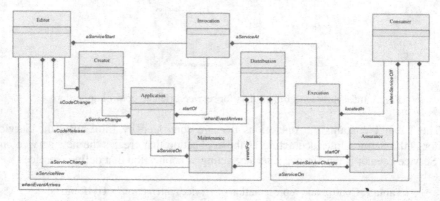

Fig. 4. InfoEntity with linked data for service-oriented control operations

Another example of service composition by using linked data is when a service invocation arrives, as signalled in the form of one or more application events, "whenEventArrives", "eServiceNew and *aServiceChange*" or "", the Invocation Service detects these events as indication of a data variation, and then instantiates the initialization of the service by instantiating the operation *"aServiceStart"*. The next phase to be performed is the execution of the service. Any location-specific parameters are defined by the *"locatedIn"* aggregation.

The service Execution implies the deployment of service code, as well as the possible evaluation of new data to monitor and manage the newly instantiated service. Monitoring is done using the Service Consumer manager interface, as it is the result of associations with execution. If maintenance operations are required, then these operations are performed using the appropriate applications, as defined by the *"aServiceOn"* aggregation, and completed when the set of events corresponding to the association *"whenServiceOff"* is received. Any changes required to the service code and/or polices for controlling the service are defined by the events associated with the *"whenServiceNew"* and *"aServiceChange"* associations.

The real sense of this representation defines how the descriptions and concepts are contained in information units and described as operations and services triggered. Event are used to control virtual infrastructures as part of control operations by using the proposed linked data mechanism. The formal description of the services, operation and linked targets has been done by using OWL grammar. The reason for using OWL is to provide a number of standard language tools for parsing and editing the data. This enables new adopters to use a tool or set of tools that is best suited to their needs. OWL is used to define the set of concepts and constraints imposed by the information model over which it defines instances.

The RDF-Schema (RDFS) [33] language emerged as a set of extensions to provide increased semantics of RDF by providing basic ontological modelling baseline, like classes, properties, ranges and domains. Finally this extensive use of XML, and the resulting use of RDF extensions, aimed to improve the expressiveness of this ontology.

4.2   Virtual Infrastructure Approach

A functional diagram for functional architecture supporting the meta-Ontology with its functional components as initial approach is depicted in Figure 5. This diagram shows the interactions between the main components. The architecture controls with a certain high level domain-based view, how service composition is performed, thus the control of the behavior is considering like added value functionalities using high level and formal representations expanding operations in other service application domains.

The depicted architecture use monitoring information to manage service operation and instructions with ontology-based information models using linked data mechanisms. Based on previous implementation experience [26]; [27] this architecture allows adaptability and dynamism to cloud services with the advantages of incorporating performance information from applications and inputs from users in form of events by using InfoEntity Class.

This functional approach offers the advantage in functionality to orchestrate system behaviour using data from business, infrastructure, and other constituencies beyond of the previously defined in the data model. By using this approach the formal models representing data can be translated and integrated as information, machine-based learning and reasoning engines are used to make the correlation between data models. In this particular approach we translate data from application-specific data models into a cloud application- and infrastructure control-neutral forms to facilitate its integration with other types of cloud systems. The key difference in this architecture relies in the usage of *semantics* to perform decision processes.

Performance values are translated into events to be co-related with the system's behaviour and then learned to make the system react when the same events occur. Linked events trigger and control each set of independent related events thus certain level of autonomy is achieved. The service composition process involves analyzing the triggering events expressed in an appropriate interoperable language via service coordination and decision-making integration, and matches them to service management and control level available [22] with the difference of this component using semantic descriptions to co-relate events with particular kind of conflicts that must be identified and evaluated. A detailed Semantic-Based Service Control Engine (S2CE) and its components as part of Functional Architecture is out of the scope of this paper, however implementation results are being analyzed and interaction between different domain events tested successfully, Details can be found in [15].

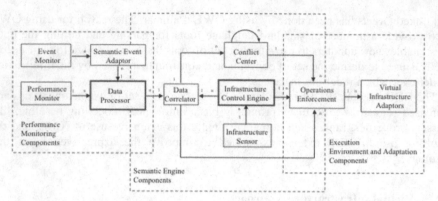

Fig. 5. Semantic-based service control engine - functional compoments

5   Background on Cloud Service Composition

The service composition is a complex process; it implies the identification of service features and elements, as well as it implies the possible evaluation of operation and functionality before the new service can be composed. Thus it can be regulated by semantic rules where if multiple operations are required, then these operations are performed using the appropriate applications, as defined by service composition rules and/or polices defined by the data associations.

In this sense, there are some interesting approaches, some of them following linked-data principles some others SOA principles; the commonality in all of them is the nature of information sharing between the different components or subservices. [35] introduces an approach where a matching algorithm SMA between cloud computing services of multiple input/output parameters is used. The algorithm considers the semantic similarity of concepts in specific lexical parameters. Particularly a service composition algorithm Fast-EP and the improved FastB+-EP were used as reference. Within this approach QoS information is utilized to rank the search results and it is shown based on experiment that this approach has better efficiency of service composition than other traditional approaches.

In other interesting approach [9] concentrates on cloud service provider requirements and their mappings with the cloud infrastructure resources in order to automate the service composition process. Founded on agent-based applications their propose a three-layered multi-agent system which by using self-organizing principles the agents make use of contract net protocol to evolve and generate adaptive service compositions. This approach demonstrates how by composing the incomplete information and make a comparison with available information about resources operations for generating a new service can be to allocate.

Another approach [17] make use of reusing pieces of information at a large scale can be exploited by harnessing the collaborative links. This approach concentrate more in a reference architecture and technical design for representation platform where sharing information and search for related pieces can act as the link generator which foster the reuse of service pieces for composition and the integration of cloud services.

Further activities have been proposed [6] more in the sense of what cloud computing can offer for new services definition rather than for re-using of services that are suitable to host new enterprise services. But while these different approach concentrates on offerings new services, even if they provide clear benefit to particular corporations, it is limited the capacity of what offers they have for composing services. The fact that applications or service systems cannot post the information they can offer as a sub-service does not help to scale or generate new enterprise enriched services. From this point of view and as an inherent feature in cloud systems, service composition is restricted or limited. However this last has not to be understood as a weakness, it is a particular and specific service-goal orientation in how to cope with the service definition and their requirement. It is just meaning that in cloud systems, it simply means composition is limited for regulations or policies and not for computing resources. In this paper we just concentrate on describing the alternatives and not to compare one or other design approach.

6 Conclusions

Information sharing is a crucial activity to satisfy the requirement in convergence service and communication systems. Implications for composing services and virtual infrastructures management are still under research (service composition in cloud). We have studied and demonstrated how formal representations of linked data can be used for modelling service compositions for cloud infrastructures.

In cloud infrastructures (virtual machines) high demands of information interoperability and of data link are demanded to satisfy service discovering and services composition requirements being controlled by diverse, heterogeneous systems and thus make more dynamic the perform of cloud-based system.

We have demonstrated with this formal representation we can use linked data mechanisms to construct relations between different functions and where diversely same information is used to for example describe different systems performance in form of information unites like data instances or likewise to compose management operations in the form of events used an by controlling virtual infrastructures.

The approach presented in this paper is based on studying linked data mechanisms and looking for alternatives to solve the complex problem of sharing information. Particular focus has been described into cloud environments as cloud computing is one of the most hot areas not only from a service composition and technology deployment perspectives else from a infrastructure and service management too.

Remaining research challenges regarding information model extensibility and information dissemination conduct our attention to continue our activity towards virtual infrastructure management, perform more cloud service control experiments and look for full linked data representations for service composition in cloud environments.

Acknowledgements. This work has been partially supported by the European Commission under contract number FP7-ICT-2011-7-287305 (OpenIoT), by Science Foundation Ireland under grant number SFI/08/CE/I1380 (Lion-II) CSET-DERI Centre for Science, Engineering and Technology - Digital Enterprise Research Institute. This work is a contribution to SFI-FAME SRC (Federated, Autonomic Management of End-to-End Communications Services - Scientific Research Cluster) Science Foundation Ireland SFI 08-SRC-I1403.

References

1. Architecture Design Project for New Generation Network, http://akari-project.nict.go.jp/eng/index2.html
2. Blumenthal, M., Clark, D.: Rethinking the design of the Internet: the end to end arguments vs. the brave new world. ACM Transactions on Internet Technology 1(1) (August 2001)
3. Clark, D., et al.: NewArch: Future Generation Internet Architecture, NewArch Final Technical Report, http://www.isi.edu/newarch/
4. Clean Slate Program, Stanford University, http://cleanslate.stanford.edu
5. Decker, S., Hauswirth, M.: Enabling Networked Knowledge. In: Klusch, M., Pěchouček, M., Polleres, A. (eds.) CIA 2008. LNCS (LNAI), vol. 5180, pp. 1–15. Springer, Heidelberg (2008)
6. Deloitte Technical Report on Cloud Computing "A collection of working papers Demystifying Clouds: Exploring Cloud and Service Grid Architectures" by Thomas B Winans and John Seely Brown, Deloitte (2009)
7. Feldmann, A.: Internet clean-slate design: what and why? ACM SIGCOM Computer Communication Review 37(3) (2007)
8. Gruber, T.: Towards "Principles for the Design of Ontologies Used for Knowledge Sharing". International Journal of Human-Computer Studies 43(5/6), 907–928 (1995)
9. Gutierrez-Garcia, J.O.: Self-Organizing Agents for Service Composition in Cloud Computing. In: 2010 IEEE Second International Conference on Cloud Computing Technology and Science (CloudCom), pp. 59–66 (2010), Digital Object Identifier: 10.1109/CloudCom.2010.10
10. Hauswirth, M., Pfisterer, D., Decker, S.: Making Internet-Connected Objects readily useful. In: Interconnecting Smart Objects with the Internet Workshop, Prague (2011)
11. Horridge, M., Knublauch, H., Rector, A., Stevens, R., Wroe, C.: A Practical Guide to Building OWL Ontologies using the Protégé-OWL Plugin and CO-ODE Tools Edition 1.0, Manchester University (August 2004)
12. Irish Future Internet Forum, http://www.futureinternet.ie
13. Kalinichenko, L., Missikoff, M., Schiappelli, F., Skvortsov, N.: Ontological Modeling. In: Proceedings of the 5th Russian Conference on Digital Libraries, St. Petesburg, Russia (2003)
14. Katsiri, E.: Middleware Support for Context-Awareness in Distributed Sensor-Driven Systems. PhD Thesis, Engineering Department, University of Cambridge. Also published as Technical Report n.620, Computer Laboratory, University of Cambridge (February 2005)
15. Keeney, J., Conlan, O., Holub, V., Wang, M., Chapel, L., Serrano, M.: A Semantic Monitoring and Management Framework for End-to-end Services. In: Proceedings 12th IFIP/IEEE International Symposium on Integrated Management, Dublin, IE, May 23-27 (2011)
16. Moore, E.: Policy Core Information Model-Extensions. IETF Request for comments (RFC 3460) (January 2003), http://www.ietf.org/rfc/rfc3460.txt
17. Motahari-Nezhad, H.R., Li, J., Stephenson, B., Graupner, S., Singhal, S.: Solution Reuse for Service Composition and Integration. In: Proceedings on SERVICES 2009 Proceedings of the 2009 Congress on Services – I, pp. 400–407. IEEE Computer Society, Washington, DC (2009) ISBN: 978-0-7695-3708-5
18. NSF-funded initiative to rebuild the Internet, http://www.geni.net/
19. OWL Ontology Web Language, http://www.w3.org/2004/OWL
20. OWL-s, http://www.daml.org/services/owl-s/

21. SFI-FAME SRC- Scientific Research Cluster: Federated Autonomic Management of End to End Communications Sesrvices, http://www.fame.ie/

22. Schönwälder, J., Straub, F.: Next Generation Structure of Management Information for the Internet. In: Proceedings of 10th IFIP/IEEE DSOM International Workshop, Zürich (1999)

23. Serrano, J.M.: Applied Ontology Engineering in Cloud Services, Networks and Management Systems, 222 p. Springer Publishers, Hardcover (2012) (to be released on March 2012) ISBN-10: 1461422353, ISBN-13: 978-1461422358

24. Serrano, M., Strassner, J., ÓFoghlú, M.: A Formal Approach for the Inference Plane Supporting Integrated Management Tasks in the Future Internet. In: 1st IFIP/IEEE ManFI International Workshop, Long Island, NY, USA, June 1-5 (2009)

25. Serrano, J.M.: Management and Context Integration Based on Ontologies for Pervasive Service Operations in Autonomic Communication Systems. PhD Thesis, UPC (2008)

26. Serrano, J.M., Serrat, J., Strassner, J.: Ontology-Based Reasoning for Supporting Context-Aware Services on Autonomic Networks. In: 2007 IEEE/ICC Intl. Conference on Communications, Glasgow, Scotland, UK, June 24-28 (2007)

27. Serrano, J.M., Serrat, J., O'Sullivan, D.: Onto-Context Manager Elements Supporting Autonomic Systems: Basis & Approach. In: 1st IEEE MACE International Workshop as Part of ManWeek 2006, Dublin, Ireland, October 23-27 (2006)

28. SID - Shared Information Data model, http://www.tmforum.org/InformationManagement/1684/home.html

29. Sloman, M.: Policy Driven Management for Distributed Systems. Journal of Network and Systems Management (1994)

30. Strassner, J., Foghlú, M.Ó., Donnelly, W., Agoulmine, N.: Beyond the Knowledge Plane: An Inference Plane to Support the Next Generation Internet. In: IEEE GIIS 2007, July 2-6 (2007)

31. Strassner, J.: Policy Based Network Management. Morgan Kaufmann (2004) ISBN 1-55860-859-1

32. TMF, The Shared Information and Data Model – Common Business Entity Definitions: Policy, GB922 Addendum 1-POL (July 2003)

33. W3C Website, http://www.w3c.org/rdf

34. Westerinen, A., Schnizlein, J., Strassner, J.: Terminology for Policy-Based Management. IETF Request for Comments (RFC 3198) (November 2001)

35. Zeng, C., Guo, X., Ou, W., Han, D.: Cloud Computing Service Composition and Search Based on Semantic. In: Jaatun, M.G., Zhao, G., Rong, C. (eds.) CloudCom 2009. LNCS, vol. 5931, pp. 290–300. Springer, Heidelberg (2009)

Enterprise Architecture Executable Patterns: Enterprise Architecture – Replacing Models with Executable Patterns

Thomas A. Tinsley

Tinsley Innovations
http://www.SelfServiceIT.com

Abstract. Using executable Enterprise Architecture patterns replaces the traditional engineering approach of using models to guide software development. This is accomplished by combining Enterprise Architecture principles, ontology reasoning, and Service Component Architecture into an executable environment. Creating this environment is the motivation for the OTTER project (Ontology Technology That Executes Real-time). This environment sets the Enterprise Architecture as the foundation for component service development and execution. In using this environment, Enterprise Architecture cannot be overlooked or bypassed during information systems development. This results in reducing the complications of application integration and data-sharing which reduces costs and problems.

Protégé [1] is used to define the layers of the Enterprise Architecture. These layers are mapped to Service Component Architecture standards to provide real-time execution of processes. Information access and service component access are both provided by OTTER using OWL data expressions. This use of OWL data expressions is an alternative to using XML web services for service access and SQL for relational database access.

Keywords: OWL, Protégé, Service component architecture, Service data object, Data access service.

1 Introduction

Replacing the engineering approach of modelling used in Enterprise Architecture is a major initiative addressed in stages. A prototype is the first step towards this goal and is embodied in the OTTER project that is presented in this paper.

The OTTER prototype integrates the proven application of Enterprise Architecture (EA) principles, OWL business definitions, and Service Component Architecture (SCA) [5] as shown in Fig. 1. This integration turns the otherwise static models of EA into executable patterns. This is done by bridging the class expression language of OWL to the Service Data Object standard of Service Component Architecture (SCA).

Applying Enterprise Architecture as a foundation for development and execution was not found in other publications after months of research. The OTTER project was then initiated to prove the validity of this approach.

---

[1] This work was conducted using the Protégé resource, which is supported by grant LM007885 from the United States National Library of Medicine.

A. Fred et al. (Eds.): IC3K 2011, CCIS 348, pp. 294–306, 2013.
© Springer-Verlag Berlin Heidelberg 2013

Fig. 1. Executable Enterprise Architecture

The real value of OTTER is in the layered extensibility of the executable EA patterns. Through extensions, business principles can be captured and very specific business applications can be defined. The *include ontology* capability of OWL provides the support for layering the architecture with each layer defined as a separate OWL ontology.

OWL provides an outstanding language for defining things and axioms that can be tested through reasoning. This capability has been proven by its application in multiple industries. This also makes it an excellent choice for defining an Enterprise Architecture.

Many major providers of infrastructure software have adopted SCA as their component model. These providers include IBM, Oracle, and SAP. They have adopted this standard across all of their infrastructure products.

The OTTER prototype creates a bridge between OWL and SCA by mapping OWL class expressions to the SCA standard of Service Data Objects (SDO). By bridging at this technology level, business service messages and component access to data is

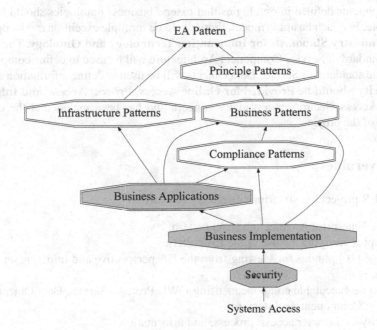

Fig. 2. Layered Ontologies

accomplished using class expressions. This replaces the need for XML to define Web Services and SQL to access data repositories.

The executable capability of the EA pattern and the patterns that extend it change these static models to implemented services. These implementations can be applied and reused by multiple applications.

The EA Pattern provides the classifications and high-level property definitions for all of the components within the business ontology. An overview of how the models are layered is shown in Fig. 2. The shaded patterns only contain individuals. The double-lined octagons indicate multiple ontologies while the single-lined octagon indicates only a single ontology.

The principles and rules for operating specific vertical lines of business would be defined by extending the EA Pattern. Principle patterns such as Accounting, Marketing, and Finance could be leveraged by multiple vertical businesses. Multiple vertical lines of business models can be combined for automatic integration.

Access to the systems would be given based upon the defined authentication and authorization defined by the security pattern.

2   GOALS

The OTTER project has a primary goal of transforming static EA models into executable patterns in a secure environment. This primary goal is accomplished by applying the following:

1. **Provide the Infrastructure for Real-time Execution.**
   Components defined in ontologies that extend business ontologies should be executable. For each business model, there could be multiple execution model options.
2. **Use Industry Standards for Information Technology and Ontology.** The industry standard of Service Component Architecture will be used to define components and the standard for Service Data Objects will be used to define information access.
3. **Security Should be Provided for Online Access, Process Access, and Information Access.** Security is a primary concern and has been included at the earliest stage of the project.

3   Deliverables

The OTTER project has six primary delivery objectives:

1. A core Enterprise Architecture (EA) pattern.
2. Examples verifying the use of the EA pattern.
3. Protégé [16] plugins for viewing from the EA perspective and initiating an HTTP server.
4. Prototype executable components using OWL Protégé, Service Data Objects, and Service Component Architecture.
5. Security for browser access, process, and information.

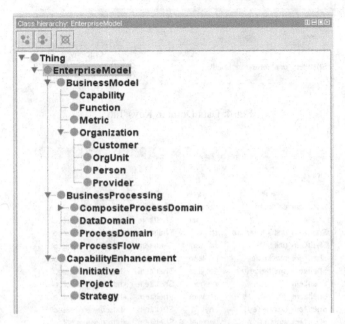

Fig. 3. Enterprise Architecture Pattern

3.1   A Core Enterprise Architecture (EA) Pattern

The EA Pattern currently used in the prototype has the high-level class structure defined in Protégé as shown in Fig. 3. The Enterprise Model is divided into three major categories. The first one, Business Model, contains the information about the business capabilities, functions, metrics, and organization. The second category, Business Processing, describes the processes used by the business, the data used by the processes, the flow of information, and the composite processes that may or may not be automated. The third category, Capability Enhancement, includes the strategies, initiatives, and projects to enhance the capabilities of the organization.

3.2   Example Verifying the Use of the EA Pattern

As a test to validate the EA Pattern, the published form of the Association of Retail Technology Standards (ARTS)[3] model was used as the base model to create multiple OWL files.

1. This model includes the high-level processes of Store, Distribution, and Home. Within these processes are ten sub-processes and forty-two information flows.
2. The data model includes nineteen subject areas and the subject areas include forty-seven sub-subject areas.
3. Additionally, twenty composite components are defined that referenced the processes, flows, and data.

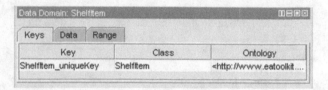

Fig. 4. Data Domain Keys Tab

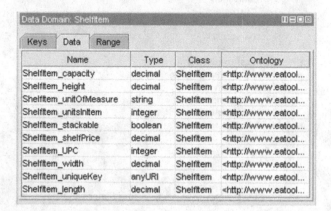

Fig. 5. Data Domain Data Tab

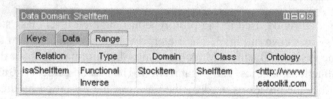

Fig. 6. Data Domain Range Tab

3.3 Protégé Plugins

There are four plugins implemented for EA viewing. They are included in Protégé under the tab "EA Pattern".

Data Domain. All data properties associated with the selected class are shown in this plugin. This includes the data and object properties of super classes. There are tabs for keys, data properties, and range object properties as shown below in the examples in Fig. 4, Fig. 5, and Fig. 6.

Process Domain. This view shows the processes using two tabs. One shows the sources of information for a process as shown in Fig. 7 and the other tab shows the consumers of information from the process shown in Fig. 8.

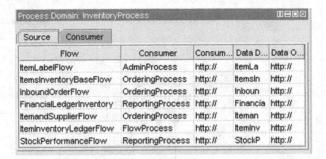

Fig. 7. Process Domain Sources Tab

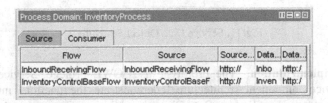

Fig. 8. Process Domain Consumers Tab

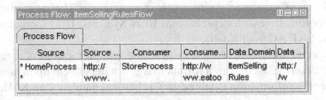

Fig. 9. Process Flow Tab

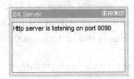

Fig. 10. Server View Tab

Process Flow. This view only has one tab that shows the sources and consumers of an information flow. This is demonstrated in Fig. 9.

EA Server. This view includes a log of alerts for the HTTP server shown in the example in Fig. 10.

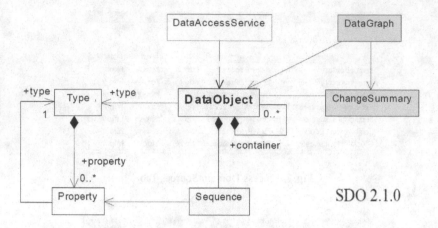

Fig. 11. Service Data Object Structure

Prototype Implementation. The Prototype uses OWL Protégé, Service Data Objects, and Service Component Architecture. It demonstrates the functional integration of OWL with the standards of Service Data Objects and Service Component Architecture.

Service Data Objects. Service Data Objects (SDO) is a specification for an application interface for accessing data to be used by programmers. The specification was included as part of the Service Component Architecture specifications. These specifications were developed in coordination with experts from BEA Systems, IBM Corporation, Rogue Wave Software, SAP AG, and Xcalia.

The SDO specification was created to describe complex business data structures. The intent is to describe the data in a format that is not dependent upon how the data is stored. Whether the information is stored in a relational database, an XML document, or other types of structures should not make any difference to the programmer.

In order to read and write different data formats, the programmer uses a Data Access Service (DAS) component that supports the specific format being processed. For example, the programmer would use a relational database DAS to access data in a relational database. To access an XML file, the programmer would use an XML DAS. In OTTER, a DAS is used to access the individuals in the business ontologies.

Standards for Programmatic Data Graph Manipulation. In the SDO specification, the API for Java is shown in Fig. 11. This diagram from the specification shows how a Data Object is defined by properties and types. It also shows how a Data Object can be a container for multiple Data Objects.

By definition, an SDO is a data graph. Since all structures in OWL are data graphs, the use of SDO to view and change the content of an OWL ontology can be accomplished in a direct manner.

SDO Serialization. The metadata and content of an SDO is serialized in OTTER by using the JSON (Javascript Object Notation) standard. This provides the serialization needed to make HTTP service requests and to return an SDO to a browser.

Ontology Data Access Object. The prototype includes an OWL Data Access Service (DAS) with the following capabilities:

- **Get_JSON.** Returns a JSON string of the data within a specified data graph.
- **Get_DataGraph.** Creates an SDO data graph from a class expression and populates it with the selected individuals.
- **Create_DataGraph.** Creates an SDO data graph from a class expression without any individuals.
- **Load_JSON.** Populate the individuals in a selected data graph from the JSON content.
- **Get_JSON.** Returns a JSON string of the data within a specified data graph.

These capabilities are accomplished by using Protégé APIs to describe the metadata for each data object. This is accomplished using an OWL class expression and following a stepwise process.

1. For each class in the expression, identify both the data properties and the object properties for that class. These properties must also be included in the expression.
2. The root classes of the data graph are identified using the reasoner.

This process produces a metadata tree structure of the data graph. Using this metadata, the individuals from the ontologies or from a JSON string can be loaded into the data graph. The programmer can then access the structure using the property names as prescribed by the SDO standard.

Example of SDO from a Class Expression. In Protégé, a class expression is used to identify a subset of individuals. In OTTER, the class expression is used to identify all of the individuals within the object properties in the class expression. This results in a data graph of sets rather than a single set. In the EA Pattern, "Person" is a class that has a data property of "hasUserID" and an object property of "inOrganization". The "Organization" class has the data property of "hasOrgUnitID." Using these classes and properties, the following can be passed to the OWL DAS as a query:

```
query=Person and (hasUserID some string) and
    (inOrganization some (OrgUnit and hasOrgUnitID some string))
```

The process extracted the following properties for the query:

```
Data Property List
    hasOrgUnitID
    hasUserID

Object Property List
    inOrganization
```

Using the individuals returned from the query and the property lists, the OWL DAS yields the following tree structure:

```
Datagraph content:
  Individual= "person1"
    property:inOrganization
      Individual= "Org1"
        property:hasOrgUnitID
          value="Organization1"
      Individual= "Org2"
        property:hasOrgUnitID
          value="Organization2"
    property:hasUserID
      value="admin"
  Individual= "person2"
    property:inOrganization
      Individual= "Org3"
        property:hasOrgUnitID
          value="Organization3"
    property:hasUserID
      value="anonymous"
```

As the results show, "person1"is in two organizations, "Organization1" and "Organiza-
tion2" and has a user ID of "admin". The individual, "person2", is in one organization,
"Organization3".

Currently, the prototype will not add properties when the "inverse" operation is used
in the query. The workaround is to always provide an inverse property in the ontology and use it in the query. In the example given, the individuals in the "Organization"
class have an object property of "hasPeople" with an inverse of "inOrganization". The
individuals of "person1" and "person2" do not actually have the property of "inOrganization" defined directly, but are rather determined by the reasoner as inferred properties.
Inferred properties are handled in the same way by the OWL DAS as direct properties.

Security for Browser Access, Process, and Information. Security functions to restrict
access of individuals to processes and information. This makes security straight-forward,
since OWL ontologies are primarily about providing restrictions to defined information.

In the prototype, the security information is in a separate OWL ontology. It includes
the business model security information and may exist as multiple included ontologies.
These ontologies can be secured separately to provide the privacy to protect the security
information.

In the prototype, the browser access is supported by HTTP basic authentication to
get the user id and password. At the time of logon, the OWL DAS is used to retrieve the
individual in the ontology with the user id and password provided. If they are found, access is approved. In the EA Pattern, security access is centered upon the "Organization"
class. The OntoGraf in Fig. 12 shows the property links.

To provide security access for a process, a "Person" can only access an "Assembled
Component" when it is used by a "Function" provided by an "Organization" that has
people that includes this "Person".

Information access is provided by defining Access Control Lists (ACL). This is accomplished by using the property "hasSecurityAccess" within the "EnterpriseModel".
The "hasSecurityAccess" property has four sub-properties: "canCreate", "canDelete",

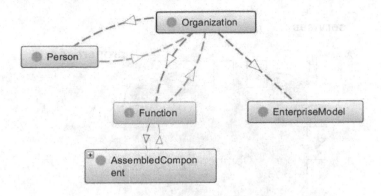

Fig. 12. Security Classes

"canRead", and "canUpdate". Each of these sub-properties is defined as a list of ontologies. When a query is made by the OWL DAS, only the authorized read ontologies are included as the ACL.

Following the standards for processing an SDO, when a changed or new data graph is applied back to the ontology, the "canCreate", "canDelete", and "canUpdate" ontology lists will be used as the ACL.

Service Component Architecture. The Service Component Architecture specification defines the APIs for accessing and constructing service components. The SCA specification was developed in coordination with experts from BEA Systems, IONA Corporation, IBM Corporation, SAP AG, Sun, and Tibco Software. In SCA, components are defined in the specification as shown Fig. 13. Each component has properties, services, and references to other services.

Components can also be assembled from other components. These components can communicate within the same technical domain. This is shown in the diagram from the specification in Fig. 14. OTTER implementation of SCA supports the loading of java components and initializing their service interfaces according to their ontology definition. Initialization includes creating the data graphs used for the request and response of each service. When a person logs on to OTTER, a session is created that maintains the person's authorized service and ontology access.

An HTTP "get" to a service will respond with a JSON containing the request metadata. This information can be used to construct a request to the service. The HTTP "post" is used to send the request and receive the response. The response to a "post" returns both the metadata of the data graph response and the data within the data graph.

3.4 Work Outstanding

The OTTER project is an active project with continuing work as listed below:

- **Patterns**
    - Include patterns to demonstrate principle patterns.
    - Incorporate the SOA Ontology published by the Open Group. [15]

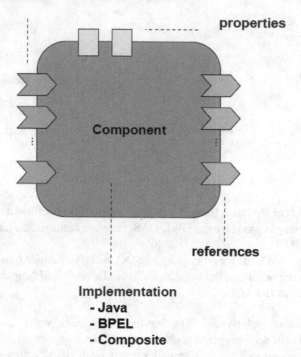

Fig. 13. SCA Component

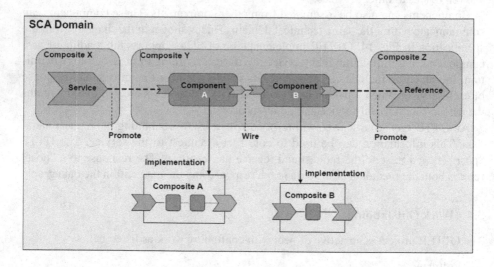

Fig. 14. SCA Domains and Assembly

- **Prototype Enhancements**
  - Support all OWL data values in SDO. (The current support includes string, integer, and double.)
  - Extend the SDO to write changes back to the ontology.
  - Add browser-based model visualizations utilizing 2d and 3d graphics.
- **Coordination**
  - Publish the prototype for review.
  - Discussion with existing Enterprise Architecture organizations to establish a process.

4   Conclusions

Mapping OWL class expressions to SDO is possible. The OTTER prototype implementation proves that the concept is sound. The metadata for an SDO can be defined from an OWL class expression. Individuals can then be loaded into the SDO data graph structure directly from the ontology for data access or from JSON for service access.

Using class expressions to define data graphs can replace the use of XML for service definition and SQL for data access. This has been demonstrated in the prototype through the implementation of a Data Access Service (DAS) for ontologies in Protégé. DAS is used in SCA to create and access SDO data graphs.

The next step of writing individuals and their properties back into Protégé will provide the proof that OWL and SDO are compatible.

The value of the EA Pattern as a base ontology proved to be a requirement for implementing access to the ontologies. The value of defining the components of SCA using OWL provided a simplified interface for services.

This prototype is intended for concept demonstration only. If all aspects of the prototype are successful, another project will be initiated to provide an industrial-strength implementation. This implementation will require a real-time reasoner that can evaluate each change made to the ontology for accuracy. The reasoner will update only the existing inferred properties that have changed.

There were comments from those attending IC3K 2011 on the technical and ontology aspects of the OTTER project. The feedback received on the technical aspects of the concept have been positive. This is in contrast to the scepticism given on the potential of ontology sharing and reuse. This scepticism is founded upon the limited amount of reuse that currently takes place. Most ontologies are developed within a unique perspective of a domain for specialized processing. The anticipation is that having Enterprise Architecture as the perspective will encourage greater sharing and reuse.

With this mapping of OWL to SCA, the Enterprise Architecture pattern can be the foundation for service component development and execution. This will result in reduced expenses, less project time, and fewer errors. This could result in a paradigm shift in the quality of information systems.

References

1. Adams, M., et al.: Service Data Objects for Java Specification, Version 2.1.0 (2006),
   `http://www.osoa.org/display/Main/`
   `Service+Data+Objects+Specifications`
2. Allemang, D., et al.: FEA Reference Model (FEA RMO) GSA OSERA Deliverable Version
   1.0 (2005) (last update: January 2005)
3. ARTS Data Model. n.d., National Retail Federation,
   `http://www.nrf-arts.org/content/data-model`
4. Baclawski, K., et al.: Finding a Good Ontology: The Open Ontology Repository Initiative.
   In: Semantic Technology Conference (June 24, 2010)
5. Beisiegel, M., et al.: SCA Service Component Architecture Assembly Model, SCA Version
   1.00 (2007),
   `http://www.osoa.org/display/Main/`
   `Service+Component+Architecture+Specifications`
6. Chandrasekaran, B., et al.: Ontology of Tasks and Methods,
   `http://www.cis.ohio-state.edu/lair`
   & `http://www.swi.psy.uva.nl/usr/richard/home.html`
7. Guizzardi, G., et al.: Grounding Software Domain Ontologies in the Unified Foundational
   Ontology (UFO): The case of the ODE Software Process Ontology. Federal University of
   Espirito Santo (UFES), Vitoria, Brazil (2006)
8. Kleb, J., et al.: A Protégé 4 Backend for Native OWL Persistence. Fraunhofer IITB & FZI
   Research Center for Information Technologies at the University of Karlsruhe
9. Knublauch, H.: Protégé OWL Programmers Guide (2010),
   `http://protegewiki.stanford.edu/wiki/`
   `ProtégéOWL_API_Programmers_Guide` (last update: June 30, 2010)
10. Knublauch, H., et al.: Weaving the Biomedical Semantic Web with the Protégé OWL Plugin,
    `http://protege.stanford.edu`
11. Krotzsch, M., et al.: Semantic Web Modeling Languages Lecture III: More OWL. In: ESS-
    LLI 2009, Bordeaux (2009)
12. MacGregor, W., et al.: Information Security: An Ontology of Identity Credentials Part1:
    Background and Formulation. NIST National Institute of Standards and Technology, NIST
    Special Publication 800-103 Draft (October 2006)
13. Mrohs, B., J.Sc. (CS) et al.: WL-SF A Distributed Semantic Service Framework, Fraunhofer
    Institute for Open Communication Systems
14. Semy, S., et al.: Toward the Use of an Upper Ontology for U.S. Government and U.S. Military
    Domains: An Evaluation. Mitre Technical Report (September 2004)
15. The Open Group, 2010. Technical Standard, Service-Oriented Architecture Ontology. Pub-
    lished by The Open Group (October 2010)
16. Stanford Center for Biomedical Informatics Research (BMIR) at the Stanford University
    School of Medicine, `http://protege.stanford.edu`
17. Tinsley, T.A.: Enterprise Architects Masters of the Unseen City, Tinsley Innovations (2009)

Identifying Services from a Service Provider
and Customer Perspectives

Carlos Mendes, João Ferreira, and Miguel Mira da Silva

Instituto Superior Técnico, Technical University of Lisbon,
Avenida Rovisco Pais, Lisboa, Portugal
{carlos.mendes,joao.p.ferreira,mms}@ist.utl.pt

Abstract. Despite the remarkable growing of the services industry in the world economy, the services quality is still affected by gaps identified two decades ago. One of these gaps occurs when the service provider has a perception of what the customer expects that diverges from the real expected service. This difference can be caused by a poor service identification process and, more precisely, by who should be included in this process. Current solutions to identify services still have drawbacks, since they are not customer driven, are web services driven or lack specific processes. In this paper, we propose a service identification process based on the Design & Engineering Methodology for Organizations (DEMO). The proposal was evaluated by comparing two lists of services provided by a Human Resources department: one based on a description given by the head of the department and another based on the customers that use the department services. The differences between the two lists show the gap between the customers' expectations and the provider perceptions of those expectations. We conclude that both client and service provider should be included in the service identification process.

Keywords: Service Identification, Enterprise Ontology, DEMO, Service Catalog.

1   Introduction

Over the past few years, organizations have faced the challenge of providing services efficiently to their clients, whether they are enterprises or other departments within the same organization. Besides this constant pressure, there has been another growing business unit demand for new services and higher service levels [1]. Nowadays services mean jobs and growth, but the companies that have been leading the charge lack a strong conceptual foundation [2]. This lack contributes to the gaps [3] that reduce the services quality for, without a solution to specify it, it is difficult for the service providers and their customers to align their expectations about the services quality.

We believe that Enterprise Ontology (EO) [4] is a theory that can be the beginning of this needed theoretical background. Therefore, we are using EO and the corresponding methodology, the Design & Engineering Methodology for Organizations (DEMO), as the foundation for our most recent proposals [5].

A. Fred et al. (Eds.): IC3K 2011, CCIS 348, pp. 307–322, 2013.

In this paper, we show how to identify the possible mismatch between what a provider thinks it provides to its customers and what is actually supplied to customers, thus emphasizing the importance of a correct identification, definition and documentation of services, for instance, in a Service Catalog. The design and development of a Service Catalog involves multiple activities such as the correct identification of the services that will be defined in this document. However, nowadays, there are no formal methods that allow to correctly perform this step [6].

Our study was conducted using the Design Science Research Methodology (DSRM) that aims at creating and evaluating Information Technology (IT) artifacts intended to solve identified organizational problems, in which these artifacts may extend the knowledge base or apply existing knowledge in new innovative ways. In this paper, we present a research that is focused in the second option (i.e. apply existing knowledge in new innovative ways). This research method includes the following phases, described in detail in [7]: problem identification, objectives definition, design and development, demonstration, evaluation and communication.

This paper is structured as follows. We start by describing the problem concerning the identification of services in Section 2. Next, we provide a brief overview of the literature on ITIL and CMMI-SVC service models, on current methodologies to identify services and on DEMO methodology (Section 3). Subsequently, we explain our DEMO-based proposal to identify services (Section 4) and present an example, based on a real organization, where we applied our proposal (Section 5). We evaluated the obtained results using an empirical method and employees' feedback (Section 6) and we explain what we have learned from this project (Section 7). Finally, we present our conclusions (Section 8).

2   Problem Statement

This section corresponds to the problem identification and motivation phase of DSRM. While the Service Catalog is an element that is increasingly important for organizations, its implementation poses several risks [8]. One of the problems currently affecting organizations, and that this document proposes to solve, is the struggle to be able to identify what services they provide to achieve business goals and outcomes. Despite the effort made in recent years, a unified methodical approach for service identification has not yet been discovered [6]. The incorrect identification and definition of IT services, due to their structural role, will affect other processes since they are largely based on these services leading to a Service Catalog that is ineffective and provides no real value to any part of the organization [1].

If the services are not identified, defined and catalogued, or due to an ineffective management of customer expectations, overpromising and others, the communication about the service will not match the actual service delivery. Consequentially, the customers may perceive the services differently from what is communicated by the provider [9].

This paper is focused on finding evidences on who should be included in the service identification process because there seems to exist a belief among the service providers that they can identify their own services without theirs clients feedback.

3   Related Work

Since this document focuses on the identification of services, it is important to define what a service is. We give a brief overview of the multiple existent definitions of service and by DEMO methodology, used in our proposal. Next, we analyze both ITIL and CMMI-SVC, each containing processes that deal with Service Catalogs and service identification as well as two of the most studied service identification approaches by researchers: Business Components Identification and Service Oriented Architectures. Finally, we analyse DEMO, used in our proposal.

3.1   Definition of Service

The term "service" has many different definitions, depending on the context in which it is being used [10]. Several definitions of service are based on technology. Some definitions of electronic services (or e-services) use the Internet and/or workflow as a conduit to new revenue or task completion. A web service has been described as an aggregation of functionality published for use. However, these definitions raise some concerns: they are focused on technology and they tend to ignore conventional services, in what is called "web service tunnel vision". It is important to notice that services are not just about technology and also include nontechnology aspects [11].

In the service marketing literature, there is a wide range of definitions of what a service entails. Usually, a service is defined as an intangible set of benefits or activities that are sold out by one party to another. Its main features are: intangibility, heterogeneity, simultaneously produced and consumed and perishability [9].

In [12] the definition of service is based on the standard transaction pattern proposed in Dietz's DEMO [4]. Though a service has many similarities with a transaction in Enterprise Ontology, they are not equal: A service is a universal pattern of Coordination and Production acts, performed by the executor of a transaction for the benefit of its initiator, in the order as stated in the standard pattern of a transaction. When implemented it has the ability:

- to get to know the Coordination facts produced by the initiator;
- to make available to the initiator the Coordination facts produced by itself.

This definition of service just given is a very generic one, since it holds for two kinds of providers: human actors and IT systems. Services executed by human actors and IT systems only differ in the way they are implemented; human services are implemented by human beings, whereas IT services are implemented by IT systems [12]. These systems assist human actors in their activities; therefore parts of a human service may also be executed by IT systems.

3.2   Service Identification Techniques

As said, although a formal and unified method to identify services has not been discovered yet, some approaches have been proposed over the years to solve this issue.

Information Technology Infrastructure Library (ITIL) [13] and Capability Maturity Model Integration for Services (CMMI-SVC) [14] provide a set of best practices for service provider organizations. While ITIL aims at controlling and managing all aspects of IT related operations, CMMI-SVC is a maturity model and a process improvement approach for general service providers. Even with different focus, both have specific areas which deal with Service Catalogs and, consequently, with service identification. However, they mostly deal with what processes should be implemented, and not so much with how they can be implemented, not providing a specific process to identify services.

In order to promote software reuse, Component-based Software Engineering (CBSE) techniques have been adopted to ease the development of large-scale complex information systems [15]. As described in [16], there are three kinds of service identification techniques, based on components: Domain Engineering based methods, CRUD Matrix based methods and Cohesion-Coupling based Clustering Analysis methods. Instead of identifying services directly, these identify the components and, since each component provides services to the exterior, they claim we can also identify its services. Nevertheless, a specific process to perform this task is not presented.

Service Oriented Architecture (SOA) is an architectural approach for designing, architecting and delivering enterprise applications that support business operations as a set of meaningful services. Many researches are suggesting various methodologies to guide the migration to SOA, each with its own approach to service identification. In [17] there is a review of some SOA methodologies such as SOMA [18] and SOAF [19]. These methodologies provide a solid basis to achieve SOA, but they do not describe all phases very thoroughly or clearly. Moreover, they are technology-based, focusing on web services.

3.3   DEMO

Design & Engineering Methodology for Organizations (DEMO) is a methodology for modeling, (re)designing and (re)engineering organizations and networks of organizations [4]. DEMO aims to develop high-level and abstract models of the construction and operation of organizations, independently of their actual implementations, by focusing on the communication patterns between human actors, i.e., models of organizations from a responsibility and communication oriented perspective. The theory that underlies this methodology is called Enterprise Ontology and consists of four axioms: Operation, Transaction, Composition and Distinction.

The Operation axiom states that an organization consists of human beings, in their role of social individuals or subjects, who achieve their goals by performing acts. A subject fulfilling an actor role, which is defined as a particular *amount* of authority and responsibility, is called an actor. An actor performs two kinds of acts: Production acts (P-acts) and Coordination acts (C-acts). On the one hand, by performing P-acts, the actors contribute to bringing about the goods or services that are provided or delivered to the environment of the organization. On the other hand, by performing C-acts, actors enter into and comply with commitments and agreements towards each other regarding the performance of P- acts. The result of successfully performing a P-act is a Production fact or P-fact, and the same applies to a C-act, creating a Coordination fact or C-fact.

The Transaction axiom states that C-acts are performed as steps in a universal pattern, called transaction, to successfully complete a P-act. Each transaction distinguishes two actor roles: the initiator, who starts the transaction and might complete it, and the executor, who is responsible for the performance of the P-act and the creation of the respective P-fact. A transaction evolves in three phases: the Order phase (O-phase), the Execution phase (E-phase) and the Result phase (R-phase). In the Order phase, the initiator and the executor negotiate about the intended result of the transaction (P-fact that the executor is going to create); in the Execution phase, the P-fact is produced by the executor; and finally, in the Result phase, the initiator and the executor negotiate and discuss the result of the transaction (P-fact actually produced).

The Composition axiom describes how these transactions can interact. According to this axiom, any transaction is either enclosed in some other transaction, initiated by an external party to the organization or a self-activated transaction. If there is an enclosed transaction, an information dependency usually exists between the enclosing and the enclosed transaction.

The Distinction axiom acknowledges three human abilities called Performa, Informa and Forma which are exerted both in C-acts and P-acts. The Forma ability concerns the form aspects of communication and information (Datalogical layer); the Informa ability is related to the content aspects of communication and information, fully abstracting from the form aspects (Infological layer); the Performa ability involves the creation of new, original things, directly or indirectly by communication (Ontological layer). This last ability is considered as the essential human ability for doing business, and is the one in which DEMO focuses on.

We just gave a short summary of the Enterprise Ontology and discussed the parts relevant for our service identification proposal. A complete overview of the theory is available in book [4] and many others publications (List: http://www.demo.nl/publications/).

4   Proposal

This section and the previous one correspond to the design and development step of DSRM.

In order to solve the problem described in Section 2, we propose to use the DEMO methodology to identify the services provided by organizations, or their departments, to their customers. Since best practices cannot help us, because they are too general and do not specify how to identify services, and the already existent methods which allow this task are focused on technology, we decided to use DEMO and the definition of service based on this methodology, because it provides abstract and high-quality models and has a strong theoretical foundation [20]. Moreover, this methodology focuses on what really matters, i.e. the business layer, not considering the implementation details which are secondary to clients.

We propose to use the following steps, already defined by DEMO in [4], to find the services of an organization:

• **Enterprise Description:** A textual description which summarizes the actions performed by the service provider to fulfill the customer's requests;

• **Performa-Informa-Forma Analysis:** Identification of the three kinds of human abilities (Performa, Informa, and Forma) performed in the context of the organization, according to the Distinction axiom, using the text of the previous step;

• **Coordination-Actors-Production Analysis:** The identified Performa items are split into C-acts/facts, P-acts/facts and actor roles who perform those acts, according to the Operation axiom;

• **Transaction Pattern Synthesis:** Identification of each transaction type, and the corresponding result, based on the identified acts/facts, according to the Transaction axiom;

• **Result Structure Analysis:** Check if there are any dependencies between the identified transaction types. Generally, these dependencies occur when the executor of a transaction is the initiator of another (inner) transaction, as already explained in the Composition axiom;

• **Actor Transaction Diagram/Service Identification:** Identification of the initiator and executor actor roles of each transaction type. When this mapping between transactions and actor roles is complete, it is possible to identify the services provided by the studied organization;

These steps are based on the EO explained in the last section and will be further detailed in the following one. According to the DSRM, this set of steps is the artifact that we will employ to identify services; we are using already existing knowledge (DEMO methodology) to show how to identify them.

5   Demonstration

In order to show the feasibility of the DEMO methodology for service identification, we will explain its relevant notions on the basis of a small real-life example, which will allow to increase the practical relevance of our study and to obtain an in-depth insight into how DEMO can assist in the service identification process. This experiment aimed to provide evidence against the importance of well identified services (for instance, in a Service Catalog). For that, we decided to study a Human Resources (HR) department and to identify the services it provided to the rest of the organization. The study focused on a European private company, leader in the wines and spiritual beverages distribution, which we will call from now on Company X. The head of the HR department (Gabriela) described the functions and actions she had to take to perform her job. In addition, we also interviewed an employee of the Marketing department (Rosario), in order to identify what interactions she had with the HR department, i.e. what services she thought the HR department provided her. So, the final objective is to prove that the services a provider thinks it offers differ from the perceived services for a customer. This section corresponds to the demonstration phase of DSRM.

5.1   Service Provider Perspective

Firstly, we describe the actions undertaken by the head of the HR department. The starting point to fulfill this task is called Enterprise Description and is characterized by producing a text which summarizes the actions performed by the service provider, such as the presented below. This text should be based in all the available documentation. Due to space limitations, we will just present the text after applying the first two analyzes.

When the text is written, one should read it carefully and try to recognize and distinguish between the Ontological, Infological and Datalogical actions described, as referred in the DEMO's Distinction axiom (see Section 3.3). This step is called Performa-Informa-Forma Analysis. To do that, we should define a notation to differentiate those actions: in this example, we have highlighted the text, using Black, Dark Gray and Light Gray colors to identify, respectively, the Ontological, Infological and Datalogical actions.

The next step concerns the identification of C-acts/facts, P-acts/facts and actor roles, using the Performa (Ontological) items identified in the previous step. We also have considered a notation to differentiate between them, similar to the one used in the Operation axiom: square brackets "[" and "]" to identify actor roles, brackets "(" and ")" to identify C-acts/facts and angled brackets "<" and ">" to identify P-acts/facts. This step is called Coordination-Actors-Production Analysis and here there is a reduction of the complexity, relatively to other methodologies, because from now on, we will only be considering the Ontological actions identified in this step. The result of applying these two analyses to the original text is presented below:

The Human Resources (HR) department of Company X is responsible for the development of the monthly payroll, management of the vehicles distribution, infrastructure management, training of the various employees, recruitment of new employees, and insurance, among others. It is constituted by 3 other employees: Vitor, Patricia and Luisa. Vitor deals essentially with fleet management, post office and banks, Patricia assists the Finance department and General Manager(she is a personal assistant of these 2 areas), and Luisa is a receptionist who deals with phone calls and Proof of Deliveries (PODs). They both report to Gabriela, head of HR.

The recruitment process of Company X starts when both the [HR] and the General Manager agree that there is a (need) to <hire> [new employees]. This need can be obtained from previous feedback given by the responsible for each department. The recruitment can be internal or external. In case of internal recruitment, such as internal staff turnover, someone in a specific area can be <moved> to another one. However, in case of external recruitment, it is possible to use the support given by the [universities] to <hire> a [new trainee] and give him/her the necessary <training>, or, through advertisements or [temporary employment companies], <hire> [someone already established on the market] and with experience. When there is a need to hire a new trainee, the HR department of Company X contacts some specific universities, according to the function and pre-requisites to perform that function. After this first contact, some universities proceed to the selection of resumes, while others send them all to Company X, which will select the most promising candidates. Next, these candidates are directly contacted by Company X to schedule the first interview with HR. If the candidate is accepted into the next stage of the recruitment process, a second *interview meeting is scheduled* with the head of the department where the function will take place. Finally, in case of satisfactory performance, the candidate is accepted. This recruitment process based on interviews is similar to all candidates, whether they are graduates or people with work experience.

About the fleet management, the [HR] is responsible for the <rental> of certain types of vehicles, which will be used by employees with determined functions. This rental is made to an [external renting company], and HR has to deal with distance control, accidents, and further expenses related to these vehicles (highway, fuel). They negotiate a contract that has a certain time limit, and includes various options such as maximum distance that can be travelled. It is also necessary to <make an insurance> for the vehicles. After receiving the vehicles, one must regularly check if they have (mechanical) problems and if they occur, the car must be <taken> to a [workshop] for repair. Exchanging vehicles between drivers to guarantee a balanced use of each must also be considered. When the contract is about to end, the vehicle is <taken> to [inspection] to check if everything is fine.

The <insurance> is also related to HR. For instance, there are several types of insurance: life insurance, health insurance, vehicle insurance and others. In case of a trainee, the work accidents insurance is the only one to be triggered, if it is a temporary contract, it includes health and work accidents insurance, and if it is a permanent contract, it includes life, health and work accidents insurance. In both cases, the HR communicates with the insurer and Social Security to deliver tax and documents.

The payroll is determined using the budget that was established for that year. The monthly salary for each employee is calculated and pre-determined from the company politics, and it is affected by absences, product discounts, among other factors which will be uploaded to the software that calculates each month's salary. To determine every factor, the head of each department must inform the HR.

The [HR] <establishes> telecommunications contracts with specified [operators] to ease the communication of the employees and checks if these contracts are fulfilled by the operating company which affects the monthly payment that has to be taken. HR is also responsible for the development of the employees' vacation map, control of the documented internal politics of the company and check if they are being carried out by employees, and occupational medicine to ensure that some employees meet certain requirements to perform some actions.

At the end of every year, the performance appraisal process is executed, during which the job performance of each [employee] is <evaluated>. The [HR] develops specific forms for each department, which must be filled in by the employees until a certain date (Self evaluation). After this phase, the head of each department gathers the feedback given by their employees and then schedules meetings with all members to discuss the performance during the last year.

When the performance appraisal process has finished, it is time to check if [someone] (needs) a particular <training session> in a specific subject. When this need is identified, one tries to identify a group of employees that also needs the same training. After that, it is necessary to plan the training session. This way, the HR starts by checking the availability of the employees, possible dates to execute it, text books, proofs of participation and other logistics steps. In case of internal training, the HR verifies who, inside the company, has the know-how to develop that training session, and then, the chosen employee will be held responsible for developing the module. In *case of external sessions, the HR also needs to* contact an accredited external company *to execute the training session.*

The infrastructure management *(office equipment, chairs, tables) and overall function of the headquarters is conducted by the [HR]. When employees* (need) office equipment*, they* ask Vitor for it*. Then, he will* <order> *this equipment in the office* store*. When the equipment arrives, the* store sends it to Company X and it is delivered to the employees*. The [HR] also* <deals> *with the logistics of the company's events (Christmas dinner, company day and* answers the employees' general doubts*.*

After these analyses, it is time to define the existent transactions in this text, by clustering the identified C-acts/facts and P-acts/facts, in what is denominated by Transaction Pattern Synthesis. The Transaction axiom can be helpful in this step, because it guarantees that each P-act/fact or C-act/fact previously found corresponds to a complete transaction. Then, for each identified transaction type, the result type (i.e., the Production fact created) should be correctly and precisely formulated, which can be achieved by uniquely identifying an entity, using variables. This result is represented in the table below, called Transaction Result Table (see Table 1).

Table 1. Result table – HR point-of-view

Transaction Types	*Result Types*
T01 – Hire a new employee	R01 – Employee E has been hired
T02 – Rent a vehicle	R02 – Vehicle V has been rented
T03 – Repair vehicle	R03 – Vehicle V has been repaired
T04 – Inspect vehicle	R04 – Vehicle V has been inspected
T05 – Insure a vehicle	R05 – Vehicle V has been insured
T06 – Insure an employee	R06 – Employee E has been insured
T07 – Establish communication contracts	R07 – Communication contract T has been established
T08 – Evaluate job performance	R08 – Job performance J has been evaluated
T09 – Give training session	R09 – Training session S has been given
T10 – Fulfill equipment requests	R10 – Request R has been fulfilled
T11 – Order and receive office equipment	R11 – Office equipment O has been received
T12 – Organize company events	R12 – Event E has been organized

After defining the transaction types and the respective result types, one must check if there are any dependencies between the transactions/ P-facts (results), as the Composition axiom describes. This step is called Result Structure Analysis and can be executed by carefully reading the text one more time. The following dependencies were found:

- There is a dependency between T01 and T06: after hiring a new employee, the HR department has to insure him/her. T06 is mandatory;

• There is a dependency between T02 and T05: when a vehicle is rented to an external renting company, the HR department is responsible for insuring that same vehicle. T05 is mandatory;

• T10 depends on T11: in order to fulfill the employee's general requests, the HR department contacts the Office Store to order the necessary equipment. T11 is optional;

• There are no dependencies involving the remaining transactions.

After identifying the transaction types, its results and dependencies, it is time to determine the environment surrounding the HR department, considering the organization context.

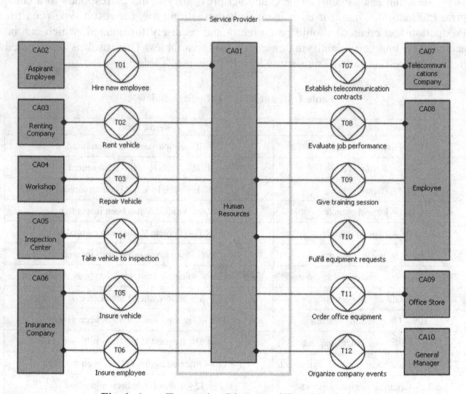

Fig. 1. Actor Transaction Diagram – HR point-of-view

The first model to be developed is called the Actor Transaction Diagram (ATD), represented in Figure 1. In this type of diagrams, a transaction is represented using a symbol, more specifically a diamond in a disk, which contains the respective combination of C-acts and a P-act. Each transaction is connected to two gray boxes, representing the initiator and executor actor roles. The initiator is connected to the transaction symbol using a solid line, while the executor is connected to the transaction using a solid line ending in a black square. These gray boxes refer to composite actor roles, i.e. elements whose exact structure is not known. All the environmental elements, i.e. elements outside the organization that we are studying, are represented with gray boxes for that reason. This also means that we can represent the studied organization with a gray box when referring to the kernel of the organization, which can be further specified using elementary actor roles (represented by white boxes). In this example, for simplicity reasons, the HR department has been considered as composite actor role. The gray-lined rectangle in the back represents the boundary under consideration.

As depicted in Figure 1, and following the service definition based on EO, we can identify four services provided by the Human Resources department of Company X, according to Gabriela: hire a new employee (T01), give a training session (T09) to Company X's employees, fulfill requests of the employees related to general office equipment (T10) and finally, organize Company X's events to its employees (T12).

5.2   Customer Perspective

Until now, we have thoroughly described all steps to find the services from the HR's point-of-view. We will now describe these steps more briefly in order to find the services provided by the HR department, but now from the Marketing department's point-of-view (customer). After collecting the textual description and applying the first two analyzes, Performa-Informa-Forma and Coordination-Actors-Production Analyzes, we have obtained the following text:

According to Rosario, when she was hired to work at Company X, there wasn't any Human Resources (HR) department yet. She was <hired> with the assistance of an external company. She had a first interview with [Vera Ribeiro] (Marketing dept.) and an agent of the external company, followed by a second meeting to discuss the working conditions and sign documents. The first week on the job was called Integration Week, when the new employees can understand what is the done by the various Company X departments.

Nowadays, regarding the <evaluation process>, [Rosario] receives emails from the HR containing the forms that must be used in the self-evaluation phase, as well as the internal procedures and politics of the company. About the training sessions, all the logistics <is arranged> by the [HR], as well as the management of the number of training hours. The head of the department is responsible for informing the HR that a specific training might be needed.

Every month, she receives the phone bill. Each employee has a limit that he/she can spend monthly, and if the limit is overreached, then the [employee] has to <pay> the difference.

To <schedule> her vacation, [Rosario] needs to send her proposal to Vera Ribeiro, who analyzes and approves it or not. Then, Vera Ribeiro sends the proposal to Vera Martins (Marketing dept.) to guarantee a consensus between the whole departments. From here on, the HR is informed.

When she wanted to change her NIB code, she had to call HR to let them know about the new one and complete the necessary documents. The same thing happened when [she] (wanted) to <include> another beneficiary in her insurance.

In order to correctly calculate her salary, in case of absences, she has to deliver some documents to the HR, and then this department proceeds to the correct discounts *In case of expenses the same procedure must be followed. When these aspects are known, the* software automatically calculates the output.

Problems with office equipment <are dealt> *with by [Vitor]. When she* wants to send a letter, *she provides it to Vitor, who is now in charge of delivering it at the postal offices.* Letters received at Company X are delivered to the respective recipients by Patricia.

After identifying the Ontological activities and the involved actor roles, we can identify each transaction and its result, on the Transaction Pattern Synthesis (see Table 2).

From Rosario's point-of-view, there are no dependencies among the identified transactions. The explanation for this occurrence is due to the fact that Rosario works at the Marketing department, so, she does not know how the HR actually performs those transactions. This step corresponds to the Result Structure Analysis.

Table 2. Result table – Marketing point-of-view

Transaction Types	Result Types
T01 – Hire a new employee	R01 – Employee E has been hired
T02 – Evaluate job performance	R02 – Job performance J has been evaluated
T03 – Give training session	R03 – Training session S has been given
T04 – Pay telecommunications invoice	R04 – Invoice I has been paid
T05 – Schedule vacations	R05 – Vacation schedule S has been developed
T06 – Change insurance status	R06 – Employee E has changed its insurance status I
T07 – Fulfill equipment requests	R07 – Request R has been fulfilled

The ATD from the point-of-view of the Marketing department is represented on Figure 2. As we can see, it is possible to identify three services provided by the HR department of Company X, according to Rosario: give a training session (T03), change the status of the insurance (T06) and finally, fulfill requests of the employees related to general office equipment (T07).

After analyzing both points-of-view, we can conclude that there are similarities and differences between them. On the one hand, we found two services which are similar, "Give training session" and "Fulfill equipment requests". On the other hand, there were also some mismatches: while in the HR's point-of-view we identified "Hire new employee" and "Organize company events", from the Marketing's point-of-view we found "Change insurance status". These differences will be discussed in the following sections.

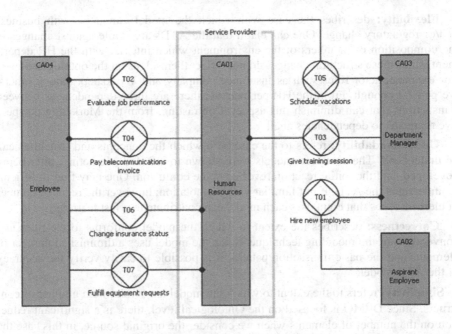

Fig. 2. Actor Transaction Diagram – Marketing point-of-view

6   Evaluation

After applying our proposal to identify the services provided by the HR department, from a HR's point-of-view and a customer point-of-view, we can evaluate the obtained results. This way, we used the Moody & Shanks Framework [21] to assess the quality of the Actor Transaction Diagrams and collect the HR and customer's feedback about the results. The evaluation phase of DSRM is achieved in this section.

In order to evaluate the quality of the produced models, we used the Moody & Shanks framework, which has been empirically tested and is composed by a set of eight quality factors:

- **Completeness:** refers to whether the model contains all user requirements. This factor is influenced by the person who describes the organization or department which is being studied. Thus, in the HR perspective, if the person who described the department has forgotten to mention some details about its functions, as consequence, it would not be possible to identify all the transactions / services. From the Marketing perspective, it is not possible for Rosario to identify all the services provided by the HR, because it depends on what are the main interactions between the HR and Marketing. For instance, it is normal that the Marketing employee did not have any idea about the organization of events (which was identified as a service by the HR itself), because in her daily functions, she had never been involved in this kind of transaction;
- **Integrity:** indicates the extent to which all business rules have been included in the model. The ATD does not express any business rules, but it was possible to gather the dependencies between transactions;

- **Flexibility:** describes the ease with which the model can cope with business and/or regulatory change. One can claim that the ATDs are stable against changes in the composition of members of the environment which interact with the HR department, since these structural changes do not reflect themselves on the ontological level. For example, actor roles such as Insurance company and Workshop, among others, are general enough. From the HR perspective, there are three dependencies between transactions that can diminish this aspect. Contrasting, from the Marketing perspective there are no dependencies at all;

- **Understandability:** refers to the ease with which the concepts and structures can be understood. The produced models were shown to HR and Marketing, after being developed, and the initial reaction revealed some confusion. One may find it difficult to understand the ATD, if not familiar with the notation; however the reduced number of element types that composes each model can contribute to a fast learning;

- **Correctness:** describes the extent to which the model conforms to the rules and conventions of the modeling technique. Since the model uses a diminished number of elements and the basic transaction pattern, it is possible to easily verify the accuracy of the ATD model;

- **Simplicity:** refers to the extent to which the model has a minimum number of constructs. Since DEMO is focused on the Ontological level, there is a significant reduction on the number of elements when we consider the original source, in this case the description, which has several Datalogical and Infological aspects. For instance, the ATD of the HR perspective has twelve transactions and ten actor roles, while the ATD of the Marketing perspective has seven transactions and four actor roles;

- **Integration:** concerns the extent to which the two different models can be integrated or compared to each other. This quality factor depends on the textual description provided by both parties. Despite the differences which were found and explained, there were also some common points (two services / transactions and several actor roles);

- **Implementability:** denotes the ease with which the data model can be implemented within the time, budget and technology constraints. In DEMO models, we are dealing with the Performa (Ontological) elements performed by actor roles, which are implementation independent. So, it is possible to conclude that DEMO models are not the best way to represent the implementation details.

The evaluation of the obtained models and results was also performed by asking for the HR and Marketing employees' opinion. In general, the provided feedback was rather positive, because they understood the obtained services. Nevertheless, a general complaint was that our list of services did not include all the functions the employees perform. This is due to the already referred feature of DEMO being implementation-independent, not considering the Infological and Datalogical activities. If one revises the textual descriptions of the HR, from both points-of-view, it is possible to conclude that the majority of the actions are included on the Forma and Informa categories (Datalogical or Infological, respectively). These actions have great potential to be automated by IT systems [4]. No Actor role performing Forma or Informa activities can ever be completely automated, but the effort that these subjects have to make in fulfilling the role can be reduced to a minimum.

7   Lessons Learned

After applying the proposal, we understand how DEMO can reduce the complexity of enterprise models: by layering it in three parts, and focusing only on the part which refers directly to the creation of new original facts (Ontological layer). Despite these advantages, there are also some downsides. To be able to apply DEMO, there is a need for a textual description, about the organization that is being studied and the surrounding environment, as input for the identification process. This text is written in natural language, by someone who has some insight about the tasks performed by the organization, which can lead to misunderstandings due to the lack of expressivity.

One also has to consider that the models produced by DEMO just contemplate the Ontological aspects performed by employees. This way, actions which are categorized as Infological or Datalogical are not directly included in these models. So, in order to have a complete overview of the actions performed by an actor, one has to consider those three kinds of abilities. This experiment revealed that in some situations it would be useful to model the most relevant infological and datalogical transactions.

An important lesson that we managed to prove is that different people have different notions of what is performed by one another. This is well showed considering the different textual descriptions and, consequently, the different services we obtained from the HR and Marketing's points-of-view. This difference represents the gap between the customers' expectations and the service provider's perception of those expectations. This gap was first identified two decades ago [3] and the fact that, nowadays, it still occurs in companies proves the importance of the service identification process. Companies still struggle to identify and manage their services and this kind of problems is motivating a new field of research: Service Science.

8   Conclusions

In this paper we have stressed the problems related with the service identification using DEMO, namely the issues that the text description can raise. We have concluded that the textual description of the organization is a critical step in the service identification process. Having it wrong may lead to gaps between the customers' expectations and the provider's perceptions of those expectations, such as in Company X.

We conclude that both client and service provider should be included in the service identification process and not only the service provider – as we have seen in the service industry so many times. There seems to exist a belief among service providers that they can alone identify their own services, and that their clients should have a minor role in this process. With the results of this paper, we hope to demonstrate that this is a wrong approach. The services are the connection point between providers and clients and having it poorly defined is the first step to not satisfy customers' expectations regarding services. Moreover, the client should be included in the beginning of the service identification process (in our case, the definition of the enterprise description) and not only in the end of the process.

As future work we intend to overcome the problems related to the text description. We will evaluate the possibility of using graphical representation of business processes, such as BPMN models, as input for the DEMO analyses and synthesis.

References

1. O'Loughlin, M.: The Service Catalog - A Practitioner Guide. Van Haren Publishing (2009)
2. Chesbrough, J., Spohrer, H.: A research manifesto for service science. Communications of the ACM 49 (2006)
3. Parasuraman, Zeithaml, V.A., Berry, L.L.: A conceptual model of service quality and its implication for future research. Journal of Marketing 49 (1985)
4. Dietz, J.: Enterprise ontology - theory and methodology. Springer (2006)
5. Mendes, C., Mira da Silva, M.: DEMO-Based Service Level Agreements. In: Snene, M. (ed.) IESS 2012. LNBIP, vol. 103, pp. 227–242. Springer, Heidelberg (2012)
6. Hubbers, J., Ligthart, A., Terlouw, L.: Ten ways to identify services. The SOA Magazine 8 (2007)
7. Peffers, K., Tuunanen, T., Rothenberger, M.A., Chatterjee, S.: A Design Science Research Methodology for Information Systems Research. Journal of Management Information Systems 24 (2008)
8. Hubbert, E., O'Donnell, G.: Service catalog: your prerequisite for effective IT service management. In: Forrester (ed.): Infrastructure & Operations Professional, Forrester (2009)
9. Zeithaml, V.A., Bitner, M.: Service Marketing. McGraw-Hill, New York (1996)
10. O'Sullivan, J.: Towards a precise understanding of service properties. Queensland University of Technology (2006)
11. Jones, S.: Toward an acceptable definition of service. Journal IEEE Software 22 (2005)
12. Albani, Terlouw, L., Hardjosumarto, G., Dietz, J.: Enterprise Ontology Based Service Definition. In: 4th International Workshop on Value Modeling and Business Ontologies, Amsterdam, The Netherlands (2009)
13. Bon, J.: Foundations of IT service management based on ITIL v3. Van Haren Publishing (2007)
14. CMMI for Services, version 1.3.: SEI - Carnegie Mellon University (2010)
15. Fan-Chao, M., Den-Chen, Z., Xiao-Fei, X.: Business Component Identification of Enterprise Information System. In: IEEE International Conference on e-Business Engineering (2005)
16. Wang, Z., Xu, X., Zhan, D.: A survey of business component identification methods and related techniques. International Journal of Information Technology 2 (2005)
17. Terlouw, L., Dietz, J.: A framework for clarifying service-oriented notions. In: Enterprise Modeling and Information Systems Architecture, vol. 5. German Informatics Society (2010)
18. Arsanjani: SOMA: a method for developing service-oriented solutions. IBM Systems Journal 47 (2008)
19. Erradi, Anand, S., Kulkarni, N.: SOAF: an architectural framework for service definition and realization. In: IEEE International Conference on Services Computing (2006)
20. Huysmans, P., Ven, K., Verelst, J.: Using the DEMO methodology for modeling open source software development processes. Information and Software Technology 52 (2010)
21. Moody, G., Shanks, D.: Improving the quality of data models: empirical validation of a quality management framwork. Information Systems 28 (2003)

Metropolitan Ecosystems among Heterogeneous Cognitive Networks: Issues, Solutions and Challenges

Salvatore F. Pileggi, Carlos Fernandez-Llatas, and Vicente Traver

ITACA-TSB, Universidad Politécnica de Valencia, Valencia, Spain
{salpi,cfllatas,vtraver}@itaca.upv.es

Abstract. Cognitive Networks working on large scale are experimenting an increasing popularity. The interest, by both a scientific and commercial perspective, in the context of different environments, applications and domains is a fact. The natural convergence point for these heterogeneous disciplines is the need of a strong advanced technologic support that enables the generation of distributed observations on large scale as well as the intelligent process of obtained information. Focusing mostly on cognitive networks that generate information directly through sensor networks, existent solutions at level of metropolitan area are mainly limited by the use of obsolete/static coverage models as well as by a fundamental lack of flexibility respect to the dynamic features of the virtual organizations. Furthermore, the centralized view at the systems is a strong limitation for dynamic data processing and knowledge building.

Keywords: Cognitive networks, Semantic technologies, Distributed computing, Sensor networks.

1 Introduction

Cognitive Networks [1] working on large scale are object of an increasing interest by both the scientific and the commercial point of view in the context of several environments and domains.

In fact, during the last years, research activities about local phenomena and their correspondent impact on global phenomena have been object of great interest inside the scientific community as well as in the context of public and private research institutions. Concrete environments (Figure 1) could depend by the research scope/goal and they can significantly vary for size, amount and kind of information, involved actors, etc. An ideal scenario in this sense is a metropolitan area that provides a complex heterogeneous ecosystem in which humans, machines and the environment are constantly interacting.

Common research activities at metropolitan area level are mainly focused on the study of climatic or environmental (e.g. chemical or natural element presence or concentration) phenomena and of human behavior (behavioral patterns, traffic, noise, etc.).

A. Fred et al. (Eds.): IC3K 2011, CCIS 348, pp. 323–333, 2013.

The study of these phenomena, first of all, interests the citizens (or concrete collectives) because it can be a complex and exhaustive feedback in order to improve the quality of life or to provide specific services for the interested collectives (allergic people for example). The evolution of these phenomena in the medium and large period, as well as its social impact, is object of great interest in the context of different domains and disciplines.

The natural convergence point for these heterogeneous disciplines is the need of a strong advanced technologic support that enables the generation of distributed observations on large scale as well as the intelligent process of the obtained information (Figure 1).

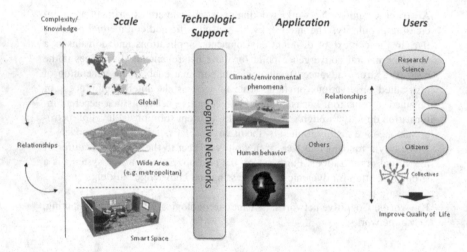

Fig. 1. Cognitive Networks as technologic support

Existent solutions at level of metropolitan area are mainly limited by the use of obsolete/static coverage models as well as by a fundamental lack of flexibility respect to the dynamic features of the most modern virtual organizations. Furthermore, the centralized view at the systems is a strong limitation for dynamic data processing and knowledge building. Finally, the heterogeneous nature of data and sources implies complex model for data representation and the related knowledge has to be analyzed according to several perspectives (e.g. local knowledge, domain, cross-domain).

This paper would exhaustively discuss the impact of the application of semantic technologies to high scale cognitive network, enabling semantic ecosystems among heterogeneous structures and information.

The paper is logically structured in two main parts. The first part has mainly the goal to define and characterize semantic ecosystems in relation with real environments. It also deals the main limitations currently existing for the massive dissemination of cognitive networks working on metropolitan scale. A short overview about the most innovative solutions for each one of the key technologic aspects featuring cognitive networks will be provided as well as a short analysis about related business models. Finally, in the last section, the impact of the semantic technologies application is analyzed, as key factor for the improving of the interoperability level among heterogeneous networks, sub-networks and data. Furthermore, the capabilities of knowledge building and intelligent analysis of data can be strongly improved.

2   Cognitive Networks: From Science to Reality

A classification of Cognitive Networks could be hard to be proposed because there is an intrinsic relation between the application and these complex systems. According to a simplified view at cognitive networks, in the context of this paper we will consider a cognitive system composed of four independent but interrelated functional/logic layers:

• *Data Sources.* Cognitive Networks implicitly work on large scale. A first and simple way to classify and categorize them is the conceptualization of the data sources. The modern society implicitly or explicitly provides great amounts of information. This information has to be "captured". We could basically distinguish between generated data (normally observations by sensors) and available data (data already existing in some electronic system). This second class of information is experimenting a constantly increasing of popularity mainly due to the social focus [2] that is becoming always more popular in large scale systems and applications. In the context of this work, generated data will be mostly referred, even if the most of approaches and solutions discussed are source-independent topics.

• *Data Representation.* Considering the heterogeneity of the available data from multiple data sources, the representation of data has a strong importance for the capabilities of systems. Semantic annotations and, more in general, ontologies could play a key role in this context as well as any other model/technology that allow advanced model for data interoperability.

• *Knowledge Building.* The main feature of the data representation is the need to provide a basic knowledge. This is not a limitation but, on the contrary, an added level of flexibility. In fact, the knowledge (understood as high-level information) building is provided by a further layer that directly works on the top of basic data. This model can assure the convergence of local knowledge to a unique virtual knowledge environment.

• *Application.* This is the only layer visible for the final user. Knowledge is represented, filtered and elaborated according to the application purposes.

Figure 2 proposes the formalization of the model using a Semantic Network [3]: semantic relations among concepts are built through a directed or undirected graph consisting of vertices, which represent concepts, and edges.

Evidently, this model is knowledge-oriented and it implicitly assumes the existence of a distributed communication infrastructure that implements the functional core of the system.

In this section, first the semantic ecosystems will be defined and characterized both with their relationships with real environments. Later, the current approaches to concrete solutions and related limitations are analyzed. Therefore, the main section scope is the definition of a generic reference scenario for semantic ecosystems in relation with their technologic and economic sustainability. As it will be discussed, there is, at the moment, a significant gap between theoretical models and their concrete application.

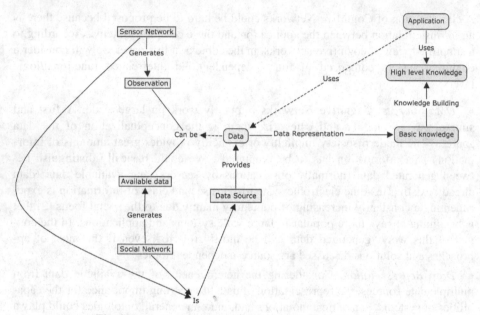

Fig. 2. Formalization of the model using Semantic Networks

2.1   Semantic Ecosystems among Heterogeneous Cognitive Networks

In the context of this work, a metropolitan or urban [4] ecosystem (Figure 3) is defined as a large scale ecosystem composed of the environment, humans and other living organisms, and any structure/infrastructure or object physically located in the reference area.

An exhaustive analysis of environmental and social phenomena is out of paper scope. Just considering that we are living in an increasingly urbanized world. From recent studies, it appears that this tendency will be probably followed also in the next future. It is a commonly accepted assumption that further increases in size and rates of growth of cities will no doubt stress already impacted environments as well as the social aspect of the problem.

Considering this tendency is hard to be controlled or modified, there are a great number of interdisciplinary initiatives, studies and researches aimed to understand the current impact of the phenomena as well as to foresee the evolution of it.

These studies have, evidently, a scientific focus, but they also could be of interest in the context of the everyday life. In fact, modern cities change their structure and physiology in function of human activities that constantly act as inputs for the feedback system. It is easy to imagine the great number of services that could improve the quality of life of citizens (or collectives) with a deep knowledge of the environment.

As mentioned, the study of the human activities, of the environmental and climatic phenomena is object of interest in the context of several disciplines and applications. All these studies are normally independent initiatives, logically separated researches and, in the majority of the cases, results are hard to be directly related. This could appear a paradox: interest phenomena happen in the same physical ecosystem, involving the same actors but the definition of the dependencies/relationships among atomic results are omitted even if they are probably the most relevant results.

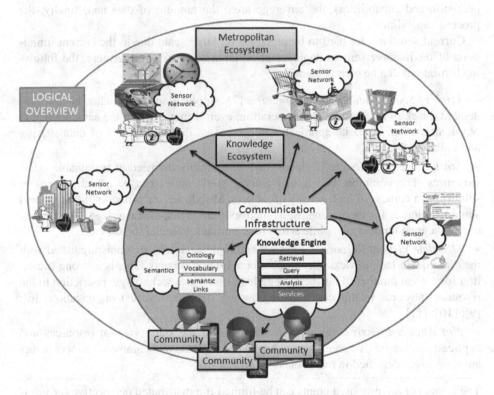

Fig. 3. Logical overview at Semantic Ecosystems

The common point is the need of great amounts of heterogeneous data, normally generated on large scale [5]. They can be "simple" measurements or complex phenomena, sometimes hard to be detected. This overall approach according heterogeneous model has a strong impact especially in the representation and processing of the information.

Summarizing, at now urban ecosystems have a directly equivalent logic concept by a knowledge perspective but its realizations are mainly knowledge environments than effective knowledge ecosystems.

2.2 Current Approaches and Limitations

The normal technologic support for enabling knowledge environment is the cognitive network [1] that assumes a physical infrastructure (sensors) able to detect interest information or phenomena and a logic infrastructure able to process the sensor data (knowledge building) eventually performing actions, responses or complex analysis.

The parameters that can potentially affect the "quality" of the applications or studies are mainly the sensor technology (constantly increasing in terms of reliability, precision and capabilities), the coverage area, the amount of data and, finally, the process capabilities.

Current solutions are hard to be proposed on large scale due to the current limitations of the massive sensors deployment on large scale [6]. Furthermore, the following limitations can be clearly identified:

• *Lack of Social View at the Information* [7]. Applications and studies have a fundamental lack of interaction and cooperation, even if they are part the same logic and physical context. A social approach could increase the possibilities of data sharing and collaboration.

• *Static Coverage Models*. High coverage areas imply the need of sustainable infrastructures. The common models that assume static nodes can be a high expensive solution in a context of high density of sensors. Mobile sensor networks [8] could be a suitable solution for environments, as metropolitan areas, characterized by the presence of a great number of mobile actors (e.g. humans, vehicles, bikes, etc.).

• *Obsolete View at Resources*. Physical and logic networks are undistinguished with the consequent lack of flexibility in distributed environments. This is a strong limitation for a great number of business scenarios as well as a technologic restriction to the resource (physical in this case) sharing among structured virtual organizations [6]; [9]; [10]; [11].

• *Not Always Effective Business Models*. Due to the static view at resources and applications, innovative scenarios are hard to be realized and common business actors are hard to be identified in real contexts.

The impact of the proposed points can be limited if a distributed perspective for infrastructures and information is assumed. Due to the heterogeneous features of the data source and information, the interoperability plays a key role for the effective realization of the model.

In the next section, a distributed approach for the main infrastructure is described both with the most advanced solutions based on semantic interoperability that allow a social perspective for the knowledge. Also the analysis for the knowledge building process based on the application of the last generation contextual semantic is proposed.

3   The Impact of Semantic Technologies: Distributed Approach

The previous section propose an abstract model for semantic ecosystems as a possible evolution of cognitive networks to a distributed approach that should allow the enablement of complex logic ecosystems in a context of flexibility and economic sustainability. This conclusion is mainly motivated by the objective difficulty of modeling virtual organizations using centralized models as well as by the low level of interoperability that currently characterizes heterogeneous systems.

Distributed solutions (Figure 4) objectively improve the flexibility of architecture but they require a high level of interoperability among systems especially if they are not part of the same social and economic context.

This section would discuss the benefits introduced by semantic technologies as general solution for improving the interoperability and as key support for the processing of heterogeneous data (knowledge building).

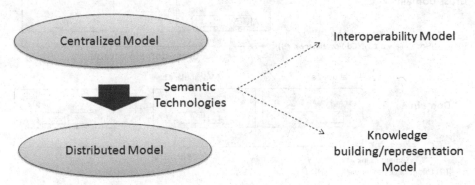

Fig. 4. Towards distributed approaches

3.1   Semantic Interoperability

Considering a distributed sensor domain, the key issue is the evolution of the Sensor Web model to the Semantic Sensor Web that, in practice, assumes systems interchanging semantic information on the top of the common functional interoperable layer [12].

The current semantic model for the web is affected by several problems. These open issues, such as ambiguities and performances, are object of an intense research activity that is proposing several solutions as simplification or particularization of the main model.

In order to enable effective working systems on large scale, a simplified model of the semantic web is considered [13]. It assumes semantic reasoners operating over three interrelated semantic structures (Figure 5):

• *Ontology* as in common semantic environment, it has to represent data and knowledge at different levels.

- *Shared Vocabulary*. It could be a contextual structure that represents an "agreement" in order to avoid possible ambiguities and semantic inconsistencies inside semantic ecosystems.

- *Semantic Link*. Additional structures that should link concepts from different ontologies and concepts from vocabularies. These structures can directly relate concepts from different ontologies and they can indirectly build contextual semantic environments.

As showed in Figure 5, the Ontology is a semantic structure normally associated to a local knowledge environment. Concepts from different ontologies can be related at domain level through semantic links to vocabularies concepts.

In the example represented in Figure 5, the concepts c1 and c6 are equivalent to the concept c3 at domain level and so, at this level, they are also equivalent to each other.

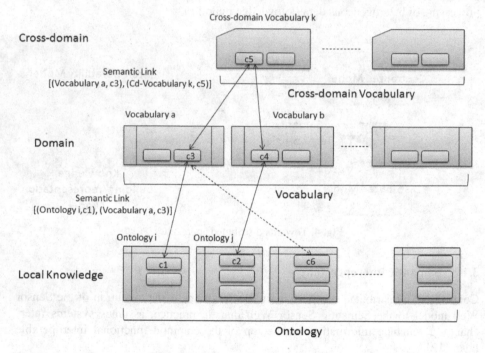

Fig. 5. Interoperability model schema

This schema could be an exhaustive model for the great part of logic environment associated to a concrete domain. But the heterogeneous features of semantic ecosystems force the knowledge environment to work in a multi-domain context. This last aspect need a further semantic layer (Figure 5): *Cross-domain Vocabularies* are defined in order to relate concepts from different domains through semantic links.

In the example of Figure 5, the concepts c3 and c4 are equivalent to c5 at global level. This also implies that c1 (linked to c3) and c2 (linked to c4) are equivalent to c5 and that they are equivalent to each other at global level.

A short analysis of the model proposed in Figure 5 first of all puts in evidence the hierarchical structure of the semantic knowledge building according to an increasing level of abstraction.

On the other hand, the semantic model is completely open and assures, through semantic links to higher concepts, a high level of expressivity and interoperability without forcing standard data models.

This last aspect has a critical importance at application level where models, rules and relationships need integrations, particularizations and extensions in function of concrete applications and domains.

3.2   Knowledge Building

This second support is the natural complement to the first one in order to provide systems with the capability of building abstracted knowledge on the base of basic sensor data on the model of [14].

The main challenge is the generalization of this approach on large scale and considering heterogeneous environment. As showed in Figure 6, a local knowledge schema (Ontology) is composed of two kinds of concepts:

• *Low-level Concepts*: They have a mean only in the context of their local knowledge environment. The main consequence is the lack of any class of semantic link. In practice, they are low-abstracted information that normally is "visible" only in the local system.

• *High-level Concepts*: They are a set of concepts that naturally complete the previous one. In fact, they are high-abstracted concepts that have evidently a local mean but also a domain and/or global mean.

A deeper analysis of the structure (Figure 6 on the right) allows the definition of semantic layers inside the main structure:

• *Data Source*. Set of low-level concepts that represent the data-sources (sensors or any other kind of physic/human data source).

• *Data*. As the previous one but representing data.

• *Core*. Abstracted layer composed of semantic rules that relate low-level and high-level concepts. Due to its critical role, this is the key layer in the semantic structure.

• *Domain-specific Layers*. Any set of high-level concept required in the context of concrete domains and applications.

The main advantage introduced by the schema is the possibility to have a common ground for data source and data representation, as well as a clearly defined set of standardized high level abstracted concepts.

Also the core part of the ontology, that has the goal of building the knowledge of basic data, is an ad-hoc component of specific applications. In the context of an ideal semantic ecosystem, any class of information (basic data or abstracted knowledge) can be correctly interpreted in the context of the owner system as well as inside other systems socially connected.

Local Knowledge

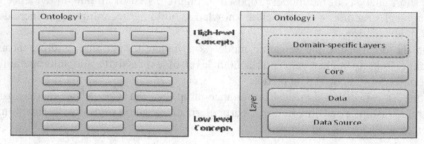

Fig. 6. Local knowledge model

4   Conclusions

The power of collecting and relating heterogeneous data from distributed source is the real engine of high-scale cognitive networks.

The economic sustainability, as well as the social focus on the great part of the applications, determines the need of an innovative view at networks and architectures on the model of most modern virtual organizations.

These solutions require a high level of interoperability, at both functional and semantic level. The current "Semantic Sensor Web" approach assures a rich and dynamic technologic environment in which heterogeneous data from distributed source can be related, merged and analyzed as part of a unique knowledge ecosystem.

References

1. Thomas, R.W., DaSilva, L.A., MacKenzie, A.B.: Cognitive networks. In: 2005 First IEEE International Symposium on New Frontiers in Dynamic Spectrum Access Networks, DyS-PAN 2005, November 8-11 (2005)
2. Wang, F.-Y., Carley, K.M., Zeng, D., Mao, W.: Social Computing: From Social Informatics to Social Intelligence. IEEE Intelligent Systems 22(2), 79–83 (2007)
3. Borgida, A., Sowa, J.F.: Principles of semantic networks: explorations in the representation of knowledge. Morgan Kaufmann Pub. (January 1991)
4. Urban ecosystem, Wikipedia,
   http://en.wikipedia.org/wiki/Urban_ecosystem
5. Akyildiz, I.F., Su, W., Sankarasubramaniam, Y., Cayirci, E.: A survey on Sensor Network. IEEE Communication Magazine 40(8), 102–114 (2002)
6. Pileggi, S.F., Palau, C.E., Esteve, M.: Enabling Wireless Sensor Network within Virtual Organizations. In: Prasad, A., Buford, J., Gurbani, V. (eds.) Future Internet Services and Service Architectures. River Publishers (2011)
7. Rahman, M.A., El Saddik, A., Gueaieb, W.: Building Dynamic Social Network From Sensory Data Feed. IEEE Transactions on Instrumentation and Measurement 59(5), 1327–1341 (2010)
8. Pileggi, S.F.: A multi-domain framework for Wireless Vehicular Sensor Network. In: Proceedings of International Conference on Ultra Modern Telecommunications and Workshops (ICUMT 2009), St. Petersburg, Russia, October 12-14 (2009)

9. Foster, I., Zhao, Y., Raicu, I., Lu, S.: Cloud Computing and Grid Computing 360-Degree Compared. In: Grid Computing Environments Workshop, GCE 2008 (2008)
10. Mell, P., Grance, T.: Draft NIST Working Definition of Cloud Computing v14, Nat. Inst. Standards Technol. (2009),
    http://csrc.nist.gov/groups/SNS/cloud-computing/index.html
11. Foster, I., Kesselman, C., Tuecke, S.: The Anatomy of the Grid: Enabling Scalable Virtual Organizations. International J. Supercomputer Applications 15(3) (2001)
12. Pileggi, S.F., Palau, C.E., Esteve, M.: Building Semantic Sensor Web: Knowledge and Interoperability. In: Proceedings of the International Workshop on Semsntic Sensor Web (SSW 2010), Valencia, Spain (October 2010)
13. Pileggi, S.F., Fernandez-Llatas, C., Traver, V.: A Semantic Layer for Embedded Sensor Networks. ARPN Journal of Systems and Software 1(3), 101–107 (2011b)
14. Pileggi, S.F.: A Semantic Environment for Data Processing in Embedded Sensor Networks. In: Proceedings of the International Workshop on Semantic Interoperability (IWSI 2011), Rome, Italy (January 2011)

Code Quality Cultivation

Daniel Speicher

University of Bonn, Computer Science III
dsp@acm.org

Abstract. Two of the meanings of the word "cultivation" that are rather unrelated show a strong dependency, when applied to the domain of code quality:

The existing code in an evolving software system could be seen as the soil in which new code and new functionality is growing. While working this "soil" developers benefit from unobtrusively presented automatic feedback about the quality of their code. There are tools that verify the correct usage of good code structures ("design pattern") and other tools that highlight improvement opportunities ("bad smells").

As design patterns and bad smells are usually presented and discussed separately it has not been observed, that they partially contradict with each other. We will show that even well chosen design patterns can lead to bad smells. Thus, design quality is relative, which does not mean that it is arbitrary. Design quality knowledge has to be rendered more precisely. We suggest to co-evolve the quality knowledge specifications together with the code in a process of cultivation. Bad smell definitions can then easily be extended by taking existing design patterns into account.

When the design knowledge is cultivated together with the code, specific knowledge like typical method names can be incorporated. A case study explored unjustified "intensive coupling"-smells in ArgoUML: While a previously suggested generic structural criterion identified 13% unjustified warnings, taking the specific names into account, identified 90%.

Keywords: Code quality, Design pattern, Bad smell, Natural odor, Logic meta-programming, Case study, Knowledge evolution.

1 Introduction

Good software design improves maintainability, evolvabilty and understandability. As any maintenance or evolution step requires the developer to understand the software reasonably good, understandability is the most crucial of these qualities. Therefore it can not be a goal to develop detection strategies for design flaws that a developer does not need to understand to use them. How could a criterion for understandability be meaningful, if it is not understandable itself? Developers should know and understand the detection strategies they use. In this paper we want to argue, that in addition detection strategies should know more of what developers know. It is essential for automated design flaw detection to be adaptable to respect developers knowledge found in the design.

Detection strategies for design flaws need to be generic, as they are meant to apply to many different systems. On the other hand software solutions are at least to some part

A. Fred et al. (Eds.): IC3K 2011, CCIS 348, pp. 334–349, 2013.

specific. If there were no need for specificity, it would not require many developers to build software. This does not yet say, that generic quality criteria are inappropriate, as there might be - and probably are - some general principles that should be met always. What it does say, is that developers answer to specific design challenges, so that there is at least some probability that there are good reasons to make a different choice in a specific situation than one would make while discussing design in general. We will present such (moderately) specific situations.

We wrote this article on the background of established Refactoring literature. With a broader acceptance of object-oriented programming at the end of the last century programmers needed advice to build systems with high maintainability and evolvability. Today the catalog of signs for refactoring opportunities ("bad smells") [1] developed by Beck and Fowler as well as the catalog of proven design solutions ("design pattern") [2] developed by Gamma, Helm, Johnson and Vlissides are common software engineering knowledge.

Marinescu devised in [3] a method to translate the informal descriptions of bad smells [1] and object-oriented design heuristics [4] into applicable so called detection strategies. The method first decomposes the informal descriptions into parts, which are then translated into an expression built of metrics and comparisons with thresholds. These expressions are then composed into one rule with the help of the elementary logical operators. Marinescu and Lanza elaborate in [5] how developers should systematically work through the bad smells (here called disharmonies) and step by step resolve cohesion problems ("identity disharmonies") then coupling problems ("collaboration disharmonies") and finally problems in the type hierarchy ("classification disharmonies"). Developers are in general well equipped with this book and established reengineering und refactoring literature [6], [1], [7].

Overview. The rest of the paper is organized as follows. Section 2 will discuss, how the same design fragment can be a well respected design pattern and an instance of a well known bad smell. As the choice of the design pattern is expected to be the result of trade-offs, the smell detection should be enabled to take the knowledge about the pattern into account and ignore this smell instance. Section 3 therefore develops our approach based on logic meta-programming. We represent the Java code base as Prolog facts, on which we define detection strategies and design pattern as predicates. Finally we explain, how we suggest to incorporate the knowledge about pattern into the knowledge about smells. Section 4 sketches the twofold "cultivation" process we envision. Quality knowledge specifications provide feedback to the developers and developers refine quality knowledge where precision is useful. Section 5 finally reports about a case study, that gives an example where taking developer intentions into account strongly increases the precision of a detection strategy. Section 6 collects some related work about design pattern definition, bad smell detection, logic meta-programming. Section 7 summarizes our contributions.

2   Visitors Tend to Have Feature Envy

Objects bring data and behavior together. This is at the core of object-orientation and allows developers to think of objects in programs similar as of real objects. The

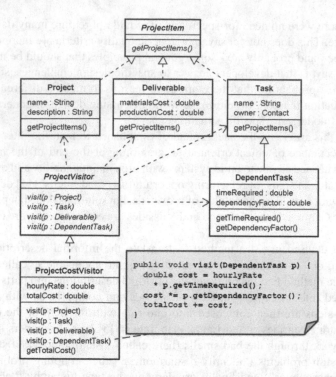

Fig. 1. A simple example of the *visitor* pattern: The `ProjectItems` build a tree structure via the `getProjectItems()` method. The `ProjectCostVisitor` implements the single responsibility to calculate the total costs of a project. He accesses the data of the `ProjectItems` to fulfill this responsibility.

deviation from this ideal are the smells *data class* and *feature envy*. A class is a *data class* if the class mainly has data and only little behavior. A method in a class has *feature envy*, if it operates for the major part on data of another class. Still in some situations one wants to separate the data and the behavior of one conceptual object into two or more technical objects, which can result in both smells.

The *visitor* pattern[1] as in Fig. 1 places functionality that operates on the the data of certain objects (*elements*) in separate classes (*visitors*). One reason for this separation is, that the *elements* build a complex object structure and the functionality belongs rather to the whole structure than to single *elements*. Another reason might be, that the functionality is expected to change more frequently and/or is used only in specific configurations. Since the functionality in the *visitor* accesses the data of the *elements*, this intended collaboration could falsely be identified as *feature envy*.

Before we continue with the discussion of other design patterns, we want to review this judgement again, as it is not impossible that the authors of the original pattern catalog suggested imperfect solutions and better design patterns would not have a flaw

---

[1] This example was taken with some minor modifications from (accessed in 09/2011):
http://www.java2s.com/Code/Java/
Design-Pattern/VisitorPattern1.htm

Table 1. Data-behavior separation: Roles in the pattern that can develop a smell as a consequence of strong data-behavior separation

Pattern	Data Class	Feature Envy
Flyweight	Extrinsic State	Flyweight
Interpreter	Context	Expression
Mediator		Concrete Mediator
Memento	Memento	Originator
State	Context	Concrete State
Strategy		Concrete Strategy
Visitor	Element	Concrete Visitor

like *feature envy*. Indeed there have been thorough discussions about the *visitor* pattern and a variety of alternatives had been suggested [8], [9], [10]. Yet, the criticism was never about *visitors* accessing data of *elements* but only about the dependencies from the *abstract visitor* to *concrete elements* as it implies that all *visitors* need to be changed, once a *concrete element* is added. The *visitor* is a good choice, if the improved cohesion or maintainability of the functionality realized by the *visitor* outweighs the cohesion of parts of the functionality in the *visitor* with the *elements*.[2] That is: sometimes *feature envy* is rather a "natural odor" than a bad smell of a *visitor*.

There are variations of the same data-behavior separation in other design patterns, as listed in Tab. 1. We separate data into an extra object, if we want other objects to share this data. So does the *extrinsic state* in the *flyweight* pattern and the *context* in the *interpreter* pattern. As a result the classes which use this data (*flyweight* in *flyweight*, *expressions* in *interpreter*) develop *feature envy*. In the *memento* pattern some data of the *originator* is stored in a separate *data class* called *memento*. We separate behavior from the data, if we want to change it dynamically as we do by exchanging one *concrete state* for another in the *state* pattern. The more data is left and accessed in the *context* class, the stronger the *feature envy* will be. Finally, if we want to let *colleague* classes interact with each other without knowing about each other the *concrete mediator* might operate on the data of a few of the *colleagues*.

The separation of data and behavior is not the only possible consequence of design pattern. Some pattern give certain roles (*factory, builder, singleton, façade, mediator*) such a central responsibility that they naturally tend to have strong afferent coupling (*shotgun surgery* in the sense of [5]). Roles that are meant to coordinate or configure other objects (*concrete factory, builder* and its *director, façade, concrete mediator, concrete visitor, context* in *state*) naturally tend to have strong efferent coupling and might even attract high complexity. Types that are inserted into existing hierarchies for the purpose of a pattern (*adapter, composite, abstract decorator, concrete decorator*) tend to extend the interface disproportional or to ignore the inherited features. Finally the level of indirection that is added by some roles (*adapter, abstraction* in *bridge, abstract*

---

[2] The trade-off is a direct consequence of the object-oriented principle that behavior belongs to exactly one object. The trade-off dissolves in languages that have other means to combine data and behavior: With AspectJ for example the cohesive behavior can be described in a "*visitor*"-aspect that introduces behavior to all *elements* by inter-type declarations [11].

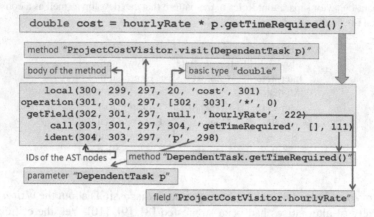

Fig. 2. Fact representation of the first line of the method `visit(DependentTask)` in `ProjectCostVisitor` and the referenced nodes of the AST

decorator, façade, proxy, concrete mediator, context in *state*) would without the pattern be considered as the smell *middle man* defined in [1].

3   Logic Based Code Analysis

The specifications of quality knowledge are defined on a representation of Java source code as Prolog facts (3.1), on which we define graph specifications ("structures", 3.2) as a basis for smell (3.3) and pattern definitions (3.4). Finally these smell definitions can easily be adapted to take developer intentions expressed as patterns into account (3.5).

3.1   Fact Representation of Object-Oriented Code

Our prototypes are implemented as plug-ins for the development environment Eclipse. They contain a builder that runs everytime the Java builder is run and translates the complete Java Abstract Syntax Tree (AST) of a program into a representation in Prolog facts. In this AST all references are resolved. All language elements of Java 5 are represented. Fig 2 shows the facts representing the first statement in the method `visit(DependentTask)` in `ProjectCostVisitor`. Each fact represents a node of the AST. The first parameter of a fact is a unique ID of this node. The second parameter is a back reference to the parent node. The third parameter references the enclosing method. The remaining parameters contain attributes as well as references to the child nodes. It is very easy to build more abstract predicates based on this fact representation. We will use many predicates without defining them as we hope that the names of the predicates convey their meaning. Here are two example definitions:

```
method_contains_call(M, C) :- call(C, _, M, _, _, _, _).
call_calls_method(C, M)    :- call(C, _, _, _, _, _, M).
```

3.2   Structures

The structure behind smells and metrics and the structure of design pattern can be implemented similarly. The smell *feature envy* analyses how strongly a method operates on fields of another class. This can be seen as the analysis of a graph (illustrated in Fig. 3 (a)) with nodes for the method, the own and foreign attributes and edges for the access relation. The pattern *visitor* can be seen as graph (partially illustrated in Fig. 3 (b)) with nodes for the single *abstract visitor* and the single *abstract element* and a few nodes for the *concrete visitors* and *concrete elements*. As there can be at least a few of these graphs in the program, we want to be able to distinguish them. Therefore we call one node of each graph which belongs to only one of these graphs the *corner* of it. The corner of the graph for *feature envy* would be the method under consideration. The corner of the graph for the *visitor* can be the *abstract visitor*.

Definition 1 (Structure Schema). *A structure schema consists of (1) one unary predicate that tests whether a program element is the* corner *of a structure, (2) some binary predicates that tests whether given the corner of a structure, another program elements plays a certain* role *in the structure with this corner, (3) and some ternary predicates that tests whether given the corner, a* relation *between two other program elements that play a role in this structure exists.*

Definition 2 (Structure Instance). *Given a structure schema and a program element that fulfils the corner predicate, we call this program element the* corner of the structure [instance], *where the structure consists of (1)* role players, *which are all program elements annotated with the names of the role predicates of the structure schema, for which the role predicate is true, if we use the corner as a first argument and the role player as a second argument; (2)* relation *player duos, which are all ordered pairs of program elements annotated with the name of a relation predicate of the structure schema, for which the relation predicate is true, if we use the corner as a first argument, the first element of the pair as second argument, and the second element of the pair as a third argument.*

3.3   Smell Detection Strategies

Detection strategies as defined by Marinescu and Lanza in [5] are elementary logical formulas built of comparisons of metrics with corresponding thresholds. To illustrate their idea and our implementation we will explain the detection strategy for the smell *feature envy* top-down. The detection strategy reads:

```
feature_envy(M) :-
    feature_usage_structure(M),
    access_to_foreign_data(M, ATFD),       ATFD >  2,
    locality_of_attribute_access(M, LAA),  LAA  <  0.3,
    foreign_data_provider(M, FDP),         FDP  =< 5.
```

Here the first goal verifies that M is a method for which we can calculate the *feature envy*. The rest is built of three metrics *Access To Foreign Data* (The number of directly or indirectly accessed fields in a foreign class), *Locality of Attribute Access* (The ratio to which, the accessed fields are from the own class) and *Foreign Data Provider* (The number of foreign classes from which the methods accesses a field) are calculated. These metrics simply count the number of certain relations in the structure, as the source code shows:

```
access_to_foreign_data(M, V) :-
  count(F, method_accesses_foreign_field(M, M, F), V).
locality_of_attribute_access(M, V) :-
  count(F, method_accesses_own_field(M, M, F),     AOF),
  count(F, method_accesses_foreign_field(M, M, F), AFF),
  Value is AOF / (AOF + AFF).
foreign_data_provider(M, V) :-
  count(C, method_accesses_foreign_class(M, M, C), V).
```

The metrics build on the role (*own class, own field, foreign class, foreign field*) and relation (*method accesses own field, method accesses foreign field*) definitions of the feature envy structure as illustrated in Fig. 3. The first lines of the definition read like follows:

```
feature_usage_structure(S) :-
  source_method(S), not(abstract(S)).
method(S, M) :-
  feature_usage_structure(S), S = M.
own_class(S, C) :-
  method(S, M), method_is_in_type(M, C).
own_field(S, F) :-
  own_class(S, C), type_contains_field(C, F), private(F).
[...]
method_accesses_foreign_field(S, M, F) :-
  method(S, M),
  foreign_field(S, F),
  method_accesses_field(M, F).
[...]
```

3.4 Lightweight Pattern Definition

To make our smell detection pattern aware, we need only a very lightweight pattern definition. The structure for the pattern consists of the roles *visitor, concrete visitor, method in concrete visitor, visited element* and *field in visited element*. The complete definition reads like follows:

```
visitor_pattern(P) :-
  declared_as_visitor(P).
visitor(P, V) :-
  visitor_pattern(P), P = V.
concrete_visitor(P, C) :-
  visitor(P, V), sub_type(C, V), not(interface(C)).
```

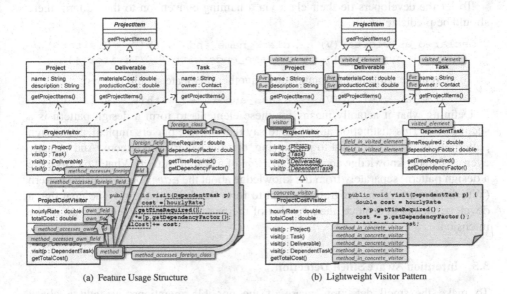

(a) Feature Usage Structure   (b) Lightweight Visitor Pattern

Fig. 3. The instance of the visitor pattern overlayed with: (a) The *feature usage structure* for the method `visit(DependentTask)` in the `ProjectCostVisitor`. (b) The lightweight pattern definition for the *visitor* pattern with the corner `ProjectVisitor`. "five" is an abbreviation for "field in visited element".

```
method_in_concrete_visitor(P, M) :-
    concrete_visitor(P, C), type_contains_method(C, M).
visited_element(P, E) :-
    visitor(P, V), type_contains_method(V, M),
    method_has_parameter(M, R), parameter_has_type(R, E).
field_in_visited_element(P, F) :-
    visited_element(P, E), type_contains_field(E, F).
```

Note that we impose very little constraints on the elements and relations within the pattern. The idea is to focus in a first step on the identification of the elements and relations, describing the extension of the design pattern. The predicates are means to capture the intended extension of the developer under the assumption that he expressed it well enough. Typically developers use standard names[3] or name parts at least for some of the role players in a pattern and we found that all other role players can be identified starting from one of these. Alternatively one could require that one or a few role players are annotated with the role.

---

[3] It is reasonable to use the naming conventions used in [2] even if identifiers are not in English. At least, one can expect some consistency in the naming of types and methods if design patterns are used intentionally. This names are meant to be configured by predicates like the ones we give here.

To let the developers tie their class via a naming convention to the pattern, there should be predicate like:

```
declared_as_visitor(V) :- class_name_ends_with(V, 'Visitor'),
  not(sub_type(V, P), class_name_ends_with(P, 'Visitor')).
```

To tie it to the pattern via an annotation, another predicate should be used:

```
declared_as_visitor(V) :- class_annotated_with(V, 'Visitor').
```

Defining what it actually means for these elements to form a design pattern is a second step. This distinction allows us to say that the role players implement a design pattern incorrectly in contrast to just saying, that the pattern is not there. And we may even evolve our knowledge about the *intension* of a design pattern (What it means to be a design pattern) separately from the knowledge about the *extension* of a design pattern (Which elements of the program a meant to form the design pattern.) The goal that a developer wants to achieve, the *intention* of the pattern, should be seen as part of the intension.

3.5   Intention Aware Smell Detection

To make the smell detection aware of the possible intentions, we add a check into the respective predicates at the earliest possible place, i.e. as soon as the variables are bound. For this purpose we define predicates `intended_field_access`, `intended_method_call`, `natural_odor`. Here is the adapted code for the relation *method accesses foreign field* of the *feature usage structure* and the *data class*:

```
data_class(C) :-
  named_internal_type(C),
  not(natural_odor(data_class, C)),
  weight_of_class(C, WOC), WOC < 3.34,
  [...]
method_accesses_foreign_field(S, M, F) :-
  method(S, M),
  foreign_field(S, F),
  not(intended_field_access(M, F)),
  method_accesses_field(M, F).
```

Given this adaptation and the definition of the pattern, it is easy to make the smells specifications ignore the intended field access and the natural odor:

```
natural_odor(data_class, Element) :-
  concrete_element(_, Element).
intended_field_access(M, F) :-
  visitor_pattern(P),
  method_in_concrete_visitor(P, M),
  field_in_visited_element(P, F).
```

The smell *data class* will now ignore any class that plays the role of a *concrete element* in the *visitor* pattern. In the calculation of the *feature usage structure* all accesses from a method in a *concrete visitor* to a *field in visited element* will be ignored. That is, feature envy towards other classes will still be detected.[4,5]

4   Conflicts Stimulate Knowledge Evolution

We suggest to use our technology for an alternating "cultivation" process of code reviews (quality improving) and code structure documentation (quality knowledge improving). For the process of quality improvement with respect to generic criteria a detailed guideline can be found in [5]. For the process of improving the explicit quality knowledge, we gave a first suggestion here. Currently we expect unjustified smell warnings to be the strongest stimulus to increase the design knowledge.[6] In addition pattern mining tools can be used to identify pattern instances [13]. Making the design explicit allows to ignore the smells on certain elements, to use specific thresholds or to adapt the metrics to it. It provides the additional benefit of a defined structure that can be verified.

The ad-hoc solution to avoid unjustified warnings by manually excluding them e.g. with an annotation[7] is not desirable: Many locations would need to be annotated. The trade-off that led to the decision is still implicit. Every time the code changes, the reasons for ignoring the smells would need to be revisited. Such an annotation could be seen as a dependency of the general and rather stable definition of the smell to the concrete and relatively unstable code. Dependencies towards unstable concrete code increase the burden of maintenance.

In the longer run, the design knowledge needs reviewing as well. We would suggest to make further expectations explicit and verify them. For example there should be no dependencies from any *concrete element* in the *visitor* pattern to any *concrete visitor*. Such an constraint can be added as predicate to the pattern schema. Another expectation is, that the *element* classes can stay much more stable than the *visitors* do. This expectation could be added to the structure schema as an absolute or relative upper bound on the change frequency of the artifact in the repository. The development environment could regularly check the change frequencies such providing an third feedback cycle.

---

[4] Having to adapt all the different relations and smell definitions is not desirable. An aspect-oriented adaptation would be very helpful here and Prolog is very well suited to be enhanced with Aspects.

[5] Another way to adapt the smells is to use thresholds that depend as well on the roles an element plays. We will not discuss this obviously considerable option.

[6] This guarantees that the design knowledge is relevant to the developer and not only specified because of an ideal of completeness. Exman discussed for the area of software reuse from Web sources ("Websourcing") in [12] that knowledge is necessary incomplete. The incompleteness is even desirable as completeness is computational expensive and the ideal of completeness might lead to inconsistencies, especially as the specified knowledge has to be interpreted. He demonstrates that relevance can be a good replacement for completeness.

[7] Similar to the Java 5 annotation @SuppressWarnings to suppress compiler warnings.

5 Evaluation

5.1 Smell in Pattern

Sebastian Jancke implemented the detection strategies defined in [5] as well as a few more as part of his diploma thesis [14]. He analyzed different open source projects and conducted a successful user study. Tab. 2 lists some of the instances of natural odors he found.

Table 2. Natural odors found in open source software projects

Smell	Role/Concept	Source Code
Feature Envy	Concrete Visitor	JRefactory 2.6.24, SummaryVisitor
Feature Envy	Concrete Strategy	JHotDraw 6, ChopBoxConnector
Middleman	Abstract Decorator	JHotDraw 6, DecoratorFigure
Law of Demeter Violation	Embedded DSL	Google Guice 2.0
Shotgun Surgery	(stable) API	[everywhere]

Although we assume that the idea to adapt smell definitions based on knowledge about design intentions is convincing, we are planning larger code studies to explore the relevance of our observations further. The following case study focuses on the aspect of our approach that we consider to be least convincing: To identify structure instances we suggest to refer to design knowledge that the developer has made explicit in names of methods, classes, packages or with annotations. As we elaborate in [15] the meaning of some names can be seen as the structure in which they are used.

5.2 Taking Advantage of Expressed Intentions: Creation Methods

In [5, Ch. 6] "Collaboration Disharmonies" Lanza and Marinescu reference design knowledge in a way, that is unique in the book. The two strategies *intensive coupling* (many dependencies to a few other classes) and *dispersed coupling* (dependencies to many other classes) contain besides conditions about the coupling an additional condition "Method has few nested conditionals" measured by $MAXNESTING > 1$, where $MAXNESTING$ is the maximal nesting depth of blocks in the method. The authors motivate this condition as follows:

> "Additionally, based on our practical experience, we impose a minimal complexity condition on the function, to avoid the case of configuration operations (e.g., initializers, or UI configuring methods) that call many other methods. These configuration operations reveal a less harmful (and hardly avoidable) form of coupling [...]." [5, p.121]

This motivation references the concept of operations (methods) for configuration. Although this concept is not precisely defined, the description gives every experienced OO programmer a first operational impression about it. We wanted to test this statement with a little case study using the current version of the same source code that was used in [5][8]. To discuss this statement we call methods with $\text{MAXNESTING} = 1$ *flat* and present three hypotheses:

Flat methods are configuration methods. (1)

Coupling in flat methods is no design problem. (2)

Coupling in configuration methods is no design problem. (3)

Table 3. Methods in ArgoUML: The maximal nesting within the method and the classification into configuration, test and other methods influences whether the method has *intensive* or *dispersed coupling*

	All Methods				Intensive Coupling				Dispersed Coupling			
Max. Nesting	config	test	other	Σ	config	test	other	Σ	config	test	other	Σ
1	772	880	6416	8068	19	57	26	102	27	143	32	202
2	318	110	2468	2896	88	97	19	204	60	24	213	297
> 2	152	53	2128	2333	209	234	29	472	52	34	568	654
Σ	1242	1043	11012	13297	316	388	74	778	139	201	813	1153

A quick view at the source code showed that the hypothesis (1) is wrong. Many of the flat methods are obviously test methods and some are neither test nor configuration methods. It turned out that the code was well designed enough, so that we could rely on naming conventions instead. Exploring twenty randomly and a few systematically chosen methods we found that in ArgoUML methods with names starting with "init", "create", "build" or "make" are configuration methods and methods with names starting with "test" or containing the term "Test" in their name or in the name of the enclosing class are test methods. Given this two name based rules and the detection strategies for *intensive coupling* and *dispersed coupling* (without the condition of the methods being flat) we were able to classify the methods. The result is presented in Table 3.

On the first view Hypothesis (1) seems to be backed by the data, as many ($772/1242 = 62\%$) configuration methods are indeed flat and most ($1090/1242 = 88\%$) have nesting not bigger than 2. Unfortunately these configuration methods build only a small fraction ($772/8068 = 10\%$ or $1090/10964 = 10\%$) of the methods with limited nesting, so that hypothesis (1) is not true. Still, there are many ($8068/13297 = 61\%$) flat methods, so that excluding them from further analysis can improve performance. We further observe, that the major part of the methods with *intensive coupling* are configuration methods ($316/778 = 41\%$) and test methods ($388/778 = 50\%$). The major part ($781/1153 = 68\%$) of the methods with *dispersed coupling* are other methods with a nesting of at least 2.

---

[8] http://argouml.tigris.org/, accessed in 06/2011. The 13 projects listed in the file "argouml-core-projectset.psf" were used and all 13297 non abstract methods of the 2083 named classes were analyzed. [5] used the version from 10/2004. An spreadsheet with our data can be downloaded from our website http://sewiki.iai.uni-bonn.de/private/daniel/public/ic3k2011

We still have to discuss whether it is safe to ignore flat methods, configuration methods and test methods. The coupling in configuration methods is indeed hardly avoidable as all the decoupled classes need to be instantiated and connected somewhere. That is, the coupling in some specific methods is a natural consequence of the overall decoupling effort across the system.[9]

To test Hypothesis (3) that coupling in configuration methods can be ignored, we reviewed the 13 configuration methods with the highest coupling intensity $(22 - 116)$ that have one of the smells: Even they are clearly understandable and the coupling is not harmful. The same is true for the 13 test methods with the highest coupling intensity $(28 - 68)$. To challenge the nesting condition, we reviewed the 13 methods with highest coupling intensity $(10 - 16)$ within the other methods with no nesting, but with one of the smells. Our impression was not that clear as in the two cases before, but still the coupling did not require any refactoring[10]. Therefore we see no reason to reject Hypothesis (2).

To summarize, we expect all smells in configuration methods and all test methods to be false positives. The same is true for all methods with no nesting, i.e. Hypothesis (2) and (3) are plausible. The nesting condition reduces the smell results by $102/778 = 13\%$ or $202/1153 = 18\%$ while ignoring configuration and test methods reduces the results by $704/778 = 90\%$ or $340/1153 = 29\%$. So, if smell detection can use other information than structure (e.g. naming conventions) to identify configuration methods and test methods, the number of false positives can be strongly reduced. Here is how we would implement the corresponding adaptation:

```
natural_odor(intensive_coupling, E) :-
  configuration_method(E).
configuration_method(M) :-
  declared_as_configuration_method(M).

declared_as_configuration_method(M) :-
  source_method(M), method_name(M, N),
  member(P, ['init', 'create', 'build', 'make']),
  prefix(P, N).
```

6 Related Work

Wuyts coined the term "logic meta-programming" in his Ph.D. thesis [16]. He presented the Smalltalk Open Unification Language ("SOUL"), a programming language for logic meta-programming on Smalltalk programs. His thesis introduced the term "co-evolution" for the consistent evolution of program elements, supported by specified consistency conditions. We suggest to consider the evolution of the meta-programs as well, as shortcomings in the meta-programs may be only observed over time.

Hajiyev, Verbaere, and de Moor presented in [17] the source code querying tool CodeQuest. This system uses Datalog queries that are guaranteed to terminate. These

---

[9] Even if the responsibility for configuration is "extracted" into XML configurations.

[10] Indeed half of them turned out to be a sort of configuration method again. So restricted (!) to the methods with smells the nesting condition is a reasonable heuristic.

queries are translated into efficient SQL queries. As far as we presented our queries they are very close to Datalog queries and our suggestions could directly be used with CodeQuest. In particular we did not have any negation in a recursive cycle so that our code is stratified. The reasons why we currently prefer Prolog over Datalog - like the meta-circularity of Prolog - are beyond the scope of the paper. Kniesel, Hannemann, and Rho [18] showed that that the performance of this and our platform is comparable.

As the structures we define can be seen as typed graphs, approaches analysing the code as graph structures are as well appropriate implementation means. An example would be the TGraph approach by Ebert, Riediger, Winter, and Bildhauer [19] with its mature query language GReQL [20].

Riehle suggested to model design pattern as systems of roles and relations in Ph.D. thesis [21]. As a case study he documented all the design pattern in JHotDraw 5.1.[11] The role modeling approach could be seen as the current default approach for design pattern modeling and is used in [13] as well.

Brichau, Kellens, Castro, and D'Hondt describe in [22] IntensiVE a tool suite build on SOUL. They demonstrate that structural regularities like design pattern as well as smell definitions can easily be defined as queries (here called "intensions"). Again, we would expect it to be straightforward to implement our approach with this tool.

Moha, Guéhéneuc, Le Meur, and Duchien present in [23] the Software Architectural Defects Specification Language ("SADSL"). They analyse the design defects "Swiss army knife", "Functional decomposition", "Blob", and "Spaghetti code" and demonstrate that they can specified with SASDL. Detection algorithms are generated from this specification.

7 Contributions

This paper showed that automated bad smell detection should take developer intentions into account, as different structural quality criteria are appropriate, depending on these intentions. To illustrate this point we discussed a well known design pattern and how it is still good design even if it shows a smell. These intentions are often already expressed in the code, but not yet available to the automated analysis. We presented a technological and conceptual framework that allows to combine the perspectives of structures that should be avoided (bad smells) and structures that are useful building blocks (design pattern). We presented our approach to implement structures based on logic meta-programming and explained how smell detection can be made aware of existing structures in code. We suggested to use our technology for an alternating "cultivation" process of code review (quality improving) and code documenting (quality knowledge improving). A case study showed that the precision of smell detection can be increased if the design knowledge that developers already made explicit is utilized instead of guessing developer intentions from structural properties.

---

[11] Versions of JHotDraw from the last ten years can be found online starting from http://www.jhotdraw.org/. As this code was implemented to demonstrate the use of design pattern and is considered as well designed it is very often used in reverse engineering case studies.

Acknowledgements. The author wants to thank everyone who made this research possible. Sebastian Jancke developed the "smell detection in context" plug-in, an earlier version of the prototype presented here. My colleague Jan Nonnen, current and former diploma students as well as former participants of our labs contribute to the evolution of "Cultivate", our platform for code quality evaluation and coding culture evolution. Tobias Rho, Dr. Günter Kniesel and further students develop "JTransformer", our platform that makes logic meta-programming for Java possible and comfortable. Finally the author wants to thank Prof. Dr. Armin B. Cremers for detailed discussions on earlier versions of this work and for supporting it.

References

1. Fowler, M., Beck, K., Brant, J., Opdyke, W., Roberts, D.: Refactoring: Improving the Design of Existing Code. Addison-Wesley (1999)
2. Gamma, E., Helm, R., Johnson, R.E., Vlissides, J.M.: Design Patterns: Elements of Reusable Object-Oriented Software. Addison-Wesley (1995)
3. Marinescu, R.: Detection Strategies: Metrics-Based Rules for Detecting Design Flaws. In: Proceedings of the 20th IEEE International Conference on Software Maintenance (ICSM 2004), Washington, DC, USA (2004)
4. Riel, A.J.: Object-Oriented Design Heuristics. Addison-Wesley (1996)
5. Lanza, M., Marinescu, R.: Object-Oriented Metrics in Practice. Springer (2006)
6. Demeyer, S., Ducasse, S., Nierstrasz, O.: Object Oriented Reengineering Patterns. Morgan Kaufmann (2002)
7. Kerievsky, J.: Refactoring to Patterns. Pearson Higher Education (2004)
8. Martin, R.C.: Acyclic Visitor. In: Pattern Languages of Program Design 3, pp. 93–103. Addison-Wesley Longman Publishing Co., Inc. (1997)
9. Nordberg III, M.E.: Default and Extrinsic Visitor. In: Pattern Languages of Program Design 3, pp. 105–123. Addison-Wesley Longman Publishing Co., Inc. (1997)
10. Palsberg, J., Jay, C.B.: The Essence of the Visitor Pattern. In: Proceedings of the Twenty-Second Annual International Computer Software and Applications Conference, COMPSAC 1998, pp. 9–15. IEEE (1998)
11. Nordberg III, M.E.: Aspect-Oriented Dependency Management. In: Filman, R.E., Elrad, T., Clarke, S. (eds.) Aspect-Oriented Software Development, pp. 557–584. Addison-Wesley, Boston (2005)
12. Exman, I.: Knowledge Incompleteness Considered Desirable. In: Fraga, A., Llorens, J. (eds.) Proc. of 2nd Workshop on Knowledge Reuse (KREUSE 2009) Hosted by ICSR 2009, Falls Church, VA, USA (2009)
13. Kniesel, G., Binun, A.: Standing on the Shoulders of Giants - A Data Fusion Approach to Design Pattern Detection. In: 17th International Conference on Program Comprehension (ICPC 2009), Vancouver, Canada (2009)
14. Jancke, S.: Smell Detection in Context. Diploma Thesis, University of Bonn (2010)
15. Speicher, D., Nonnen, J., Mügge, H.: How many realities fit into a program? - Notes on the meaning of meaning for programs. In: Proceedings of the Second International Workshop on Software Knowledge (SKY 2011), Paris, France (2011)
16. Wuyts, R.: A logic meta-programming approach to support the co-evolution of object-oriented design and implementation. PhD thesis, Vrije Universiteit Brussel (2001)
17. Hajiyev, E., Verbaere, M., de Moor, O.: *codeQuest:* Scalable Source Code Queries with Datalog. In: Thomas, D. (ed.) ECOOP 2006. LNCS, vol. 4067, pp. 2–27. Springer, Heidelberg (2006)

18. Kniesel, G., Hannemann, J., Rho, T.: A Comparison of Logic-Based Infrastructures for Concern Detection and Extraction. In: Proceedings of the 3rd Workshop on Linking Aspect Technology and Evolution (LATE 2007), New York, NY, USA (2007)
19. Ebert, J., Riediger, V., Winter, A.: Graph Technology in Reverse Engineering. The TGraph Approach. In: Proc. 10th Workshop Software Reengineering. GI Lecture Notes in Informatics (2008)
20. Ebert, J., Bildhauer, D.: Reverse Engineering Using Graph Queries. In: Engels, G., Lewerentz, C., Schäfer, W., Schürr, A., Westfechtel, B. (eds.) Nagl Festschrift. LNCS, vol. 5765, pp. 335–362. Springer, Heidelberg (2010)
21. Riehle, D.: Framwork Design, A Role Modeling Approach. Dissertation, ETH Zürich (2000)
22. Brichau, J., Kellens, A., Castro, S., D'Hondt, T.: Enforcing Structural Regularities in Software using Intensive. Sci. Comput. Program. 75, 232–246 (2010)
23. Moha, N., Guéhéneuc, Y.-G., Le Meur, A.-F., Duchien, L.: A Domain Analysis to Specify Design Defects and Generate Detection Algorithms. In: Fiadeiro, J.L., Inverardi, P. (eds.) FASE 2008. LNCS, vol. 4961, pp. 276–291. Springer, Heidelberg (2008)

Misbehavior Discovery through Unified Software-Knowledge Models

Iaakov Exman

Software Engineering Department, Jerusalem College of Engineering,
POB 3566, Jerusalem, 91035, Israel
iaakov@jce.ac.il

Abstract. UML statecharts are a widely accepted standard for modeling software behavior. But, despite the increasing importance of semantics for software behavior, semantics has been treated within UML as mere reasoning add-ons. We propose fully integration of UML statecharts with behavioral knowledge obtained from novel behavioral ontologies into a Unified Software-Knowledge model. These unified models have two important characteristics: first, misbehaviors are explicitly represented; second, behavioral ontologies generate graphs isomorphic to UML statecharts, by construction. This approach is applicable to run time measurements, to check the actual software behavior correctness and efficiency. Measurement discrepancies may trigger knowledge discovery mechanisms to update the unified models. The approach is illustrated with statechart examples from the domain of GOF software design patterns.

Keywords: Unified Models, Software, Knowledge, Behavioral Ontology, Nano-ontology, Eventuality, Misbehavior.

1   Introduction

UML – the Unified Modeling Language (see e.g. [1] and the OMG standard [2]) – is a de facto standard modeling language used for software development. It consists of separate structure and behavior diagrams with logic OCL add-ons which have been used for reasoning about class and state diagrams.

Semantics has been increasingly important for understanding and testing software behavior. Nevertheless it is not directly represented in UML diagrams – see e.g. the discussion by Selic [3] on the Semantics of UML.

An example of a problem of relevance to semantic application is checking by runtime measurement the actual behavior correctness of a software package vis-à-vis the declared behavior. For instance, if an observer design pattern is indeed correct, most of the time it actually behaves as designed, since its messages are lost with low probability.

This work proposes, as an improvement of the solution to the run-time measurement problem, the direct integration of semantics into UML statecharts, resulting in a Unified Software-Knowledge behavioral model.

A. Fred et al. (Eds.): IC3K 2011, CCIS 348, pp. 350–361, 2013.
© Springer-Verlag Berlin Heidelberg 2013

This approach is illustrated by statecharts and a novel type of behavioral ontologies from the domain of GOF[1] software design patterns [4].

In the remaining of the paper we illustrate UML statecharts in the context of design patterns (section 2), discuss explicit misbehaviors for measurement and reasoning (section 3), introduce novel behavioral ontologies (section 4), propose integration of ontologies' knowledge within statecharts (section 5), and conclude with a discussion (section 6).

2   Behavioral UML: Design Pattern Statecharts

Statecharts are not the only behavioral diagrams in UML. There are also sequence diagrams – with an explicit time axis, useful to describe specific scenarios – and collaboration diagrams.

Statecharts [5] were chosen in this work – in preference to other behavioral diagrams – because they encompass all possible behaviors for a given system, instead of just specific scenarios. For instance, a single statechart can cover all possible calls of a mobile phone. One would need different sequence diagrams to describe scenarios, such as a successful call, a busy line, a voice-mail message left, etc.

A statechart contains states graphically represented by rounded rectangular nodes, linked by labeled edges standing for transitions. Thus, a statechart is essentially a type of labeled graph. This is an important feature for integration of knowledge with software statecharts, as will be described in section 5.

2.1   Design Pattern Statecharts

While statecharts are in widespread use for system development, since they were published in the late eighties by Harel [5], they were not systematically applied to describe GOF design patterns. In particular, the GOF book [4] describes design patterns by UML class and sequence diagrams, without statecharts.

The reasons for the conspicuous absence of statecharts from design pattern descriptions are an interesting question per se. We show elsewhere [6] how to systematically build statecharts for GOF patterns. For any design pattern the basic question is how to determine the relevant states and transitions. Here we provide statecharts for sample patterns, to illustrate this paper's ideas.

A statechart for the Singleton design pattern [4] is seen in Figure 1. This and subsequent examples in this work are taken from the domain of GOF design patterns.

The Singleton design pattern, depicted in the statechart of Fig. 1, perfectly characterizes the declared behavior of a singleton: either zero or exactly one instance for a class. Since the Singleton design pattern is defined in terms of numbers of instances, states are chosen to have a fixed number of instances. Consequently, transitions between states are events that change the number of instances.

---

[1] GOF – Gang Of Four – is the well-known nickname of the authors of the book that introduced design patterns in software engineering, viz. Gamma, Helm, Johnson and Vlissides (ref. [4]).

Another example, is the Observer pattern [4]. This pattern has two roles: a subject that may change and observers that are notified of the subject change, ask for the subject new state, and are updated. For this design pattern, we set transitions to be messages among roles, similarly to the reasoning that leads to the state diagrams of a TCP connection (TCP is the Transmission Control Protocol within the TCP/IP family of communication protocols, see e.g. ref. [7]). Then, states are characterized by variable values in between message transmissions.

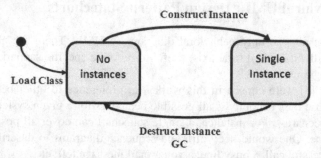

Fig. 1. UML Statechart of the Singleton design pattern – It displays two states and transitions between states. The left arrow, labeled "Load Class", stands for an initial transition. GC means Garbage Collection, a way to destruct the single instance.

In Figure 2 one sees a partial statechart representing the ***notify-update*** cycle of the subject and a respective observer.

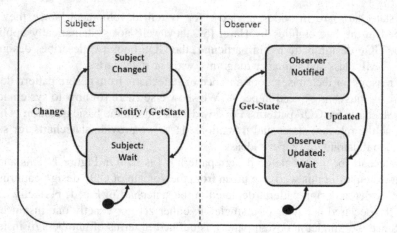

Fig. 2. Observer pattern partial statechart – the states separated by the dashed line work in parallel: the states of the subject and of the observer are independent. For the perfect Observer pattern the ***notify-update*** cycle is always obeyed.

The perfect Observer complies with the notify-update cycle, i.e. messages are always in this order: first the subject notifies the observers, then each observer requests the subject state by a Get-State message, finally each observer is updated.

Observers can also be attached or detached to the subject as needed. The latter actions are not represented in the above statechart.

3   Misbehaviors for Run-Time Measurements

In order to take advantage of actual run-time measurements – observing whether an implemented pattern complies with the perfect declared behavior – we need a more general model that takes into account that things can go wrong.

3.1   The Need for Explicit Misbehaviors

Misbehaviors are conceivable, perhaps possible behaviors – states and transitions – that are not part of the intended *perfect* design of a system.

There are two main reasons for adding explicit misbehaviors to a statechart:

1. *Measurement* – to represent actual behaviors, which may show up in laboratory measurements of the real system; these behaviors may be either acceptable or demand correction of the actual system to fit the perfect design;

2. *Reasoning* – to allow setting of conditions and reasoning about state and transition semantics.

3.2   Design Pattern Statecharts with Misbehaviors

We use the above design pattern statecharts to illustrate the addition of misbehaviors.

A more general model of the Singleton pattern – including possible misbehaviors – is seen in the example statechart of Figure 3. This statechart has an added state with K=2 instances, contradicting the definition of a singleton.

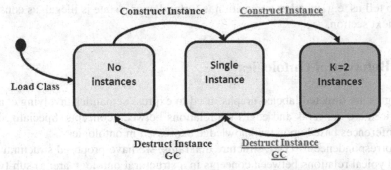

Fig. 3. UML Statechart of the Singleton design pattern with misbehaviors – It has an additional illegal state (K=2 instances) and forbidden transitions to/from this state (*Construct Instance* and *Destruct Instance* in bold underlined, red colored)

Such statechart with misbehaviors is not unique. For instance, it could have additional transitions from the illegal state (K=2 instances) to other states, enabling creation and destruction of additional instances. The number of additional misbehaviors has an unbounded value K.

Next, we see in Figure 4. a statechart of the Observer pattern with a sample misbehavior.

The illustrated misbehavior may occur because messages may be lost in an actual system, where the subject and observers reside in different machines communicating by TCP/IP messages.

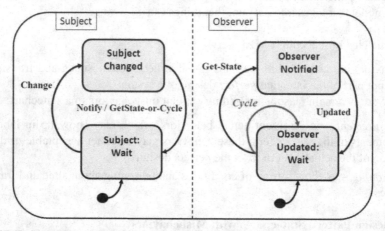

Fig. 4. UML Statechart of the Observer design pattern with misbehavior – It has an additional forbidden transition within the r.h.s. Observer state: the *Cycle* transition (in bold italics, red colored). If the *GetState* message is processed correctly as in the perfect statechart, the Observer changes its state to "Observer Notified". Otherwise, the Observer state is not changed represented by the new *cycle* (red colored) arrow.

In this context – i.e. statecharts containing misbehaviors – semantics enters the stage to tell us, e.g. whether a transition is forbidden or a state is illegal, as considered in the next section.

4   Behavioral Ontologies

Ontologies are directed labeled graphs[2] used to express semantics involving concepts. Its vertices are concepts and edges are relations between concepts. Specialized tools make inferences based upon the knowledge expressed in ontologies.

In correspondence to UML structure diagrams, we have proposed structural ontologies. Typical relations between concepts in a structural ontology are: a) sub-type; b)

---

[2] Graphs – as noted by Weisstein [8] – have two different meanings: 1- a graphical plot; 2-an abstract mathematical entity. We refer here to the second meaning.

composition; c) domain dependent relations. Structural ontologies are extensively discussed elsewhere.

Here we describe in detail a novel type of ontologies – behavioral ontologies – to deal with behavior.

In particular, these behavior ontologies are nano-ontologies viz. very small ontologies with restricted numbers of concepts. The justification for the reduced size of nano-ontologies is reasoning efficiency, especially when used in conjunction with real-time measurements.

4.1 Behavioral Nano-Ontologies

A software behavioral ontology is a novel type of ontology explicitly referring to software behavior concepts (i.e. states and transitions), with kinds of relations referring to these and relevant concepts.

The single most important new concept in a behavioral ontology – besides states and transitions – is "*eventuality*". Eventuality in natural language means a "possible occurrence or outcome". We shall refer to concepts of either "transition eventuality" or "state eventuality", meaning the a priori knowledge one has about the eventual occurrence of a transition or the outcome of being in a state.

Eventuality is composed of three concepts:

a) *Authority* – is a Boolean concept; the model developer – the authority – decides that a transition is either "forbidden" or "allowed"; similarly, a state is either "legal" or "illegal";

b) *Probability* – this is the probability of occurrence of a transition or that an outcome is a given state; a probability can be expressed numerically or qualitatively, say it is either low or high probability;

c) *Constraints* – specific domain dependent constraints; say in the real time domain, a transition may not occur before a certain timeout.

It is important to point out that *authority* and *probability* are not redundant concepts. A forbidden transition does not necessarily imply zero probability. A forbidden transition may have a finite non-zero probability of occurrence even if it is desirable to have a low probability in a particular case (see e.g. the discussion of the Observer design pattern below).

Figure 5 displays a software behavioral nano-ontology for transitions. Its graph represents the "transition eventuality".

The behavioral ontology for states is identical, except that one should substitute "transition" by "state" and "transition eventuality" by "state eventuality".

The authority notion is illustrated by the Singleton pattern: in Fig. 3 there were both legal and illegal states. For the singleton pattern all transitions not within the perfect declared behaviors are forbidden. In this particular case, they actually have zero probability.

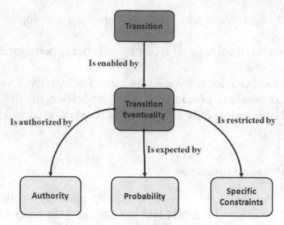

Fig. 5. Transition Eventuality Behavioral Nano-ontology – This graph displays the top concepts for transitions between states. The central concept of *"Transition Eventuality"* is itself composed of three concepts: authority, probability and specific constraints.

The notion of finite probability can be illustrated by the Observer pattern. If the subject and observers are not in the same machine, they communicate by IP messages (IP is the Internet Protocol, also within the TCP/IP family of communication protocols, see e.g. ref. [7]). Since IP messages may travel through different routes and sometimes are lost, the Observer pattern cycle is not always perfect. Thus there is a low but finite probability of changing message orders.

It should be stressed that, even with mistakenly ordered messages, as long as these occur with low probability, we could still accept such a program as a correct implementation of the Observer pattern (see e.g. ref. [9] on the measurement of Observer patterns).

Note that the probability of a *misbehavior*, say in the Observer example, is a statement about the pattern correctness. On the other hand, a probability of an allowed transition is a statement concerning the specific behavior of a certain system and not about the whole pattern.

Nano-ontologies – such as the transition eventuality in Figure 5 – usually consist of the top sub-graph of larger ontologies. For instance, in Figure 6 one can see a possible extension of the above nano-ontology, in which each of the three sub-concepts of the transition eventuality is refined into additional sub-sub-concepts.

In Figure 6, authority has three sub-concepts:

a) temporary – meaning that the authority is not permanent, but limited to bounded periods;

b) domain restricted – the authority is limited to a sub-domain of the system containing the transitions;

c) delegate – the authority is delegated and maybe removed.

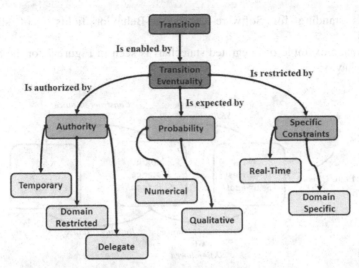

Fig. 6. Extended Transition Eventuality Behavioral Ontology – it displays one additional level of sub-concepts for transitions between states, below the top nano-ontology of Fig. 5

Probability has two sub-concepts:

a) numerical – quantitative with values between 0 and 1; this probability condition may be either an equality (e.g. probability = 0) or an inequality (e.g. probability > 0.8).

b) qualitative – with a discrete set of values, say high and low.

The Specific constraints concept also has two sub-concepts:

a) real-time – related to strict time constraints;

b) domain specific – related to the domains relevant to the software system in case.

The resulting ontology could be conceivably extended even further, with additional lower levels e.g. detailing the domain specific constraints.

5   Integration of Software and Knowledge in a Unified Model

By construction, behavioral ontologies originate a one to one correspondence between a *transition* or a *state* and their respective "*eventuality*" (in Fig. 5).

Thus for each statechart entity – either state or transition – there is a behavioral ontology eventuality: one obtains a strict isomorphism between the software only UML statechart and the Unified Model graph obtained by labeling states and respective transitions by their corresponding eventualities. There is no reason to keep them separated: their integration is desirable by reasoning and efficiency considerations.

5.1   Behavioral: Integrated Statecharts

Practical integration of statecharts with behavioral ontology labels is quite straightforward. Semantic information is added as a new kind of attribute fields –

SKB-fields standing for Software-Knowledge-Behavior fields – of states and transitions.

A schematic example of integrated statechart is seen in Figure 7 for the Singleton with misbehaviors.

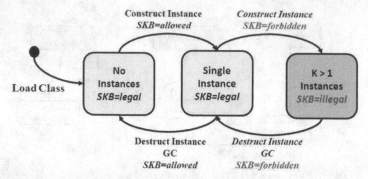

Fig. 7. Integrated Statechart model of Singleton pattern with misbehaviors – This statechart schematically displays legal/illegal states, allowed/forbidden transitions within integrated semantic SKB-fields.

5.2   Misbehavior Discovery with Unified Models

An approach to misbehavior discovery by comparing run-time measurements with unified behavioral models is outlined as follows:

1- *Obtain the Unified Model* – This model contains states and transitions for the declared behaviors of the chosen pattern. It also contains additional states and transitions representing expected misbehaviors. The SKB fields label the states and transitions with the a priori knowledge one has about the pattern, i.e. the eventualities for states and transitions.

2- *Perform Measurements* – Once one has obtained the unified model, one performs measurements of the actual behavior of a software system that implements the pattern.

A typical measurement, that one may perform for the Observer pattern, is the frequency of notify messages within a subject transition preceding GetState messages from its observers.

3- *Compare Measurements with Model* – One compares the measured frequencies with the a priori probabilities in the Unified Model. One might conclude that: a- the model is accurate and the implementation correctly reflects the model; b- the implementation is faulty; c- the model is inaccurate and it should be updated.

4- *Update Model with Discovery* – If the model is inaccurate it is updated with the discovered knowledge.

In practice, to use the Unified models that were proposed one typically needs a knowledge base, independent of the specific case under test to be measured. Since the knowledge base is a relatively small addition to a standard statechart, our claim is that Unified models should speed-up the solution of a run-time measurement problem.

6   Discussion

The main argument for software and knowledge integration is that one actually starts software development from fundamental concepts about the system, in other words one starts from structural and behavior knowledge. One should not lose sight of knowledge, when the software is gradually developed.

6.1   Validation

The run-time measurement problem refers both to the overall correctness of the design pattern behavior, and to the efficiency of the pattern behavior. For instance, suppose one expects a low probability of occurrence of certain transitions due to message losses, and the resulting measurements indicate a higher than expected probability. This points out to a system needing redesign.

To validate both correctness and efficiency of the approach, one needs a software tool that is able to fetch/store data from/to the Unified Model, and also is able to interact with the measurement or simulation environment.

For instance, Teplitsky and Exman have described (in ref. [9]) a system to simulate the behavior of the Observer pattern. According to their results one could set for the allowed transitions in this pattern a probability threshold of $p=0.8$. The referred tool – described in a forthcoming paper – would enable actual comparison of the latter threshold with a statistically representative set of measurement results.

6.2   Related Work

There are a few sets of ontologies related to GOF design patterns, most of them structural. For instance, reference [10] contains examples in OWL of a sub-set of GOF patterns. Another set by Kampffmeyer – the design pattern intent ontology [11] – also deals systematically with GOF patterns. Its purpose is more of a classificatory nature, aiming to help the user to find the most adequate design pattern that fits to one's development needs.

There exist also proposals for statechart specific patterns – differing from the GOF ones. Such a proposal by Yacoub and Ammar is found in ref. [12].

The closest kinds of work to ours are the relatively rare application of ontology-based approaches to software behavior. Huang et al. in ref. [13] describe such an application to malware behavioral analysis. Barcellos et al. in ref. [14] deal with software process behaviors within organizations.

The approach proposed here has similarities to Model Checking techniques (see e.g. [15]) using Kripke structures instead of unified statecharts, to make inferences about systems. For example, it describes systems by graphs in which states are labeled with atomic propositions which are true in these states. System models – e.g. of a microwave oven – contain both allowed and 'undesirable' transitions. On the other hand, Model Checking does not require either a running system or measurements.

OCL – the Object Constraint Language – has been used for reasoning about class diagrams (see e.g. Clavel et al. [16] and Queralt and Tenient [17]) and about state diagrams (see e.g. Flake and Muller [18]).

6.3   Future Work

One should extensively investigate a variety of software systems of practical interest. This may lead to a different characterization for behavioral ontologies and the SKB-fields. There could be values of a different nature or even small data structures within states.

Concerning the inference techniques, an important issue is applicability. We mean the kinds of inferences that can be done with the integrated models. One should also consider different forms of time representation, say the operators of temporal logic [15] and explicit real-time constraints.

6.4   Main Contribution

The main contribution of this work is the proposal of Unified *software knowledge* behavioral models. In particular, as a knowledge discovery technique, it is important to model misbehaviors, as well as allowed/legal behaviors. Unified models represent an alternative to disconnected diagrams in an already cluttered UML zoo of diagrams.

References

1. Rumbaugh, J., Jacobson, I., Booch, G.: The Unified Modeling Language Reference Manual, 2nd edn. Addison-Wesley, Boston (2004)
2. UML – Unifying Modelling Language – Version 2.4 Beta 2, OMG – Object Management Group Specification (March 2011), Web site:
   http://www.omg.org/spec/UML/2.4/
3. Selic, B.: On the Semantic Foundations of Standard UML 2.0. In: Bernardo, M., Corradini, F. (eds.) SFM-RT 2004. LNCS, vol. 3185, pp. 181–199. Springer, Heidelberg (2004)
4. Gamma, E., Helm, R., Johnson, R., Vlissides, J.: Design Patterns – Elements of Reusable Object-Oriented Software. Addison-Wesley, Boston (1995)
5. Harel, D.: On Visual Formalisms. Comm. of the ACM 31(5) (1988)
6. Exman, I.: Systematic Construction of Statecharts for Design Patterns (to be published)
7. Kurose, J.F., Ross, K.W.: Computer Networking – A Top-Down Approach, 5th edn. Pearson, Boston (2010)
8. Weisstein, E.W.: Graph, From MathWorld – A Wolfram Web Resource, Web Site:
   http://mathworld.wolfram.com/Graph.html
9. Teplitsky, M., Exman, I.: Measuring Behavioral Software Design Patterns. In: Proc. IEEEI-2006, 24th IEEE Convention of Electrical & Electronics Engineers in Israel, Eilat, Israel, pp. 384–388 (2006)
10. Design Pattern Ontologies,
    http://www.cs.uwm.edu/~alnusair/ontologies/patterns.html

11. Kampffmeyer, H.: The Design Pattern Intent Ontology – Finding the Pattern you need. VDM Verlag Dr. Muller, Saarbrucken (2007)
12. Yacoub, S.M., Ammar, H.H.: A Pattern Language of Statecharts. In: Proc. PLoP 1998 Conference (1998)
13. Huang, H.D., Chuang, T.Y., Tsai, Y.L., Lee, C.S.: Ontology-based Intelligent System for Malware Behavioral Analysis. In: Proc. WCCI 2010 (2010)
14. Barcellos, M.P., Falbo, R.A., Rocha, A.R.: A Well-founded Software Process Behavior Ontology to Support Business Goals Monitoring in High Maturity Software Organizations. In: 14th IEEE Int. Enterprise Distributed Object Computing Conf. Workshops, pp. 253–262 (2010)
15. Clarke Jr., E.M., Grumberg, O., Peled, D.A.: Model Checking. MIT Press, Cambridge (1999)
16. Clavel, M., Egea, M., de Dios, M.A.G.: Checking Unsatisfiability for OCL Constraints. Elec. Comm. EAAST 24, 1–13 (2009)
17. Queralt, A., Teniente, E.: Reasoning on UML Class Diagrams with OCL Constraints. In: Embley, D.W., Olivé, A., Ram, S. (eds.) ER 2006. LNCS, vol. 4215, pp. 497–512. Springer, Heidelberg (2006)
18. Flake, S., Mueller, W.: Formal semantics of static and temporal state-oriented OCL constraints. Softw. Syst. Model (2003)

Structured Knowledge: An Universal Indexing System Approach

Anabel Fraga, Juan Llorens, and Karina Robles

Universidad Carlos III de Madrid, Departamento de Informática,
Av. Universidad 30, Leganés, 28911 Madrid, Spain
{llorens,afraga,krobles}@inf.uc3m.es

Abstract. Knowledge is one of the main assets that humans have, the knowledge achieved in one area may be applied in another different area; all that you need is to remember it and adapt it to the new area or problem. If we apply this concept in computer science, knowledge could be a powerful asset to store (as remember) and reuse (as adapt). Knowledge could be structured using different kinds of Knowledge Organization Systems (KOS), but in all the cases the metamodel is important to be known in order to match with consistency the diverse kinds of knowledge. If it is possible to index any kind of knowledge stored as a KOS in a repository, it means that knowledge coming from diverse sources could be merged in a unique repository. The merge activity is important in the Reuse process because it makes possible to trace different pieces of knowledge at the end it will be retrieved improving the Reuse process and reducing costs at last. A method for indexing structured knowledge is described as well as the algorithms and practical examples in case of the metamodel describing the knowledge is not available.

Keywords: Universal indexing, Knowledge management, Structured information, Metamodel generation, KOS.

1 Introduction

In previous research [1] the major problems found in the systematic reuse have been shown. A new perspective of reuse going back to the origins has been shown as well [1], it allows the process of improving retrieval techniques and methods, dropping investments costs, including traceability in the process and fully integrated into the software development process. Retrieval is one of the major *lost* activities in the reuse process. Diverse proposals arise in order to solve it, and diverse repositories and libraries are supporting storage and retrieval. Domain Analysis is another example trying to solve the retrieval issue, but in this case cost is really high because everything must be modeled a priori for further retrieval. It is a problem in the industry because it is almost forgot, reuse is not applied because of costs and ROI is low or negative in some cases.

Diverse storage and retrieval proposals are present nowadays in terms of unified structures or repositories helping the retrieval and representation tasks, but indexing is

A. Fred et al. (Eds.): IC3K 2011, CCIS 348, pp. 362–373, 2013.

still on top, it depends on the kind of information. An indexer for each type of information is needed, so each type could be retrieved. And here a problem is foreseen: for each type of information an indexer must be developed. If an indexer for any kind of information could be developed then the process of reuse could be really improved reducing cost at last.

Following this research line, the main focus of it is the study of techniques, rules and development of a universal indexer, a transformer autonomous of the kind of information.

Universal Reuse [1] is the notion of knowledge reuse independently of the kind of information, the user that demands the need or even the context where it must be reused Knowledge is an asset important for everyone, particular or company, and it has a peculiar characteristic, it is an asset that is possible to reuse in different situations, by diverse people demanding the reuse. It requires a special treatment or process in order to be reused in any context with the organization.

The reminder of this paper is structured as follows: Section 2 explains the Universal Reuse Representation Schema, Section 3 explains the Universal Reuse Indexing schema, Section 4 explains an overview of the metamodel generation method based on the structured information data, and finally Section 5 includes some conclusions.

2   Universal Reuse Representation

Knowledge is very difficult to accumulate, be sought and be integrated for new needs. One of the basic problems with different types of knowledge is that reusers do not always get what they need from repositories, for reasons that have to do in part with how repositories are created, in part with not up-to-date retrieval techniques, and with almost not existing solutions for smart merging and integrating knowledge within other knowledge. This is a big part of the *window* to be covered by the Knowledge Reuse area [2][3][4][5][6]. A well modelled knowledge implies a well retrieved knowledge later, and the time spent classifying knowledge will imply less time in the retrieval process. So, modelling is a challenge because it must be universal in the assumption of a universal reuse.

The Knowledge Organization Systems (KOS) domain is a formal and well-studied area aiming to create accurate structured knowledge, where Ontologies have a fundamental role. Ontologies play an important role in Knowledge Reuse as they could serve as repositories or even be reusable assets as well [7] [8] [9] [10].

The Universal indexing process could be defined as a process of transforming all sort of knowledge into a format compatible with a universal representation model. So far, not losing information is a very difficult task. First, some rules for transforming information must be implemented and these rules must take into consideration a representation model able to keep any kind of knowledge. One of the main problems to deal with is to index or transform knowledge without loss of information. For that, the rules must be a complete set of transformation rules.

The indexing process deals with:

• A metamodel of the knowledge to be represented must be always accessible, if not it must be extracted somehow. The metamodel provides the structure of the instances and its semantics, providing also the knowledge in an upper level.

• The creation of transformation rules that makes the process in an automatic way, using as input the information and its metamodel.

For us, the structured information can be defined as information that have a data model (metamodel) that explains unambiguously (explicitly) and entirely (completely) the contents that its creator intended to [6]. It means the metamodel and its content must be represented in the same schema in order to be linked and retrieved later on.

2.1   Universal Schema

An universal schema is important for the success of the reuse process. This scheme must be able to keep the information and its metamodel representation, if both levels of representing information could be kept in the same scheme, then we could call it as an Universal Schema. The problem to be solved in the case of modelling is that we need to model any kind of information. For that, we must have a generic metamodel information. Diverse kind of schemas for representing information are available, we choose RSHP because it is a very suitable schema for this research. RSHP [11] [12] is an advantage for this research because any kind of information can be represented on it.

RSHP stands for *RelationSHiP* [13], and it was designed to jointly represent all different types of information using a common meta-model that allows all possible information models to be stored in the same repository, in order to enable traceability between the represented information artifacts.

The philosophy of RSHP is based on the ground idea that knowledge is in essence, related facts, and therefore, it is necessary to bring the relationship itself to the highest priority of a representation model. As a result of this premise, in order to represent information in the RSHP representation model, the main description element to be found within the container of the information to be represented should be the relationship, this relationship is in charge of linking concepts [11] [12].

The advantage here is that RSHP could help us to represent the complete set of information, also some stages in the reuse process is solved thanks to the use of RSHP (i.e. retrieval). If we can solve the representation in RSHP of any kind of information some process like retrieval are solved by reciprocal, because RSHP has a retrieval process solved for the repository and information that it manages (see Figure 4). And in the case that each stage of the general reuse process could be solved in the side of the RSHP schema, then it will be solved for the process itself. Thanks to this schema, knowledge could be stored in a universal repository. These are the advantages of using the RSHP schema, which provides an easy answer to the doubt we had regarding the schema selection. It means, instead of using as for instance RDF because this XML (eXtensible Markup Language) based technology does not provide a retrieval system by default.

3   Universal Reuse Indexing

If the indexing activity is limited to structured information, but not free text which is managed by Natural Language Processing techniques (NLP); then it is possible to design an algorithm for the achievement of this task (Universal Indexing).

3.1   Indexing Process

The indexing process could be defined as a process of transforming knowledge into a storage model without loss of information. So far, not losing information is a very difficult task. First, some rules to transform information must be implemented and these rules must consider a storage model able to keep any kind of knowledge.

As proposed, RSHP is a well suited storage schema and it helps in the process of preserving information due to its inherent properties as a generic representation model. One of the main problems to deal with is to index or transform the knowledge without loss of information. For that, the rules must be a complete set of transformation rules.

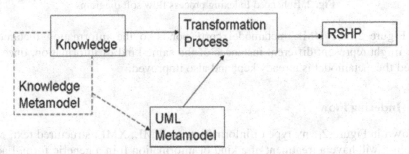

Fig. 1. Indexing process flow diagram

As shown in Figure 1, the information to be indexed needs the existence of its metamodel, in order to keep the meta-relationships between the concepts thanks to the transformation process that follows rules, it transforms the information and its metamodel into the universal repository schema. Thanks to the metamodel, the relationships could be extracted and also information could be semantically treated. At the end, a semantic retrieval is achieved, useful for adapting to another context afterwards.

The metamodel could be represented in UML or it could be extracted if not available using the XML files structure.

Even more, the indexing process could be expanded using auxiliary indexers, as for example NLP indexers (see Figure 2). Each kind of information has its own indexer (T', T''…. and so on), the indexing process is attached to its information kind, and it means an indexer is needed for each of them at the end in at the present, but using the Universal Indexing System this problem could be avoided.

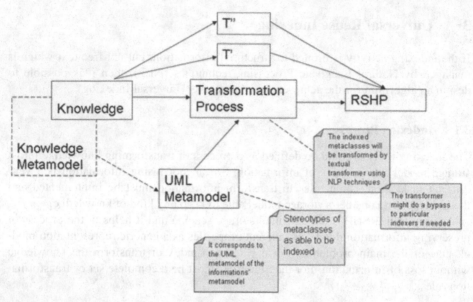

Fig. 2. Enhanced Indexing process flow soft diagram

In Figure 2, the UML metamodel corresponds to the information metamodel, which might represent different instances of the same kind of information, once it is indexed the metamodel is already kept and also improved.

3.2   Indexing Flow

As shown in Figure 2, any type of information as: UML, XML, structured text, code, and so on; will have a treatment of a kind of information Ii in a generic form. Each Ii must be stored and later on retrieved for adapting it to a new context.

As shown in Figure 3, the universal indexing must transform any kind of information (Ii), without loss of it.

For that, a set of rules has been proposed, these rules are implemented in the universal transformer, and it is called Tu. Each Ii requires a metamodel for later transformation, without a metamodel it won't be possible to obtain the relationships between information. Thanks to the metamodel, information relationships are extracted and semantic knowledge is kept in the repository.

Each Ii is a structured information representation, so it could be kept in XML which is widely used for structured information interchange.

The metamodel for each Ii will be represented using UML as modeling language. UML could be represented in a structured schema called: XMI (XML Metadata Interchange). And it will be used in the universal transformer as input.

So far, not losing information is a very difficult task. First, some rules to transform information must be elucidated and these rules must consider a storage model able to keep any kind of knowledge. RSHP is a well suited storage schema and it helps in the process of preserving information due to its inherent properties as a generic representation model. On the other hand, a user should aid in the definition of the information's metamodel. If it is not useful for the final user then there is no need to keep it, so at the end the information will be transformed at all from the user side of view and the user best interest.

A metamodel is required for indexing the information; it is needed in order to know the structure followed by the information. Furthermore, the transformers must support any version or tool application of XMI (XML Metadata Interchange), with independence of the UML version or the Tool providing the XMI file. Thant's the reason why XMI transformers are needed. For that purpose, transformers for XMI files have been generated, using XSLT (Extensible Stylesheet Language Transformations) as technology for transforming XML files to a standard version used in this research: UML 2.1.

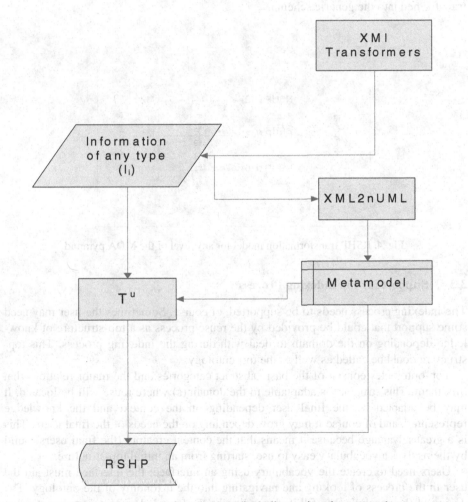

Fig. 3. Universal Indexer process (including main sub processes)

Independently of the kind of information, if we analyze the MDA levels, it is possible to transform information from any level to RSHP[14]. It means that the model and its metamodel could be transformed or indexed in the generic or universal schema for representing any kind of information, as RSHP. As shown in Figure 4, levels M0, M1 and M2 are represented following the MDA approach, but it also could be generalized to apply M3, if exists on another research requirement.

RSHP is a schema that brings the possibility to represent any kind of information, and its relationships, due to its general representation. For instance, if the reuse process has a need for reusing information in the M0 or M1 level of the MDA pyramid (see Figure 4), just because it is reused in the transformation side (M0, M1) inside RSHP (M0 or M1) then it is reused as well in the M0 or M1 levels at the MDA pyramid side by correspondence. It is possible because the information has been fully transformed into the generic schema.

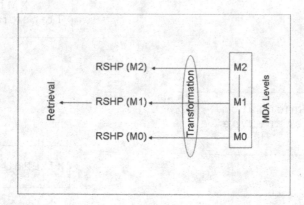

Fig. 4. RSHP transformation model for any level of the MDA pyramid

3.3   Support for the Indexing Process

The indexing process needs to be supported, of course. Sometimes the user may need some support that could be provided by the reuse process as a top-structure of knowledge depending on the domain to deal with during the indexing process. This top-structure could be called as well as the top-ontology.

Top-ontologies consist of the basic abstract categories and the major relations that link them. This structure is adaptable to the domain(s) where reuse will be located. It may be adapted by the final user depending on the context and the knowledge represented, and of course it may grow depending on the needs of the final users. This is a great advantage because it means that the content creators (the final users) build by themselves a vocabulary easy to use, starting from an initial top-structure.

Users need to create the vocabulary using an interface; the interface must aid the user in the process of looking and navigating into the taxonomy of the ontology. The design of the vocabulary to fill it, depending on the knowledge domain to reuse, must be a top-ontology that will make possible to trace any kind of knowledge.

The vocabulary and types of vocabulary in the top-structure will be treated as entities. Entities are just types of information, concepts that are treated as nouns in the taxonomic structure. That's a lot of pre-coordinated work, and it doesn't mention how much maintenance goes into the vocabulary structure.

If the final user is changing and re-structuring the taxonomy day by day depending on the needs s/he has, then at the end of the day the users will finally have a good domain ontology representation for the company/enterprise. The top-structure is ultimately a structure of entities or abstract objects organized by users depending on their needs that aids as final goal in a richer retrieval results.

4   Metamodel Generation Method Based on the Structured Information Data

If the metamodel is not available, a reverse engineering process is needed. It is called XML2nUML inside the universal indexer process. The metamodel is extracted using XML information as foundation. The metamodel could be described by a human, but automatic generation is designed in order to be independent for better validation.

Table 1. Indexer algorithm in pseudocode

Algorithm: Indexer(Metamodel, Information_source[1])
1. For each Information_source corresponding to *Metamodel*:   *// Initialization*   2. If not(Existing_Metamodel)   *// Parse Metamodel*   3.   Nodes := Parse_Metamodel(nodes in   Information_source)   4.   Load_XML_structures_in memory(Nodes)   5.   GenMetamodel := Generate_Metamodel(Information_source)   6. Else if (Existing_Metamodel)   7.   GenMetamodel := Load_Metamodel(Metamodel)   8. endIf   *//In case of a new metamodel, it is generated to be univocal, reused and extended*   9. If isNew(GenMetamodel)   10.   IdMetamodel := Save_Metamodel(GenMetamodel)   11. Else   12.   IdMetamodel := Retrieve_Metamodel_Id(GenMetamodel)

---

[1] Each information_source corresponds to each Ii (see Figure 3).

```
13. endIf

   //The metamodel is loaded

14. LoadedMetamodel := Load_metamodel _inMemory(IdMetamodel)
15. Parse_content(Information_source, LoadedMetamodel)

   // Parse   Instances
```

The method in the pseudocode algorithm called GenMetamodel (base for XML2nUML) responsible for generating the metamodel could be explained in simple rules, following some of the Jensen's principles [15] as follows:

1. The namespaces of the XML file will be transformed as Models in UML.

2. The main namespace will be the main Model element.

3. Each element in the XML file will be transformed as a Class, and it will be part of the main Model element.

4. If an XML element has XML attributes, each XML attribute will be transformed as attribute of the UML Class.

5. If an XML element has sub-elements, each sub-element will be treated as a Class and related with its parent Class as an UML composition relationship between classes. Also part of the main Model element.

6. An association is created if available attributes as HREF, IDREF, ID between the elements.

4.1   Metamodel Example

We provided an example in the domain of orchestral instruments to illustrate the idea of metamodel generation.

For instance, if an XML document contains the information regarding a KOS, for instance type Thesauri:

```
<?xml-stylesheet type="text/xsl" href="thesauri.xsl"?>
  <Domain>
    <DomainName>Orchestral Instruments</DomainName>
    <Descriptor>
      <Name>A</Name>
      <ScopeNote>Musical note</ScopeNote>
      <Synonym>La</Synonym>
    </Descriptor>
    <Descriptor>
      <Name>Wolfgang Amadeus Mozart</Name>
    </Descriptor>
...

    <Relation>
```

```
      <Descriptor1>Wolfgang Amadeus Mozart
</Descriptor1>
      <Descriptor2>Tuba</Descriptor2>
      <AssociationKind>
        <Association
 AssociationName="has written for">
</Association>
      </AssociationKind>
    </Relation>
    <Relation>
      <Descriptor1>Orchestra Instruments
</Descriptor1>
      <Descriptor2>Wood Block</Descriptor2>
      <RelationKind>Hierarchy</RelationKind>
    </Relation>
    <Relation>
      <Descriptor1>Wood Block</Descriptor1>
      <Descriptor2>Undetermined Wood
</Descriptor2>
      <AssociationKind>
<Association AssociationName="belongs to">
</Association>
      </AssociationKind>
    </Relation>
    <Relation>
      <Descriptor1>Wind</Descriptor1>
      <Descriptor2>Woodwind</Descriptor2>
      <RelationKind>Hierarchy</RelationKind>
    </Relation>
</Domain>
```

The resulting schema extracted without the available metamodel is as follows:

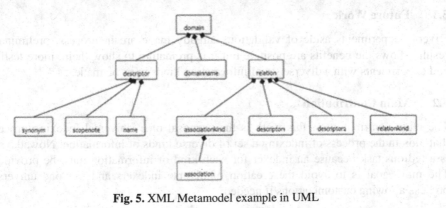

Fig. 5. XML Metamodel example in UML

The example shown in Figure 5 is the corresponding metamodel extracted from one XML file that correspond to a particular information kind. It is interesting to notice that an analogous metamodel may be extracted from a completely different XML information source.

Additional relationships between elements could not be automatically extracted because a misunderstanding in semantic could be generated. The user must include additional relationships thanks to the metamodel defined by s/he at the moment of creating the contract for reuse.

5   Conclusions

The classical systematic reuse process failed in the industry environment because of the huge investment needed to be accomplished by practitioners. Low or negative ROI ratios became one of the key problems for its wide-spreading. Aside ad-hoc reusers also gained a certain level of success but the accomplishment level is low, reuse is only applied to code, dlls and components, and the practice of this reuse has been chaotic.

Industry would get worth of dealing with any kind of knowledge, in any context, and by any user: anything, anywhere, and anybody. For that reason, we offer the concept of UKR as an open door to get all the benefits of theoretical reuse avoiding the well-known drawbacks of systematic and ad-hoc reuse.

The universal knowledge reuse process has to deal with: a universal representation model, a universal indexer, a universal retrieval and adaptation activities, a universal accessing, knowledge visualization and a universal reuse metric. All of these activities have to face the issue that each one could be applied to any kind of knowledge, in any context, and each activity might be required by any user. In this paper the Indexing system is presented and extended, providing its algorithms, problems and facing solutions.

The universal indexing system explained in this paper provides a method, algorithms and examples of generating a metamodel based on the structured knowledge available. Problems as locating a suitable universal schema and the retrieval method are also explained in the schema selection (RSHP) which provides a suitable option for this case.

5.1   Future Work

Diverse experiments aside of validations already made are in process, preliminary results shows the benefits are positive but it is premature to show them, more testing and experiments with a diverse set of information kinds must be made.

5.2   Main Contribution

The main contribution of this work is the proposal of a Universal Indexing process that aids in the process of indexing a set of diverse kinds of information. Nowadays it is a tedious task because an indexer for each kind of information must be provided. The main goal is to avoid the creation of diverse indexers and use one universal process allowing customizations if needed.

Acknowledgements. Many thanks also to Prof. Dr. Iaakov Exman for reading this paper and giving us valuable improvements.

References

1. Fraga, A., Llorens, J.: Universal Knowledge Reuse: anything, anywhere, and anybody. In: Proceedings of the International Conference of Software Reuse. International Workshop on Knowledge Reuse, KREUSE 2008/ICSR 2008, Beijing, China (2008) ISBN: 978-84-691-3166-4
2. Knowledge acquisition and retrieval apparatus and method. US Patent Issued on (August 26, 2003), http://www.patentstorm.us/patents/6611841-description.html (last visited on November 15, 2008)
3. Davis, R., Shrobe, H., Szolovits, P.: What is a Knowledge Representation? AI Magazine 14(1), 17–33 (1993)
4. Markman, A.B.: Knowledge Representation. Lawrence Erlbaum Associates (1998)
5. Brachman, R.J.: What IS-A is and isn't. An Analysis of Taxonomic Links in Semantic Networks. IEEE Computer 16(10) (October 1983)
6. Sowa, J.F.: Knowledge Representation: Logical, Philosophical, and Computational Foundations. Brooks/Cole, New York (2000)
7. KOS: Knowledge Organisation Systems, http://www.db.dk/bh/lifeboat_ko/concepts/knowledge_organizat ion_systems.html (last visited November 15, 2008)
8. Bechofer, S., Goble, C.: Thesaurus construction through knowledge representation. Data & Knowledge Engineering 37, 25–45 (2001)
9. Hill, et al.: Integration of Knowledge Organization Systems into Digital Library Architectures (2002), ASIST SigCR – http://www.lub.lu.se/SEMKOS/docs/Hill_KOSpaper7-2-final.doc. (last visited March 15, 2009)
10. Janée, G., Ikeda, S., Hill, L.: ADL Thesaurus Protocol v1.0. (2002), http://www.alexandria.ucsb.edu/thesaurus/protocol http://nkos.slis.kent.edu/2002workshop/janee.ppt (last visited March 15, 2009)
11. Llorens, J., Morato, J., Genova, G.: RSHP: An information representation model based on relationships. In: Damiani, E., Jain, L.C., Madravio, M. (eds.) Soft Computing in Software Engineering. STUDFUZZ, vol. 159, pp. 221–253 (2004), Available for reviewers in http://www.ie.inf.uc3m.es/llorens/ICSR.zip
12. Lloréns, J., Fuentes, J.M., Prieto-Diaz, R., Astudillo, H.: Incremental Software Reuse. In: Morisio, M. (ed.) ICSR 2006. LNCS, vol. 4039, pp. 386–389. Springer, Heidelberg (2006)
13. Llorens, J., Morato, J., Genova, G.: RSHP: An information representation model based on relationships. In: Damiani, E., Jain, L.C., Madravio, M. (eds.) Soft Computing in Software Engineering. STUDFUZZ, vol. 159, pp. 221–253. Springer (2004), Available for reviewers in ftp://www.ie.inf.uc3m.es/llorens/ICSR.zip
14. Bézivin, J.: On the Unification Power of Models. SOSYM 4(2), 171–188 (2005)
15. Jensen, M., Moller, T., Pedersen, T.: Converting XML Data to UML Diagrams For Conceptual Data Integration. In: 1st International Workshop on Data Integration over the Web (DIWeb) at 13th Conference on Advanced Information Systems Engineering, CAiSE 2001 (2001)

Part III

Knowledge Management and Information Sharing

The Role of External Information Awareness
and Proactiveness of Innovation Strategy in Employees'
Innovation Behavior: A Psychological Perspective

Jing Tang, Loo Geok Pee, and Junichi Iijima

Department of Industrial Engineering and Management,
Tokyo Institue of Techonology, Tokyo, Japan
{tang.j.aa,peelg.aa,iijima.j.aa}@m.titech.ac.jp

Abstract. Due to innovation is highly knowledge intensive, employees' innovation behavior plays a central role in knowledge creation and distribution in organizations. Therefore, in knowledge management initiatives, it is important to encourage employees' IB, which involves developing, promoting, judging, distributing and implementing new ideas at work. From a psychological perspective, this study applies the theory of planned behavior (TPB) to better understand employees' IB, and also extends TPB by considering the effects of two unexamined yet important organizational factors: external information awareness (EIA) and proactiveness of innovation strategy (PIS). Results from a survey of employees in Japanese organizations indicate that EIA and PIS are positively related with employees' attitude towards innovation, subjective norm about innovation, and perceived behavioral control to innovation, which is, in turn, significantly influence employees' IB. Employees' attitude, subjective norm, and perceived behavior control mediate partially the effects of EIA and completely the influence of PIS to employees' IB. These findings provide directions for more efficient employees' IB encouragement, by focusing on improving perceived behavior control, EIA and PIS.

Keywords: Employees' innovation behavior, External information awareness, Proactiveness of innovation strategy, Theory of planned behavior.

1   Introduction

In the last decade, the fast emergence and diffusion of information technology led to the customer's need becoming a trend of diversified, which is, in turn, stressed the importance of innovation to competitive advantage and long-term survival for companies. Recently, innovations have often occurred in the cumulative and diffuse process of knowledge [14]. But interactivity and complexity of innovation both within and across organizations advance new challenges in exploration and exploitation knowledge, which is one cardinal foundation of "organizational innovative potential" [41]. Growing emphasis on innovation through knowledge management, innovation behavior (IB) of knowledge workers, which involves developing, promoting, judging, distributing, and implementing new ideas, is the primary source for organizational

A. Fred et al. (Eds.): IC3K 2011, CCIS 348, pp. 377–392, 2013.
© Springer-Verlag Berlin Heidelberg 2013

innovation [20]; [37]; [41]. Turgoose [43] suggests that the acceptance rate of ideas suggested by employees positively influences organizational performance. The process research of knowledge management and innovation also emphasizes the importance of employees' innovation behavior. It suggests that innovation in organizations is a "relay race" based on successful connection of individual innovations along continuous stages from new idea and knowledge initiation to implementation [11]; [23]; [41]. Hence, employees' innovation is indispensable for organization success and it is important to understand individual employees' innovation behavior.

In this study, innovation behavior is defined as employees' behavior "directed towards the initiation and intentional introduction of new and useful ideas, processes, products, or procedures" at work [11]. In order to encourage employees' innovation behavior, prior research has tried to identify antecedents of employees' innovation behavior. Examples include leadership, work groups relationship, multifunctionality of jobs, organizational knowledge structure, and external work contacts [11]; [13]; [37]. These studies provide many advices for managers, but few studies have focused on the psychological analysis of employees' innovation behavior. Scott and Bruce [37] suggest that, at individual level, employees' innovation behavior is a primary response to cognitive meaningful and feasible interpretation of situations, which is more integrative, rather than to the situations per se. So there is a path model of individual innovation from situational and personal characteristics to psychological factors, and to behavior sequentially. Hence, we consider a structured study of employees' innovation behavior's psychological antecedents is important, because it helps to improve management efficiency by focusing on those factors related to the more effective psychological antecedents of employees' innovation instead of paying attention to everything. In this paper, the theory of planned behavior (TPB), which is a well-conceived psychological model, is used to explain employees' innovation behavior. Attitude towards innovation (ATT), subjective norm about innovation (SN) and perceived behavioral control to innovation (PBC) are expected to influence it.

The organization provides basic conditional environment for individual behaviors, so organizational factors have potential influence on employees' innovation behavior [11]; [25]; [37]. In this study, we consider two factors that have been neglected in prior research: external information awareness (EIA) and proactiveness of innovation strategy (PIS). Now, in the knowledge-based economy, accessing knowledge external to organizations offers potential technology and market opportunities to innovate. Meanwhile, tight customer-supplier relationships could also raise employees' consciousness of the importance of innovation. From the resource-based view (RBV), proactive strategy is positive with firm performance when proactiveness bolsters firms to develop some competitive advantages [5]. As the highly-qualified innovative employee is treated as one important strategic resource for firms in terms of knowledge creation and diffusion to keep long-term competitive advantage [28], there is a lack of research considering the relationship between proactiveness of innovation strategy and employees' innovation. In this study, we posit that external information awareness and proactiveness of innovation strategy could prompt employees to engage in innovation behavior and improve the explanatory power of TPB for employees' innovation behavior.

As previous empirical research on innovation behavior has been mainly conducted in western countries, this study addresses the gap by collecting data in Japan which is among the most innovative but own specific different culture [14]. In general, Japanese companies view innovation as an overall collaboration rather than a task for limited teams [15]. It is therefore interesting to examine what motivates Japanese employees to engage in innovation behavior.

In sum, the research questions addressed in this study are:

RQ1: What are the psychological factors influencing employees' innovation behavior?

RQ2: Does organizational external information awareness and proactiveness of innovation strategy influence employees' innovation behavior?

Based on a questionnaire survey in Japan, we found that employees' attitude, subjective norm, and perceived behavioral control are positively related to their innovation behavior as predicted by TPB. More interestingly, they partially mediate the effects of external information awareness and completely mediate the influence of proactive innovation strategy. This study potentially contributes to research and practice in several ways. First, this is the first study to apply TPB to study employees' innovation behavior. The findings indicate that TPB is suitable for understanding the behavior. Second, we examine how the factors in the TPB mediate the influences of organizational factors on employees' innovation behavior. This provides explanations for how organizational factors influence employees' innovation behavior. Together, these findings offer insights into how innovation behavior may be promoted in organizations. Third, this is the first study to examine the innovation behavior of employees in Japanese companies, which is a highly relevant but understudied context.

2   Conceptual Background

2.1 Innovation Behavior and Theory of Planned Behavior

As innovation is about both newness and profit, it is not just only related to idea generation, but also the application and commercialization of new ideas [37]. According to the research of De Jong and Den Hartog [11], employees' innovation behavior refers to the behavior "directed towards the initiation and intentional introduction of new and useful ideas, processes, products, or procedures" at work. Prior research on innovation behavior has identified many individual and organizational antecedents or motivators from different perspectives, such as self-confidence, problem-solving style, leadership, work group relationship, job autonomy, organizational knowledge structure, and organizational support [11]; [13]; [37].

However, few prior studies have focused on the psychological antecedents of employees' innovation behavior. As behavior is a result of rational decision of individuals based on the judgment of "perceived" existence of related preconditions, the associated psychological process and psychological factors are important [1]. So, "perceived" psychological antecedents are more directly related to innovation behavior rather than other organizational and environmental factors. It addresses the importance of psychological analysis of employees' innovation behavior.

This study use the theory of planned behavior to provide explanations of psychological influences on employees' innovation behavior, because of its high predictive power in predicting various employees' behaviors [21]; [32]; [45]. TPB is a deliberative processing model in which individuals make behavioral decisions based on careful consideration of available information [1]. It posits that human behavior is preceded by intention formation and that intention is determined by individuals' attitude, subjective norm, and perceived behavioral control. Since this study is cross-sectional, we focus on actual behavior rather than intention. Attitude is a personal evaluation or interest about performing the target behavior by an individual. Subjective norm reflects the individual's perception of social influence and pressure from relevant social constituents such as peers and superiors about the necessity to perform the target behavior. Perceived behavioral control reflects the perceived existence of necessary facilitators (e.g. time, ability) to successfully perform the target behavior. In other words, perceived behavioral control is an assessment of the ability to overcome possible obstacles for performing the target behavior [1].

2.2 External Information Awareness

External information is an important driver of innovation that provides signals of market and technological trend and extends limited internal innovation capability [10]. In a volatile environment where customer needs and technology changes rapidly, organizations need to maintain strong relationships with their environmental constituents in their innovation endeavor. It has been emphasized that firms should openly "use external ideas as well as internal ideas", especially those from key customers, suppliers, competitors, research organizations and market to accelerate innovation. Chesbrough [8] names it as open innovation. These suggest that it is important for organizations to have strong external information awareness.

External information awareness refers to the extent to which organizations track best performers, main competitors and technologies in the industries, and maintain contact with suppliers, customers, and the government to gather information from the external environment [31]; [44]. In firms with active network to access both internal and external knowledge and expertise, employees' awareness and access of external knowledge and knowledge sharing among employees will be strengthened also [9]. Even many scholars certified empirical linkage between external information awareness and innovation performance on the organizational level [42], but few prior studies have considered the potential influence from external information awareness to employees' innovation behavior for its capability to bolster employees' external information and knowledge access. This study provides new insights by examining how external information awareness influences innovation behavior through affecting employees' attitude, subjective norm, and perceived behavioral control.

2.3 Proactiveness of Innovation Strategy

Innovation strategy guides organizations' innovation endeavor [30]; [35]. An important aspect of innovation strategy is proactiveness. Proactiveness "implies taking initiative, aggressively pursuing ventures, and being at the forefront of efforts to shape the environment in ways that benefit the firm," which is opposite with reactiveness [24]. In other words, proactiveness of innovation strategy refers to the organization's quickness to innovate and to introduce new products or services. According to the resource-based view (RBV), proactiveness is posited to be positively related to firm performance when firms some develop competitive advantages from proactive strategy, while a reactive strategy of innovation is considered to be not effective to keep long-term success in a dynamic and sophisticated environment [5].

A proactive innovation firm is likely to be a leader rather than a follower [30]. Slater [39] identified four types of innovation strategy: early market innovator, early adopter, mainstream market, and conservationist (late majority and sluggards). Early market innovators are those firms which "appreciate innovation for its own sake" and continuously focus on discovering new needs of customers. Early adopters are those that are sensitive to new market trends and actively "adopt and use innovation to achieve a revolutionary improvement". Firms in the mainstream market are those that are sensitive to innovation risks and prefer to conduct a "mature" innovation that already confirmed by the market and with low risk. Conservationists are those firms that are highly conservative or averse to innovation. As innovative employees and tacit innovation processes are considered as hard-to-imitate strategic resources to generate competitive advantages [28], the potential linkage between proactiveness of innovation strategy and employees' innovation behavior can deepen our understanding of the positive effect of proactiveness on firm performance. This research considers this potential linkage.

3   Proposed Model and Hypotheses

3.1 The Effects of Attitude, Subjective Norm and Perceived Behavioral Control

Attitude is a person's evaluation or interest about performing the target behavior by an individual, which is strongly related to the perception of behavior-associated outcomes and "the strength of these associates" [2]. Based on the "principle of compatibility", employees' innovation behavior should be anticipated by their attitude toward innovation [2]. That is to say, as a general rule, employees tend to do innovation when they view it as beneficial or favorable, as it has high possibility of increasing their job efficiency and reputation in the workplace. Lee and Wong [26] points out the positive relationship between attitude and performance of R&D workers. Williams [46] also certifies the empirical link between attitude towards divergent thinking, which is "an integral process in creativity", and employees' creation. In addition, innovators are often with higher attitude toward innovation than non-innovators [33]. So we hypothesize that:

Hypothesis 1a: Employees' attitude toward innovation is positively associated with their innovation behavior.

Subjective norm reflects an individual's perception of social encouragement and pressure from relevant social referents such as peers and superiors about the necessity to perform the target behavior [2]. When key social referents in the workplace seem to all "suggest" employees to conduct innovation behavior, they are likely to feel pressured to engage in innovation. For example, Amabile [3] considers that leaders' expectations are important for employees' creative work, and Amo [4] indicates that the perceived opinion of "important others" as managers and colleagues influence health-care workers' innovation behavior. Meanwhile, CEOs' commitment toward innovation indicates the importance of innovation in firm's development strategy and customers' new service or product requirements and expectations and they compel employees to innovate continuously. In addition, from the process view of innovation, the implementation of new innovation ideas demands heavily on the engagement of these key social referents. Then we hypothesize that:

Hypothesis 1b: Employees' subjective norm about innovation is positively associated to their innovation behavior.

Perceived behavioral control reflects the perceived existence or absence of necessary non-volitional facilitators (e.g. time, ability) to successfully perform the target behavior [2]. Facilitators such as opportunities for innovation, freedom to innovate, and resources provided by organization are important, as they provide the basic "physical" preconditions for employees to carry out innovation. Hence, the existence of these facilitators is another independent factor which will be considered when employees make a rational decision of conducting innovation behavior. In support, it has been found that perception of organizational innovation support and resource supply, which is an important part of PBC, has strong positive effects on employees' innovation behavior [37]. Accordingly we postulate that:

Hypothesis 1c: Employees' perceived behavioral control to innovation is positively associated with their innovation behavior.

3.2 The Effects of Organizational Context

The Effects of External Information Awareness. Nowadays, with increased globalization, innovation requires firms utilize both internal and external innovation sources to advance their R&D capability [8]. High external information awareness means organizations tend to be highly open to environment to absorb external knowledge or gain complementary resources. Since these external sources accesses compensate the lack of internal ability, external information awareness enhances the employees' perception of innovation success and support innovative initiatives, especially for employees in innovation-adopter firms. Meanwhile, external information awareness will improve employees' perception of innovation necessity. So, in organizations with strong external information awareness employees will develop a positive attitude towards innovation.

Hypothesis 2a: External information awareness is positively associated with employees' attitude towards innovation.

As external information awareness could also improve CEO and senior managers' perceived necessity to innovate, they will tend to persuade and require employees to do innovation through assigning more innovation-related tasks and giving more innovation rewards. At the same time, employees may also feel more innovative pressure from direct contact with external stakeholders such as customers. In addition, closely cooperation among employees are always needed in the innovation project corresponding to external customer requirements, so the innovation pressure from colleagues will be enhanced sequentially. Then, we posit that:

Hypothesis 2b: External information awareness is positively associated with employees' subjective norm about innovation.

External information awareness is associated with a wider knowledge and technology base to innovate. Employees are likely to get more innovation support and freedom in extrovert firms. External information about customers, suppliers and competitors is a trigger of employees' innovation to provide innovation hints, and external resources may also extend employees' innovation capability. In addition, tight customer-producer relationships speed up the feedback of innovation. So external information awareness will do not only improve employees' self-efficacy beliefs about their capability to do innovation, but also directly provide technology and market opportunities for employees' innovation behavior.

Hypothesis 2c: External information awareness is positively associated with employees' perceived behavioral control to innovation.

Hypothesis 3: External information awareness is positively associated with employees' innovation behavior.

The Effects of Proactiveness of Innovation Strategy. Proactiveness of innovation strategy refers to an organization's quickness to innovate and the speed to introduce new products or services according to new market opportunities [30]. Proactiveness of innovation strategy reflects the high priority of innovation inside organizations. In proactive organizations, employees' proactive innovation is more appreciated than in reactive organizations. According to the theory of organizational alignment, rewards system should be aligned with the strategic goals and values, so innovators are expected to receive more formal or informal organizational rewards [36]. Hence, employees will develop more positive attitude toward their innovation behaviors.

Hypothesis 4a: Proactiveness of innovation strategy is positively associated with employees' attitude towards innovation.

Innovation strategy directly reflects administrators' expectation of employees' work, and the highlight of innovation management. Managers in organizations with proactive innovation strategy are likely to focus more on continuous generation and implementation of new ideas actively to react to new market trends quickly. Consequently, employees are likely to feel more pressure from social referents to innovate.

Hypothesis 4b: Proactiveness of innovation strategy is positively associated with employees' subjective norm about innovation.

Proactive organizations invest more in R&D and human capital than reactive ones [5], so they are likely to provide more support (i.e., money, times, and opportunities) for employees' innovation. With the aim to innovate quickly, proactive organizations are likely to give faster feedback, and implement employees' idea bravely.

Hypothesis 4c: Proactiveness of innovation strategy is positively associated with employees' perceived behavioral control to innovation.

4   Survey Instrument Development

The proposed model was assessed with data collected in a survey. The questions related to attitude, subjective norm, perceived behavioral control and employees' innovation behavior were adapted from prior studies applying TPB [6]; [16]; [29]. Attitude was assessed with four questions: "...engaging in innovation behavior is enjoyable", "...innovation behavior is valuable", "...innovation behavior is beneficial", and "...innovation behavior is favorable". Subjective norm was measured in terms of perceived innovation encouragement and pressure from CEOs, supervisors, colleagues, and customers (e.g., "...receive innovation encouragement and competitive pressure from customers to innovate"). Perceived behavioral control was assessed in terms of perceived existence of resources for innovation (e.g., technology, financial support), opportunities for innovation, freedom to innovate, and feedback, such as "there are many opportunities for employees to innovate in my company" and "employees in my company are given the freedom to innovate at work". Employees' innovation behavior was measured with four questions related to frequency of innovation, time spent on innovation, activeness in innovation, and participation in innovation projects. Examples include "...innovate actively" and "...spend significant time innovating at work". Measures for external information awareness were developed based on prior studies [22]; [31]; [40] and focuses on the capture and sharing of information about market trend, government policy, customers, suppliers, competitors, and strategic partners such as research and development institutes and consultants. All items were measured on a five-point Likert scale anchored by "strongly disagree" (1) and "strongly agree" (5).

Proactiveness of innovation strategy was measured by four multiplicative measures. The multiplicative measures were based on the forms of proactive innovation strategy identified by Slater [39]: early market innovator, early adopter, mainstream market, and conservationist (late majority and sluggards). Early market strategy is perceived when firms continuously consider expressed and latent customers needs in conducting innovation to create new market trends (item PIS1). Early adoption strategy is perceived when firms are sensitive to follow new market trends that created by competitors (item PIS2). Mainstream strategy is perceived when firms prefer to capture external market trend, but postpone implementing innovation until it becomes mature inside organization (item PIS3). Conservation strategy is perceived when firms develop non-active attitude toward innovation (item PIS4). Among these designs, early market strategy is the highest level of proactiveness of innovation strategy, while conservation strategy is the lowest one. These four items form a Guttman-type scale. Actually, these four strategies are on a development continuum for firms to accommodate environment changes. Firms mainly using early market strategy should also relatively emphasize early adoption strategy in order to timely alter wrong market expectations. Firms mainly using early adoption strategy should also considering early market strategy and mainstream strategy to capture high potential first-mover advantages and avoid high financial risks. Similarly, firms mainly with mainstream strategy will consider early adoption strategy and conservation strategy, and firms mainly with conservation strategy will try mainstream strategy under some situations.

Four control variables: industry, firm size, department, and job position were included in this research. Industry was measured as a categorical variable indicated by respondents as either from manufacturing (1) or non-manufacturing sectors (0). Firm size was measured by the number of employees. Department was measured as a categorical variable indicated by respondents as either from IT-related (1) or non IT-related departments (0). Job position was measured by the hierarchical level of respondents from employee (1) to department head (3).

5   Data Collection and Analysis

A survey was conducted in Japanese companies with the support of the Japanese Innovation Management College in late 2010. The questionnaire was sent to 1,819 employees listed in the database of NTT DATA Corporation. We received 127 completed responses and the response rate is 7 percent. Most of responses are from large organizations with more than 1000 employees (64.6 percent). The majority of respondents work in manufacturing sector (70 percent), but also some in non-manufacturing sector as distribution service, transportation and finance. Among the respondents, 74.8 percent of responses are from corporate development department. In terms of job position, 48.8 percent are section heads, and 32.3 percent are non-managerial staff.

5.1 Tests of Measurement Model

The proposed model was assessed using Partial Least Squares (PLS) analysis. We employed Smart PLS version 2.0 and the Bootstrap resampling method with 1000 resamples. All scales show high internal consistency and reliability. The Cronbach's alpha estimates for attitude, subjective norm, perceived behavioral control, employees' innovation behavior and external information awareness shown in Table 1 were all above the recommend threshold of 0.70. In structural equation modeling (SEM), composite reliability (CR) is also used to value the reliability of constructs, and the suggested threshold of it is 0.70 [7]. In Table 1, all CRs of constructs are above 0.85. In addition, the loadings of each item to constructs are also significant at $p<0.001$.

Convergent validity is assessed by average variance extracted (AVE) and factor analysis. In Table 1, all AVEs are above the recommended acceptable value of 0.50 [7]. The exploratory maximum-likelihood factor analysis with Equamax rotation supports our proposed evaluation of constructs. Five corresponding factors are extracted. Next, an acceptable individual reliability of item is shown by the item loadings to their related constructs being above 0.70. In our study, the loadings of each item to constructs in the sample are all above the recommended benchmark of 0.70 [7].

Table 1. Psychometric properties of constructs and construct correlations

Construct	Cronbach's Alpha	AVE	CR	Construct Correlation				
				IB	ATT	SN	BC	EIA
Employees' Innovation Behavior (IB)	.89	.75	.92	**.87**				
Attitude towards Innovation (A)	.92	.81	.94	.54	**.90**			
Subjective Norm about Innovation (SN)	.80	.63	.87	.48	.44	**.79**		
Perceived Behavioral Control to Innovation (PBC)	.83	.66	.88	.54	.57	.41	**.81**	
External Information Awareness (EIA)	.84	.62	.89	.36	.25	.26	.24	**.79**
Proactiveness of Innovation Strategy (PIS)*	-	-	-	.37	.38	.39	.40	.19
* The Cronbach's alpha, AVE, and CR of PIS is not computed as it is measured with a Guttman scale								

The discriminant validity demonstrates the difference of construct measures in the research model. Results of comparing square root of AVEs and constructs correlation coefficients support the adequate discriminant validity of our questionnaire. In construct correlation part of Table 1, bold numbers in the diagonal are the square roots of AVE, while off-diagonal numbers are Kendall's tau correlation coefficients among constructs. Kendall's tau correlation coefficient is better measure of correlations of ordinal variables, which can be interpreted as same as Pearson correlation coefficient [27]. In Table 1, none of the constructs correlation coefficients is bigger than the corresponding square roots of AVE, which means all constructs are more correlated with their own measuring items than with any other constructs.

A test of multicollinearity was also conducted. The highest variance inflation factors (VIF) is 2.29, which is well below the threshold value of 3.3, suggesting that multicollinearity is unlikely to be a problem for our data [12].

5.2 Tests of Structural Model

Figure 1 shows the results of the structural model. All hypotheses were supported at 0.05 level. Among the control variables, only industry had a significantly effects on employees' innovation behavior. All related factors explained 71% of the variance in employees' innovation behavior. Among three psychological predictors, perceived behavioral control has stronger influence on the behavior than the other two factors. Furthermore, external information awareness and proactiveness of innovation strategy explain about 30% in employees' attitude, subjective norm, and perceived behavioral control. But the effect of external information awareness on employees' innovation behavior is significant, while that of proactiveness of innovation strategy is not.

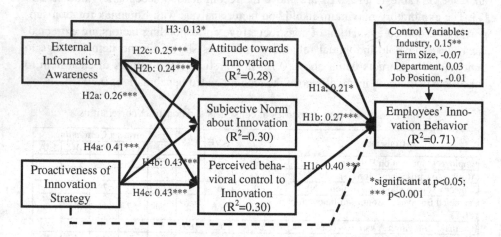

Fig. 1. Results of structural model

Table 2. Mediation test statistic

Independent Variable	Mediation Test Statistic	Mediator		
		Attitude towards Innovation (ATT)	Subjective Norm about Innovation (SN)	Perceived Behavioral Control to Innovation (PBC)
External Information Awareness (EIA)	Sobel	2.17*	2.44*	2.73*
	Aroian	2.12*	2.40*	2.69*
	Goodman	2.23*	2.48*	2.76*
Proactiveness of Innovation Strategy (PIS)	Sobel	2.45*	3.66***	3.53***
	Aroian	2.42*	3.63***	3.50***
	Goodman	2.49*	3.69***	3.57***
* significant at p<0.05, *** p<0.001				

6   Discussion

6.1 Implications for Research

As employees own limited ability to process all kinds of stimuli around them, they use affective and cognitive representations of related information to handle the complexity [14]. Although previous studies have examined employees' innovation behavior [11], there are few empirical studies on the psychological analysis of employees' innovation. This research applied the theory of planned behavior to empirically understand employees' innovation behavior. Results of the survey support all our hypotheses and help us to answer our two research questions. Attitudes, subjective norms, and perceived behavior control are three primary psychological factors influencing employees' innovation behavior. External information awareness and proactiveness of innovation strategy, as two important organizational factors is positively related to employees' innovation behavior through improving three mediating psychological factors.

The first contribution of this study is to examine the power of TPB model in explaining the innovation behavior of employees. While prior research only focuses on the influence of some psychological factors, TPB provides a strong theoretical structure to this study to understand the effects of psychological factors to employees' innovation behavior. Through comparing path coefficients of the three psychological antecedents, our results show that perceived behavioral control to innovation has the strongest influence on employees' innovation among the three antecedents. Therefore, future studies may examine ways to improve perceived behavioral control.

The second contribution of this paper is examining the effects of two organizational factors: external information awareness and proactiveness of innovation strategy on employees' innovation behavior. In empirical studies of TPB, only individual psychological factors are considered and the potential influences of other factors, especially organizational factors, have been neglected. But in practical management, those organizational factors are likely to be more controllable by managers than individual factors. Hence, this study addresses a limitation of prior research. TPB also suggests that organizational factors may influence the way that employees perceive things or actions, and, as a result, affect behavior [2]. Similarly, we have shown that the effects of external information awareness and proactiveness of innovation strategy are mediated through attitude, subjective norm and perceived behavioral control. Understanding this mediating relationship is important because it empirically demonstrates of the underlying mechanism through which organizational factors influence employees' behavior. Our results also suggest that external information awareness is also an important trigger for employees' innovation, when both of them can improve psychological motivation to innovate. This may also help to explain the importance of external information, and the positive relationship between proactive innovation strategies with firm performance, when employees' innovation behavior becomes an important strategic resource to gain and maintain competitive advantage.

Third, this study is among the first to collect data from Japanese companies. Based on its unique continuous innovation strategy, Japan owns a big market share in some industries like automobiles and electronics [14]. Recent studies on employees' innovation behavior focusing on the effect of factors such as job design and leadership have mainly been conducted in the western countries as the United Stated and Netherlands [11]; [34]; [37]; [38]. However, Japan is generally considered to be culturally different from these countries in terms of social collectivism, privilege preference, seniority-based social status, tolerance of hierarchy and risk aversion [19]. It is therefore interesting to examine whether their findings apply to Japan. This study suggests that the findings of studies in other countries are possible to apply to Japan, but the careful reconsideration based on Japanese culture is also needed.

6.2 Implications for Practice

Our results show that attitude, subjective norm, and perceived behavioral control are important to predict employees' innovation behavior. Among them, perceived behavioral control has stronger effect on employees' innovation behavior than the others. Thus, in order to effectively encourage employees' innovation behavior, managers may play supporting role rather than deciding and persuading role to increase employees' perceived controllability and self-efficacy for innovative behavior. Some ways include providing innovation freedom, innovation opportunities, innovation-related resources, and training to employees.

This research focuses on two important organizational factors: external information awareness and proactiveness of innovation strategy. Our results suggest managers to improve external information awareness of their companies, for its positive influence to employees' attitude, subjective norm, perceived behavioral control and directly to the innovation behavior. So, organizations should establish better relationships with external innovation partners, and share the captured information within organizations. Existing knowledge management technology may be helpful for its capability to capture information from external environment and share them within organizations.

The potential influence of proactiveness of innovation strategy may also be important in practice. Therefore, organizations should firstly emphasize to be an innovation leader rather than an innovation follower, and change to be research-oriented. Then they also should generate and access a wide range of new ideas and bravely invest in the quick implementation of them to capture new opportunities. Although there is a high risk in proactive innovation strategy, its benefits to increase employees' innovation behavior also need to be taken into account. As recent research considers human resources and business processes to be unique resources to gain competitive advantages, an innovation leader will be difficult to be copied and surpassed by an innovation follower.

6.3 Limitations and Future Research

The findings in this study should be interpreted in view of its limitations. First, most of the respondents are from the manufacturing sector. More studies of other sectors are needed to assess the proposed model. Second, this study focuses on Japanese companies, so there may be some geographical or cultural specificity and the findings may not generalize to other settings. It may be interesting to assess the proposed model in other countries, especially those with different culture compared to Japan, like China, and the United State. Third, only subjective measures have been used in our research. As self-reports may contain some presentational biases [17], future research may consider using objective measures of employees' innovation behavior. Other than addressing the limitation of this study, in order to better understand the phenomenon, future research may further explore other organizational and environmental factors to extend our proposed model. Examples include organizational structure, risk-taking tendency, job and business process orientation, and environmental dynamism.

7   Conclusions

The TPB-based psychological analysis of employees' innovation behavior deepens our understanding of employees' innovation behavior by considering the effects of two organizational characteristics: external information awareness and proactive innovation strategy. Our finding suggests a more effective way to encourage employees' innovation is to ensure the availability of innovation sources, as knowledge, time and opportunity. While both external information awareness and proactive innovation strategy own high motivational influence of employees' innovation, some ideas and technologies from external customers, research institutions or other sectors are likely to directly initiate employees' innovation. So considering organizational factors will improve the explanatory power of TPB in real world. In addition, it suggests external information awareness as an important characteristic of an innovative organization. Meanwhile, managers should the benefits of proactive innovation strategy on employee's innovation behavior, as well as its potential risks.

References

1. Ajzen, I.: The Theory of Planned Behavior. Organizational Behavior and Human Decision Process 50(2), 170–211 (1991)
2. Ajzen, I.: Attitudes, Personality, and Behavior, Maidenhead. Open University Press, England (2005)
3. Amabile, T.M.: A Model of Creativity and Innovation in Organizations. Research in Organizational Behavior 10, 123–167 (1988)

4. Amo, B.W.: Employee Innovation Behavior in Health Care: the Influence from Management and Colleagues. International Nursing Review 53(3), 231–237 (2006)
5. Arogon-Correa, J.A., Sharma, S.: A Contingent Resource-based View of Proactive Corporate Environmental Strategy. The Academy of Management Review 28(1), 71–88 (2003)
6. Bock, G.W., Zmud, R.W., Kim, Y.G.: Behavioral Intention Formation in Knowledge Sharing: Examining the Roles of Extrinsic Motivators, Psychological Forces, and Organizational Climate. MIS Quarterly 29(1), 87–111 (2005)
7. Chin, W.W., Marcolin, B.L., Newsted, P.R.: A Partial Least Squares Latent Variable Modeling Approach for Measuring Interaction Effects: Results from a Monte Carlo Simulation Study and Voice Mail Emotion/Adoption Study. In: DeGross, J.I., Jarvenpaa, S., Srinivasan, A. (eds.) Proceedings of the 17th International Conference on Information Systems, pp. 21–41 (1996)
8. Chesbrough, H.W.: Open Innovation: The New Imperative for Creating and Profiting from Technology. Harvard Business School, Boston (2003)
9. Cohen, W.M., Levinthal, D.A.: Absorptive Capability: a New Perspective on Learning and Innovation. Administrative Science Quarterly 35(1), 128–152 (1990)
10. Cooper, R.G., Kleinschmidt, E.J.: Benchmarking the Firm's Critical Success Factors in New Product Development. Journal of Product Innovation Management 12(5), 374–391 (1995)
11. De Jong, J.P.J., Den Hartog, D.N.: How Leaders Influence Employees' Innovation Behavior. European Journal of Innovation Management 10(1), 41–64 (2007)
12. Diamantopoulos, A., Winklhofer, H.M.: Index Construction with Formative Indicators: an Alternative to Scale Development. Journal of Marketing Research 38(2), 269–277 (2001)
13. Dorenbosch, L., van Engen, M.L., Verhagen, M.: On-the-job Innovation: the Impact of Job Design and Human Resource Management through Production Ownership. Creativity and Innovation Management 14(2), 129–141 (2005)
14. Fagerberg, J., Mowery, D.C., Nelson, R.R.: The Oxford Handbook of Innovation, 1st edn. Oxford University, New York (2005)
15. Forrester, R.H.: Capturing Learning and Applying Knowledge: an Investigation of the Use of Innovation Teams in Japanese and American Automotive Firms. Journal of Business Research 47(1), 35–45 (2000)
16. Fishbein, M., Ajzen, I.: On Construct Validity: a Critique of Miniard and Cohen's Paper. Journal of Experimental Social Psychology 17(3), 340–350 (1981)
17. Gaes, G.G., Kalle, R.J., Tedeschi, J.I.: Impression Management in the Forced Compliance Situation: Two Studies using the Bogus Pipeline. Journal of Experimental Social Psychology 9, 491–501 (1978)
18. Goodman, M.B.: Executive Summary: High Visibility, High Risk, High Reward. In: 1989 Professional Communication Conference, pp. 18–20, 245–248 (1989)
19. Hofstede, G., Hofstede, G.J.: Cultures and Organizations: Software of the Mind. McGraw-Hill, New York (2004)
20. Jassen, O.: How Fairness Perceptions Make Innovative Behavior More or Less Stressful. Journal of Organizational Behavior 25(2), 201–215 (2004)
21. Jimmieson, N.L., Peach, M., White, K.M.: Utilizing the Theory of Planned Behavior to Inform Change Management: an Investigation of Employee Intentions to Support Organizational Change. Journal of Applied Behavioral Science 44(2), 237–262 (2008)
22. Kaufmann, A., Todtling, F.: How Effective is Innovation Support for SMEs? An Analysis of the Region of Upper Austria. Technovation 2(3), 147–159 (2002)
23. King, N., Anderson, N.: Managing Innovation and Change: a Critical Guide for Organizations. Thomson, London (2002)

24. Knight, G.: Entrepreneurship and Marketing Strategy: the SME under Globalization. Journal of International Marketing 8(2), 12–32 (2000)
25. Krueger, Jr., N.: The Cognitive Infrastructure of Opportunity Emergence. Entrepreneurship, part II, pp. 185–206 (2007)
26. Lee, L., Wong, P.K.: Individual Attitudes, Organizational Reward System and Patenting Performance of R&D Scientists and Engineers. M. P. R. Archive, (ed.) vol. 595, p. 44 (2006)
27. Lee, R., Klobas, J., Tezinde, T., Murphy, J.: The Underlying Social Identities of a Nation's Brand. International Marketing Review 27(4), 450–465 (2010)
28. Lieberman, M., Montgomery, D.: First Mover Advantages. Strategic Management Journal 9(1), 41–58 (1998)
29. Lin, H.F., Lee, G.G.: Perceptions of Senior Managers towards Knowledge-sharing Behavior. Management Decision 42(1), 108–125 (2004)
30. Lumpkin, G.T., Dess, G.G.: Clarifying the Entrepreneurial Orientation Construct and Linking it to Performance. Academy of Management Review 21(1), 135–172 (1996)
31. Mendelson, H.: Organizational Architecture and Success in the Information Technology Industry. Management Science 46(4), 513–529 (2000)
32. Morris, M.G., Venkatesh, V., Ackerman, P.L.: Gender and Age Differences in Employee Decision about New Technology: an Extension to the Theory of Planned Behavior. IEEE Transactions on Engineering Management 52(1), 69–84 (2005)
33. Pizam, A.: Psychological Characteristics of Innovators. European Journal of Marketing 6(3), 203–210 (1972)
34. Pieterse, A.N., Knippenberg, D.V., Schippers, M., Stam, D.: Transformational and Transactional Leadership and Innovative Behavior: the Moderating Role of Psychological Empowerment. Journal of Organization Behavior 31(4), 609–623 (2010)
35. Saleh, S.D., Wang, C.K.: The Management of Innovation: Strategy, Structure, and Organizational Culture. IEEE Transactions on Engineering Management 40(1), 14–21 (1993)
36. Sender, S.W.: Systematic Agreement: a Theory of Organizational Alignment. Human Resource Development Quarterly 8(1), 23–40 (2007)
37. Scott, S.G., Bruce, R.A.: Determinants of Innovation Behavior: a Path Model of Individual Innovation in the Workplace. The Academy of Management Journal 37(3), 580–607 (1994)
38. Scott, S.G., Bruce, R.A.: Following the Leader in R&D: the Joint Effect of Subordinate Problem-solving Style and Leader-member Relations on Innovative Behavior. IEEE Transactions on Engineering Management 45(1), 3–10 (1998)
39. Slater, S.F., Mohr, J.J.: Successful Development and Commercialization of Technological Innovation: Insights Based on Innovation Type. The Journal of Product Innovation Management 23(1), 26–33 (2006)
40. Souitaris, V.: External Communication Determinants of Innovation in the Context of a Newly Industrialized Country: a Comparison of Objective and Perceptual Results from Greece. Technovation 21(1), 25–34 (2001)
41. Swan, F., Newell, S., Scharbrough, H., Hislop, D.: Knowledge Management and Innovation: Networks and Networking. Journal of Knowledge Management 3(4), 262–275 (1999)
42. Tambe, P.B., Hitt, L.M., Brynjolfsson, E.: The Extroverted Firm: How External Information Practices Affect Productivity. In: 2008 International Conference on Information Systems (2009)

43. Turgoose, C., Thacker, C., Adams, M., Carmichael, C., Gray, M., Hall, L., Todd, C.: Innovation in Manufacturing SMEs in South Yorkshire. Innovation Advisory Service, Sheffield (2000)
44. Von Hippel, E.: The Sources of Innovation. Oxford University Press, New York (1988)
45. Wiethoff, C.: Motivation to Learn and Diversity Training: Application of the Theory of Planned Behavior. Human Resource Development Quarterly 15(3), 263–278 (2004)
46. Williams, S.D.: Personality, Attitude, and Leader Influences on Divergent Thinking and Creativity in Organizations. European Journal of Innovation Management 7(3), 187–204 (2004)

Determining the Collaboration Maturity of Organizational Teams: A Study in the Automotive Industry

Imed Boughzala[1] and Gert-Jan de Vreede[2,3]

[1] Télécom Business School, Paris, France
[2] The Center for Collaboration Science, University of Nebraska at Omaha, Nebraska, U.S.A.
[3] Systems Engineering Department, Delft University of Technology, Delft, The Netherlands
imed.boughzala@it-sudparis.eu, gdevreede@unomaha.edu

Abstract. Many researchers argue that the quality of collaboration directly affects the quality of an organization's outcomes and performance. This paper reports on the first field application of a Collaboration Maturity Model (Col-MM) through an automotive field study. This model was empirically developed during a series of Focus Group meetings with professional collaboration experts to maximize its relevance and practical applicability. Col-MM is intended to be sufficiently generic to be applied to any type of collaboration and useable to assess the collaboration maturity of a given team holistically through self-assessments performed by practitioners. The purpose of the study reported in this paper was to apply and evaluate the use of the Col-MM in practice. The results should be of interest to academic researchers and information systems practitioners interested in collaboration maturity assessment. The paper further serves as a starting point for future research in this area.

Keywords: Collaboration, Collaboration Quality, Performance, Collaboration Technology, IT, Maturity Model, Virtual Team, Chief Knowledge Officer.

1   Introduction

Organizations form to create value and products that individuals cannot create alone [29]. To ensure their organizational performance and competitive advantage it is thus critical for organizations to achieve successful collaboration [8]; [18]. In today's increasingly unstable and competitive socio-economic environment, trends like globalization and increased product and service complexity have pushed organizations to use more distributed, cross-disciplinary, cross-culture, virtual teams [7]. In this context, the quality of collaboration directly affects the quality of an organization's outcomes and performance [23]; [1]. This means that the disposition and capabilities of an organization's work force to collaborate will directly affect organizational performance, productivity and profitability [17]; [18].

It is important for organizations to assess the quality of the collaboration in their teams. This will enable them to identify measures to improve collaboration by better selecting and designing the appropriate collaboration technologies (IT/IS) and

A. Fred et al. (Eds.): IC3K 2011, CCIS 348, pp. 393–406, 2013.

therefore to improve the management of their virtual teams and projects. This requires organizations to answer questions such as: Under what conditions do teams collaborate better? Are there different levels of collaboration quality that can be recognized and that teams should aim for? To what extent should management styles be taken into account? Which role should collaboration technologies play to foster effective collaboration? How can we measure the impact of collaboration on organizational performance?

Several studies propose models and methods for collaboration assessment from different points of view: collaboration processes (see e.g. [32]; [12] or collaboration technologies and their usage (see e.g. [10]; [31]; [20]. One way to assess the overall collaboration quality of teams is through maturity model approaches used extensively in quality assurance for product development [15]. Using these types of models as an assessment instrument helps an organization to identify best practices and trouble spots, and to stimulate discussion among practitioners to initiate activities for continuous improvement [14]. However, few efforts have been reported on using maturity models to assess collaboration. Those that have are limited in that they apply only to certain domains or just cover a few phases of the project life cycle [11].

This paper reports on the first field application and evaluation of a Collaboration Maturity Model (Col-MM) in an automotive industry for assessing a virtual team distributed in two European countries. The purpose of the study reported in this paper was to apply and evaluate the use of the Col-MM in practice. Col-MM was empirically developed through a Design Science perspective approach [21] during a series of Focus Group meetings with professional collaboration experts to maximize its relevance and practical applicability [4]. It was intended to be sufficiently generic to be applied to any type of collaboration, virtual or not (e.g. project teams, organizational teams, cross functional/organizational teams, inter-organizational team, or communities of practice) and useable to assess the collaboration maturity of a given team holistically by practitioners for conducting self-assessments.

The remainder of this paper is structured as follows. We first introduce the methodological background related to maturity models in general and the Col-MM in particular. Next, we report on the application and the evaluation of the Col-MM in a field study in the automotive industry. Last, we discuss the appropriateness and usefulness of Col-MM, followed by our conclusions which summarize the limitations of this study and present future research directions.

2   Background

Maturity, literally meaning 'ripeness', describes the transition from an initial to a more advanced state, possibly through a number of intermediate states [15]. The fundamental underlying assumption of maturity models is that a higher level of maturity will result in higher performance. Maturity models reflect the degree to which key processes or activities are defined, managed, measured, and executed effectively. They typically describe the characteristics of an activity at a number of different levels of performance as described in Fraser et al. [14]:

"At the lowest level, the performance of an activity may be rather ad hoc or depend on the initiative of an individual, so that the outcome is unlikely to be predictable or reproducible.

As the level increases, activities are performed more systematically and are well defined and managed. At the highest level, 'best practices' are adopted where appropriate and are subject to a continuous improvement process" (p.1500).

2.1   Maturity Models

In the literature, collaboration has been defined in different ways [25]; [5]; [3]. In the context of this study, we define collaboration as a process in which two or more agents (individuals or organizations) share resources and skills to solve problems so that they can jointly accomplish one or more activities. During this process, the agents communicate with each other to coordinate their tasks. Based on this definition of collaboration, we define collaboration maturity as a team's current maximum capability to collaborate where team members effectively communicate, reach shared understanding, and adjust their tasks and behaviors to produce high quality outcomes.

The main objective of our work is to introduce a new collaboration maturity mo-del that addresses some of the limitations described above. This model aims to holistically assess the collaboration maturity of a (virtual) team that uses several collaboration technologies. However, its applicability is not limited to a particular form of collaboration and the model can be used for different settings. Further, it supports the development of recommendations in form of an action plan to reach improved project management, collaboration performance and quality of collaboration outcomes.

The Col-MM was designed during a design science study in which we cooperated with a Focus Group consisting of professional collaboration experts. These experts included 15 Chief Knowledge Officers (CKOs) from companies of different sizes (ranging from 500 – 1,000 to 100,000-200,000 employees; including 10 multinational firms) in different sectors (including Automotive, Software, Audiovisual, Civil engineering, and Telecommunications), holding at least a master-level degree (four held a PhD) from different areas, and having at least 10-14 years of work experience with 50% of them having 5-9 years as a CKO. Their average age was 48 and 73% of them were male (see Table 1). They were accustomed to meet in the context of a business association to share their best practices regarding methods, techniques and tools in the collaboration and knowledge management area.

The involvement of the experts group enabled us to combine relevance and rigor by meeting a business need with applicable knowledge and so to maximize the resulting artifacts' relevance and applicability. The experts expressed the following critical requirements for the Col-MM:

- **Resource Efficient:** The Col-MM should be fast to complete.

- **Rich Data:** The Col-MM should report on different points of view and concerns from the workplace, using both quantitative and qualitative data.

- **Limited Need for Further advanced Data Analysis:** The supporting tool should provide integrated support for results interpretation.

- **Self-assessment:** Practitioners should be able to apply the Col-MM themselves.

- **Constructive Learning:** The Col-MM should promote team building and organizational learning rather than control and sanction.

After a series of meetings with the Focus Group in which several initial versions of the Col-MM were presented and pilot results were shared, the first full version of

Col-MM was completed [4]. The Col-MM consisted of a number of artifacts including: The Col-MM structure that describes the collaboration areas of concerns (topics) and their related criteria; The Col-MM questionnaire that includes questions, levels of rating and mathematical equations for analysis; The Col-MM method that (a) defines the steps and provides guidance on how to run the Col-MM questionnaire in the field, and (b) supports the development of recommendations; and the Col-MM tool which is a customized MS Excel application that represents the implementation of the above artifacts, and enables the execution of a concrete assessment by enabling the collection and analysis of quantitatively and qualitatively questionnaire data. It provides different presentations of results (e.g. individual and team spider diagrams, comparison curves, and cloud matrices) and the results' report generation.

Table 1. Expert focus group characteristics

Sector	Firm size	Firm type	Age	Gender	Background	Level	Years of work experience	Years of work as CKO
Automotive	100,000 – 200,000	M	45-49	M	Industrial Design	MSc	20-24	0-4
Automotive	10,000 – 20,000	M	50-54	M	Mechanical Eng.	MSc	25-29	5-9
Automotive	5,000 – 10,000	M	50-54	F	Math Eng.	PhD	25-29	5-9
Software Edition	50,000 – 100,000	M	40-44	M	Computer Science	MSc	15-19	0-4
Software Edition	1,000 – 5,000	N	35-39	M	Ergonomics	PhD	10-14	0-4
Audiovisual / Communications	1,000 –5,000	N	55-59	F	Human Resources	MBA	30-34	10-14
Civil Engineering	5,000 – 10,000	M	50-54	M	Cognitive Sciences	MSc	25-29	5-9
IS/IT Consulting	5,000 – 10,000	M	45-49	M	Management	MBA	20-24	0-4
IS/IT Consulting	1,000 –5,000	M	40-44	M	IT	MSc	15-19	0-4
Research and Development	1,000 – 5,000	N	50-54	F	Management	PhD	25-29	5-9
Aeronautic	10,000 – 20,000	M	50-54	M	Knowledge Eng.	MSc	25-29	5-9
Telecommunications	100,000 – 200,000	M	40-44	M	Management	MBA	15-19	0-4
Telecommunications	1,000 – 5,000	N	45-49	F	IT	MSc	20-24	0-4
Energy	100,000 – 200,000	N	50-54	M	Knowledge Eng.	PhD	25-29	5-9
Transport and Freight	10,000 – 20,000	M	55-59	M	Cognitive Sciences	MSc	30-34	5-9
10 Sectors	15 firms	Multi-national (10) National (5)	35-39 (1) 40-44 (3) 45-49 (3) 50-54 (6) 55-59 (2)	F (4) M (11)		PhD (4)		

The Col-MM distinguishes between four maturity levels: Ad-hoc, Exploring, Managing, and Optimizing. At the Ad-hoc level, teams are collaboratively immature. Individuals have many difficulties to communicate effectively, to reach shared understanding, and to adjust their tasks and behaviors to produce high quality outcomes together. At the Exploring level, teams are well aware of their weaknesses in terms of collaboration quality. Individuals try work together to produce valuable outcomes, but are faced with many collaboration challenges. Some initiatives to address these are attempted but without major impacts. At the Managing level, individuals are able to produce collaborative outcomes of good quality. They have overcome many challenges to collaborate productively, but there still is room for improvement. At the Optimizing level, teams are collaboratively mature. Teams work together optimally and accomplish high quality collaborative outcomes. Furthermore, they engage in critical self-reflection and continuous improvement efforts.

The Col-MM explores the maturity of a given team holistically from different perspectives related to collaboration. The following perspectives, or areas of concerns, were considered essential by the participants in the Focus Group meetings [4]:

Table 2. Col-MM areas of concerns and criteria

Areas of concern	Criteria
Collaboration Characteristics	1. Collaboration object 2. Collaboration depth 3. Working mode 4. Interaction intensity 5. Collaboration forms 6. Formalization of relationships 7. Commitment and availability of individuals 8. Collaboration boundaries
Collaboration Management	9. Collaboration goal 10. Management style 11. Decision-making 12. Leadership endorsement 13. Rewarding 14. Collaboration progress
Collaboration Process	15. Collaboration framework 16. Resources sharing 17. Awareness 18. Conflicts management 19. work methodologies and technologies
Information and Knowledge Integration	20. Information collection 21. Information structuring 22. Information access 23. Knowledge validation 24. Knowledge reusing 25. Knowledge creation

• **Collaboration Characteristics:** This covers the characteristics of the collaboration.

• **Collaboration Management:** This covers the way in which collaboration processes and activities are managed.

• **Collaboration Process:** This covers how actors perform collaboration on a daily basis.

- **Information and Knowledge Integration:** This covers how actors manage the information and knowledge required for productive collaboration.

For each area of concern, a number of criteria were defined (see Table 2). These criteria represent the topics for a questionnaire (Col-MM questionnaire). Each criterion is represented by an item that is evaluated on a 4-point scale. To support the respondents, the levels of each criterion are described briefly, with examples wherever possible. An example of a criterion item is provided in Figure 1. When a respondent cannot answer, no score is recorded. The calculation of points provides the level of maturity (Ad hoc (<20), Exploring (20-50), Managing (50-80) and Optimizing (80-100)).

COLLABORATION STEERING
Collaboration goal

Why we collaborate? What is the purpose of the collaboration?

Level 1   Undefined
The collaboration aim is not defined. Nobody knows the purpose of collaboration.

Level 2   Blur
The collaboration aim is unclear. Few people know exactly why they work together.

Level 3   Clear
The collaboration aim is clear, but it is not necessarily very precise for everyone.

Level 4   Precise
The collaboration aim is clear. Everybody knows exactly the purpose.

Fig. 1. Example of criterion in Col-MM

In essence, the Col-MM is structured as a library of criteria. Sometimes, not all criteria are relevant. So, an organization can decide which criteria fit its particular context. It can also decide to expand the set of criteria. Also, for some organizations certain criteria may be more important than others. In such situations, it is possible to assign different weights to the criteria.

The Col-MM method defines the steps to perform the analysis. Figure 2 summarizes the seven main steps in the method. At the scoping step, the purpose of the Col-MM analysis is defined according to the organizational context and the business strategy. The boundaries of the analysis (i.e. organizational scope) are defined before starting. The reasons for performing the assessment should be communicated to all concerned individuals and teams. It is important to present this as a holistic team assessment to improve overall organizational performance, rather than as an individual retrospective evaluation.

The data collection is performed through individual and/or collective interviews based on the Col-MM questionnaire (quantitative data). The selection of individuals should be representative of the target team. During interviews, qualitative observations should be collected to enrich the analysis and gain a deeper understanding of any perception differences that may exist. After the data collection, a first quantitative data analysis is performed using the Col-MM tool (This tool allowed for quantitative and qualitative data collection during the interviews and to analyze quantitative data automatically. It also supports different presentations of results and the results' report generation.).

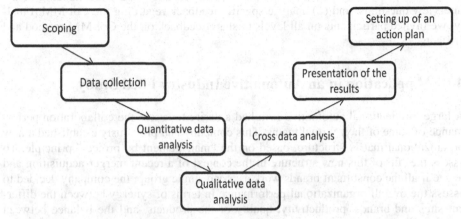

Fig. 2. The seven steps in the Col-MM method

The first quantitative analysis presents individual perceptions about the collaboration maturity of the team. It also helps to identify critical perception differences concerning the different criteria. The qualitative data analysis (using a content or thematic analysis based on the interview statements) helps to get a more in-depth understanding of these perception differences for each criterion or group of criteria (area of concern). Follow-up discussions and consensus building efforts could be carried out for relevant scores, in order to settle on an acceptable assessment. The cross analysis may yield additional interpretations by combining criteria for specific measurements of capabilities according the goal of the assessment.

3   Method

The Col-MM was developed following Hevner et al.'s (2004) design science approach. In this paper we will not report on the development of the Col-MM but only on its first field application and evaluation to demonstrate the model's practical feasibility and utility. This study therefore answers Hevner et al.'s Design Evaluation Framework recommendation for the use observational methods (2004 p. 86). Our role as researchers was limited to the organization and execution of (group) interviews, the analysis of collected interview data, and the gathering of participants' feedback regarding Col-MM. Our interventions during the study were only aimed at supporting the organization in achieving its goals in the project. The researchers had no personal stake in the project, neither with the problem situation nor with the solutions that were to be explored. The primary motivation for the client organization to involve the researchers was its desire to assess and improve the collaboration in a number of its key teams.

Research data was collected from both quantitative and qualitative sources to enable a rich understanding of the application of the Col-MM in practice. First, while observing the different activities in the study, we kept notes of incidents, remarks and events that conveyed critical information. Second, the (group) interview results were analyzed to gain insight into (1) the participants' reaction and understanding of the interview questions, and (2) analyze specific feedback regarding the Col-MM. Finally, we invited participants on all levels to share feedback on the Col-MM method and artifacts.

4   Application in an Automotive Industry Filed Study

A large multinational automotive firm had a desire to assess the collaboration performance of some of their virtual teams. This company had previously established a new organizational matrix structure, based on the "management by project" principle. To assess the 'fit' of this new structure in the context of a recent merger-acquisition and to see if all the constituent brands work as a one single group, the company decided to assess the overall organizational performance in terms of synergy between the different sites and brands, productivity, quality of the products, and the balance between product diversity and process complexity. The collaboration maturity assessment was part of this larger organizational performance assessment.

As a first step it was decided to apply the Col-MM to measure the collaboration maturity of one virtual team distributed over two European countries (two sites) with different cultures, different work habits, and different management styles. This virtual team was in charge of the "Engine After Treatment System" (EATS) that was part of a larger development project of a new diesel engine that was taking place under the responsibility of a business unit distributed over three countries. The leading site in this project will be referred to as site A below.

4.1   Field Study Steps

The field study was performed over the course of 5 months. It was applied and followed the Col-MM method steps (Figure 2).

At the scoping step, the purpose of the Col-MM analysis was defined according to the company's requirements. As presented to the stakeholders, the objectives of the Col-MM analysis were to check:

• If the organization had adequate capabilities to effectively support high quality collaboration.

• If collaboration technologies were well selected and configured.

• If collaboration technologies provided were effectively used.

• If there were critical issues related to cultural differences (national, organizational, technological, etc), given the merger-acquisition context.

The data collection was performed through individual and/or collective interviews based on the Col-MM questionnaire (quantitative data). The selection of respondents was done with the assistance of the firm' CKO according to criteria such as job position, responsibility, process step intervention, and working experience. All respondents had similar levels of education (MSc degree) but from different engineering fields: mechanical, electronic, electric, and industrial. The Col-MM questionnaire was sent to the respondents before the meeting with an introduction of the company's Col-MM objectives. Anonymity and confidentiality of the treatment of the responses were formally assured. Nine individual interviews were conducted face-to-face in the respondents' native language in the two European countries (3 in the site A and 6 in the site B). Each interview lasted about 90 minutes. During the interviews the Col-MM tool was used for data collection, followed by a first quantitative data analysis. This analysis presented individual perceptions about the collaboration maturity of the team. It also helped to identify perception differences concerning the different criteria. All interviews were recorded for further qualitative data analysis, consisting of content/thematic analysis. This analysis helped to get a more in-depth understanding of these perception differences for each criterion or group of criteria (area of concern). Two collective interviews (one for each site) were conducted to examine these perception gaps on some criteria. Follow-up discussions and consensus building efforts were carried out for relevant scores, in order to settle on an acceptable assessment. The cross analysis yielded additional interpretations by combining criteria for specific measurements of capabilities according to the focus of the assessment, such as project management, knowledge management, IT adequacy, value creation, and organizational learning.

The last step of the Col-MM method concerns the definition of an action plan. This plan was included in the report. An initial version of the report was sent to the respondents to solicit any corrections before the final report was prepared. A final presentation to the company's top management reported the results and recommendations in the form of a list of suggested future actions.

4.2   Findings

The findings were reported as observations and discussions of the different recorded perceptions related to the Col-MM criteria and topics. Examples of findings reported to top management according to the four Col-MM areas of concern include:

- **Collaboration Characteristics:** There were virtually no differences between the different sites in terms of their perceptions regarding the nature of collaboration. We found similar understandings of collaboration goals and team members' commitment for both sites. This may have been facilitated by the technical subject matter that the team members in the different sites had to collaborate on; this created a common language and hence understanding.

- **Collaboration Management:** We noticed different perceptions between the two sites with respect to project management style and decision-making (hierarchical management vs. consensual management). Site B respondents felt unfairly rewarded compared to site A. They felt that because site A had the project lead, its employees always had an advantage.

- **Collaboration Process:** We noticed that site B respondents had less awareness about different collaboration approaches to enhance the team's performance. Because of their positions and responsibilities in the process, they focused more on their individual contribution to the overall process rather than on developing collaborative relationships. We also noticed differences in terms of conflict management by the leadership in each site: Conflict management in the site A was based on consensus while in site B it was based on hierarchical decision making and negotiation.

- **Information and Knowledge Integration:** We found different perceptions regarding information access. For site A respondents, access to information was not organized as well as they wished. Information was very distributed and access should be simplified. We found consistent perceptions between the two sites regarding collaborative knowledge creation; both sites felt this process was well organized.

Through the qualitative data analysis we found that some cultural differences between sites appeared to be related more the organizational culture rather than to the national culture. For example, the balance between private and professional life appeared to be different. Also, there was a different brand identity: Site B respondents felt they were still belonging to their original brand (i.e. from before the merger) rather than to the group of brands. We also found different work attitudes. Site B respondents were more reactive compared to Site A respondents who were more proactive. According to some respondents, this was because of their position in the project. Possible explanations could be related to their contracts type (tenure status) and social protection.

The general findings reported can be summarized as follows:

- Collaboration was mainly based on "individuals' goodwill" as for example related to resource sharing and knowledge management.

- The team was not as collaboratively mature as was expected – they were at the Exploring Level (36 points). Because of the asymmetric collaboration awareness between the two sites, their collaboration is mostly of a coordination nature. This makes it difficult to further improve the quality of their outcomes.

- The new matrix structure did not resolve all problems with respect to the imbalance between responsibility and authority.

In the final report, various recommendations were proposed, including:

- Make collaboration a clear strategic goal in all project management initiatives.

- Re-think the management of collaboration (steering) and provide training for managers.

- Nominate full-time facilitators for collaboration.

- Take into account diversity aspects related to culture.

- Make explicit recognitions for the contributions of every actor toward effective collaboration.

After six months, we learned that three of the suggested recommendations were followed up with concrete actions:

- The first recommendation was clearly mentioned in the company's project management standard.

- Following the third recommendation, one full time collaboration facilitator was assigned to each business unit.

- Following the fifth recommendation, a 'collaboration capability' criterion was added to the annual individual performance assessment.

5   Discussion

During the application of the Col-MM during this field study, we gathered various experiences and feedback regarding the appropriateness and usefulness of Col-MM. According to the respondents, the Col-MM analysis was satisfactory and correctly represented their perceptions. It focused on real collaboration problems and allowed traditionally 'unspoken issues' to surface. They were also satisfied with the feedback provided to top management and the subsequent actions that were taken related to the assessment's recommendations. According to the operational managers, the results were relevant. Further, most of them felt able to reuse the Col-MM by themselves in the future. According to the top managers, the study was satisfactory in terms of results and recommendations, as they confirmed and reinforced some of their own perceptions. This allowed them, for example, to focus more on the organizational culture than on national culture and to understand the problems related to the project-based new organizational structure.

We also received feedback and recommendations from the respondents on the Col-MM questionnaire such as the possibility to review some criteria and questions. The respondents stated that some criteria were a little difficult to understand. Also, the nuances between levels of responses were sometimes subjective or difficult to distinguish. In addition, they proposed to add some criteria such as culture, work experience, and practice diversity, and to rename some areas of concern such as "collaboration readiness" instead of "collaboration characteristics". Finally, they suggested putting a stronger focus on *virtualness* (i.e. the extent to which a process can be virtualized [27]) and collaboration technology rather than on information and knowledge integration. Interestingly, this was contrary to the wishes expressed by the focus group. However, since the Col-MM is developed as a library of criteria, the review of the Col-MM structure according to a specific context is possible and therefore the respondents' suggestions can be easily accommodated. In terms of execution, most respondents expressed that they preferred the use of collective rather than individual interviews as this would enable a faster application of the Col-MM process.

Based on the experiences and feedback from this field study, we observe the following regarding the extent to which Col-MM meets the requirements as proposed by the Focus Group experts:

- Resource Efficient: Col-MM appears to be resource efficient. A total of 36 hours were spent: 1.5 hours for the assessment preparation, 16.5 hours for the engineering interviews, 3 hours for the CKO interviews, 3 hours for the top management interviews and 12 hours for the analysis and report preparation. We feel that this is a modest and reasonable effort in terms of resources spent.

- Rich Data: The combined use of quantitative and qualitative data analysis enabled richer findings. We felt that qualitative observations enabled us to better uncover and interpret the various points of views expressed by the respondents through the Col-MM questionnaire.

- Limited Need for Further Advanced Data Analysis: The analysis needs in the field application were limited and the Col-MM tool provided sufficient support (among others the report generation).

- Self-assessment: The operational managers expressed confidence that they could perform future applications of the Col-MM themselves.
- Constructive Learning: The respondents' feedback shows that when the Col-MM study is carefully communicated, participation can be effective and generate discussions on real problems that further facilitate the acceptance of proposed solutions. In this respect, anonymity and confidentiality seem to be crucial. This was confirmed by feedback from the participants.

6   Conclusions

In this paper, we report on the first field application and evaluation of an initial version of a new collaboration maturity model, Col-MM, to assist in the assessment of teams' collaboration performance. The Col-MM was developed prescriptively to meet a real business need as expressed by 15 CKOs and others experts that are regularly confronted with collaboration performance challenges. Our contribution is both theoretical and practical as we propose a model, an application method, a supporting tool, and empirical evidence of their application. Our experiences show that the Col-MM can be applied in a resource-efficient fashion and yields results that are useful for organizations.

However, there are limitations related to this work with respect to the Design Science Evaluation Framework. First, our empirical evidence is based on several pilot studies but only a single field application. Further field studies have to be executed to expand the evaluation of the Col-MM artifacts and to further enhance the Col-MM. Particular care will have to be taken to ensure that Col-MM can take into account all levels of collaboration and all collaboration processes in an organization in different settings. This cannot be achieved by just expanding the number of criteria as this will overly complicate the use of the model. Second, at this stage, the Col-MM cannot yet be used to investigate a correlation between collaboration maturity levels and organizational/team performance. However, it provides a first step into this direction.

We recommend several directions for future research to enhance the current version of Col-MM. First, the model has to be applied in different types of organizations for different types of teams. The experiences from these applications will assist in the further development and evaluation of the Col-MM artifacts. Second, organizational and team performance measures have to be developed to enable an analysis of the relationship between collaboration maturity and organizational productivity. Fourth, from a behavioral science perspective, some further confirmatory studies should be performed using Structural Equation Modeling (Bollen, 1989) to validate the correlation between these variables (i.e. Col-MM constructs and performance).

References

1. Banker, R., Bardhan, I., Asdemir, O.: Understanding the impact of collaboration software on product design and development. Information Systems Research 17, 4 (2006)
2. Bollen, K.A.: Structural Equations with Latent Variables. Wiley (1989)

3. Boughzala, I.: Ingénierie de la collaboration: théories, technologies et pratiques, Paris, Hermès (September 2007)
4. Boughzala, I., de Vreede, G.J.: A Collaboration Maturity Model: Development and Exploratory Application. In: Proceedings of the 45th Hawaii International Conference on System Sciences, Hawaii (January 2012)
5. Briggs, R.O., Kolfschoten, G.L., de Vreede, G.J., Dean, D.L.: Defining Key Concepts for Collaboration Engineering. In: Garcia, I., Trejo, R. (eds.) Proceedings of the 12th Americas Conference on Information Systems, Acapulco, Mexico, pp. 121–128 (August 2006)
6. Chiesa, V., Coughlan, P., Voss, C.: Development of a technical innovation audit. Journal of Product Innovation Management 13(2), 105–136 (1996)
7. Chudoba, K.M., Wynn, E., Lu, M., Watson-Manheim, M.B.: How virtual are we? Measuring virtuality and understanding its impact in a global organization. Information Systems Journal 15, 279–306 (2005)
8. Clark, K.B., Fujimoto, T.: Product development performance: Strategy, organization and management in the world auto industry. Harvard Business School Press, Boston (1991)
9. Crosby, P.B.: Quality is Free. McGraw-Hill, New York (1979)
10. Damianos, L., Hirschman, L., Kozierok, R., Kurtz, J., Greenberg, A., Holgado, R., Walls, K., Laskowski, S., Scholtz, J.: Evaluation for Collaborative Systems. ACM Computing Surveys (CSUR) 31(2), Article 15 (1999)
11. Daoudi, J., Bourgault, M.: What Do We Know About Collaborative Maturity? In: 4th Annual Sprott Doctoral Symposium, April 19-20 (2007)
12. Den Hengst, M., Kolfschoten, G., Dean, D.L., Chakrapani, A.: Assessing the quality of collaborative processes. In: Proceeding of the 39th Annual Hawaii International Conference on System Sciences. IEEE Computer Society (2006)
13. Dooley, K., Subra, A., Anderson, J.: Maturity and its Impact on New Product Development Project Performance. Research in Engineering Design 13, 30–29 (2001)
14. Fraser, P., Farrukh, C., Gregory, M.: Managing product development collaborations – A process maturity approach. Proceedings of the Institution of Mechanical Engineers 217(11), 1499–1519 (2003)
15. Fraser, P., Moultrie, J., Gregory, M.: The use of maturity models / grids as a tool in assessing product development capability. In: IEEE International Engineering Management Conference, Cambridge, August 19-20 (2002)
16. Fraser, P., Moultrie, J., Holdway, R.: Exploratory studies of a proposed design maturity model. In: 8th International Product Development Management Conference, University of Twente, Holland, June 11-12 (2001)
17. Frost, Sullivan: Meetings Around the World: The Impact of Collaboration on Business Performance, White paper (2007),
http://newscenter.verizon.com/kit/collaboration/MAW_WP.pdf
(last accessed: April 3, 2011)
18. Hansen, M.T., Nohria, N.: How to build collaborative advantage. MIT Sloan Management Review 46(1), 22–30 (2004)
19. Harter, D.E., Krishman, M.S., Slaughter, S.A.: Effects of Process Maturity on Quality, Cycle Time, and Effort in Software Product Development. Management Science 46(4), 451–466 (2000)
20. Herskovic, V., Pino, J.A., Ochoa, S.F., Antunes, P.: Evaluation Methods for Groupware Systems. In: Haake, J.M., Ochoa, S.F., Cechich, A. (eds.) CRIWG 2007. LNCS, vol. 4715, pp. 328–336. Springer, Heidelberg (2007)
21. Hevner, A.R., March, S.T., Park, J., Ram, S.: Design Science in Information Systems Research. MIS Quarterly 28(1) (2004)

22. Hsieh, P.J., Lin, B., Lin, C.: The Construction and Application of Knowledge Navigator Model (KNM TM): The Evaluation of Knowledge Management Maturity. Expert Systems with Applications 36, 4087–4100 (2009)
23. Jordan, M.H., Field, H.S., Armenakis, A.A.: The relationship of group process variables and team performance—A team-level analysis in a field setting. Small Group Research 33, 121–150 (2002)
24. Lebrun, J., Gosset, P., Pallot, M., Roux, P., Gandelot, D., Vannier, M.: FREE Capability Assessment Framework (FREE-CAF), Vol. 1: Overview, Deliverable 132-1, ESPRIT Project 23286 FREE (1998)
25. Levan, S.K.: Travail collaboratif sur Internet: Concept, méthodes et pratiques des plateaux projet. Vuibert, Paris (2004)
26. Macbeth, D., Ferguson, N.: Partnership sourcing: an integrated supply chain management approach. Pitman Publishing, London (1994)
27. Martins, L.L., Gilson, L.L., Maynard, M.T.: Virtual teams: what do we know and where do we go from here? Journal of Management 30(6), 805–835 (2004)
28. McGrath, M.E. (ed.): Setting the PACE in Product Development: A Guide to Product and Cycle-Time Excellence. Butterworth-Heinemann, Oxford (1996)
29. Mintzberg, H.: The Structuring of Organizations. Prentice-Hall, Englewood Cliffs (1979)
30. Paulk, M., Curtis, B., Chrissis, M., Weber, C.: Capability maturity model for software, Version 1.1.Technical Report CMU/SEI-93-TR-024 ESC-TR-93-177 (February 1993)
31. Pinelle, D., Gutwin, C.: Task analysis for groupware usability evaluation: Modeling shared-workspace tasks with the mechanics of collaboration. ACM Transactions on Computer-Human Interaction (TOCHI) 10(4), 281–311 (2003)
32. Pinsonneault, A., Kraemer, K.L.: Middle management downsizing: An empirical investigation of the impact of information technology. Management Science 43(5), 659–679 (1997)
33. Ramasubbu, N., Krishnan, M.S., Kompalli, P.: A process maturity framework for managing distributed development. IEEE Software, 80–86 (May/June 2005)
34. Sander, P.C., Brombacher, A.C.: Analysis of quality information flows in the product creation process of high-volume consumer products. International Journal of Production Economics 67(1), 37–52 (2000)

Open Source Tools for Enterprise 2.0:
A Feature-Based Adaptable Analysis

Bettina Schauer, Michael Zeiller, and Robert Matzinger

Information Technology and Information Studies,
University of Applied Sciences Burgenland,Campus 1, Eisenstadt, Austria
{bettina.schauer,michael.zeiller,
robert.matzinger}@fh-burgenland.at

Abstract. When introducing Enterprise 2.0 tools to support knowledge workers working together on cognitive tasks and sharing information, companies and organizations face the problem of choosing the right tools from a huge market of systems. In particular for SMEs, open source tools for Enterprise 2.0 offer a good alternative to commercial systems, but the diversity of systems makes this marketplace quite confusing. Therefore we present a study of the growing market for Enterprise 2.0 systems that focuses entirely on systems available under an open source license. We use a set of 97 individual features and criteria that are grouped around the central functionalities of communication, coordination, collaboration and connection, to analyze the suitability of a representative sample of open source Enterprise 2.0 tools for the average knowledge worker. The evaluation matrix can be easily adapted to get more specific evaluation results for more particular company requirements. Our results show that there are many technically mature solutions with a broad range of functionality available from the market of open source tools for Enterprise 2.0.

Keywords: Enterprise 2.0, Electronic Collaboration, Knowledge Sharing, Collaboration Systems, Social Software, Open Source.

1   Introduction

Enterprise 2.0, the use of emergent social software platforms within companies, or between companies and their partners and customers, has rapidly gained momentum since this term was coined by McAfee [1] in 2006. Although Enterprise 2.0 incorporates more aspects than simply applying social software or comprehensive Enterprise 2.0 software systems, the software tools supporting collaboration processes have a great impact on the utilization of Enterprise 2.0 concepts in enterprises and organizations [2]. A lot of vendors – big, well-known market actors as well as small startup companies – offer numerous tools and systems. In addition market analysts, journals and online communities provide market reports and analyses to support potential customers to identify the appropriate Enterprise 2.0 system.

The basic concept of Enterprise 2.0 can be applied to all kinds and sizes of organizations. Small and medium size enterprises (SME) use social software platforms just

A. Fred et al. (Eds.): IC3K 2011, CCIS 348, pp. 407–422, 2013.

like large enterprises for sharing information and knowledge among their employees. They need maximum but specific functionality and community support at a reasonable price and workload. Besides small vendors, niche players and big software companies that offer highly flexible but complex and sometimes costly commercial platforms, there are also several Free and Open Source software tools available. Open source Enterprise 2.0 tools often provide a good alternative solution especially for SMEs and not-for-profit organizations to support team collaboration in a cost-effective way [3].

Market reports on Enterprise 2.0 by market analysts typically target large enterprises and therefore rarely cover open source tools [4-5]. Individual reports on specific open source tools often appear in journals and in several channels within the social software, Enterprise 2.0 or open source community. However, substantiated analyses and comparisons of open source tools can hardly be found. In this paper we present a market analysis of the growing market of Enterprise 2.0 systems that focuses entirely on those that are available under an open source license. These Enterprise 2.0 systems have to support and enable communication, coordination, collaboration and connection [6], [2]. Consequently, only those open source tools that support – at least partially – these four primary interaction processes of enterprise collaboration are included in our study. We introduce features and criteria that can be applied to assess Enterprise 2.0 systems and present detailed results for five selected open source Enterprise 2.0 systems.

In Section 2 we discuss Enterprise 2.0 and team collaboration and present related work on open source Enterprise 2.0 systems. Section 3 identifies and characterizes Enterprise 2.0 software and analyses the Enterprise 2.0 marketplace. Section 4 presents the applied assessment methodology. Section 5 provides detailed results of a number of open source Enterprise 2.0 systems. In section 6, we summarize key findings on open source systems for team collaboration based on the sample of evaluated Enterprise 2.0 systems. Finally, Section 7 concludes this paper.

2   Enterprise 2.0

McAfee [1] first introduced the term *Enterprise 2.0* in his trend-setting paper "Enterprise 2.0: The Dawn of Emergent Collaboration" where he discusses how companies can benefit from Web 2.0 technologies to support knowledge workers. McAfee [7] defines Enterprise 2.0 as "... the use of emergent social software platforms within companies, or between companies and their partners or customers". The utilization of social media within enterprises and organizations has significantly increased since then. For example, weblogs may be used as project logs or to communicate between the CEO or head of marketing and the company's customers. Knowledge workers of organizations create shared knowledge bases with the help of wikis. Microblogging applications allow for easy communication among team members. Thus, social media help employees and team members to work together on cognitive tasks and share information and knowledge.

2.1   Enterprise 2.0 Systems

All definitions of the term Enterprise 2.0 have in common that they refer to the use of social software or other web-based technologies to support enterprises and organizations. For example the AIIM defines Enterprise 2.0 as a system of web-based technologies that provide rapid and agile collaboration, information sharing, emergence and integration capabilities in the extended enterprise [8].

Software to support communication, cooperation and collaboration within teams of (knowledge) workers has been used for decades under the terms Groupware, Group Support Systems and Computer Supported Cooperative Work (CSCW) [9], [10]. Ellis et al. [11] define groupware as "computer-based systems that support groups of people engaged in a common task (goal) and that provide an interface to a shared environment".

One of the basic functionalities of Enterprise 2.0 systems is the support of electronic collaboration (short: e-Collaboration) among team members. e-Collaboration systems, i.e. information systems to support collaborative work, are "software for supporting communication, coordination and cooperation between people processes in groups" [6]. Therefore software and tools to support the ideas and concepts of Enterprise 2.0 can be found as Enterprise 2.0 software, groupware, CSCW systems and e-Collaboration systems.

To describe the characteristics of Enterprise 2.0 systems, McAfee [1] uses the acronym SLATES, which indicates six key components of Enterprise 2.0 technologies: Search – Links – Authoring – Tags – Extensions – Signals. Hinchcliffe [12] extends SLATES by four elements: Freeform – Social – Network-oriented – Emergent (resulting in the mnemonic FLATNESSES). While many authors (e.g., [9], [11]) categorize groupware and CSCW according to three basic interaction modes – communication, coordination, cooperation – Cook [2] modifies and extends this approach to four primary functions of social software and Enterprise 2.0: communication – cooperation – collaboration – connection (4C). We will follow this approach based on Cook's 4Cs to organize our feature-oriented analysis of Enterprise 2.0 systems.

2.2   Related Work

The market of Enterprise 2.0 tools and e-Collaboration systems offers a large variety of features and configurations. Riemer [6] presents a study which structures the range of available systems into system classes using cluster analysis. He applies the classification process to a sample of 94 systems that are used to derive four main system classes. Xu et al. [13] present a survey on asynchronous collaboration tools, i.e., systems with limited functionality. Büchner et al. [14] analyze seven commercial and open source Enterprise 2.0 tools based on 51 core Enterprise 2.0 services (in 13 service categories) to examine which concepts and services are supported by these tools. Comparative studies, which analyze open source tools exclusively, can hardly be found [3].

Market analysts like Gartner Research [4], Forrester Research [5] or Real Story Group [15] tend to cover the entire market for collaboration platforms and social software vendors. However, they focus on vendors with strong market presence and diverse functionality. Consequently, small niche players and open source tools are rarely included. These market reports, for example, evaluate the ability to execute and the completeness of vision to identify niche players, visionaries, challengers and leaders [4]. Functionality and features of Enterprise 2.0 tools are represented in these studies only to a subordinate extent.

3   e-Collaboration Systems

This paper discusses Enterprise 2.0 software that supports collaboration among team members. Thus we focus on those software tools that are commonly denoted as e-Collaboration systems and tools that cover the same functional range. From the large variety of available systems, we narrow our analysis to those systems available under an open source license (OSS).

Refining and enhancing Riemer's definition [6], we define e-Collaboration systems as software for supporting and enabling communication, coordination and collaboration between people in shared projects, processes and teams within organizations and for cross-organizational use. According to Cook's 4Cs model [2], comprehensive e-Collaboration systems should cover all types of social interaction in collaborative team processes – communication, coordination, collaboration and connection.

3.1   e-Collaboration Marketplace

The Enterprise 2.0 and e-Collaboration systems marketplace is highly dynamic and diverse and consists of heterogeneous system classes. It is made up of various tools with different levels of support of Enterprise 2.0 features. In his map of the 2009 Enterprise 2.0 marketplace, Hinchliffe [16] arranges more than 70 major products along two dimensions *Enterprise Capability* and *Support for Core Enterprise 2.0 Features*. He clusters them into three categories *Established Software Firms & Incumbent Players Territory – Enterprise 2.0 Sweet Spot – Open Source, Startup, Web Co. Territory*.

An evaluation report on enterprise collaboration software by the Real Story Group [15] analyses Functional Business Services and Technology Services (application services, administrative & system services), Vendor Intangibles and Universal Scenarios for 27 products. These products are organized into six categories: platform vendors (4), social software suites (7), wikis (5), blogs (3), white-label community services (4) and public networks (4).

Market analysts like Gartner Research or Forrester Research typically analyze in their reports only vendors with significant market presence. Koplowitz [5] discusses the collaboration platform products of 11 vendors and organizes them into three categories: leaders, strong performers, contenders. In the 2010 version of their Magic Quadrant for Social Software in the Workplace Gartner discusses 23 vendors after assessing their market presence and the functional capabilities of the products [4]. Based on their evaluation criteria concerning *Ability to Execute* (7 criteria) and *Completeness of Vision* (8 criteria) they identify three leaders, two challengers, seven visionaries and eleven niche players. Besides these major reports, listings on Enterprise 2.0, E-Collaboration systems and groupware can be found in (online) journals, in Wikipedia and in various Enterprise 2.0 and open source communities.

3.2   OSS Tools for Team Collaboration

A large variety of open source tools for team cooperation and collaboration exists which are referred to as groupware or e-Collaboration systems. According to our definition of e-Collaboration systems, only those tools will be part of a detailed analysis that support all four basic types of social interaction (full support or partial support per interaction process, but all types have to be supported). Applying this limitation, we eliminate the vast number of single function open source tools, e.g. all those wikis, weblogs, chats, video conferencing tools, project management tools, content management tools, tagging or bookmarking solutions, etc., that offer only a limited number of features according to their system class, but do not cover the entire spectrum of functions for team collaboration.

Based on a detailed market analysis in which we analyzed the functional range, we set up a long-list of 15 open source tools to be included in our study:

- Alfresco Share
- dotProject
- EGroupware
- IGSuite
- Liferay Portal / Social Office
- MindTouch Core
- more.groupware
- Novell Open Workgroup Suite
- NullLogic Groupware
- OpenGroupware
- phpGroupWare
- PHProjekt
- Simple Groupware
- TUTOS
- Zimbra Collaboration Suite

Other open source tools that are often labeled as groupware, too – like Kolab, netOffice, Web Collab, Plone or Scalix – have not been included because of their focus on project or content management or simply because their scope of operation does not meet our requirements for e-Collaboration systems.

Due to reasons of presentation in consequence of lack of space in this paper we had to reduce the long-list to a shortlist containing only five e-Collaboration systems:

- Alfresco Share, Community v3.4.0
- EGroupware, v1.4
- Liferay Portal, Community Edition v6.0.5 CE
- Simple Groupware, v0.701
- TUTOS, v1.7

These products have been chosen to represent the major types of collaboration tools on the market including tools focusing on content-based collaboration (Alfresco Share), content sharing (Simple Groupware), and project management (i.e., focus on coordination; TUTOS) as well as connection-oriented tools (Liferay), and groupware style tools (EGroupware). Thus, these five tools fairly represent collaboration solutions on the open source marketplace. Despite this limitation to a representative number of Enterprise 2.0 systems, the presented feature list and evaluation criteria can be easily applied to all tools in the long-list and to new emerging tools.

4   Evaluation Methodology

The evaluation methodology used to assess the open source e-Collaboration systems consists of the following three steps [17]:

1. Selection of e-Collaboration systems
2. Definition of evaluation criteria
3. Assessment

The basis for all three steps of the evaluation are the primary interaction processes for collaboration defined by Riemer [6] and Cook [2]. Tasks in electronic collaboration can be assigned to one of the following primary interaction processes, which are also called the 4Cs (following Cook [2]):

- Communication
- Coordination
- Collaboration
- Connection

In the first step of the evaluation process we select a number of e-Collaboration systems for the assessment (see Section 3.2). Due to space limitations, results will be presented in Section 5 for members of the shortlist only.

The second step of the evaluation process covers the definition of appropriate evaluation criteria. Functionalities provided by e-Collaboration systems were assigned to each of the primary interaction processes, which resulted in a feature list grouped into the 4Cs. In addition, two more categories were identified to be important for the evaluation of open source tools: *cross-sectional features* as well as *administration & technology*. These features were selected on the basis of a detailed study of literature on electronic collaboration, CSCW software and groupware (e.g., [6], [11]), of related reports and studies such as [3], [4], [5] and [14] which were complemented by the results of a number of interviews with experts in the field of CSCW and electronic collaboration plus personal expertise of the authors. For a detailed description of the respective features, refer to the following sections and [17].

For the assessment of the e-Collaboration systems according to their strengths and weaknesses with respect to the 4C categories, weightings were assigned to the features representing their relative importance within each category. Since related studies and market reports (e.g., [3], [14]) often do not provide weightings, we based ours on the experiences of experts (e.g., derived from interviews with consultants) plus personal expertise of the authors (see Section 5 for details of features and weightings).

The evaluation was not undertaken for a specific use case, but for a standard team collaboration scenario. This scenario covers typical activities of a team of collaborating knowledge workers, such as creating knowledge by sharing and structuring ideas, preserving knowledge by creating and editing documents together, sharing knowledge by providing central access to information for all team members as well as coordinating collaboration activities. However, in the case of applying the proposed method to a specific use case, the weightings of all features can be adjusted correspondingly to quickly get a more specific evaluation result that may support an individual process of choosing the right e-Collaboration tool.

Due to the application of a standard team collaboration scenario in our work, all categories were supposed to be equally important. Thus, we do not compute an overall sum of all features to get a single rating number. Each system was evaluated according to whether it offered a certain feature, which subsequently increased the total score within each category. The assessment resulted in a score for each category with a maximum of 100% per category.

5   Enterprise 2.0 Open Source Tools Profile

In the following subsections the features of the 4C categories, the category cross-sectional features as well as administration & technology, with a special emphasis on features relevant for choosing an open source system, are presented. The tables show the features, the assigned weightings for each feature and scores for the analyzed tools in the shortlist (see Section 3.2). The five e-Collaboration systems got the according score if the feature was supported (●) or not (○).

5.1   Communication

This category covers tools for asynchronous and synchronous communication as well as contact management and social presence. Among the asynchronous tools, the functionality of email with connection to an external email server was considered to be the most important as email is still the most widely used means of communication. An integrated and convenient solution for managing contacts was also regarded as highly important. Social presence features, which are often integrated in instant messaging services, have gained more and more significance for unified communication as they allow users to choose the right means for contacting other team members.

In the category communication, Liferay supports most of the important features. All systems support email using external email servers. Social presence showing the status of a team member as well as his/her availability is supported by Alfresco, EGroupware and Liferay, thus facilitating unified communication.

5.2   Coordination

Category coordination comprises features for task management, project management and workflows as well as the coordination of appointments and meetings. A group calendar providing an overview of the appointments of all team members was regarded as the most critical feature for efficient and effective coordination, followed by the management of tasks. Some systems provide sophisticated functionalities for project management, depending on whether the system was designed to support project management or was developed out of a project management tool and combined with collaborative features. Simple workflows are also supported by some e-Collaboration systems. However, for the full integration of workflows that also allows for designing complex processes, additional software packages are required in most cases.

Table 1. Features of category communication

Category Communication	Weight	Alfresco Share	EGroup-ware	Liferay	Simple Group-ware	TUTOS
Total score	100,00	50,00	63,00	85,50	50,50	45,00
Asynchronous	**50,00**					
Internal email	2,50	○	●	●	●	●
External email server	17,50	●	●	●	●	●
Discussion forum	7,50	●	○	●	●	○
Weblog	10,00	●	○	●	○	○
Microblogging	2,50	○	○	○	○	○
News	5,00	○	○	●	○	○
Comments	5,00	○	●	●	○	●
Synchronous	**15,00**					
Instant messaging	3,00	○	●	●	●	○
Broadcast	0,75	○	○	○	○	○
Desktop telephone conference	0,75	○	○	○	○	○
Video conference	3,00	○	○	○	○	○
Web conference	2,25	○	○	○	○	○
Telephone (audio)	3,75	○	○	○	○	○
Video telephone	1,50	○	○	○	○	○
Contact management	**20,00**	○	●	●	●	●
Social presence	**15,00**	●	●	●	○	○

In the category coordination, EGroupware gets the highest score followed by TUTOS. EGroupware provides a fully developed calendar and task management system including features for software development projects such as bug tracking, feature requests and patches as well as timesheets and resource management. EGroupware also supports the management of multiple projects.

TUTOS also supports the management of software development projects including Scrum, invoice, risk management, bug tracking and the administration of installations. Even though these features are not really necessary for standard team collaboration scenarios, the consequence of this emphasis on project management are extensive functionalities for coordination tasks.

5.3   Collaboration

Within the category collaboration, wikis are regarded as a widely used, suitable and thus important means of shared content production. A fully developed administration of shared content is crucial for a system to adequately support team collaboration. Creating documents out of the shared workspace proved to be a significant feature for e-Collaboration systems to be integrated into daily work routines. Workspaces supporting this functionality have got a higher chance of being used like the desktop. On the other hand, e-Collaboration systems that provide only up and downloading of documents risk being used as a repository for documents instead of supporting active collaboration.

Table 2. Features of category coordination

Category Coordination	Weight	Alfresco Share	EGroup-ware	Liferay	Simple Group-ware	TUTOS
Total score	100,00	53,50	92,50	48,00	52,50	67,50
Surveys	5,00	O	●	●	●	O
Workflow management	**10,00**					
Modeling of workflows	2,50	O	●	O	O	O
Ad hoc workflows	6,00	●	O	O	O	O
Workflow templates	1,50	O	O	O	O	O
Project management	**20,00**					
Gantt charts	5,00	O	●	O	O	●
Work package coordination	4,00	O	●	O	O	●
Assign work packages to employees	3,00	O	●	O	O	●
Status work packages	3,00	O	●	●	O	●
Resource management	4,00	O	●	O	O	●
Multi project mgmt.	1,00	O	●	O	O	●
Task management	**25,00**					
Task coordination	10,00	●	●	●	●	●
Assign tasks to employees	7,50	●	●	●	●	●
Status of tasks	7,50	●	●	O	●	●
Appointment coordination	**30,00**					
Group calendar	18,00	●	●	●	●	●
Personal calendar	4,50	●	●	●	●	●
Search for free dates	7,50	O	●	O	O	O
Meeting coordination	**10,00**					
Coordination of participants	5,00	O	●	O	O	O
Coordination of documents	5,00	O	●	O	O	O

In the category collaboration, Liferay gets the highest score followed by Alfresco and Simple Groupware. While Liferay supports the collaborative organization of content and meta data with the help of social tagging, bookmarking and cataloguing, Alfresco´s strength is the administration of shared content. Simple Groupware offers several features for dealing with shared content and enables editors to create (wiki) text, spreadsheets and graphics in the workspace, but does not support tagging.

5.4   Connection

The category connection provides the presentation of the user profiles as well as social features. Profiling allows users to present personal expertise, and people search supports finding the right person to contact. e-Collaboration systems with an emphasis on connection provide a short profile of the author together with contact details for every piece of content.

Table 3. Features of category collaboration

Category Collaboration	Weight	Alfresco Share	EGroup-ware	Liferay	Simple Group-ware	TUTOS
Total score	100,00	70,25	58,25	75,50	68,00	18,50
Shared content production	**20,00**					
Wiki	10,00	●	●	●	●	○
Whiteboard	3,00	○	○	○	○	○
Synchronously shared documents	4,00	○	○	●	○	○
Shared ideas / brainstorming	3,00	○	○	○	○	○
Working together on the same objects	**15,00**					
Social tagging	6,75	●	○	●	○	○
Social bookmarking	6,00	●	●	●	●	○
Social cataloguing	2,25	○	○	●	○	○
Administration of shared content	**50,00**					
Document sharing	7,50	●	●	●	●	●
Image sharing	2,00	●	●	●	●	●
Video/audio sharing	2,00	●	●	○	●	○
Restricted access for content	4,00	●	○	●	●	○
Restricted access for folder	2,50	●	●	●	●	○
Check in/check out	5,00	●	○	●	●	○
Up- & download	5,00	●	●	●	●	●
Versioning	4,00	●	●	●	●	●
Archiving	2,50	○	●	●	●	○
Folder / shared folder	7,50	●	●	●	●	○
Content tagging	4,00	●	●	●	○	○
Folder tagging	1,50	●	○	○	○	○
Personal site	2,50	●	○	●	●	○
Creating documents out of the shared workspace	**15,00**					
Text document	5,25	○	●	○	●	○
Spreadsheet	1,50	○	○	○	●	○
Presentation	1,50	○	○	○	○	○
Graphics	0,75	○	○	○	●	○
…using MS Office	6,00	○	○	○	○	○

Liferay´s emphasis is on the social connection of team members. The system provides activity tracking on blogs, message boards and wikis with a Facebook-like activity wall. Furthermore the social networking aspect is supported by the possibility to view the friends of a team member as well as their activities together with the team member´s profile. All systems provide user profiles and the possibility to search for people.

5.5   Cross-Sectional Features

Cross-sectional features cover all functionalities that do not only belong to a single category, but support some or all of them. Among these features, configurable areas for users were regarded as the most important, as they allow for the designing of a workspace that suits the specific requirements of the team without the need for an administrator to customize the workspace. Also the "pull" instead of the "push" way of getting informed about news is integrated into this category by the ability to get newsfeeds and set alerts. The "pull" metaphor is one of the essential elements when trying to cope with the information overload that is mainly caused by emails and to filter only relevant information. Equally important for getting the right pieces of information in time are complex search functionalities, which incorporate various sources of content and also people profiles. As knowledge work becomes more and more flexible and location independent, the synchronization of the e-Collaboration system with mobile devices is regarded as an important aspect for today's working conditions.

Table 4. Features of category connection

Category Connection	Weight	Alfresco Share	EGroup-ware	Liferay	Simple Group-ware	TUTOS
Total score	100,00	50,00	58,00	86,00	50,00	50,00
People search	25,00	●	●	●	●	●
People profiling	25,00	●	●	●	●	●
People tracking	8,00	O	●	●	O	O
Social networking	20,00	O	O	●	O	O
Visualization of the network	9,00	O	O	O	O	O
Network analysis	5,00	O	O	O	O	O
People tagging	8,00	O	O	●	O	O

Liferay and EGroupware almost fully support all cross-sectional functionalities and thus provide high integration of the features of all other categories.

Table 5. Features of category cross-sectional features

Category Cross-sectional Features	Weight	Alfresco Share	EGroup-ware	Liferay	Simple Group-Ware	TUTOS
Total score	100,00	76,00	94,00	96,00	75,00	28,00
Newsfeeds	10,00	●	●	●	●	O
Syndication	2,00	O	●	●	O	O
Personalization	8,00	●	●	●	●	O
Dashboard	5,00	●	●	●	O	O
Configurable areas	15,00	●	●	●	●	O
Mashup	2,00	O	●	●	O	O
Alerts	10,00	●	●	●	O	O
Tracking	3,00	●	●	●	O	●
Rating	3,00	O	O	●	O	O
Ranking	3,00	O	O	●	●	O
Filtering	4,00	O	●	O	●	O
Handheld delivery	10,00	O	●	●	●	O
Documentation	4,00	●	●	●	●	●
(Online) Help	6,00	●	●	●	●	●
Simple search	5,00	●	●	●	●	●
Complex search	10,00	●	●	●	●	●

5.6   Administration and Technology

The category administration & technology covers features which are mainly important for the administrator of the e-Collaboration system. With respect to administration the definition of user groups and roles turned out to be highly significant.

Furthermore, the smooth integration of the e-Collaboration system into the existing system landscape was regarded as a determining factor. Finally, the possibility to customize the system according to specific user requirements was identified as an important aspect, too.

Additionally, two groups of features especially relevant for open source e-Collaboration systems were identified: support & extendibility as well as installation. Regarding support & extendibility, the activity of the community serves as an indicator about how actively the system is further developed and how much support can be expected. Plugins & API denotes the availability of a collection of extensions and plugins and the existence of an API that allows programming of self-written plugins. Regarding installation, we distinguish between three options: *Repository installation* refers to package installation from the standard repositories of common Linux distributions (checked with Debian/Ubuntu); *Download and config.* refers to installation via download and configuration from a simple config file or web interface (requiring only a running Apache/PHP/MySQL environment). While these options are regarded as rather easy to be installed, the third category *advanced installation* refers to more complex installation procedures, including having to (re-)compile the code from the sources, or being based on specific database configuration or server installation, etc.

Table 6. Features of category administration & technology

Category Administration & Technology	Weight	Alfresco Share	EGroup-ware	Liferay	Simple Group-ware	TUTOS
Total score	100,00	64,80	92,40	86,50	75,20	35,50
Monitoring	3,50	○	●	○	●	○
Reporting	5,60	○	●	●	○	○
Scalability	5,60	●	●	●	○	○
Configuration	7,00	●	●	●	●	●
Customization	10,50	●	●	●	●	●
Designer toolkit	5,60	○	○	●	○	○
Integration	10,50	○	●	●	●	●
Backup/recovery	4,20	●	●	●	●	○
User Management	**17,50**					
User groups/roles	14,00	●	●	●	●	○
Directory	3,50	●	●	●	●	●
Support and Extendibility	**18,00**					
Active community	10,00	●	●	●	●	○
Plugins, API	8,00	●	●	●	●	○
Installation	**12,00**					
Repository installation	6,00	○	●	○	○	○
Download and config. installation	4,00	○	●	○	●	●
Advanced installation	2,00	●	○	●	○	○

Liferay and EGroupware provide fully developed administration facilities. Concerning support and extendibility for further developing the system, apart from TUTOS, all systems have got an active and large community and offer a convenient way of extending the existing features. While EGroupware, Simple Groupware and TUTOS can be rather easily installed with only basic system administration skills,

Alfresco and Liferay belong to the installation type *advanced installation* which implies that a skilled administrator is required to set up the system.

6   Review

Table 7 summarizes the scores out of 100% as the total score for each category that the e-Collaboration systems in the shortlist obtained in the categories communication, coordination, collaboration and connection as well as cross-sectional features and administration & technology. This table combines the lines "total score" of Table 1 to 6 in a single table including an arithmetic mean in each category. Table 7 does not show an overall score for each of the e-Collaboration systems (e.g. computed as an arithmetic mean or a weighted sum), as the proposed evaluation is based on the standard team collaboration scenario briefly described in Section 4. It should thus provide an overview of the results in the various categories, so that the actual selection by a specific SME can be based on the categories relevant for the use case under consideration.

Table 7. Evaluation of five open source e-Collaboration systems

	Alfresco Share	EGroupware	Liferay	Simple Group-ware	TUTOS	*Average*
Communication	50 %	63 %	86 %	51 %	45 %	*59 %*
Coordination	54 %	93 %	48 %	51 %	68 %	*63 %*
Collaboration	70 %	58 %	76 %	68 %	19 %	*58 %*
Connection	50 %	58 %	86 %	50 %	50 %	*59 %*
Cross-sectional features	76 %	94 %	96 %	75 %	28 %	*74 %*
Administration & Technology	65 %	92 %	87 %	75 %	36 %	*71 %*

Alfresco provides very well supported and integrated document and content management features and thus got the highest scores in the categories collaboration and cross-sectional features. Alfresco belongs to the well-known e-Collaboration systems. It is widely used, well supported and it stands out due to many reference installations.

EGroupware scores very high in the category coordination as it supports many project management features and also special functionalities for software development projects. The social aspect is not a strength of EGroupware and the linking between content and people providing the content is not as transparent as in Liferay for example, which puts a special emphasis on the social aspect and thus got the highest score in the category connection. Even though EGroupware seems to have been developed for collaboration in software development projects, it is easy to use and provides many other collaborative features apart from project management. TUTOS on the other hand also scores high in the category coordination as it supports the development of software projects from acquisition to installation, but falls short concerning the other categories and thus is not really suited for standard team collaboration.

Although Simple Groupware is developed only by a small company and does not seem to have a big community, it appears technically mature. Simple Groupware scores with its ability for managing content in enterprise, project and personal spaces.

Due to limitations in usability and lack of project management capabilities, Simple Groupware is well suited only for small and medium size teams.

The studied systems differ in installability, support, extendibility and maturity, which is typical for open source systems. However, some of them are also supported commercially for those not wanting to hassle with technical issues themselves (in our study: Alfresco, EGroupware, Liferay).

The evaluation of open source E-Collaboration systems among standard closed source systems showed that there are many technically mature open source solutions. As the assessment in this paper shows, a differentiated view on open source E-Collaboration systems is necessary. In addition, many open source systems provide complex functionalities, offer support and training so they can definitely be regarded as a serious alternative to closed source platforms. Market readiness for open source systems is also confirmed by market analysts such as Gartner Research [4] or Forrester Research [5], who list open source systems such as MindTouch together with well-known proprietary solutions.

7   Conclusions

Small and medium size enterprises as well as small organizations, e.g. not-for-profit organizations, are looking for e-Collaboration systems at a reasonable price and workload to support their needs in communication, collaboration, and information sharing among their team members. Open source Enterprise 2.0 tools have shown to offer sufficient functionalities to fulfill these requirements. These tools have strengths in one or more basic interaction processes of electronic collaboration – just like comparable closed source systems have; and like these systems, they also show weaknesses because they do not support all features we identified to be important for Enterprise 2.0 tools. When choosing an open source e-Collaboration system, special attention should be paid to installation, support and operation options offered by open source systems, especially since many vendors of commercial systems offer easy deployment options like SaaS. However, our analysis of the various installation options provided for open source tools shows that also a number of open source E-Collaboration systems are easy to install and maintain.

This study provides enterprises – especially SMEs – with a comprehensive set of evaluation criteria focusing on communication, coordination, collaboration and connection. In contrast to related studies evaluating open source tools for electronic collaboration, like [3], our criteria cover the entire spectrum of collaboration features and represent state-of-the-art features of modern Enterprise 2.0 systems. Especially with respect to the category connection our evaluation approach includes features for social networking which are not covered by comparable assessments. In comparison to the evaluation of E-Collaboration systems presented by Büchner et al. [14], which focuses on content-centric collaboration, our report also covers features for communication and coordination. We do not intend to present a market study on all major open source Enterprise 2.0 systems such as attempted by Spath et al. [3], but to introduce an easy to implement evaluation method. Since other studies and reports like [14] include either none or only a small number of open source systems, SMEs will benefit from these results that meet their limited resources.

Enterprises and organizations planning to implement Enterprise 2.0 software will find a market analysis of relevant open source tools and an evaluation method that will help them to identify an appropriate collaboration tool. The systematic evaluation approach including a set of nearly 100 individual features in six distinct categories can be applied to various types of Enterprise 2.0 and e-Collaboration systems to provide a sound assessment. Furthermore, by adapting the weights of the evaluation matrix, our evaluation approach can easily be specialized for the requirements profile of a particular company or user group.

Future work will include commercial platforms to allow for a comparison of open source software and commercial tools. An even more detailed analysis of selected collaboration tools will rate individual features not in a binary mode – at a scale of "supported" ● or "not supported" ○, but will apply performance ratings on an appropriate multilevel scale.

Acknowledgements. This work was funded by FFG / Österreichische Forschungsförderungsgesellschaft mbH in grant No. 821111 "eCollaboration 2.0: Collaboration Tools und Social Media für Teamarbeit in KMUs".

References

1. McAfee, A.: Enterprise 2.0: The Dawn of Emergent Collaboration. MIT Sloan Management Review 47(3), 21–28 (2006),
   http://sloanreview.mit.edu/the-magazine/articles/2006/
   spring/47306/enterprise-the-dawn-of-emergent-collaboration/
   (retrieved on May 1, 2011)
2. Cook, N.: Enterprise 2.0 – How Social Software Will Change the Future of Work. Gower, Aldershot (2008)
3. Spath, D., Schimpf, S., Kugler, A. (eds.): Webbasierte Open Source-Kollaborations plattformen. Fraunhofer IRB Verlag, Stuttgart (2007)
4. Drakos, N., Mann, J., Rozwell, C.: Magic Quadrant for Social Software in the Workplace. Gartner (2010),
   http://www.gartner.com/
   DisplayDocument?id=1456713respectively (retrieved on May 1, 2011)
5. Koplowitz, R.: The Forrester Wave™: Collaboration Platforms, Q3 2009. Forrester Research Inc. (2009)
6. Riemer, K.: The Market for E-Collaboration Systems – Identification of System Classes Using Cluster Analysis. In: Österle, H., Schelp, J., Winter, R. (eds.) Proceedings Fifteenth European Conference on Information Systems, pp. 346–357 (2007)
7. McAfee, A.: Enterprise 2.0, version 2.0. Andrew McAfee's Blog - The Business Impact of IT (2006),
   http://andrewmcafee.org/2006/05/enterprise_20_version_20/
   (retrieved on May 1, 2011)
8. AIIM: What is Enterprise 2.0? (n.d.),
   http://www.aiim.org/What-is-Enterprise-20-E20
   (retrieved on May 1, 2011)
9. Koch, M.: CSCW and Enterprise 2.0 – towards an integrated perspective. In: Proceedings 21st Bled eConferencee Collaboration, pp. 416–427 (2008)

10. Koch, M., Gross, T.: Computer-Supported Cooperative Work – Concepts and Trends. In: Proceedings 11th Conference of the Association Information and Management (2006)
11. Ellis, C.A., Gibbs, S.J., Rein, G.L.: Groupware – Some Issues and Experiences. Communications of the ACM 34(1), 38–58 (1991)
12. Hinchcliffe, D.: The state of Enterprise 2.0. Enterprise 2.0 Web Blog, ZDNet (2007), http://blogs.zdnet.com/Hinchcliffe/?p=143 (retrieved on May 1, 2011)
13. Xu, J., Zhang, J., Harvey, T., Young, J.: A Survey of Asynchronous Collaboration Tools. Information Technology Journal 7(8), 1182–1187 (2008)
14. Büchner, T., Matthes, F., Neubert, C.: A Concept and Service Based Analysis of Commercial and Open Source Enterprise 2.0 Tools. In: Proceedings International Conference on Knowledge Management and Information Sharing, pp. 37–45 (2009)
15. Real Story Group: The Enterprise Collaboration & Community Software Management Evaluation Report (2010), http://www.realstorygroup.com/ (retrieved on May 1, 2011)
16. Hinchcliffe, D.: Assessing the Enterprise 2.0 marketplace in 2009: Robust and crowded. Enterprise 2.0 Web Blog, ZDNet (2009), http://blogs.zdnet.com/Hinchcliffe/?p=598 (retrieved on May 1, 2011)
17. Schauer, B., Zeiller, M., Riedl, D.: Reviewing the E-Collaboration Marketplace – A Survey of Electronic Collaboration Systems. In: Proceedings ICE-B 2010 International Conference on e-Business, pp. 69–75 (2010)

Assessing the Impact of In-government Cooperation Dynamics: A Simluation-Based Systems Inquiry

Lawrence John[1], Patricia M. McCormick[2], Tom McCormick[2],
Greg McNeill[3], and John Boardman[4]

[1] Analytic Services, Inc., Arlington, VA, U.S.A.
larry.john@anser.org
[2] Alpha Informatics, Ltd., Manassas, VA, U.S.A.
{pmccor,tmccor}@inforead.com
[3] ExoAnalytic Solutions, Inc, Mission Viejo, CA, U.S.A.
mceillg@teamseas.com
[4] School of Systems and Enterprises, Stevens Insitute of Technology, Hoboken, NJ, U.S.A.
john.boardman@stevens.edu

Abstract. We coined the term "government extended enterprise" (GEE) to describe sets of effectively autonomous government organizations that must cooperate voluntarily to achieve desired GEE-level outcomes. A GEE is, by definition, a complex dynamical system of systems (SoS). Our continuing research investigates the proposition that the interaction of four "canonical forces" affects both internal GEE cooperation and SoS-level operational effectiveness, changing the GEE's status as indicated by the "SoS differentiating characteristics" detailed by Boardman and Sauser. Three prior papers have described the concepts involved, postulated the relationships among them, discussed the n-player, iterated "Stag Hunt" methodology applied to execute a real proof-of-concept case (the U.S. Counterterrorism Enterprise's response to the Christmas Day Bomber) in an agent-based model, and presented preliminary conclusions from testing of the simulation. This paper adds key insights gleaned from additional in-depth review of relevant literature and data analysis.

Keywords: System of Systems, Extended Enterprise, Game Theory, Prisoner's Dilemma, Stag Hunt, Agent-based Modelling, Complex Dynamical Systems, Institutional Analysis, Organizational Norms, Organizational Behaviour, Systems Thinking.

1 Introduction

On Christmas Day 2009, 19-year old Farouk Abdulmutallab and a few supporters exposed fatal flaws in an extended enterprise of at least "1,271 government organizations and 1,931 private companies" with a combined budget in excess of $75 billion [1]. Yet, according to the findings of the Senate Select Committee on Intelligence [?], this leviathan failed because its members *chose not to share critical information* that would have foiled the plot—they chose to not cooperate.

A. Fred et al. (Eds.): IC3K 2011, CCIS 348, pp. 423–439, 2013.
© Springer-Verlag Berlin Heidelberg 2013

We believe the discipline of systems engineering—specifically, system of systems (SoS) engineering—has both the ability and the responsibility to help future decision makers understand why this happened and how to prevent similar failures. A systems engineer might describe such a problem:

A heterogeneous network of autonomous nodes, each with its own "private" goals, exists collectively to serve one or more "public" goals. The nodes must cooperate to produce preferred SoS-level outcomes, but the network underperforms due to a lack of internal cooperation—intentionally or not, some nodes place their private goals ahead of the network's goals.

Examples of networks fitting this description abound—in industry (e.g., standards consortia, corporate alliances), the non-profit sector (e.g., collections of community service organizations), the military (e.g., Services trying to jointly field capabilities or conduct operations), and government at all levels (e.g., cabinet departments or legislative committees with overlapping jurisdictions, collections of international, federal and/or state organizations working in one more domains). Yet decision makers appear to lack sound, theory-based approaches and methods to generate, promote and sustain the required level of cooperation in enterprises like them. This may be especially true for networks of high-level government organizations, where there is no analog to "market forces" to punish recalcitrant members.

2   Background

Serious thought about cooperation began with the classical philosophers. Thomas Hobbes, David Hume and Adam Smith interpreted, enriched and passed it through Thomas Malthus to Charles Darwin. From there, one spur proceeds through Emile Durkheim, William Hamilton and Robert Trivers to contemporary sociology and evolutionary biology, while another flows through David Ricardo, Francis Edgeworth and Leon Walras to contemporary neoclassical economics [3, pp. 3-4].

Our first paper [4] discusses cooperation-related concepts from many disciplines to investigate two propositions: 1. a set of "canonical forces" (Sympathy, Trust, Fear and Greed) affects the dynamics of SoS operating under conditions of need and uncertainty; and 2. understanding these forces may enable network leaders to address SoS performance issues caused by component systems' failure to cooperate.

Our unit of analysis is the "Government Extended Enterprise" (GEE). Our definition of a GEE extends concepts from [5] and [6] to include "the entire set of collaborating [entities], both upstream and downstream, from [initial inputs] to [end-use decisions, policies and actions], that work together to bring value to [the nation]." Thus, GEEs are sets of relatively autonomous government enterprises that must achieve enough "propensity to cooperate" [7] to produce the voluntary cooperation that is a prerequisite for coordinated action. GEEs face the "Hobbesian paradox" of social dilemmas—cooperative action pursued by self-interested individuals [8].

2.1 Central Concepts

Our understanding of cooperation has evolved based on continuing review of work by researchers in engineering, the behavioural and social sciences, business and management, and the life sciences (the reference list alone exceeds the total space allotted for this article; a critical review is in the works). We have arrived at the following definition of *cooperation*: "activity one actor carries out deliberately that benefits one or more other actors individually or collectively, potentially at a cost to itself and with no guarantee of a direct benefit to itself." While classical theorists have long held altruistic cooperation antithetical to Darwin, recent experiments in biology show cooperation as fitness-enhancing even in some lower species [9-11].

In [4], we postulated four *canonical forces*: Sympathy, Trust, Fear and Greed. Sympathy, is the "fellow feeling" that brings individuals together [12-15]. Trust is a three-part attitude (*a* trusts *b* with respect to *x*) [16], [17] reflecting "one's willingness to be vulnerable to another's actions with the belief that the other will perform as expected" [18-20] and is viewed as critical to primitive man's survival as a species [21] [22]. Fear is "the cognition of an expected deprivation" [23]. Fear "induces ... focus on events that are especially unfavourable" [24 citing 25] and erodes SoS cohesion by causing components to advance their private goals in preference to the network's public goals. In GEEs it is anxiety about being viewed as a failure, of incurring a business or political loss or cost, or loss of control [26]. Greed for success, power, budget or influence [15], [27], [28] also increases focus on private goals, reducing propensity to cooperate, and reinforcing the effects of Fear. Extending [16], we now view the forces not as general conditions (i.e., actor *x* experiences level of force $\{S, T, F, G\}$ all the time), but as relationships of the form "in situation *n*, actor *x* experiences level of force toward actor *y* about subject a.

Our work also postulates two conditions to which GEEs are subject because they are open systems [29], [30]. They are: Need, primarily for resources, which in this case involves the competitive federal budget process [31] and Uncertainty, which produces Fear, with concomitant Risk and potential for profit [32-37].

2.2 SoS Differentiating Characteristics

We leverage the five characteristics by which Boardman and Sauser [38], [39] differentiate systems of systems from systems of components: Autonomy (A), both a component system's native ability to make independent choices (an "internal" system-level property conveyed by its nature as a holon), and more importantly, the fact that other members of the system of systems "respect" this ability by permitting the component to exercise it; Belonging (B), a direct reflection of the components' recognition of a shared mission or shared (but not merely coincident) interests; Connectivity (C), "the agility of structure for essential connectivity in the face of a dynamic problematique that defies prescience" [39, pp. 158-159]; Diversity (D), "noticeable heterogeneity; having distinct or unlike elements or qualities in a group," [39, p. 157] that reflects the impact of the law of requisite variety [40] on systems of systems; and Emergence (E): the ability to "match the agility of the problematique" by adding new responses based on "auxiliary mechanisms for anticipation" [39, pp. 160-161].

Our prior paper [4] discusses the postulated relationships between the forces and these characteristics in detail, summarizing them in a table that uses a five-point nominal scale to indicate both how strongly "positive," "neutral," or "negative" a force is with respect to a characteristic, and whether the characteristic requires or is inimical to the force. These values drive the cooperative model selected by the agents in the simulation described in Section 6, below, and support tracking the resulting chain of causality. Table II in the same paper describes the relationships between the levels of the Boardman-Sauser characteristics and a component's "Cooperation Model"—cooperate or "co-opetate" (attributed to Novell founder Ray Noorda). Table III uses a five-point nominal scale to indicate the postulated impact of changes in the Boardman-Sauser characteristics levels on the "cooperation model" each agent uses in the game, expressed as the force "favouring" or "disfavouring" a choice. Our work captures and measures these changes and the resulting SoS-level behaviour. The same paper discusses the impact of two other potentially important factors: History of Behaviour and Leadership.

3   Theory and Approach

Recent research [38], [39], [41-45] has posited and attempted to quantify how collections of systems that should work together become more manifestly SoS as their levels of the characteristics rise. We theorize that in action situations that demand cooperation, assuming increasing the level of cooperation improves the operational performance of the SoS, each organization's Probability of Cooperation with an emerging coalition is the result of the interaction of the proposed forces, each organization's principles-based strategy and a set of behavioural factors.

Informed by noted cooperation scholars [7], [22], [32], [46-54] our methodology applies game theory in an agent-based simulation of a complex adaptive system and a real-world case (see [55] for a detailed description). Our approach centres on a "Stag Hunt" game [56], [57] that treats information in the GEE as a "common-pool resource" [48] and established payoff dirven (Hicks) and risk dominant (Nash) equilibria that correspond with the GEE's public and private goals. The GEE cannot succeed if key nodes fail to cooperate by sharing information in ways that meet the requirements in the unclassified Executive Summary of the SSCI report [2].

3.1   Hypotheses and Assumptions

Testing has led us to refine the previously declared set of hypotheses [48]. Given a SoS ("S")—the GEE—comprising Executive Agent "a_1" and autonomous components "a_2" though "a_n", operating under conditions of uncertainty and with knowledge of each others' history of behavior with respect to themselves:

1. Hypothesis 1. an's levels of Probability of Cooperation with a1, will be:

    (a) positively correlated with a_n's level of Risk Tolerance,

    (b) positively correlated with a_n's level of Sympathy and Trust with respect to a1,

    (c) positively correlated with a_1's History of Behavior,

(d) negatively correlated with a_n's level of Greed

(e) negatively correlated with a_n's level of Fear.

2. Hypothesis 2. S's level of Belonging will be:

(a) positively correlated with S's level of Sympathy (where the value of Sympathy is the median of the values for S's members)

(b) positively correlated with S's level of Trust, (where the value of Trust is the median of the values for S's members)

(c) negatively correlated with S's levels of Greed (where the value of Greed is the median of the values for S's members)

(d) negatively correlated with S's level of Fear (where the values of Fear is the median of the values for S's members).

3. Hypothesis 3. S's level of EE Belonging, will be positively correlated with key components' aggregate Probability of Cooperation.

4. Hypothesis 4. S's level of EE Connectivity, will be positively correlated with key components' aggregate Probability of Cooperation.

5. Hypothesis 5. S's level of EE Diversity will be positively correlated with key components' aggregate Probability of Cooperation.

6. Hypothesis 6. S's level of EE Emergence will be positively correlated with key components' aggregate Probability of Cooperation.

Continuing research has led us to add the assumption that all of the player's "coope-rate or defect" decisions must comply with the letter and intent of U.S. law and policy to our prior list [48, Section III.B]. This enables us to explicitly incorporate the deon-tic component of social decision making—rules about what one must, may not and should do [46], [58], [59].

3.2   Validation and Data Analysis Process

After testing has been completed, a Review Panel—a set of experts with long expe-rience as both operators and executives in the relevant organizations—will provide data on force levels for each situation. This will aid validation by accounting for agent-based models' sensitivity to initial conditions [60], [61].

Data analysis centres on the use of non-parametric statistics processes. These are appropriate when using agent-based models because one cannot make useful *a priori* assumptions about the distributions of the data.

3.3   Sample Case

Our second paper [55] provides a detailed explanation of the sample case, which cov-ers an 18-month period comprising five decision points where the SSCI Report found that components of the GEE could have foiled the attack by sharing information they already possessed. Figure 1 illustrates the SSCI's core operational finding (example, at Event #1)—the solid blue arrows represent expected information flows (i.e., full information sharing) while dashed blue arrows represent flows that did not occur.

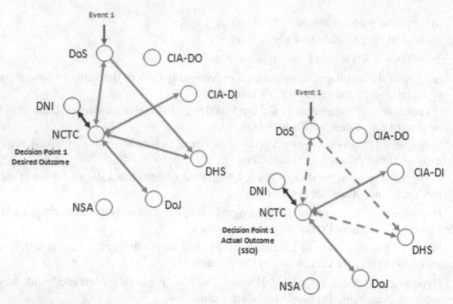

Fig. 1. Desired and actual Information Flows for Decision Point #1

3.4   Agent-Based Simulation

Our second paper [55] also provides a detailed description of our computational agent based simulation, which is similar to a process used in a wide variety of domains in the physical and social sciences, including studies of cooperation [49], [62] and complex adaptive systems [49], [61]. We eliminate the potential impact of signalling issues [49] by assuming that all choices are made simultaneously, approximating the impact of effective administrative information security procedures.

We chose the Stag Hunt over the more widely used iterated Prisoner's Dilemma [7], [46], [62], [63] because it provides two equilibria that can be viewed as "satisfactory"—a risk-dominant (Nash) equilibrium satisfies private goals, while a payoff-dominant (Pareto-optimal Hicks) equilibrium satisfies public goals—in non-zero sum games [47], [48]. Creation of viable social contracts [50-54], [57] and artificial intelligence and network switching [64-70] present similar decision making challenges.

3.5   Factors Governing Behaviour

Behaviour within and among organizations is governed by "institutional statements"—rules, norms and shared strategies [22], [32], [46], [49], informed by knowledge—the deontic, axiological and epistemological components of social decision making [58], [59]. They are embodied in a set of behavioural factors that represent key inputs to decision making, and can be described algorithmically. Our second paper [55] presents our core algorithm (Equation 1), defines the eight factors that affect an agent's Propensity to Cooperate (P_c) and details the processes by which the model leverages them. The characteristics are: F1 Level of Risk, F2 Payoff to the Sharing

Agent, F3 Payoff to the Receiving Agent, F4 History of Behaviour, F5 Risk Tolerance, F6 Perceived Level of Need, F7 Perceived Level of Damage Due to Disclosure
(a powerful analogue to "Subtractability of Flow" in common-pool resource problems
[71]), and F8 Sharing Agent's Perceived Level of Confidence in the Information.

$$P_c = (F4*F6) \ ((F1+F2+F3+F5)/4) \ ((F7+F8)/2) +3 \qquad (1)$$

3.6   Principle-Driven Strategies

Principles—the sum of an organization's values, standards, ideals, precepts, beliefs,
morals and ethics—drive the strategy that drives decisions by helping decision makers
"to establish whether a decision is right or wrong" [72]. They are the axiological
component of social decision making. Our second paper [55] describes the process by
which we leveraged aspects of Vroom's image theory (based on Maslow and Herz-
berg, cited in [73] to derive and leverage the six alternative (self-regarding, neutral or
other-regarding) principles that underlie the information sharing decision making
strategies of US CT Enterprise components. This enabled us to create a game strategy
profile [49], [62] consisting of 11 strategies and to establish weighting coefficients for
the behavioural factors used by Equation 1 to calculate P_c for each situation.

3.7   Simulation Toolset

This effort uses Systems Effectiveness Analysis Simulation (SEAS), an agent-based,
complex adaptive systems simulation that is part of the Air Force Standard Analysis
Tool Kit [74]. SEAS agents incorporate the components of social decision making by
functioning at the physical, information, and cognitive levels to maintain awareness of
their situations, and by leveraging a set of simple, principle-based behaviour rules that
incorporate the impact of norms to make decisions "on the fly." SEAS-generated data
enables us to infer the impact of the forces on the Boardman-Sauser characteristics
and sets the stage for root cause analyses.

4   Analysis

4.1   Boundary Conditions

Exploring boundary conditions (i.e., the outcomes produced by agents adopting extreme strategies) is a key step in the use of agent-based models [61]. Our initial exploration of the game matrices for agents employing a "pure" strategy (e.g., always share
if it favours the GEE, or always favour their own organization), verifies that the "Stag
Hunt" game is a good simulation for this problem. Extremely cooperation-friendly
strategies produced payoff-dominant results, and cooperation-antagonistic strategies
produced risk-dominant results. We used the results of these initial analyses to select
the applicable ranges and effects of the Decision Making Freedom Factors. Based on
tests to date, a normalized P_c of 0.8 appears to represent a "ceiling" below which
agents will always refuse to cooperate, while values above 1.225 represent a "floor"
above which agents will always cooperate. Approximately 20% of calculated values

fall in one of these two areas.   Test data indicates that these values manifest at higher force levels.

4.2   Data Analysis Process

The team conducted exploratory data analysis and a series of statistical tests to establish the presence of significant patterns within the data. We sought to determine whether the observed outcomes vary directly with the levels of the canonical forces, thereby reliably refuting the null hypotheses. We used the Mann-Whitney U test to detect differences in the distribution of results for different conditions, and the Kruskal-Wallis test as an analogue to a one-way analysis of variance.

Initial testing in early March 2010 focused on verifying that the simulation manipulates data and computes results in accordance with our design, and that the design contains no egregious errors. We chose a subset (43 cases) of the possible combinations of integer-value force levels (256 cases), to support linear regression analysis. We ran the entire five-event scenario for each strategy in 200 blocks; each block held 112 opportunities for cooperation. Table 1 is a small sample of the simulator output. A "1" in the "Share Sender?" and or "Share Receiver?" column indicates that the computed probability of cooperation resulted in that agent deciding to share ("cooperate"). "Score" is the sharing agent's payoff.

Table 1. Simulator output (Unprocessed)

Sharer	Receiver	Share Sender?	Share Receiver?	Score
DOS_CA	DHS_CBP	1	0	0
DOS_CA	NCTC	1	1	10

An Excel macro cleans and orders the data, computes summary statistics (the number of "share" decisions, as well as minimum, maximum, mean and median values for probability of cooperation and payoff by block), and transfers the ordered data to another workbook (Table 2).

Table 2. Partial simulator output (Unprocessed)

ST_0	Trust = 1 all others = 0				ST_1
#	Sender Share Opps.	Receiver Share Opps.	Score		Sender Share Opps.
1	0.678571	0.660714	139		0.696429
2	0.625	0.758929	180		0.642857
3	0.723214	0.741071	179		0.767857

The macro computes additional summary statistics (minimum, 25th percentile, average, 75th percentile, and maximum $Prob_c$ values) for each case and plots them vs. force configuration using a "box and whiskers" format.   It also counts the frequency of $Prob_c$ values in a series of ranges for plotting by force configuration and strategy in a three-dimensional "ribbon chart" format. We also use a "box and whiskers" format to plot linear regressions for the impact of forces on $Prob_c$ and line charts to plot the impact of strategies or forces on payoff.

The resulting data set is large. Our current process treats the forces as discrete variables, but we are currently analyzing comparative plots of the manually computed values of Pc versus the simulator-computed values of Probability of Cooperation based on integer values assigned to each of the force configurations to begin to illuminate the space between the data points to support interpolation in future versions of the simulation. Interpolation across strategies may be problematic.

4.3   Addressing Threats to Validity

Executing 200 Monte Carlo trials of each of force configuration (a "case") produces a statistical confidence above .95 for each set of results. Section IV of [55] briefly discusses our approach to addressing internal and external validity, face and construct validity, criterion validity, and construct validity.

5   Preliminary Results

The following are preliminary conclusions, some of which may have the potential for significant implications for GEE members or future researchers.

We have demonstrated the ability to encapsulate agents' belief systems in key model elements and leverage that encapsulation to produce internally self-consistent results. This means the experiment may offer a useful evaluation of the postulated relationship between the forces and the SoS characteristics.

The neutral strategies are, by their nature as firm, all-purpose decision making heuristics, essentially insensitive to the forces. While preliminary results demonstrate that the forces can impact decision making, the effect appears to be significant only when the decision maker's principles evidence some level of preference for public or private goals. In general when considering the forces individually, Sympathy tends to have the greatest impact, followed in descending order by Greed, Trust and Fear.

The neutral strategies produce results that are predictable, but uninteresting. Kruskal-Wallis testing indicates that some strategies produce sufficiently similar results that we can eliminate some and reduce the mass of data to be analyzed. The combination of Trust and Sympathy at Level 3 (with other forces = 0) has produced anomalous results with two strategies; further investigation is required. We have yet to evaluate the interaction effects among the forces in complicated force configurations (for example, each force at a different level), but must do so, as we expect these conditions to be relevant to the case study.

We see preliminary indications of an unexpectedly dynamic relationship between the forces and strategies. Strategies tend to dominate Level 1 forces, but Level 3 forces (and, presumably, their interaction effects) dominate non-heuristic strategies. Fear appears to play a major role only when added to other forces, but it seems to dampen the impact of many strategies. On the surface, this appears to be consistent with results reported in [75], which investigated sociologists' view that, in contrast to what game theory predicts, "players [in cooperation games] will pay attention to the size of Greed, of Fear, and of Cooperators' Gain in making their decisions while these cardinal values do not play a role in standard game theoretical predictions" [76], [77], but more detailed investigation is required.

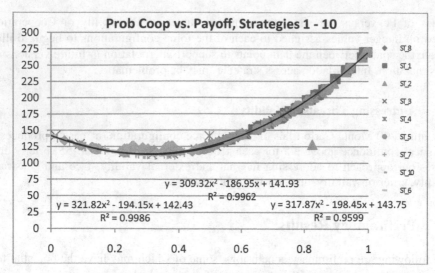

Fig. 2. Median payoff at probability of cooperation

Figure 2 plots median gross payoff as a function of probability of cooperation by strategy (10 of 11 strategies). We believe it clearly indicates that agents engaged in repeated interactions within a Stag Hunt situation will generally earn higher cumulative payoffs if they choose to cooperate. One could also infer from this that (assuming no externalities to the contrary) the same is true of one-shot Stag Hunt situations, but this is not the focus of our research. It must be noted, however, that the current version of the simulation does not feature live play of situational (Need and Uncertainty) and behavioural factors that could have a profound impact on decision making. If the plotted results persist, however, they should lead GEE members who are uncertain of whether they should choose to cooperate to do so. Because the only strategies that produce better-than-minimum payoffs from defecting are those held by agents with principles that motivate strongly against cooperation, these results indicate that GEEs may be able to safely incorporate agents with principles that are uncertain or neutral.

We believe our work highlights a key difference in the explanatory potential of the models researchers use to study human behaviour. The rational egotist model is fundamentally *prescriptive* in nature—it presumes to tell people how they *should* behave but ignores how they often actually behave. We believe it is unlikely to be useful in any but the simplest action situations and the most disciplined of actors. Models that include the possibility of prosocial behaviour are *descriptive*, precisely because they are based upon observations of how people *do* behave and actively seek to explain why. Their potential appears limited only by the depth of researchers' understanding of the factors affecting when prosocial and anti-social behaviour will manifest.

Both research and experience show that even in similar conditions and with similar goals, assumptions and objectives, different decision makers may not make the same decisions. This leads us to believe that any effort to understand human cooperation—perhaps even the possibility of cooperative behaviour by *any* actor that possesses intelligence and autonomy—must include consideration of ideas many years of research in evolutionary biology [78], cognitive neuroscience [79], the formation and operation of behavioural norms [22], [80-85] and psychology [86-93].

6   Limitations

We recognize this effort is essentially a proof-of-concept, based on a single case. That case may not generalize as fully as we hope; other cases may lack a convenient set of findings to use as a measuring stick for evaluating the relationship of the forces to the characteristics; the SSCI's root cause analysis may be flawed. It is also possible that the assumptions and abstractions we have used to simplify the problem may contain important complexities or factors we have failed to recognize and account for. For example, our model assumes that agents do not learn—they do not change decision making strategies in the course of a case. We also eliminate the effects of information transfer time, and differences in individual capabilities and authorities by assuming that when any member of an organizational element gains access to a piece of information, the entire element gains access and understanding immediately, and has the authority to act on that understanding. Moreover, the process used in our tests to date lacks explicit recognition of costs that would discount payoffs and qualify a decision to cooperate as altruistic [22]. Moreover, our treatment of history of behaviour assumes that organizational memory and individual memory are essentially the same— that is, even if the individuals representing organizations change, knowledge and reaction to history of behaviour will be the same. This idea must be tested.

The Stag Hunt is not without controversy. Skyrms wrote, "Many modern thinkers have focused on the prisoner's dilemma [as a simple exemplar of the central problem of the social contract], but I believe that this emphasis is misplaced. The most appropriate choice is not the prisoner's dilemma, but rather the stag hunt" [57]. Gintis replied that Skyrms had fallen prey to the "Folk Theorem of Repeated Games" [95], [96] whose "central weakness is that it is only an existence theorem with no consideration of how the Nash equilibrium can actually be instantiated as a social process....Rather, ... strategic interaction must be socially structured by a choreographer—a social norm with the status of common knowledge" [94]. Based on our understanding of Simon's analyses of the behaviour of individuals in organizations [27], [97], and others' [22], [49] confidence in the power of cultural norms as the "choreographer" of individual and group behaviours, we believe norms within government organizations may be powerful enough to establish the Nash equilibrium.

Earlier in the same book, Gintis also questions the need for those applying game theory to social dilemmas to eschew the "rational actor" model in favour of bounded rationality. He contends that explicitly accounting for each agent's beliefs, preferences and constraints allows for rational, self-regarding agents that operate with defined limits with respect to their knowledge and their own perspective (their utility function). We designed our model to address this by explicitly including the impact of decision maker principles, which are often based on organizational "received wisdom," and perceived constraints in the strategy formulation process.

7   Future Work

Our most recent paper [98] refers to the idea that this effort is one step on a much longer journey toward what we view as an emerging "Science of Belonging," which we now believe must work hand-in-hand with the science of cooperation. We believe the understanding that can be derived from focused inquiry into these areas will be crucial in a world full of autonomous, machine-based systems guided by software. Future work must establish how decision makers can change the levels of the forces in their extended enterprises—the specific "levers" decision makers can pull—as well as how they can accurately measure the resulting amount of change in the Boardman-Sauser characteristics. We must also ascertain the existence and impacts of other useful forces and characteristics. We must establish whether systems of systems can operate in the absence of Belonging—and a sound way to test that idea.

In pursuing this research, we are mindful that real success, especially in the realm of human activity SoS, requires that we understand cooperation dynamics so thoroughly that we can help decision makers avoid choosing solutions that create what Sterman calls *policy resistance*, "the tendency for interventions to be defeated by the response of the system to the intervention itself … [which] breeds a sense of futility about our ability to make a difference" [99, pp.504-506]. To that end, we must be conscious of Emanuel Derman's impassioned warning on the impact of confusing metaphors, models and theories [100]. We must examine our results critically by asking *why?* questions that illuminate the difference between what we *think* we see and what *is* really there, what we've thought to document and what we've left undocumented. We must examine the boundaries of both our simulation models and our underlying thought models to ferret out the implicit assumptions that, while apparently subtle, can have profound impacts by omitting important factors.

The reception to presentations of our work has thus far been positive. This convinces us it presents fertile ground for important work, including the creation of useful tools and analysis processes to advance development of cooperation theory for systems of systems. Additional research into the sensitivity of P_c, $Prob_c$, and Payoff to the level of decision maker bias embodied in the strategies (self-regarding or other-regarding) may be very illuminating. For example, do one or more of these outputs vary in a linear fashion, a step function or some other way? We are also working to discover additional sample cases that may hold promise for helping us determine if our work does generalize. We are very grateful to the University of Manchester's Carole Goble for her observation that cooperation among scientific researchers, especially across disciplines, may provide an interesting and compelling case.

References

1. Priest, D., Arkin, W. M.: Top Secret America: A Washington Post Investigation (2010), http://projects.washingtonpost.com/top-secret-america/articles/a-hidden-world-growing-beyond-control (accessed February 10, 2011)
2. Unclassified Executive Summary of the Committee Report on the Attempted Terrorist Attack on Norwest Airlines Flight 253, Senate Select Committee on Intelligence (2010)
3. Gintis, H., Bowles, S., Boyd, R., Fehr, E. (eds.): Moral Sentiments and Material Interests: The Foundations of Cooperation in Economic Life. MIT Press, Cambridge (2005)
4. John, L., McCormick, P.M., McCormick, T., Boardman, J.: Self-organizing Cooperative Dynamics in Government Extended Enterprises: Essential Concepts. In: Proceedings of the 2011 IEEE International Systems Conference. IEEE (2011)

5. Fine, C.H.: Clockspeed: Winning Industry Control in the Age of Temporary Advantage. Perseus Books, New York (1998)
6. Davis, E.W., Spekman, R.E.: The Extended Enterprise: Gaining Competitive Advantage through Collaborative Supply Chains, p. 20. Financial Times Prentice Hall, Upper Saddle River (2004)
7. Axelrod, R.: The Complexity of Cooperation: Agent-Based Models of Competition and Collaboration. Princeton University Press, Princeton (1997)
8. van Lange, P.A.M., DeCremer, D., van Dijk, E., van Vugt, M.: Self-Interest and Beyond: Basic Principles of Social Interaction. In: Kruglanski, A.W., Higgins, E.T. (eds.) Social Psychology: Handbook of Basic Principles, ch. 23. Guilford, New York (2007)
9. Bartal, I.-B.-A., Decety, J., Mason, P.: Empathy and Pro-Social Behavior in Rats. Science 334(6061), 1427–1430 (2011)
10. Langford, D.J., Tuttle, A.H., Brown, K., Deschenes, S., Fischer, D.B., Mutso, A., Root, K.C., Sotocinal, S.G., Stern, M.A., Mogil, J.S., Sternberg, W.F.: Social approach to pain in laboratory mice. Social Neuroscience 5(2), 163–170 (2010)
11. Langford, D.J., Crager, S.E., Shehzad, Z., Smith, S.B., Sotocinal, S.G., Levenstadt, J.S., Chanda, M.L., Levitin, D.J., Mogil, J.S.: Social Modulation of Pain as Evidence for Empathy in Mice. Science 312(5782), 1967–1970 (2006)
12. Hume, D.: In: Norton, D.F., Norton, M.J. (eds.) A Treatise of Human Nature, pp. 1739–1740. Oxford University Press, Oxford (2000)
13. Smith, A.: The Theory of Moral Sentiments, 6th edn. A. Millar, London (2000 [1790]), http://www.econlib.org/library/Smith/smMS1.html (accessed October 2010)
14. Ostrom, V.: Citizen Sovereigns: The Source of Contestability, the Rule of Law, and the Conduct of Public Entrepreneurship. PS: Political Science and Politics 39(1), 13–17 (2005)
15. Cohon, R.: Hume's Moral Philosophy. In: Zalta, E.N. (ed.) The Stanford Encyclopedia of Philosophy (Fall 2010 Edition) (2010), http://plato.stanford.edu/archives/fall2010/entries/hume-moral/
16. Hardin, R.: Trust. Maldin. Polity Press, MA (2006)
17. McLeod, C.: Trust. In: Zalta, E. N. (ed.) The Stanford Encyclopedia of Philosophy (Spring 2011 Edition) (2011), http://plato.stanford.edu/archives/spr2011/entries/trust/
18. Jarvenpaa, S.L., Knoll, K., Leidner, D.E.: Is anybody out there? Antecedents of trust in global virtual teams. Journal of Management Information Systems 14, 29–64 (1998)
19. Ridings, C.M., Gefen, D., Arinze, B.: Some antecedents and effects of trust in virtual communities. Journal of Strategic Information Systems 11, 271–295 (2002)
20. Kanter, R.M.: Collaborative Advantage: The Art of Alliances. Harvard Business Review on Strategic Alliances, 97–128 (1994)
21. Bowles, S., Gintis, H.: The Origins of Human Cooperation. In: Hammerstein, P. (ed.) The Genetic and Cultural Origins of Cooperation. MIT Press, Cambridge (2003)
22. Bowles, S., Gintis, H.: A Cooperative Species: Human Reciprocity and Its Evolution. Princeton University Press, Princeton (2011)
23. Parsons, T., Shils, E.A.: Toward a General Theory of Action: Theoretical Foundations for the Social Sciences, p. 134. Transactions Publishers, New Brunswick (2001)
24. Shefrin, H.: Beyond Greed and Fear, pp. 120–121. Oxford University Press, New York (2002)

25. Lopes, L.: Between Hope and Fear: The Psychology of Risk. Advances in Experimental Social Psychology 20, 255–295 (1987)
26. Van Dijk, E., DeCremer, D., Mulder, L.B., Stouten, J.: How Do We React to Feedback in Social Dilemmas. In: Biel, A., Eek, D., Gärling, T., Gustafson, M. (eds.) New Issues and Pardigms in Research on Social Dilemmas, pp. 43–56. Springer Science+ Business Media, New York (2008)
27. Simon, H.A.: Administrative Behavior: A Study of Decision-Making Processes in Administrative Organizations, 4th edn., pp. 206–207. The Free Press, New York (1997)
28. Skinner, B.F.: Science and Human Behavior, p. 361. The Free Press, New York (1965)
29. von Bertalanffy, L.: The Theory of Open Systems in Physics and Biology. Science III(2872), 23–29 (1950)
30. Wiener, N.: Cybernetics. Wiley, New York (1948)
31. Garrett, E.: Rethinking the Structures of Decisionmaking in the Federal Budget Process. Harvard Journal on Legislation 35, 409–412 (1998)
32. Ostrom, E.: An Assessment of the Institutional Analysis and Development Framework. In: Sabatier, P.A. (ed.) Theories of the Policy Process, 2nd edn. Westview Press, Boulder (2007)
33. Williams, J.: Foreword. In: Grabo, C. (ed.) Anticipating Surprise: Analysis for Strategic Warning. Center for Strategic Intelligence Research, Joint Military Intelligence College, Washington DC (2002)
34. Wohlstetter, R.: Pearl Harbor: Warning and Decision. Stanford University Press, Stanford (1962)
35. Prange, G.W.: At Dawn We Slept. McGraw-Hill, New York (1981)
36. Hansson, S. O.: Risk. In: Zalta, E. N. (ed.) The Stanford Encyclopedia of Philosophy (Fall 2011 Edition),
http://plato.stanford.edu/archives/fall2011/
entries/risk/ (2011)
37. Bernstein, P.L.: Against the Gods: The Remarkable Story of Risk. John Wiley & Sons, New York (1998)
38. Boardman, J., Sauser, B.: System of Systems – the meaning of "of.". In: Proceedings of the 2006 IEEE/SMC International Conference on System of Systems Engineering. IEEE (2006)
39. Boardman, J., Sauser, B.: Systems Thinking: Coping with 21st Century Problems, pp. 155–161. CRC Press, Boca Raton (2008)
40. Ashby, W.R.: An Introduction to Cybernetics. Chapman & Hall, London (1956)
41. DiMario, M.J., Boardman, J.T., Sauser, B.J.: System of Systems Collaborative Formation. IEEE Systems Journal 3 (2009)
42. Gorod, A., Sauser, B., Boardman, J.: System-of-systems engineering management: A review of modern history and a path forward. IEEE Systems Journal 2, 484–499 (2008)
43. Baldwin, W.C., Sauser, B.: Modeling the Characteristics of System of Systems. In: IEEE International Conference on System of Systems Engineering, pp. 1–6 (2009)
44. Epelbaum, S., Mansouri, M., Gorod, A., Sauser, B., Fridman, A.: Target Evaluation and Correlation Method (TECM) as an Assessment Approach to Global Earth Observation System of Systems (GEOSS). International Journal of Applied Geospatial Research 2(1), 36–62 (2011)
45. Baldwin, W.C.: Modeling of a Decision Mechanism for the Formation of Collaborative System of Systems. PhD Dissertation, Stevens Institute of Technology, Ann Arbor: ProQuest (2011)

46. Ostrom, E.: Understanding Institutional Diversity. Princeton University Press, Princeton (2005)
47. Pacheco, J.M., Santos, F.C., Souze, M.O., Skyrms, B.: Evolutionary dynamics of collective action in n-person stag hunt dilemmas. Proceedings of the Royal Society Biological Sciences 276(1655), 315–321 (2009)
48. Poteete, A.R., Janssen, M.A., Ostrom, E.: Working Together: Collective Action, the Commons, and Multiple Methods in Practice. Princeton University Press, Princeton (2010)
49. Gintis, H.: The Bounds of Reason: Game Theory and the Unification of the Behavioral-Sciences. Princeton University Press, Princeton (2009)
50. Binmore, K.: Game Theory and The Social Contract, vol. 1: Playing Fair. MIT Press, Cambridge (1994)
51. Binmore, K.: Game Theory and The Social Contract, vol. 2: Just Playing. MIT Press, Cambridge (1998)
52. Binmore, K.: Why do people cooperate? Politics, Philosophy & Economics 5, 81–96 (2006)
53. Binmore, K.: Rational Decisions. Princeton University Press, Princeton (2009)
54. Binmore, K.: Natural Justice. Oxford University Press, Oxford (2011[2005])
55. John, L., McCormick, P., McCormick, T., McNeill, G., Boardman, J.: Self-Organizing Cooperative Dynamics in Government Extended Enterprises: Experimental Methodology. In: Proceedings of the 6th IEEE International Systems Conference. IEEE (2011)
56. Shor, M.: Stag Hunt. In: Dictionary of Game Theory Terms, Game Theory.net (2010), http://www.gametheory.net/dictionary/games/StagHunt.html (accessed October 12, 2010)
57. Skyrms, B.: The Stag Hunt and the Evolution of Social Structure. Cambridge University Press, Cambridge (2004)
58. Stamper, R., Liu, K., Hafkamp, M., Ades, Y.: Understanding the Roles of Signs and Norms in Organizations – a Semiotic Approach to Information Systems Design. Behaviour and Information Technology 19(1), 15–27 (2000)
59. Filipe, J.B.L., Fred, A.L.N.: Collective Agents and Collective intentionality Using the EDA Model. In: Sixth Mexican International Conference on Artificial Intelligence, Special Session, pp. 211–220 (2008)
60. Windrum, P., Fagiolo, G., Moneta, A.: Empirical Validation of Agent-Based Models: Alternatives and Prospects. Journal of Artificial Societies and Social Simulation 10(2), 8 (2007), http://jasss.soc.surrey.ac.uk/10/2/8.html
61. Miller, J., Page, S.: Complex Adaptive Systems: An introduction to Computational Models of Social Life. Princeton University Press, Princeton (2007)
62. Mertikopoulos, P., Moustakas, A.L.: The Emergence of Rational Behavior in the Presence of Stochastic Perturbations. The Annals of Applied Probability 20, 1359–1388 (2010)
63. Shor, M.: Prisoners' Dilemma. In: Dictionary of Game Theory Terms, Game Theory.net, http://www.gametheory.net/dictionary/games/ PrisonersDilemma.html (accessed at October 12, 2010)
64. Permuter, H., Shamai, H.S., Somekh-Baruch, A.: Message and State Cooperation in Multiple Access Channels. IEEE Transactions on Information Theory 57(10), 6379–6396 (2011)
65. Escrig, B., Roviras, D., Paillassa, B., Panichpattankul, W.: A Framework for Cooperative Communications at the System Level. In: Second IEEE International Workshop on Technologies and Standards for Wireless Mesh Networking (2008)

66. Murphy, P., Sabrharwal, A.: Building a Cooperative Communications System. Computer, 1–19 (2007)
67. Kramer, G., Marić, I., Yates, R.D.: Cooperative Communications. Foundations and Trends® in Networking 4(3-4), 271–425 (2006)
68. Giordano, S., Urpi, A.: Self-Organized and Cooperative Ad Hoc Networking. In: Basagni, S., Conti, M., Giordano, S., Stojmenovic, I. (eds.) Mobile Ad Hoc Networking, ch. 13, pp. 355–371. Wiley-IEEE Press, Hoboken (2004)
69. Michiardi, P., Molva, R.: Ad Hoc Networks Security. In: Basagni, S., Conti, M., Giordano, S., Stojmenovic, I. (eds.) Mobile Ad Hoc Networking, ch. 12, pp. 329–354. Wiley-IEEE Press, Hoboken (2004)
70. Wolpert, D.H.: Collective Intelligence. In: Fogel, D.B., Robinson, C. (eds.) Computational Intelligence: The Experts Speak, ch. 17. IEEE, Wiley, New York (2003)
71. Ostrom, E.: Personal communication, Request for Opinion on Subtractability of Value as a replacement for Subtractability of Flow in Information Exchanges (June 25, 2010)
72. Miner, J.B.: Organizational Behavior 2: Essential Theories of Process and Structure, pp. 109–126. M. E. Sharpe, Inc., Armonk (2006)
73. Miner, J.B.: Organizational Behavior 1: Essential Theories of Motivation and Leadership, pp. 94–113. M. E. Sharpe, Inc., Armonk (2005)
74. SEAS: What is SEAS? (2010), http://teamseas.com/content/view/55/71/ (accessed October 12, 2010)
75. Ahn, T.K., Ostrom, E., Schmidt, D., Shupp, R., Walker, J.: Cooperation in PD Games: Fear, greed, and history of play. Public Choice 106, 137–155 (2001)
76. Komorita, S.S., Parks, C.D.: Interpersonal relations: Mixed-motive interaction. Annual Review of Psychology 46, 183–207 (1995)
77. Bonacich, P., Shure, G.H., Kahan, J.P., Meeker, R.J.: Cooperation and group size in the n-person prisoner's dilemma. Journal of Conflict Resolution 20, 687–706 (1976)
78. West, S.A., El Mouden, C., Gardner, A.: Sixteen common misconceptions about the evolution of cooperation in humans. Evolution and Human Behavior 32, 231–262 (2011)
79. Noë, R.: Cooperation. In: Koob, G.F., Le Moal, M., Thompson, R.F. (eds.) Encyclopedia of Behavioral Neuroscience, pp. 345–353. Elsevier (2010)
80. Parson, T.: The Social System. Routledge, New York (1951)
81. Arrow, K.J.: A Utilitarian Approach to the Concept of Equality in Public Expenditure. The Quarterly Journal of Economics 85(3), 409–415 (1971)
82. Akerlof, G.A.: The Economics of Caste and of the Rat Race and Other Woeful Tales. The Quarterly Journal of Economics 90(4), 599–617 (1976)
83. Hardin, R.: Collective Action. Resources for the Future, New York (1982)
84. Axelrod, R.: The Evolution of Cooperation, Revised Edition. Basic Books, New York (2006[1984])
85. Bicchieri, C.: Rationality and Coordination, 2nd edn. Cambridge University Press, Cambridge (1996, 1993)
86. Kahneman, D.: Thinking, Fast and Slow. Farrar, Straus and Giroux, New York (2011)
87. Kahneman, D.: Judgment and decision making: A personal view. Psychological Science 2, 142–145 (1991)
88. Kahneman, D.: A perspective on judgment and choice: Mapping bounded rationality. American Psychologist 58(9), 697–720 (2003)
89. Kahneman, D., Tversky, A.: Prospect Theory: An Analysis of Decision under Risk. Econometrica 47(2), 263–291 (1979)
90. Kahneman, D., Tversky, A.: Choices, Values and Frames. American Psychologist 39, 341–350 (1984)

91. Kahneman, D., Slovic, P., Tversky, A.: Judgment under Uncertainty: Heuristics and Biases. Cambridge University Press, New York (1982)
92. Kramer, R.M., Tenbrunsel, A.E., Bazerman, M.X. (eds.): Social Decision Making: Social Dilemmas, Social Values, and Ethical Judgments. Psychology Press, New York (2010)
93. Biel, Eek, D., Gärling, T., Gustafson, M.: New Issues and Paradigms in Research on Social Dilemmas, pp. 43–56. Springer Science+ Business Media, New York (2008)
94. Gintis, H.: Review of B. Skyrms, The Stag Hunt and the Evolution of Social Structure (2009), http://www.amazon.com
95. Aumann, R.J., Shapley, L.S.: Long-Term Competition: A Game-Theoretic Analysis. In: Megiddo, N. (ed.) Essays in Game Theory in Honor of Michael Maschler, pp. 1–15. Springer, New York (1994)
96. Rubinstein, A.: Equilibrium in Supergames. In: Megiddo, N. (ed.) Essays in Game Theory in Honor of Michael Maschler, pp. 17–28. Springer, New York (1994)
97. Simon, H.A.: Invariants of Human Behavior. Annual Review of Psychology 41, 1–19 (1990)
98. John, L., McCormick, P., McCormick, T., McNeill, G., Boardman, J.: Impact of Behavioral Forces on Knowledge Sharing in an Extended Enterprise System of Systems. In: Proceedings of the 3rd International Conference on Knowledge Management and Information Sharing. INSTICC (2011)
99. Sterman, J.D.: All models are Wrong: Reflections on Becoming a Systems Scientist. Systems Dynamics Review 18, 501–531 (2002)
100. Derman, E.: Models. Behaving. Badly. Why Confusing Illusion with Reality Can Lead to Disaster, on Wall Street and in Life. The Free Press, New York (2011)

Using Work Context and Levels of Expertise to Foster Person-to-Person Knowledge Exchange: A Concept for Connecting Knowledge Workers Based on (Socio-) Psychological Theories

Jörg Schmidl and Helmut Krcmar

Department of Informatics, Technische Universität München,
Boltzmannstr. 3, Munich, Germany
{joerg.schmidl,helmut.krcmar}@in.tum.de

Abstract. Knowledge within organizations is increasingly distributed, which raises the challenge to connect the right individuals for knowledge exchange. This contribution analyzes this challenge and proposes a concept to connect the right individuals. We first investigate relevant theoretical models namely transactive memory theory, social capital theory for knowledge exchange and a model based on socio-motivational and problem solving theory to find relevant constructs. We then analyze the relevant state-of-the-art to find that all approaches have some limitations with respect to the theoretical models. Our proposed solution builds on task histories for the matching, and we show how it can be used to determine contextual overlap and level of expertise. We then describe a case study in which our concept was employed in a three month timeframe with 93 individuals. A survey after the case study showed that our assumptions concerning the relevance and benefit of context overlap are substantiated.

Keywords: Expert Mediation, Expert Recommendation, Knowledge Networks, Yellow Page Systems, Person-to-Person Knowledge Exchange, Social Capital Theory, Transactive Memory Systems, Cognitive Motivation Theory, Problem Solving Theory.

1   Introduction

With the increase in relative importance of knowledge for the success of a company there also is a change in the role the single individual plays. Nowadays, workforce increasingly consists of knowledge workers and hence suitable support of their work becomes more important [9]. According to Drucker [10] knowledge workers are specialists in their profession who govern their work on their own adapting their performance to feedback from their environment.

However, owing to the increasing diversification and radical advancements in knowledge, knowledge workers' specialization is naturally limited to certain areas – there are no universal geniuses like Leonard DaVinci anymore. Hence an organizations is often seen as "[...] a society of knowledge workers who are interconnected by

A. Fred et al. (Eds.): IC3K 2011, CCIS 348, pp. 440–454, 2013.

a computerized infrastructure" [13] and a fundamental challenge for organizations lies in the systematic coordination of knowledge in this network [23].

Leveraging the potential of this network of knowledge workers can be done in two stereotypic ways [9]: Either, the general goal is to give knowledge workers access to as many and as diverse sources of information, including fellow knowledge workers, and assume that they will handle and integrate the information autonomously. Or, alternatively, the information delivered to the knowledge worker is governed to a larger degree by structured processes and systems.

While structured delivery is well-suited for tasks that follow a routine, pursuing a free access model assumes that knowledge workers know what information they can use, how to manage it and how to find it. However, "[…] workers may know how to use technology tools, they may not be skilled at searching for, using, or sharing the knowledge." [9] and hence if possible some structure should be imposed to guide the knowledge workers. This also applies when accessing knowledge by interacting with others that serve as source of help for a concrete challenge, i.e. in person-to-person knowledge exchange.

However, owing to specialization and the resulting diversification of knowledge in companies, finding suitable experts is a challenge, further fueled by geographic distance, time-zone differences and large pools of potential candidates typically found in larger organizations. Therefore, the limited human attention puts a natural limit to the ability to collaborate [24] while at the same time a knowledge worker's attention becomes a crucial resource that should be handled with care when searching for interaction partners [33].

Suitable solutions should therefore induce only small effort for the knowledge seeking individual and at the same time, in a global view, should limit the attention consumed for possible interaction partners. Both taken together can increase the likelihood of two knowledge workers exchanging information and hence contribute to fostering knowledge exchange in organizations.

The remainder of this contribution is structured as follows. In chapter 2 we discuss relevant theoretical models that describe person-to-person knowledge transfer processes. In chapter 3 we investigate approaches currently employed in corporate settings to facilitate person-to-person knowledge transfer. We especially relate them to the constructs of the theoretical models discussed in chapter 2 and indicate where they face challenges. Chapter 4 describes our concept for addressing the challenges found in chapter 3 that also complies with the constructs described in the theoretical models. A discussion about the conceptual design choices of the concept is dealt with in chapter 5. The subsequent chapter describes the evaluation of our concept in a case study, while chapter 7 concludes the contribution with a summary of results and an outlook to further research.

2   Theoretical Basis for Person-to-Person Knowledge Exchange

When knowledge is exchanged from person to person a social process between the two actors starts. Therefore, to understand the antecedents and constituting steps in this social process, appropriate models from sociology and psychology that describe this process need to form the basis for any further design. Guiding the design of new artifacts based on foundational theories is a consequent approach that is known as theory-driven design [7]; [6]. For our discussion we will rely on one model from social psychology, another from sociology and a third one from cognitive psychology.

2.1   Transactive Memory Systems

According to the theory's originator, transactive memory systems can be described as "[...] a set of individual memory systems in combination with the communication that takes place between individuals" [32]. Individuals use others as their "external storage" by remembering pointers to those that possess relevant knowledge, i.e. by establishing know-who, instead of remembering the procedural or factual knowledge itself. Transactive memory system (TMS) theory also encompasses processes that determine who is responsible to store new knowledge on behalf of the group and processes to later disseminate it within the group.

While the theory tries to explain knowledge transfer processes between individuals, its unit of analysis are groups that consist of individuals acquainted to each other. Especially the necessity of individuals to assume responsibility for the group necessitates a binding element that socially motivates the individual to contribute. In the original TMS theory this binding element is personal acquaintance, i.e. an established social fabric that manifests in strong relationships between a group's individuals. In non-co-located, dispersed organizations, close relationships across team or department are rare, which is why personal acquaintance may not act as a binding element.

Hence, according to TMS theory, appropriate approaches for supporting person-to-person knowledge exchange need to have a suitable surrogate for those strong interpersonal relationships that still allows to establish a form of TMS.

2.2   Social Capital Theory

Social capital (see for example [17] for an overview) relates to an individual's previously established connections to (known) others, their strength and reliability and the individual's ability to take benefit out of this network. The theory has been adapted to explain knowledge exchange processes [19] which subsequently has been applied to empirically study this process in electronic networks of practice [30]; [16]. Here, the original concept of social capital had to be relaxed. While in its original form for social capital to build up, it is necessary to know the other individual so that later one's own effort for an individual may be reciprocated by this individual, electronic networks of practice are effectively anonymous and individuals do not know each other. Therefore, in electronic networks of practice other constructs surrogate for this. Statistical evidence could be found among others for the following constructs [30]; [16]: 1) The more communication threads an individual has with others the more likely he is to contribute, which is subsumed under the concept network centrality 2) During interactions the more one can rely on a shared language the higher the willingness to contribute 3) The more an individual can identify with the network or more precisely identify with the interaction partners in this network the more likely he will contribute. Approaches that foster knowledge exchange have to adopt these three aspects.

2.3   Model Based on Problem Solving and Cognitive Motivation Theory

Olivera et al. [22] developed a model to describe how and why people contribute in distributed organizations through IT-mediated means. They argue that to understand the contribution behavior, two strands of theories have to be combined: theories of problem solving [20] and cognitive motivation theories [14]. The model distinguishes three subsequent mediating mechanisms. The first, awareness, relates to a person recognizing an opportunity to contribute. In the second, searching and matching, the individual determines whether and how his knowledge is sufficient to help another individual. The third mechanism, formulation and delivery, deals with formulating and communicating the individual's knowledge to help the other. Each of these mechanisms inflicts costs for the individual who can possibly help. The necessary over-compensation of these costs is described by constructs from cognitive motivation theory. The relevant ones are: Self-enhancement, that is fueled by e.g. liking to express one's expertise [30] and living up to one's self-identity [8]; Exchange motivations which is the equivalent to expectation of (individual or generalized) reciprocity as discussed in relation to social capital theory; Instrumental motivation which refers to rewards such as recognition.

Appropriate approaches to support person-to-person knowledge exchange should hence try to support the aforementioned three mechanisms and build upon the three means of motivation.

3   State-of-the-Art in Person-to-Person Knowledge Exchange

In light of the theoretical models describing the antecedents and constituting steps of person-to-person knowledge exchange, in this chapter we want to investigate the properties of contemporary approaches supporting this process.

There are diverse approaches and tools that quite directly or more indirectly fall under the umbrella term Knowledge Management. Binney has arranged them according to a spectrum from transactional systems to innovation-supporting systems [1] while others, e.g. [2] suggested to use the SECI model [21] to impose structure on the set of approaches and tools. Using the respective structure's dimensions and contrasting those with the situation we look at – person-to-person knowledge transfer, for complex, highly-adaptable, knowledge-intensive tasks in a distributed setting – we find that three approaches fall into a comparable category with respect to our research: Yellow Page Systems, Expert Recommender Systems, and Knowledge Networks.

3.1   Yellow Pages Systems

A yellow page system (YPS) contains lists of the individuals in an organization along with their competencies, knowledge and skills in those areas that are relevant to the organization. A knowledge seeker may search for required knowledge and will be presented with those individuals that match the request. If the system contains levels of proficiency, the result may also be ranked. The profiles contained in the system are often manually maintained, while some data may be extracted from directory services

[15] or Human Capital Management systems [12]. Also, keeping those profiles up-to-date is normally a manual process.

TMS' backbone are interpersonal relationships that in distributed settings need to be surrogated by other means. The query mechanism in YPS typically only operates on the level of expertise to find relevant matches, neglecting the previous relationship between actors and hence there is no obvious surrogate for interpersonal relationships. On the other hand, the application of social capital theory on knowledge exchange suggests that many communication threads increase likelihood to contribute. YPS have no means to increase this number nor is it typically tracked. Also, shared language, another positive influence for knowledge contribution according to social capital theory, is not part of the matching of knowledge requester and potential helper. Identification with potential others is not part of the matching, but result lists of queries may contain affiliation and other socio-metric information that may serve this purpose. In terms of socio-cognitive and problem solving theory, YPS have no means for creating awareness on the side of the potential helper for requests of a knowledge seeker – it follows a pull interaction schema. However, match of request and expertise tends to be high, if the profiles are up-to-date. YPS include no measures that facilitate the formulation of responses, though. Also, the motivation factors such as self-identity are not specifically supported by the matching delivered by YPS. Reciprocity on the other hand is often a motivator in YPS settings as the individuals are acquainted after the interaction.

Many of the aspects that theory predicts to be important for supporting person-to-person knowledge exchange are not present in YPS. Along with the relatively high effort for keeping YPS up-to-date, they seem to leave room for improvement.

3.2   Expert Recommender Systems

Expert recommender systems (ERS) can be seen as the next evolution step of YPS. Tasks such as profile generation and mediation of communication are automated. ERS help in the following way: When someone seeks an expert, he wants to know if there is an expert that can answer the user's questions, but also what level of expertise the user has and how it compares to others, if there are others that also fulfill the criteria and how the person can be reached [25]. The automation of expert determination is achieved by deriving levels of expertise in relation to queried knowledge items from sources that may be scanned for expertise evidence. Those sources can be communication-based, such as e-mail messages, document-based, such as websites or electronic documents stored on intranets or interaction-based where software usage is utilized as source of expertise evidence.

Many of the aspects that theory predicts to be important are however not present in ERS. Considering suitable surrogates for TMS' interpersonal relationships, ERS do usually not include means for this. The query mechanism in ERS typically only operates on the level of expertise to find relevant matches, neglecting the previous relationship between actors and hence there is no obvious surrogate for interpersonal relationships. Only few attempts can be found to somewhat remedy this downside e.g. [26]. On the other hand, the application of social capital theory on knowledge exchange suggests that many communication threads increase the likelihood to contribute. ERS have no means to increase this number nor is it typically tracked. Also shared language, another positive influence for knowledge contribution according to social capital theory, is typically not part of the matching of knowledge requester and potential helper, again with only few exceptions. As was true for YPS, identification with others is not part of the matching, but results of queries may contain affiliation and other socio-metric information that may serve this purpose. In terms of socio-cognitive and problem solving theory, ERS have no means for creating awareness on the side of the potential helper for requests of a knowledge seeker – it also follows a pull interaction schema. However, match of request and expertise tends to be high, if the expertise extraction mechanism fits the users' expectations. ERS include no measures that facilitate the formulation of responses, though. Also, the motivation factors such as self-identity are not specifically supported by the matching delivered by ERS. Reciprocity on the other hand is often a motivator in ERS settings as the individuals are acquainted after the interaction.

Many of the aspects that theory predicts to be important for supporting person-to-person knowledge exchange are also not present in ERS, while due to its decreased effort they appear more promising than YPS.

3.3   Electronic Networks of Practice

Electronic networks of practice (ENP) are a geographically distributed group of individuals that are engaged in a shared practice. However in contrast to other forms of knowledge networks, the group of individuals can be large, virtually limitless in size, the individuals are loosely knit, but may not know each other at all nor necessarily do they expect to ever meet face-to-face [4]; [30]. By relying on IT-mediated communication, ENP allow quick and effortless access to a broad source of expertise through a wide variety of knowledge carriers [28].

In relation to TMS' backbone – interpersonal relationships – the interactions in an ENP support the creation of weak ties. Those ties are far less reliable and pronounced than those between acquainted individuals; however, they are a suitable surrogate for distributed setting where people normally never meet face to face. In contrast to the two previous approaches, ENP do not offer explicit mechanisms to find suitable interaction partners. It is rather up to the individual to find relevant individuals or, more often, relevant outlets within the ENP, such as a forum concerned with his knowledge request. Social capital theory's indication that many communication threads are beneficial can be supported in ENP, as many of them feature mechanisms to be informed by updates in parts of the ENP, e.g. forums, which are relevant to the individual. This measure can also increase awareness of knowledge requests on the knowledge bearer's side. In ENP, shared language establishes over time and with more and more interactions. The same holds for the ability to identify with others that also increases over time while being a member of the ENP. However, the challenge of identifying with individuals without previous interaction history remains, especially since socio-metric information is often not available in ENP. Another challenge lies in matching available and requested expertise: If the requester does not know where or who to ask he is hindered and if the question is addressed to the wrong individuals, answers are unlikely and effort is wasted on the side of all affected individuals. Requests for knowledge and responses are also not associated with the work context of neither requesting nor replying individual and hence to some degree decontextualized, which may affect ease of request and response formulation. Reciprocity is often present in ENP - in its generalized form, though. Also self-identity, another motivation factor, may be reinforced by the ENP itself, as other like-minded individuals are likely to find one's interactions in the ENP.

While many of the social factors of the theoretical models can be matched onto features of ENP, there remain challenges such as facilitating the searching, finding and matching of interaction partners or limiting consumption of awareness. Table 1 provides an overview of the discussion above.

Table 1. Evaluation of contemporary approaches in light of the theoretical models

Theory	Construct	YPS	ERS	ENP
TMS	Interpersonal Relationship	-	-	+
Social Capital Theory	Increasing Network Centrality	-	-	+
	Shared Language	-	O	O
	Identification	-	-	O
Olivera et al.'s model	Creating Awareness	-	-	+
	Match of Knowledge Area	+	+	-
	Facilitated Response Formulation	-	-	-
	Self Identity	-	-	+
	Reciprocity	+	+	+

4   Task History as Binding Element

The theoretical models discussed in chapter 2 described the mechanisms that allow knowledge transfer to happen mainly from the point of view of the contributor, i.e. the knowledge bearer helping the knowledge seeker. The knowledge request was treated as a given prerequisite. However understanding the knowledge seeker's intention is relevant as well. Seid [25] analyzed which circumstances lead someone to consult an expert. First, someone might need access to information that is not documented. Second, someone might not be able to exactly specify what he needs to know, rather the dialogue with an expert acts as the process to facilitate information acquisition. Third, someone might want to utilize an expert to be more efficient. Someone with advanced expertise can handle tasks faster than novices can: Therefore, relying on the expert improves the initial individual's efficiency. Fourth, often users do not want a context-free, general piece of information but rather need a contextualized, situated interpretation of more general knowledge that the expert might poses. Fifth, someone might simply prefer relying on social interaction instead of using anonymous media like documents. Therefore, sufficient levels of expertise on the side of the potential helper is important for the knowledge seeker and hence for the knowledge transfer to start and to be successful.

On the other hand, next to being able to help, the helper needs to be willing to help and the requester needs to be willing to ask this individual for help. TMS address this aspect by stressing the importance of established interpersonal relationships. In social capital theory identification with others and use of shared language expresses this aspect. The cognitive psychology model expresses this aspect in the constructs self-identity and recognition. In a distributed setting interaction partners are unacquainted and do not have a previous history of interactions. Nevertheless, an appropriate mechanism to determine suitable interaction partners that reflects the constructs of those theoretical models is necessary.

As both, ability and willingness to help, are relevant for knowledge exchange to happen on the knowledge bearer side, and both willingness to accept help and ability to understand the offered help are necessary on the requester side, a suitable glue needs to be found. We suggest using the histories of tasks performed by and knowledge bearers as this binding element.

History of tasks refers to the tasks a knowledge worker is currently engaged in or has been performing in the recent past. Nowadays, large portions of a company's operations are supported by information systems and a large extend of knowledge work is as well. In some organizations that are customer service-focused up to 75 % of a knowledge worker's tasks are IT-supported [18]. Therefore the current and previous task context of a knowledge worker is often adequately reflected in IT system use that can be extracted from the logs that those systems create for administrative purpose.

The history of tasks can serve both purposes that we elicited to be important: ability and willingness to engage in knowledge transfer for both, knowledge seeker and knowledge bearer. The more often a knowledge worker has performed a specific task, the higher the chance that he has proficiency in performing the task. Therefore the number of times a knowledge worker performs a task, can be used to determine his proficiency related to this task – an assumption that is often taken as valid [25].

On the other hand, the history of tasks, especially its very recent or current part, gives indication of the knowledge worker's current work context. If the work context of the knowledge seeker and the one of the knowledge bearer overlap, they are more likely to engage and benefit from interaction. This assumption is backed by fundamental results from socio-psychology, with one of its clearest results being, that one likes others that are similar to oneself [35] and that we identify with those that are similar to us [27]; [29]. Other psychological results further support this aspect. Similar attitudes were shown to predict interpersonal attraction [5] and joint interest and mutual trust also correlate [34].

While not nearly as expressed, this contextual overlap is a surrogate for the interpersonal relationships that TMS has as its backbone, for a setting in which personal acquaintance is scarce or non-existent. Also the relevant constructs of social capital theory can be supported. Being in similar work context increases the ability to rely on the same task-specific terminology and hence use of shared language is possible, as is the identification with the other as reasoned above. Context overlap also translates nicely to the constructs used in the model relying on problem solving and socio-cognitive theory. When work contexts overlap, the knowledge bearer can more easily determine how his knowledge matches with the request and the likelihood that it does is higher, as the request relates to what he currently does or has done just recently. Also, being in similar work contexts facilitates formulation of responses as it is possible to rely on shared terminology. The motivational aspects of the model map to context overlap as well. Being recognized as expert is a strong motivator that is even higher when the recognition comes from individuals that are similar to one. This is in line with social comparison theory [11] that states that we want to be better than our reference group of similar peers. Similarly, self-identity is more pronounced when one can help in areas that are relevant to oneself, which applies for tasks that one is currently doing or has done just recently.

5   Expertise vs. Contextual Overlap

In the previous chapter we have argued that task history may serve the purpose of identifying levels of expertise and at the same time may act as surrogate for determining willingness to interact due to similarity in task context and hence situation and previous history. Finding suitable interaction partners essentially is a filtering task, as otherwise requests for help could just be broadcasted to all individuals in an organization. When filtering, the question shifts to determining which dimensions to filter on and which filter values to set for them. As we argued, level of expertise as well as contextual overlap are relevant and form the two dimensions we may use for filtering. Figure 1 illustrates the four different ways to configure the filtering values.

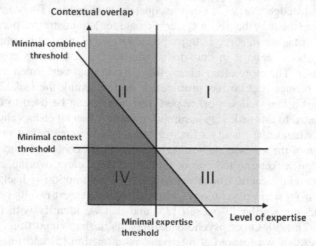

Fig. 1. Filtering thresholds operating on contextual overlap and level of expertise

One way to filter out non-suitable individuals lies in choosing only those that exhibit a minimum level of expertness. This minimum value can be absolute, e.g. only individuals who have performed a task more than ten times, or it can be relative to the knowledge seeker, e.g. only individuals that have performed a task at least five times more than the knowledge seeker. If an individual is below this threshold (rectangle II and IV), he is filtered out, while the ones that are above are eligible but we may choose to rank them according to the second dimension and only consider a fixed number of them that have most contextual overlap.

Another way to filter lies in defining a minimum value of contextual overlap to find those individuals that are possibly willing to interact with the knowledge seeker and that also can do so with only small effort because their mental models are alike. Again, the ones that are below the threshold (rectangle III and IV) are filtered out, while the remaining ones are all eligible, but we may choose to rank them according to the first dimension and only consider a fixed number of them that have highest levels of expertise. However, this may induce the problem that also arises in expert recommendation systems. Experts with much higher levels of expertise face a mental challenge when interacting with (relatively seen) laypersons. Their mental models about the relevant topics are "compressed", single facts are aggregated into larger chunks and abstractions are used to condense the relevant knowledge. To help, experts have to unpack these chunks, which is a high mental effort as they must undo their previous learning to understand the requester's problem context [3].

A third way of filtering combines the thresholds defined on the level of expertise dimension and on the contextual overlap dimension. In this case, those individuals that have insufficient expertise (rectangle III and IV) and those that do not share sufficient context (additionally rectangle II) are filtered out, while the ones in rectangle I are suitable candidates.

A final way of filtering lies in defining a combined threshold that takes into account level of expertise and contextual overlap at the same time. Individuals who are very similar to the knowledge seeker but have only low levels of expertise may be suitable, while also individuals that do not share much commonalities with the knowledge seeker but are very knowledgeable may be suitable candidates as well. Therefore, in both cases one dimension might compensate the lack in the other. However, individuals that neither have a sufficient contextual overlap nor sufficient levels of expertise (darker area in the lower left part) are filtered out.

6   Application of Concept

To evaluate the concept of using histories of tasks to foster knowledge exchange between knowledge workers, we implemented a prototype that utilized the concept and applied it in a case study. Within the timeframe of three month we sought to foster knowledge exchange among the knowledge workers that were executing knowledge-intensive tasks in an SAP system. More precisely, the 93 case study participants were tasked to design a company's organizational setup in the SAP system – a complex system configuration task with multiple options and the challenge to master the system and its interactions in addition. We offered an interface in the operational system that combined features of expert recommendation systems and knowledge networks where the overlap in task context served as the glue (see Figure 2). The similarity of tasks was determined in analogy to the discussion in chapter 4: We analyzed the history of transactions, the SAP concept of tasks, to find those individuals that have common task contexts and find those that are, relatively seen, experts whenever an individual seeks support from within a certain task context.

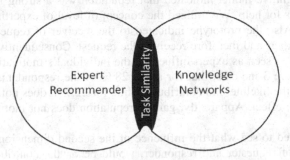

Fig. 2. Using task similarity to combine Expert Recommendation and Knowledge networks

The prototype worked as follows. Whenever an individual wanted to interact with another to find help for solving a challenging problem, he could, directly in the SAP system, call up a program. There, the individual would formulate a question and send it out without specifying recipients. In the background, the program then distributes the request to the "right" individuals, based on the current context of the requester and the features, i.e. context and level of expertise, of the receivers. Only those that are suitable (see previous chapter for filtering options) were informed that their expertise was being asked for along with the message itself. The response of the knowledge bearers was also automatically distributed, so that the individual could focus on response formulation knowing that the requester was in a similar context.

We implemented additional functionality into the prototype, e.g. a forum to collect past interactions with task-oriented structuring, facilitated message creation, notification systems for possibly relevant messages and other features. In this contribution we only focus on and describe those that deal with the core concept of using histories of tasks for determining levels of expertise, contextual overlap and their perceived value for the participants.

7   Evaluation

After the three month case study, we surveyed the participants to find out about their perceived value of the concept we applied. Among other items that related to the additional features of the prototype, we included a number of items that asked for the participants' perception of contextual overlap and level of expertise. We received 18 fully filled surveys. This corresponds to a return rate of roughly 20 % - a normal value for online surveys.

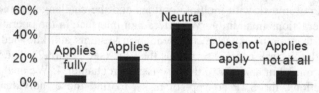

Fig. 3. Influence of being seen as expert on contribution behavior

The socio-cognitive model indicated that reputation was a strong motivator to respond to requests for help. We utilized the concept of level of expertise as one filtering dimension. Also the prototype indicated to the receiver of requests that he was determined as expert and therefore received the request. Consequently, we wanted to find out how being seen as expert influences the individual's motivation. The results as shown in Figure 3 indicate, that for about 28 % of the respondents being seen as expert increases the likeliness to contribute, while for 22 % it does not and 50 % were undecided on the effect. Apparently, gain in reputation does not motivate all participants likewise.

We also wanted to see what the influence of the second dimension we used, contextual overlap of requester and responder, would be on the contribution behavior. Figure 4 illustrates that roughly four out of ten survey participants felt that they would respond more often if the work contexts match. Interestingly, this is a higher value than for being seen as expert and suggests that contextual overlap is more important for the potential responder than the level of expertise.

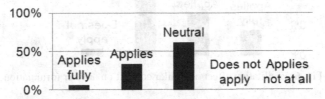

Fig. 4. Influence of context match on contribution behavior

In our argumentation in chapter four we further argued, that a shared context not only increases the willingness of individuals to respond to requests but also that the shared context facilitates the interaction. We especially argued that the shared context allows relying on shared terminology that facilitates the formulation of messages.

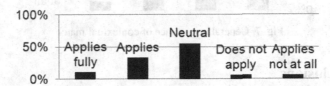

Fig. 5. Facilitation effect of similar context on request formulation

In our survey we also asked whether it is easier for the knowledge seeker to formulate requests for help, when knowing that the potential receivers will be in a similar context. As illustrated in Figure 5, 11 % of the survey respondents strongly agreed that it is easier to formulate requests in this case and another 33 % agreed, while only 6 % did not or strongly not think so, respectively. This indicates that for a large portion of participants, knowing that the receiver is in a similar work context helps them in communicating their request and starting an interaction with them.

Similarly, we wanted to determine the possible facilitation effect on the responders' side. Figure 6 shows whether the survey participants thought that knowing about the similar context of the initial requester would help them in formulating answers to the knowledge request. While half of the responses indicated that the participants were undecided whether or not this knowledge would help them in formulating answers, 28 % agreed it would and 17 % strongly agreed. With nearly half of the respondents indicating the value of knowing the contextual circumstances of the receiver as being high, this appears to be a relevant feature.

Independent of the facilitation effect and the increased willingness to interact, we also asked the survey participants whether they perceived the overlap of work context with their interaction partners as generally valuable. Figure 7 shows that while 50 % were undecided whether or not they perceived this match to be important, 33 % did and 11 % did so strongly yet only 6 % indicated that they did not. The inclusion of context overlap in the mediation mechanism therefore seems to be a suitable design choice.

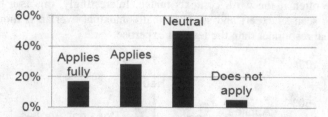

Fig. 6. Facilitation effect of similar context on answer formulation

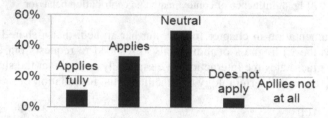

Fig. 7. General importance of contextual match

8   Conclusions

In this contribution we addressed the challenge of connecting knowledge workers to foster knowledge exchange. Starting with three theoretical models that describe the antecedents and process of knowledge exchange we determine those constructs that apply for our setting: distributed, non-acquainted knowledge workers that interact across temporal, physical and organizational borders. Subsequently, we looked at state-of-the-art approaches that support this setting and contrasted those approaches with the constructs of the theoretical models to find that the contemporary approaches do not support all constructs. We hence suggest using a different concept that relies on the history of tasks at its core. Using this concept we describe how it can be used to determine willingness and ability to support fellow knowledge worker. Then, we described how we implemented the concept in a case study with 93 individuals and describe the results we could obtain by surveying the individuals after the three month case study. We found that contextual match is at least as important for the participants as the level of expertise when interacting with other knowledge workers. Also the survey results support our expectation that formulation of messages among knowledge workers is facilitated by contextual overlap. Additionally, the participants found it important to have a work context match with their unknown interaction partners and indicated that knowing that there is a contextual overlap motivates them to contribute more. Our results appear promising, but may be substantiated by replicating the case study setting with more individuals and by using different host systems for embedding new prototypes and organizational settings to better understand its limitations and benefits.

References

1. Binney, D.: The knowledge management spectrum - understanding the KM land scape. Journal of Knowledge Management 5(1), 33–42 (2001)
2. Böhmann, T., Krcmar, H.: Werkzeuge für das Wissensmanagement. In: Bellmann, M., Krcmar, H., Sommerlatte, T. (eds.) Wissensmanagement. Symposion Publishing GmbH, Duesseldorf (2002)
3. Bromme, R., Jucks, R., Rambow, R.: Experten-Laien-Kommunikation im Wissensmanagement. In: Reinmann, G., Mandl, H. (eds.) Psychologie des Wissensmanagements. Perspektiven, Theorien und Methoden, 1st edn., pp. 176–188. Göttingen, Hogrefe (2004)
4. Brown, J.S., Duguid, P.: Knowledge and organization: A social-practice perspective. Organization science 12(2), 198–213 (2001)
5. Byrne, D.E.: The attraction paradigm. Academic Press (1971)
6. Briggs, R.: On theory-driven design and deployment of collaboration systems. International Journal of Human-Computer Studies 64(7), 573–582 (2006)
7. Card, S.: Theory-driven design research. In: Applications of Human Performance Models to System Design, pp. 501–509 (1989)
8. Constant, D., Sproull, L., Kiesler, S.: The kindness of strangers: The usefulness of electronic weak ties for technical advice. Organization Science 7(2), 119–135 (1996)
9. Davenport, T.H.: Rethinking knowledge work: A strategic approach. McKinsey Quarterly (2011)
10. Drucker, P.: The coming of the new organization. Harvard Business Review (1988)
11. Festinger, L.: A theory of social comparison processes. Human Relations 7(2), 117–140 (1954)
12. Gronau, N., Uslar, M.: Integrating Knowledge Management and Human Resources Via Skill Management. Journal of Universal Computer Science, 135–142 (2004); Proceedings of I-Know'04
13. Holsapple, C.W., Whinston, A.B.: Knowledge-based organizations. The Information Society 5(2), 77–90 (1987)
14. Kanfer, R.: Motivation theory and industrial and organizational psychology (1990)
15. Krcmar, H.: Informations manamgent. Springer, Berlin (2010)
16. Law, S.P.M., Chang, M.K.: Fostering knowledge exchange in online communities: a social capital building approach. In: Proceedings of ICIS, Paris, France, December 14-17, p. Paper 173 (2008)
17. Lin, N.: Building a network theory of social capital. Social Capital: Theory and Research, 3–29 (2001)
18. Makolm, J., Weiß, S., Reisinger, D.: Proactive knowledge management: the DYONIPOS research and use-case project. In: Proceedings of 1st International Conference on Theory and Practice of Elecontric Governance, ICEGOV. ACM, New York (2007)
19. Nahapiet, J., Ghoshal, S.: Social capital, intellectual capital, and the organizational advantage. Academy of Management Review 23(2), 242–266 (1998)
20. Newell, A., Simon, H.A.: Human problem solving. Prentice-Hall Englewood Cliffs (104) (1972)
21. Nonaka, I.: A dynamic theory of organizational knowledge creation. Organization Science 5(1), 14–37 (1994)
22. Olivera, F., Goodman, P.S., Tan, S.S.L.: Contribution behaviors in distributed environments. MIS Quarterly 32(1), 23–42 (2008)
23. Quinn, J.B.: Intelligent enterprise. A knowledge and service based paradigm for industry. Free Press, New York (1992)

24. Qureshi, S., Briggs, R., Hlupic, V.: Value Creation from Intellectual Capital: Convergence of Knowledge Management and Collaboration in the Intellectual Bandwidth Model. Group Decision and Negotiation 15(3), 197–220 (2006)
25. Seid, D.Y., Kobsa, A.: Expert finding systems for organizations: Problem and domain analysis and the DEMOIR approach. Journal of Organizational Computing and Electronic Commerce (2003)
26. Serdyukov, P.: Search for expertise. Going beyond direct evidence. SIKS dissertation series (2009)
27. Tajfel, H.T.: The social identity theory of intergroup behavior. Psychology of Intergroup Relations, 7–24 (1986)
28. Teigland, R.: Knowledge networking: Structure and performance in networks of practice. Stockholm School of Economics (2003)
29. Turner, J.C., Hogg, M.A., Oakes, P.J., Reicher, S.D., Wetherell, M.S.: Rediscovering the social group: A self-categorization theory. Basil Blackwell (1987)
30. Wasko, M.M., Faraj, S.: Why should I share? Examining social capital and knowledge contribution in electronic networks of practice. MIS Quarterly 29(1), 35–57 (2005)
31. Wasko, M.M., Faraj, S.: "It is what one does": why people participate and help others in electronic communities of practice. Journal of Strategic Information Systems 9(2-3), 155–173 (2000)
32. Wegner, D.M., Giuliano, T., Hertel, P.: Cognitive interdependence in close relationships. In: Ickes, W.J. (ed.) Compatible and Incompatible Relationships, New York, pp. 253–276 (1985)
33. Ye, Y., Nakakoji, K., Yamamoto, Y.: Understanding and Improving Collective Attention Economy for Expertise Sharing. In: Bellahsène, Z., Léonard, M. (eds.) CAiSE 2008. LNCS, vol. 5074, pp. 167–181. Springer, Heidelberg (2008)
34. Ziegler, C.-N., Golbeck, J.: Investigating Interactions of Trust and Interest Similarity. Decision Support Systems 43(2), 460–475 (2007)
35. Zimbardo, P.G.: Psychologie. Springer, Berlin (1983)

Thinking Out of the Box: Discovering the Relevance of External Context to Business Processes

Eduardo Costa Ramos, Flavia Maria Santoro, and Fernanda Baião

NP2Tec, Department of Applied Informatics, Federal University of the State of Rio de Janeiro (UNIRIO), Rio de Janeiro, Brazil
{eduardo.ramos,flavia.santoro,fernanda.baiao}@uniriotec.br

Abstract. Successful organizations are those able to identify and respond appropriately to changes in their internal and external environments. The search for flexibility is linked to the need for the organization to adapt to frequent and exceptional changes in scenarios imposed to them. Those disruptions in routine should be reflected in business processes, in a sense that processes must be adjusted to such variations, taking into account both internal and external variables, typically referred in the literature as the context of the process. In particular, defining the relevance of external context for the execution of a process is still an open research issue. We propose a method to identify and prioritize external variables that impact the execution of specific activities of a process, applying competitive intelligence concepts and data mining techniques. We have evaluated the method in a case study, which showed how the discovered variables influenced specific activities of the process.

Keywords: Business Process, External Context, Knowledge Management, Competitive Intelligence, KDD.

1 Introduction

Successful organizations are those able to identify and respond appropriately to changes in their internal and external environments. The search for flexibility is linked to the need to adapt to frequent and exceptional changes in scenarios imposed to them. Those disruptions in routine should be reflected in business processes, in the sense that the processes must be able to adjust to such variations [24]. This may be conducted with the support of Knowledge Management (KM) and Competitive Intelligence (CI) disciplines, that recommend initiatives to boost the strategic goals of the organization, dealing both with internal and external variables.

Competitive Intelligence focuses on monitoring and internalizing information from the external environment. Knowledge Management intends to collect, share and reuse internal knowledge generated in the organization. By taking internal and external environment variables into account, the organization is able to address important questions, such as how a business process was executed last time the country experienced a similar economic scenario; whether that process execution brought

A. Fred et al. (Eds.): IC3K 2011, CCIS 348, pp. 455–470, 2013.

positive results or not; which were the external environmental reasons that posed changes in previous process executions [23]. Those variables are typically referred in the literature as the context of the process. The problem of KM and CI, however, does not rely on the lack of contextual information about processes, but in its quality, content and the reasoning about it. Therefore, a major challenge for organizations is how to define the relevance of their information, more specifically, in its various contexts [18].

Context can be defined as any information that can be used to characterize the situation of an entity [6]. In a business process scenario, context is the minimum set of variables containing all relevant information impacting the design and implementation of a business process [24]. Context information could be associated to any process element, such as activities, events, or actors. Furthermore, its analysis should provide insights to identify problems and learn with the past, besides helping to make decisions.

According to Rosemann et al [25], the external context is related to elements that are part of an even broader system whose behavior is beyond the sphere of control of an organization. Those elements exist within the business network where the organization operates. Besides, the environmental context lies ahead of the business network and is related to categories such as society, nature, technology and economy. Even if this context is not in close proximity to the day-to-day business of the organization, it still represents a high impact in the way of how the organization defines and executes business processes.

Although there are few proposals that deal with context associated to business process [18]; [25]; [26], defining the relevance of external information (or external context) for the execution of a process in an organization is still an open research issue. The manipulation of all stored organizational knowledge, as well as environmental and external information, requires the application of knowledge discovery techniques so as to automatically handle and extract patterns from it. In this regard, Liebowitz [13] proposed a set of frameworks to help a project manager in conceptualizing and implementing knowledge management initiatives, and posed some important questions that need to be addressed: (i) how knowledge discovery techniques can be applied for mining Knowledge bases; (ii) how is Knowledge originating from outside a unit evaluated for internal use?; (iii) does lack of a shared context inhibit the adoption of knowledge originating from outside a unit?; (iv) how much context needs to be included in knowledge storing to ensure effective interpretation and application?

We propose a method to identify and prioritize external variables that impact the execution of specific activities of a process [23]. The proposed method applies Competitive Intelligence concepts and data mining techniques (feature selection and decision trees). We have evaluated the method in a case study, which showed how the discovered variables influenced specific activities of the process. In this paper we extend the results presented before.

This paper is structured as follows: Section 2 presents related work, while defines context and KM concepts. Section 3 explains the proposed method, which was applied to a case study discusses in details in Section 4. Section 5 concludes the paper and points to future work.

2   Context-Aware Business Processes

The concept of context has recently revealed its relevance in business process management area. Identifying, documenting and analyzing contextual issues might help to make clear how changes in the environmental setting of an organization should lead to adaptations in processes. Literature points to the importance of considering contextual information, both in the design of business processes; and also, throughout process instances execution. As a result, an important issue should be identifying contextual elements that impact the process.

A taxonomy for context, described by Saidani and Nurcan [26], which is composed of the most usual contextual information (location, time, resource and organization) aims at supporting context elicitation. Nunes et al. [18] also presented a model for context to support knowledge management within the scenario of a business process. The model developed by these authors is an ontology that establishes a representation for context elements associated with process activities. Based on this model, process instances and their context are stored and further could be re-used. The types of context elements presented are: (i) information that exist during the execution of an activity (time, artifacts), (ii) information about individuals or groups that perform an activity, (iii) information to spell out the interaction between individuals within the activity performed. Both proposals do not provide explicit methods for context elicitation and neither consider external environment context.

Rosemann et al. [25] integrate context in process modeling and define a meta-model concerned to the structure of a process, its goals, and context. They also describe a context framework where diverse context levels are depicted in layers, and a procedure to use it: (i) identify process goals; (ii) decompose process, (iii) determine relevance of context, (iv) identify contextual elements, (v) type context. Our research is directly related to the detailing of step 4 as an evidence-based task.

Another approach for bringing out context is stated by Soffer et al. [27] with the goal of learning and gradually improving business processes considering three elements: process paths, context and goals. Similar to our work, they argue that the success of a process instance can be affected not only by the actual path performed, but also by environmental conditions, not controlled by the process. Their work is based on an experience base, including data of past process instances: actual path, achieved outcome, and context information.

We propose context identification to be handled at the activity level, thus enabling process stakeholders to dynamically interfere into a specific activity result by applying previously acquired knowledge during the execution of a process. The circumstances are defined according to the external environment. External contingencies can be considered as opportunities or constraints that influence the structure and internal processes of organizations, according to Competitive Intelligence initiatives [13]. The CI implementation cycles generally include steps to identify information that should be collected. Therefore, based on [13]; [14]; [5]; [9], [22] described the CI process cycle steps to support a Context-based KM Model as shown in Figure 1.

The first step is to identify process, therefore key business processes are chosen from goals and organization strategy. Then, external variables should be identified and represented and associated to the process model through a Bus Matrix [14]. After that step, it is possible to start collecting and keeping these information through properly sources (databases, sensors, etc.). All information is stored in a repository called Organizational Memory, and a number of techniques (KDD, inferences) are applied in order to search for evidences of their impact in process instances. This might result in scenarios and recommendations, which might improve the process, either at the instance or at the model level. The process manager is able to make decisions based on that outcomes; it could possibly cause process adaptations. Then, the cycle starts in on again.

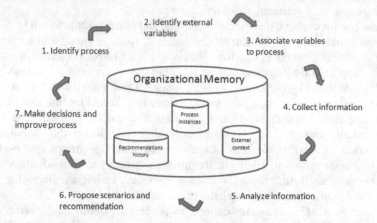

Fig. 1. Competitive Intelligence and decision-making process life cycle

The problem addressed in this paper is specifically related to steps 2 and 3 from this cycle. Next section describes a method to identify the external context, or the kind of information that generally cannot be captured in transactional systems, but from outside of the organization.

3   Method for Discovering External Context

In order to capture and use context information, it is first necessary to specify which context information has to be handled by the organization [18]. We propose a method to discover external context variables (Figure 2) that may not be part of the organizational memory elements, but can be very relevant to the organization in achieving its process goals. This method also identifies which specific activities and process outcomes are impacted by the external context variables. Once discovered, the intelligence analyst may retrieve and analyze external context variables to define scenarios and recommend actions for decision-makers. The decision-makers evaluate the previous decisions and make new decisions that can reflect on improving, creating or removing processes.

There are several methods related to the definition of information needs, e.g., questionnaire, interview and observation that are widely used in different contexts [31]. However, the most suitable methods for the definition of information at the strategic level used by competitive intelligence are Key Intelligence Topics (KIT) [9]; [10] and

Critical Success Factors (CSF). The use of a systematized or formal "management-needs identification process" is a proven way to accomplish this task [9]. Key Intelligence Topic (KIT) support specification, definition and prioritization of information needs at the strategic level of the organization. KITs are items that must be constantly monitored to guarantee business success. They should be more detailed in the form of KIQs (Key Information Questions), which are items that specify the contents of each KIT. For example, the KIT "Strategic Investment Decisions" may consist of the following KIQs: "What is the involvement of other investors in competitors?" and "What are the critical investments from competitors?" [31].

The KITs are identified through interviews with managers, asking open questions. They fall into three categories: (i) strategic decisions and actions; (ii) topics for early warning, considering threats and issues on which decision makers do not want to be surprised, and (iii) major players in the market, such as customers, competitors, suppliers and partners [9]. The technique also proposes the concept of surveillance areas, which are macroeconomic variables that impact the business sector, and that should be monitored.

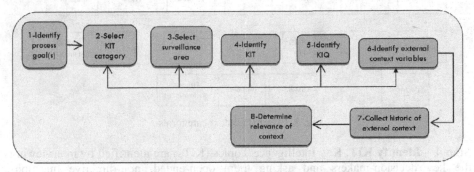

Fig. 2. Method for external context variables identification

The method steps are described as follows.

Step 1 – Identify Process Goal(s). Identify the goal related to a given process and their appropriate measures [25]. Repeat this step to identify others goals after concluding the last step.

Step 2 – Select KIT Category. Herring [9] has divided KITs into three categories: 1) Strategic Decisions and Issues, 2) Early-warning KITs, considering threats and issues on which decision makers do not want to be surprised and 3) Key player KITs (such as customers, competitors, suppliers and partners).

Step 3 – Select Surveillance Area. To define the external context variables, the steps 3 to 6 are part of a top-down approach. Top level areas must be considered to give support to the next step. A model to categorize context information would help to select those areas. The areas can be selected from any framework or a combination of them, such as Five Forces model [21], or SLEPT or STEEP Analysis [30]. In general, they are: social, technology, economic, ecology, political, legal and competitors, due to all industries are influenced by them. These forces are continually in a state of change and then should be scanned. Most research about context in business process deal with internal context, i.e. process attributes inherent to the way process is performed, to the organization of activities and internal rules. Few context categories are proposed, such as location, time, and organization environment. Our work focuses on the events that occur externally to the process, or ultimately to the organization where it runs, but somehow interfere within this process, provoking good or bad effects. There are not many proposals to categorize this kind of context information. Rosemann *et al.* [25] propose that the external layer of their model is composed of the following types of context: suppliers, capital providers, workforce, partners, customers, lobbies, states, competitors. Based on the works mentioned before, we propose the categories depicted in Figure 3. Repeat this step for each of the three KIT categories.

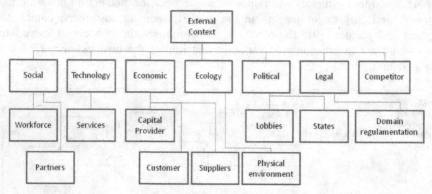

Fig. 3. External context categories

Step 4 – Identify KIT. Key Intelligence Topics (KITs) are identified by interviewing the key decision-makers and asking them open-ended, non-directive questions (Herring and Francis, 1999). An interview protocol can be very useful to ensure the consistency of results . Repeat this step for each of the surveillance area selected.

Step 5 – Identify KIQ. Key Intelligence Questions (KIQs) should be identified for each KIT. KIQs represent the information needs listed in the KIT, i.e. what the manager needs to know to be able to make the decisions. It is possible to have the same KIQ for more than one KIT. Repeat this step for each KIT selected.

Step 6 – Identify External Context Variables. Each KIQ may reference one or more external variables. These are the external context variables and are identified in this step. It is possible to have the same variable for more than one KIQ. Repeat this step for each KIQ identified in the previous step. For each process goal, the result of all the executions of steps 2 to 6 will be the final Intelligence Tree with the following columns: Process Goal, KIT category, Surveillance Area, KIT, KIQ and External Context Variable.

Step 7 – Collect Past Information of the External Context. In this step, the historic of the external context is collected and stored in the organizational memory.

Step 8 – Determine Relevance of the External Context to the Process Outcomes and to the Process Activities Outcomes. It is not feasible to store all context information that could form part of the Organization Memory. That's is why, this step helps prioritizing which context to capture and store, by classifying the variables by relevance using data mining. This step follows the KDD process of Fayyad *et al.* (1996) that is interactive and iterative, involving numerous steps with many decisions made by the user. The term Knowledge Discovery in Databases (KDD) is generally used to refer to the overall process of discovering useful knowledge from data, where data mining is a particular step in this process [7].

Several data mining problem types or analysis tasks are typically encountered during a data mining project. Depending on the desired outcome, several data analysis techniques with different goals may be applied successively to achieve a desired result [12]. Before applying the KDD process, it is necessary to develop an understanding of the application domain and the relevant prior knowledge and identifying the goal of the KDD process from the customer's viewpoint [7]. Our method uses KDD for the following KDD goal: predict the process goal and determine the relevance of the external context to the process outcomes and to the process activities outcomes to achieve the process goal defined in step 1. The KDD process steps [7] are:

Step 8.1 (Selection) - this step consists on creating a target data set, or focusing on a subset of variables or data samples, on which discovery is to be performed. In this step, the historic of the external context is associated to the process activities outcomes and to the process execution results, for the same period.

Step 8.2 (Pre-processing) - this step consists on the target data cleaning and pre processing in order to obtain consistent data;

Step 8.3 (Transformation) - this step consists on data reduction and projection: finding useful features to represent the data depending on the goal of the task. With dimensionality reduction or transformation methods, the effective number of variables under consideration can be reduced, or invariant representations for the data can be found [7].

Step 8.4 (Data Mining - DM) - this step consists on the searching for patterns of interest in a particular representational form, depending on the DM objective (usually, prediction). Many models can be created to allow comparing which one has the best accuracy for predicting the target attribute, in the case of prediction. The chosen model must easily show the relevant variables that must be scanned and what specific values may trigger some decisions.

Step 8.5 (Interpretation/Evaluation) - this step consists on the interpretation and evaluation of the mined patterns. It works as follows: from the external potential variables identified in the previous stage, a data mining technique is applied to discover which of the external variables presents the most relevant results concerning their relationships with the business process. In our work, we adopted the feature selection and the decision tree data mining techniques, since they generate adequate representations that will allow knowledge managers to understand and interpret the circumstances in which a change in an external variable impacted the results of process instances in the past and, further, which activities of the processes were impacted. We argue that this interpretation will enable the decision maker to prepare process adaptations required to handle future modifications in the external variables, or at least to quickly react to those changes in the environment, whe they occur.

4   Case Study Using Data from Open Source Projects

An explanatory case study was conducted in order to evaluate the proposed method. The reasons for selecting the case study evaluation approach are twofold: it does not require control of event behaviour and it focus on contemporaneous events [32].

The case study research question was "how to determine the relevance of variables of the external context to a business process?".

4.1   Source Forge Software Development Process Model

We applied our proposed approach in the domain of Open Source Software Development. In fact, the software development domain (especially in the case of open source development) is typically impacted by external environment, such as economical and political situations, which may interfere in the project budget, investment priority, or even in the availability of developers effectively participating in it. In such organizations, it would be valuable information for a project manager if he could anticipate the risk of a project not being concluded to the moment of deciding whether or not to authorize its beginning. Even better, if he could systematically foresee that a strategic project is bound to fail and trigger actions that will revert it and increase the chance of a success. Figure 4 presents the software process model we considered.

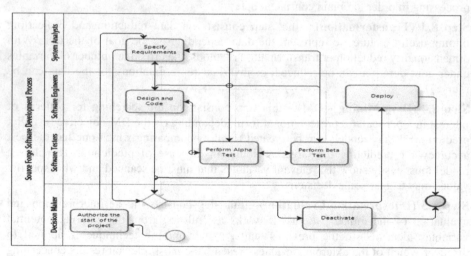

Fig. 4. Source forge software development process model

This process model was derived from Source Forge software development projects, modeled using the Bizagi Process Modeler tool [4], following the BPMN 1.2 notation [19].

For specifying this process, we considered the process log repository maintained by the Source Forge project [29]. This repository holds log data from thousands of real open software development projects, including which was the programming language used, the development status of the project at each moment, what were the operational systems compatible with the software, and the number of developers involved in the team. The development status of a project is one of six subsequential levels, from the earliest stage of production to fully developed software, according to Comino *et al.* [3]: planning, pre-alpha, alpha, beta, production/mature; additionally, the inactive status means that the project was suspended. The process activities then reflected the sequence flow of these development status and on literature [20]; [29]: "planning" is the result of the "Specify Requirements" activity, "pre-alpha" is the result of the "Design and Code" activity, "alpha" is the result of the "Perform Alpha Test" activity, "beta" is the result of the "Perform Beta Test" activity, "production/mature" is the result of the "deploy" activity, "inactive" is the result of the "Deactivate" activity.

Our process model is then a simple, yet effective, representation of a straightforward development process that works in this way: when the Decision Maker (which is typically a project manager in the organization) decides to authorize the start of a development project, the project instance is created in the Source Fourge repository. Software requirements are then specified by System Analysts, Software Engineers then design and code the software. The testing team performs alpha tests on the first published release of the software, in which there are known bugs, so as to gather user feedback. Next, the testing team performs beta tests on the published beta software version; if tests are considered successful then the decision maker decides to deploy the software, thus publishing it as a mature software to its users. Anytime, the decision maker may decide to cancel the project.

4.2   Data Set

The proposed method was applied to the Open Source (OS) projects from Source Forge projects database [29]. SourceForge (SF) thrives on community collaboration to help creating the leading resource for open source software development and distribution. With the tools it provides, 2.7 million developers create software in over 260,000 projects. SF connects more than 46 million consumers with these open source projects and serves more than 2,000,000 downloads a day [29]. SourceForge.net is the largest existing online platform providing OS developers with useful tools to control and manage software development. Project administrators register their software project on SF and provide the required information which is then available on-line [3].

The dataset we employed in our analysis consists of 1,087 projects that were hosted on SF and that had an English version and that got started after January 2005 at the "Specify Requirements" activity, and that achieved firstly one of the following activities before January 2011: "Deploy" or "Deactivate". All the 1,087 projects are aligned with the process of Figure 4. This dataset has 1 dependent variable and 10 predictors pertaining to projects. These predictors consist of 1 process outcome and 9 process activities outcomes.

For each project, the binary outcome (dependent) variable "final status" is available and indicates whether the project achieved firstly the status of "production/mature" (good projects) or "inactive" (bad projects). This dataset contains 295 bad projects and 792 good projects. It means that 27% of the 1087 projects achieved the "final status" as inactive, and 73% of them, as production/mature. In addition, this dataset has also 9 process activities outcomes available for each project, describing the total duration of the project in each process activity and the percentage it represents of the project duration. The project duration is one process outcome and represents the duration of the project from the Specify Requirements activity to the first month of one of the following activities: Deploy or Deactivate. The duration is measured in quantity of months.

In our work, we introduce new variables of the external context and relate it to the process activities and to the process execution results to support these decisions.

4.3   Application of the Method

In this explanatory case study, we applied all the 8 steps of the proposed method to define relevant external variables that influenced the project conclusion of SF projects using the dataset detailed in section 4.2 and considering the software development process defined in section 4.1. The result after applying the steps 1 to 6 of the proposed method 1 is a list of possible relevant external variables. The result applying the steps 7 to 8 is a list showing just the relevant variables among the external contexts, the activities outcomes and the process outcomes; and a decision tree showing the relation among these relevant variables.

Step 1. The goal "Conclude the software development in the Deploy activity" was considered for the process of Figure 4. This goal is achieved when the dependent variable "final status" is production/mature.

Step 2 to 6. For the defined process goal, the result of steps 2 to 6 is illustrated in Table 1. This table contains possible relevant external variables that can impact the process goal.

Table 1. (Partial) Final intelligence tree after step 6

KIT category	Surveillance Area	KIT	KIQ	External Context Variable
Strategic decisions and actions	Economic	Economic recession	What are the predictions for IT investments of public and private organizations for the next years?	**IT Investment Prediction;**
			What are the predictions for the unemployment rate for next years?	**Unemployment Rate prediction; Unemployment Rate;**
			What are the predictions for the inflation rate for next years?	**Inflation Rate prediction; Inflation Rate;**
Strategic decisions and actions/ Early-warning	Politic	IT goals of the Govern	What are the Open Source Software patterns adopted by the Govern?	**Open Source Software patterns;**

Step 7. In this step the focus is on collecting past information of the external variables defined previously. As the projects could be developed by people from different countries anywhere in the planet, we decided to simplify and assume USA as the original country for each of the 1,087 projects in the dataset. USA was chosen because it is one of the most influential countries in the global economy, as we could see in the global economy crisis of 2008 that got initiated in the USA.

An important issue to consider in this step is that sometimes it is not possible to collect past information of an external variable. In some cases, past values may simply not exist. In these cases, the external variable may be discarded or the decision maker may decide to start monitoring this variable and registering its values in a log data so as to use it in the future. In our research, we have used the historic of 2 external variables defined previously: the USA unemployment rate and the USA inflation rate [11].

Step 8. In this step we followed the KDD process [7] and we applied the Feature Selection technique to show the variables relevance, and we used Decision Tree C&RT (Standard Classification Trees with Deployment) to show explicitly the rules of the relation between the relevant external contexts, the relevant process outcomes and the relevant process activities outcomes for predicting the dependent variable "final status". This was the KDD goal.

Below, we explain how the data mining technique determined that Unemployment Rate was a relevant external context variable to the defined process goal and to one of its activity outcome. We used the STATISTICA Data Miner software [28] that uses the CRISP-DM process (CRoss-Industry Standard Process for Data Mining). According to Azevedo and Santos [1] CRISP-DM can be viewed as an implementation of the KDD process of Fayyad et al. [7]. KDD process steps:

Steps 8.1 (Selection) and **8.2 (Pre-processing).** These 2 steps were the most time consuming steps, as Mack et al. [16] already experienced. The data requirements for what is necessary as well as the data acquisition itself have been taken care of already with the data dump from SourceForge (SF). The output of the step 8.1 is the process log, the dataset that was detailed in section 4, and the output of the step 8.2 is a new dataset with the historic of the collected external contexts (step 7) associated to the process activities outcomes and process outcomes (step 8.1).

Step 8.3 (Transformation). In this case study, we run the Feature Selection of STATISTICA Data Miner [28] to automatically find and rank important predictor variables for predicting the dependent variable "final status" that discriminates between good and bad projects, as shown in Figure 5. Feature Selection (FS) technique is "the process of reducing dimensionality by removing irrelevant and redundant features" [2]; [8] reducing "the complexity of the problem, transforming the data set into a data set of lower dimensions" [17]. Figure 5 shows that among the 12 variables of the dataset created in the last step, there are 5 that have a p-value of less than 0.01, i.e., that stand out as the most important predictors variables to determine whether a project would be finalized in the production/mature or in the inactive status.

	Best predictors for categorical dependent var: status_final	
	Chi-square	p-value
Project duration	83,34835	0,000000
Specify requirements duration	75,96964	0,000000
Inflation rate	49,45075	0,000000
Unemployment rate	33,22903	0,000009
Perform beta test duration	18,63327	0,002249

Fig. 5. Best predictors variables for categorical dependent status_final ordered top to bottom on basis of lowest p-value to highest (Stratified Random Sampling)

The top 5 most relevant variables in a descending order of relevance are: 1-Project duration; 2-Specify requirements duration; 3-Inflation rate; 4-Unemployment rate; 5-Perform Beta Test Duration. Note that two of these relevant variables are process activities outcomes ("Specify requirements duration" and "Perform Beta Test Duration"); one is a process outcome ("Project duration"); and two variables ("Inflation rate" and "unemployment rate") are from the external context.

Step 8.4 (Data Mining). Decision trees are powerful tools for classification and prediction. The decision tree C&RT (Standard Classification Trees with Deployment) of Figure 6 was run using STATISTICA Data Miner [28] considering the relevant variables found in the previous step. We used the V-fold cross validation and a 30% sample of dataset for testing to assess the accuracy of the model. Based on the 1087 projects of the full dataset, initially we used a training data sample to build the decision tree (training phase), then, a testing data sample to refine and evaluate the decision tree (testing phase), and finally, we used another dataset with different projects to re-evaluate the accuracy of the decision tree (re-evaluation phase).

In the training phase the decision tree had an error rate of 19.12%; in the testing phase, 21.53%; and in the re-evaluation phase, 20%. The error rate of 19.12% (training phase) means that the decision tree C&RT can predict correctly with an accuracy of 80.80% whether a project will be finalized in the production/mature or in the inactive status. The percent of correct predictions for the bad projects (final status = inactive) is 77.44%; and for the good projects (final status = production/mature) is 82.10%.

Step 8.5 (Interpretation/Evaluation). The decision tree C&RT (Standard Classification Trees with Deployment) of Figure 6 show the relation among the relevant external contexts, the process activities outcomes and the process outcomes. Each node in the tree of Figure 6 shows the number of instances at that node and information about the distribution of the dependent variable values (final status). The root node stands for 748 instances. Below the root node there is the first split that breaks the dataset into 2 subsets, represented by nodes based on the predictor variable "Project duration". This variable is one outcome of the SF software development process. The leftmost node resulting from this split stands for 379 instances that are associated to the duration of the Perform beta test activity.

This decision tree shows that the process outcome "Project duration" is related to the Perform beta test activity by its outcome "Perform beta test duration" and that these outcomes are related to the external context "Inflation rate", as we can see in nodes 1, 2, 4, 6, 8 and 11. Node 11 clearly shows the relevance of the external variable to the Perform beta test activity. It evidences that, when the inflation rate raises below or equal 2.705 and greater than 1.67, then there is a higher probability of the projects, that have Project duration <=4.5 and Perform beta test duration <=0.5, to be deactivated, i.e., to be concluded as inactive.

For instance, the decision maker may use the decision tree when deciding whether to start the development of a new software project or not. Assuming a scenario in which the "Perform beta test" is scheduled to take less than 0.5 month, he should consider the estimated inflation rate for the month the project is supposed to be concluded. If this rate falls in the range from 1.67 and 2.705, there is a higher probability that this project will end up being deactivated. In such a situation, the decision maker may decide not to start this project, or to trigger other actions to maximize the chances of this project to be deployed. The same scenario may occur with an ongoing project. Therefore, relevant external context could be monitored so as to trigger changes during process execution.

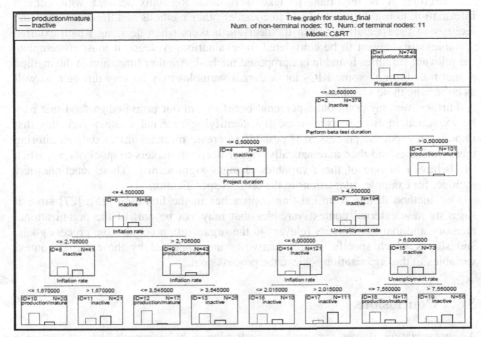

Fig. 6. Part of the decision tree C&RT (training phase) for the SF projects dataset considering the best predictor variables to the dependent variable "final status"

4.4   Result Analysis

The case study reported in this section illustrated a successful application of our proposed method, since it identified external context variables and verified their relevance towards achieving a business process goal, in a real scenario in the domain of open source software development. Moreover, the case study showed the viability of specifically identifying which process activities are impacted by those variables.

The proposed method is adequate to be applied within organizations which foster to improve their decision making processes. We interviewed 4 experts so as to gather their feedback on the method results. They recognized the value of the identified variables, but one of them considered the error rate of 24.42% too high when supporting his/her decision. Further executions of the method could incorporate the definition of an error rate threshold by the decision makers, so as to tune data mining algorithms accordingly.

An important characteristic of our method is that, as with any data mining approach, the discovered knowledge depends on the amount of detailed information available in the log. Therefore, when our approach discovers that a specific external variable is not-so-relevant, it does not mean that it is not relevant at all; instead, it means that the process log did not include enough evidences pointing to the relevance of this external variable to the historic process instances, when compared to other variables.

Therefore, it is important to take a process log into account with enough information to run our method and to consider other methods and the experience and feelings of the specialists and of the decision makers when deciding which external variables are relevant to be considered for evaluation. At least, it must contemplate the relevant variables found in our proposed method. Another limitation in the method is that transforming some KIQs into external variables may be very difficult, as well as collecting these variables.

Further, we may point out a potential benefit from our proposed method that may be exploited in further executions. By identifying external context variables that impact on a specific process, it is possible to create mechanisms to start monitoring these variables and then automatically trigger decision makers to quickly react when the behavior of one of these variables changes significantly. These reactions may include, for example, adaptations to the process specification.

Our method differs from existing approaches in the literature [25]; [27] since it suggests new external context variables that may not be part of the organizational memory and that can be very relevant to the organization achieve the process goals; and shows which specific process activities are impacted by the external context variables to the organization achieve the process goal.

5   Conclusions

An organization strives for its best performance by accomplishing its business processes. Sustainable growth requires that the organization measures, manages and improves its processes. However, to survive in an increasingly disruptive world, companies need to become adaptable and efficient, from both strategic and operational points of view. Flexibility in business processes is the ability to adapt to external changes by promoting adjusts only where they impact the process. Therefore, the decision makers within an organization need to consider external context information in order to make important decisions regarding business process adaptations. We observed that the systematic identification of relevant external context for the process, and moreover, for specific activities, in order to achieve its goals during the execution, is still a challenge.

In this paper we presented a method for supporting the identification and prioritization of variables to be considered in the context of the external environment that impacts process execution. The proposal is based on competitive intelligence and data mining techniques, and provides the process manager (decision maker) with a fact-based understanding of which are the most relevant external variables that really influenced previous process executions, among the several variables that could be taken into consideration unnecessarily. The method also shows which specific activities within the process model are impacted by these variables in relation to the organization strategic goals.

A case study illustrated the application of our method in a software development process using real data from projects of SourceForge.net. This case study showed that changes in relevant variables of the external context may fire a decision of the decision maker to quickly responding to these changes, by adapting the process specification, or creating other business rules to be followed by the business process.

The discovery of relevant external environment variables depends on the set of input data: not only the log of the variables in the external environment, but also the log of the process itself. As any data mining approach, the discovered knowledge depends on the amount of detailed information available in the log. Therefore, it is possible that the relevant variables in the external environment have not been detected; however, those ones identified have proven to be relevant, confirming the application of the method.

We are working now in applying the proposed method: in others different scenarios, such as oil&gas and risk management. The future work is to use larger samples of process logs and with more variables; as well as interviewing decision makers of the same process log organization. We also intend to refine the model evaluation of our method.

References

1. Azevedo, A., Santos, M.F.: KDD, Semma and CRISP-DM: A Parallel Overview. In: European Conference Data Mining-IADIS (2008)
2. Blum, A.L., Langley, P.: Selection of relevant features and examples in machine learning. AI 97, 245–271 (1997)
3. Comino, S., Manenti, F., Parisi, M.: From planning to mature: On the success of open source projects. Research Policy 36(10), 1575–1586 (2007), http://www.scopus.com (retrieval)
4. BizAgi Process Modeler, Version 1.6.1.0, BPMN Software (May 2011), http://www.bizagi.com
5. Cook, M., Cook, C.: Competitive Intelligence. Kogan Page Limited, London (2000)
6. Dey, A.K.: Understanding and using context. Personal and Ubiquitous Computing 5(1), 4–7 (2001)
7. Fayyad, U.M., Piatetsky-Shapiro, G., Smith, P., Uthurusamy, R.: Advances in Knowledge Discovery and Data Mining. AAAI/MIT Press (1996)
8. Guyon, I., Elisseeff, A.: An introduction to variable and feature selection. JMLR 3, 1157–1182 (2003)
9. Herring, J.P.: Key Intelligence Topics: A Process to Identify and Define Intelligence Needs. Competitive Intelligence Review 10(2) (1999)
10. Herring, J.P., Francis, D.B.: Key Intelligence Topics: A Window on the Corporate Competitive Psyche. Competitive Intelligence Review 10(4) (1999)
11. IndexMundi. USA Unemployment and Inflation rate (April 2011), http://www.indexmundi.com
12. Jackson, J.: Data Mining: a Conceptual Overview. Comm. Association for Information Systems 8(19) (2002), http://aisel.aisnet.org/cais/vol8/iss1/19
13. Jung, J., Choi, I., Song, M.: An integrated architecture for knowledge management systems and business process management systems. Computers in Industry 58, 21–34 (2006)

14. Kimball, R., Ross, M.: The Data Warehouse Toolkit. Wiley Computer Publishing, New York (2002)
15. Liebowitz, J.: A set of frameworks to AID the Project manager in conceptualizing and implementing knowledge management initiatives. Sciencedirect (2003)
16. Mack, D., Chawla, N.V., Madey, G.: Activity Mining in Open Source Software. In: NAACSOS 2005 (2005)
17. Nisbet, R., Elder, J., Miner, G.: Handbook of statistical analysis and Data Mining Applications. Elsevier Inc., California (2009)
18. Nunes, V.T., Santoro, F.M., Borges, R.B.: A Context-based Model for Knowledge Management embodied in Work Processes. Information Sciences 179, 2538–2554 (2009)
19. OMG-Object Management Group/Business Process Management Initiative. BPMN Specification Releases: BPMN 1.2 (October 2010), http://www.bpmn.org
20. PMI. A guide to the project management body of knowledge (PMBOK® Guide), 4th edn. Project Management Institute, Newtown Square (2008)
21. Porter, M.E.: How competitive forces shape strategy. Harvard Business Review (March/April 1979)
22. Ramos, E.C., Santoro, F.M., Baião, F.A.: Process Improvement Based on External Knowledge Context. In: ACIS 2010 Proceedings, Paper 34 (2010), http://aisel.aisnet.org/acis2010/34
23. Ramos, E.C., Santoro, F.M., Baião, F.A.: A Method for Discovering the Relevance of External Context Variables to Business Processses. In: International Conference on Knowledge Management and Information Sharing (KMIS), Paris, France (2011)
24. Rosemann, M., Recker, J.C.: Context-aware Process Design Exploring the Extrinsic Drivers for Process Flexibility. In: Latour, T., Petit, M. (eds.) The 18th International Conference on Advanced Information Systems Engineering. Proceedings of Workshops and Doctoral Consortium, Luxembourg, Grand-Duchy of Luxembourg (2006)
25. Rosemann, M., Recker, J., Flender, C.: Contextualization of Business Processes. International Journal of Business Process Integration and Management 3, 47–60 (2008)
26. Saidani, O., Nurcan, S.: Towards Context Aware Business Process Modelling. In: Workshop on Business Process Modelling, Development, and Support (BPMDS), Trondheim, Norway (2007)
27. Soffer, P., Ghattas, J., Peleg, M.: A Goal-Based Approach for Learning in Business Processes. In: Nurcan, et al. (eds.) Intentional Perspectives on Information Systems Engineering. Springer (2010)
28. StatSoft Inc., STATISTICA Data Miner (2010), http://www.StatSoft.com (May 2011)
29. Madey, G. (ed.): The SourceForge Research Data Archive (SRDA). University of Notre Dame (July 2011), http://srda.cse.nd.edu
30. The Times, SLEPT analysis, 100th edn. (2010), http://www.thetimes100.co.uk (last accessed April 2010)
31. Vuori, V., Pirttimäki, V.: Identifying of Information Needs in Seasonal Management. Frontiers of E-business Research, 588–602 (2005)
32. Yin, R.K.: Case Study Research: Design and Methods, 4th edn. SAGE Publications, California (2009)

Author Index